PHYSICAL GEOGRAPHY.

The subject of Physical Geography is one of which so little is generally known, that the publishers, in presenting a new and improved edition of Mrs. Somerville's excellent work, have thought it not inappropriate to introduce the following extracts from an article on its uses and applications, from the "Canada Journal of Education": —

"There is probably no study which, in comparison with its importance, has received so little attention as this. The school-boy soon wearies of learning the names and locations of continents, peninsulas, islands, capes, mountains, oceans, seas, lakes, rivers, &c., &c.; together with their comparative size, length, distance from each other, their population, navigation, character of inhabitants, varieties of animals, various productions, adding, it may be, the accompanying history of events connected with the different countries; and to what purpose? To be forgotten nearly as soon, and much more easily, than learned.

"We think that the judicious introduction of physical geography, in connection with topography, will very much increase the interest of the latter, while the knowledge it will afford, in and of itself, will exceed by far, in importance, what is usually obtained, at the present time, even in our best schools.

"Of what use is it that we know that there are certain mountains, seas, or rivers, in Europe or Asia, if we are totally ignorant of their effects upon vegetation, upon civilization, and the condition of mankind? or that the different continents are so many miles in length, and so many in breadth, if we are unacquainted with the corresponding oceanic influences and the resulting facts.

"How many scholars know why all the great deserts of the world are situated where they are, and that the physical laws are such that it is not possible that there could be anything but deserts in those places? How many know why the northern part of the Andes is almost wholly desert upon their western slope, and the southern part upon their eastern? or that, were this chain removed to the eastern side of South America, nearly the whole division would be one continuous desert?

"These things are seldom spoken of as having any connection with the study of geography, and yet it would seem that they should constitute its very foundation.

"Probably the difference in the civilization of Europe and Africa, is to be attributed more to the inland seas and gulfs, and the numerous rivers of the former, and their effects; and the absence of the same in the latter, together with other physical characteristics, than to any other causes whatever; but these things are seldom learned in the schools.

"The scholar learns the results of these causes as merely abstract facts, and remembers them about as well as he would the conclusion to a proposition in Euclid, without having been through with the demonstration.

"These things are not too difficult to be understood by the scholars in our grammar and high schools, and many of them come within the range of the lower classes.

PHYSICAL GEOGRAPHY.

BY

MARY SOMERVILLE,

AUTHOR OF THE "CONNEXION OF THE PHYSICAL SCIENCES,"
"MECHANISM OF THE HEAVENS," ETC. ETC.

A NEW AMERICAN,

FROM THE THIRD AND REVISED LONDON EDITION.

WITH NOTES, AND A GLOSSARY,

BY

W. S. W. RUSCHENBERGER, M.D.,

U. S. NAVY.

PHILADELPHIA:
BLANCHARD AND LEA.
1855.

T. K. AND P. G. COLLINS, PRINTERS.

TO

SIR JOHN F. W. HERSCHEL, BART., K.H.

&c. &c. &c.

DEAR SIR JOHN,

I avail myself with pleasure of your permission to dedicate my book to you, as it gives me an opportunity of expressing my admiration of your talents, and my sincere estimation of your friendship.

I remain, with great regard,

Yours truly,

MARY SOMERVILLE

LONDON, 29*th February*, 1848.

PUBLISHERS' ADVERTISEMENT.

THE improvements and additions embodied in the third London edition are enumerated in the Author's Preface. The additions to the last American edition have been considerably extended in this; and the Glossary of scientific and technical terms has been carefully revised. It is believed the volume is very much improved, and superior to the last London edition.

The additions which are distributed through the volume are enclosed in brackets [ ]. These, with the Glossary, add considerably to the number of pages. The American Publishers therefore believe, that it will be found well suited to the wants of the general reader, and in every particular adapted to the use of the more advanced pupils in schools.

PHILADELPHIA, *May*, 1853.

INTRODUCTION.

THE daily accumulating knowledge in every branch of science has rendered it necessary to make many additions and corrections in the third edition of this work. In doing this, the Author acknowledges her obligations to Baron Humboldt's invaluable 'Cosmos,' with Colonel Sabine's excellent notes — and to the works of M. Elie de Beaumont, Sir Charles Lyell, and Sir Henry De la Beche;[1] to the researches of Messrs. Campbell, Thomson, Strachey, and Dr. Hooker in the Himalaya; and to papers in the periodical journals of Europe, India, and America.

The Author has to express her thanks to her friend Mr. Pentland for his kindness in again superintending the passage through the press of this work during her residence abroad, and for matter hitherto unpublished, on the countries visited by him during his diplomatic missions to Bolivia and Peru; and also to M. Elie de Beaumont, the Prince of Canino, and Dr. Weddell, for valuable information on the subjects of Geology, Ornithology, and Botanical Geography.

---

[1] 'Principles of Geology, by Sir Charles Lyell,' 8vo., 1850; 'Manual of Elementary Geology,' by Sir Charles Lyell, 8vo., 1851; 'The Geological Observer,' by Sir Henry T. De la Beche, C.B., 8vo., 1851. (Republished by Blanchard & Lea.)

It was the Author's wish and her publisher's intention that this work should be accompanied by maps to illustrate the most important questions of Physical Geography; but since Mr. Alex. Keith Johnston has published an edition of his splendid 'Physical Atlas' on a reduced scale,[1] which affords all the information required, that plan has been abandoned.

The Author must also acknowledge the assistance she has received from another work recently published by the same author, his 'Geographical Dictionary,'[2] the most complete General Gazetteer that has appeared in our own or in any other language.

TURIN, *March* 4, 1851.

---

[1] 'The Physical Atlas of Natural Phenomena,' 1 vol. fol., 1847; 'The Physical Atlas, reduced for the Use of Colleges, Families, &c.' 1 vol. 4to., 1850. (Republished by Blanchard & Lea.)

[2] 'A Dictionary of Geography, forming a complete General Gazetteer of the World,' by Alex. Keith Johnston. 1 vol. 8vo., London, 1850.

CONTENTS.

CHAPTER I.

GEOLOGY.

CHAPTER XIX.

CHAPTER XX.

CHAPTER XXI.

CHAPTER XXII.

CHAPTER XXIII.

PHYSICAL GEOGRAPHY.

CHAPTER I.

GEOLOGY.

Of Physical Geography — Position of the Earth in the Solar System — Distance from the Sun—Civil Year—Inclination of Terrestrial Orbit—Mass of the Sun — Distance of the Moon — Figure and Density of the Earth from the Motions of the Moon — Figure of the Earth from Arcs of the Meridian — from Oscillations of Pendulum — Local Disturbances — Mean Density of the Earth — Known Depth below its Surface — Outline of Geology.

Physical Geography is a description of the earth, the sea, and the air, with their inhabitants animal and vegetable, of the distribution of these organized beings, and the causes of that distribution. Political and arbitrary divisions are disregarded, the sea and the land are considered only with respect to those great features that have been stamped upon them by the hand of the Almighty, and man himself is viewed but as a fellow-inhabitant of the globe with other created things, yet influencing them to a certain extent by his actions, and influenced in return. The effects of his intellectual superiority on the inferior animals, and even on his own condition, by the subjection of some of the most powerful agents in nature to his will, together with the other causes which have had the greatest influence on his physical and moral state, are among the most important subjects of this science.

The former state of our terrestrial habitation, the successive con vulsions which have ultimately led to its present geographical arrangement, and to the actual distribution of land and water, so powerfully influential on the destinies of mankind, are circumstances of primary importance.

The position of the earth with regard to the sun, and its connexion with the bodies of the solar system, have been noticed by the author elsewhere. It was there shown that our globe forms but an atom in the immensity of space, utterly invisible from the nearest fixed star, and scarcely a telescopic object to the remote planets of our system. The increase of temperature with the depth below the surface of the earth, and the tremendous desolation hurled over wide regions by numerous fire-breathing mountains, show that man is removed but a few miles from immense lakes or seas of liquid fire. The very shell on which he stands is unstable under his feet, not only from those temporary convulsions that seem to shake the globe to its centre, but from a slow almost imperceptible elevation in some places, and an equally gentle subsidence in others, as if the internal molten matter were subject to secular tides, now heaving and now ebbing, or that the subjacent rocks were in one place expanded and in another contracted by changes of temperature.

The earthquake and the torrent, the august and terrible ministers of Almighty Power, have torn the solid earth and opened the seals of the most ancient records of creation, written in indelible characters on the "perpetual hills and the everlasting mountains." There we read of the changes that have brought the rude mass to its present fair state, and of the myriads of beings that have appeared on this mortal stage, have fulfilled their destinies, and have been swept from existence to make way for new races, which, in their turn, have vanished from the scene, till the creation of man completed the glorious work. Who shall define the periods of those mornings and evenings when God saw that his work was good? and who shall declare the time allotted to the human race, when the generations of the most insignificant insect existed for unnumbered ages? Yet man is also to vanish in the ever-changing course of events. The earth is to be burnt up, and the elements are to melt with fervent heat — to be again reduced to chaos — possibly to be renovated and adorned for other races of beings. These stupendous changes may be but cycles in those great laws of the universe where all is variable but the laws themselves, and He who has ordained them.

The earth is one of thirty-two planets which revolve about the sun in elliptical orbits: of these, twenty-five have been discovered since the year 1781.[1] Mercury and Venus are nearer the sun than

[1] [The solar system consists exclusively of the sun and all the planets, satellites, and comets, whose motions are dependent upon its gravitation. It does not include the *fixed* stars. The following is believed to be a complete catalogue of the solar system:—

Name.	Date of Discovery.	Discoverer.	Place of Discovery.
Mercury	known to the ancients.		
Venus	"	"	
The earth (with one satellite)	"	"	
Mars	"	"	
Jupiter (with four satellites)	"	"	
Saturn (with seven satellites)	"	"	
Uranus (with two or more satellites)	1781	Sir Wm. Herschel.	
Ceres	June 1, 1801	Piazzi	Palermo.
Pallas	March 28, 1802	Olbers	Bremen.
Juno	Sept. 1, 1804	Harding	Lilienthal.

the earth, the others are more remote. The earth revolves at a mean distance of 95,298,260 miles from the sun's centre, in a civil year of 365 days 5 hours 48 minutes 49·7 seconds, at the same time that

Name.	Date of Discovery.	Discoverer.	Place of Discovery.
Vesta	March 29, 1807	Olbers	Bremen.
Astrea	Dec. 8, 1845	Hencke	Drissen.
Neptune	1846	Le Verrier & Mr. Adams	
Hebe	July 1, 1847	Hencke	Drissen.
Iris	Aug. 13, 1847	Hind	London.
Flora	Oct. 18, 1847	Hind	London.
Metis	April 25, 1848	Graham	Sligo.
Hygea	April 12, 1849	Gasparis	Naples.
Parthenope	May 11, 1850	Gasparis	Naples.
Clio	Sept. 13, 1850	Hind	London.
Egeria	Nov. 2, 1850	Gasparis	Naples.
Victoria	1850	Hind	London.
Irene	May 19, 1851	Hind	London.
Eunomia	July 19, 1851	Gasparis	Naples.
Psyche	March 17, 1852	Gasparis	Naples.
Thetis	April 17, 1852	Luther	Berlin.
Melpomene	June 24, 1852	Hind	London.
Fortuna	Aug. 22, 1852	Hind	London.
Massilia	Sept. 19, 1852	Gasparis	Naples.
Calliope	Nov. 16, 1852	Hind	London.
Lutitia	Nov. 18, 1852	Goldschmidt	Paris.
Thalia	Dec. 15, 1852	Hind	London.

Comets observed since January 1, 1847.

1847—I.	Feb. 6	Hind	London.
II.	May 7	Colla	Parma.
III.	Aug. 31	Schweizer	Moscow.
IV.	July 4	Mauvais	Paris.
V.	July 20	Brorsen	Scuftenberg.
VI.	Oct. 1	Miss Mitchell	Nantucket.
1848—I.	Aug. 7	Petersen	Altona.
II.	(Encke's comet) Sept. 13	Hind	London.
III.	Oct. 26	Petersen	Altona.
1849—I.	April 15.	Goujon	Paris.
II.	April 11	Schweizer	Moscow.
III.	Nov. 28	Jenkins	At sea.
1850—I.	May 1	Petersen	Altona.
II.	Aug. 29	Bond	Cambridge.
III.	(Fay's comet) Nov. 28	Challis	Cambridge, E.
1851—I.	June 27	D'Arrest	Leipsic.
II.	Aug. 1	Brorsen	Scuftenberg.
III.	Oct. 22	Brorsen	Scuftenberg.
1852—I.	May 15	Chacornac	Marseilles.
II.	July 24	Westphal	Gottingen.
III.	(Biela's comet) Aug. 25	Leechi	Rome.]

From the elements and position of the orbits of the thirteen small bodies. [namely, Flora, Vesta, Iris, Metis, Hebe, Astræa, Hygea, Parthenope, Victoria, Egeria, Juno, Ceres, and Pallas,] which revolve between Mars and

it rotates in 24 hours about an axis which always remains parallel to itself, and inclined at an angle of 23° 27′ 34·69″ to the plane of the ecliptic; consequently the days and nights are of equal length at the equator, from whence their length progressively differs more and more as the latitude increases, till at each pole alternately there is perpetual day for six months, and a night of the same duration: thus the light and heat are very unequally distributed, and both are modified by the atmosphere by which the earth is encompassed to the height of about forty miles.

With regard to magnitude, Mars, Jupiter, Saturn, Uranus and Neptune are larger than the earth, the rest are smaller, but even the largest is incomparably inferior to the sun in size; his mass is 354,936 times greater than that of the earth, but the earth is nearly four times as dense.

Though the planets disturb the earth in its motion, their form has no effect on account of their great distance; but it is otherwise with regard to the moon, which revolves about the earth at a mean distance of 240,000 miles, and is therefore so near, that the form of both bodies causes mutual disturbances in their respective motions. The perturbations in the moon's motions from that cause, compared with the same computed from theory, show that the earth is not a perfect sphere, but that it bulges at the equator, and is flattened at the poles: it even gives a value of the compression[1] or flattening. Again, theory shows that, if the earth were throughout of the same

---

Jupiter, it has been conjectured, with much probability, that they once formed the mass of a large planet which had exploded: upon this hypothesis several have actually been looked for, and found. The shooting stars which have appeared in such remarkable showers in the months of August and November, may possibly have had a similar origin, as they are believed to form a group which revolves about the sun in 182 days, in an elliptical orbit, and that in passing through the aphelion in August and November, they come in contact with the earth's atmosphere, on entering which with great velocity they become ignited and are consumed. An event so tremendous as the explosion of a world, is by no means beyond the unlimited power of steam under intense pressure.

[1] The compression of the earth is the flattening at the poles. Its numerical value is equal to the difference between the equatorial and polar diameters, expressed in feet or miles, divided by the equatorial diameter.

[The extent of compression at the poles is measured by the ratio of the difference between the equatorial and polar diameters to the equatorial diameter, which is technically termed the *oblateness*. The dimensions of the earth in miles, are as follows.

	Miles.		Diameter.
Radius at the equator	3962·6	=	7925·2
Radius at the pole	3949·6	=	7899·2
Difference of equatorial and polar radii	13·0	=	26·0
Mean radius, or at 45° latitude	3956·1	=	7912·2
Mean length of a degree	69·05		———
The fourth part of a meridian	6214·2		———]

density, it would be much less flat at the poles than the moon's motions show it to be, but that it would be very nearly the same were the earth to increase regularly in density from the surface to its centre; and thus the lunar motions not only make known the form, but reveal the internal structure of the globe. Actual measurement has proved the truth of these results.

The courses of the great rivers, which are generally navigable to a considerable extent, show that the curvature of the land differs but little from that of the ocean; and as the heights of the mountains and continents are inconsiderable when compared with the magnitude of the earth, its figure is understood to be determined by a surface at every point perpendicular to the direction of gravitation, or of the plumb-line, and is the same which the sea would have if it were continued all round the earth beneath the continents. Such is the figure that has been measured in various parts of the globe.

A terrestrial meridian is a line passing through both poles, all the points of which have their noon contemporaneously, and a degree of a meridian is its 360th part. Now, if the earth were a sphere, all degrees would be of the same length; but, as it is flattened at the poles, the degrees are longest there, and decrease in length to the equator, where they are least. The form and size of the earth may therefore be determined by comparing the length of degrees in different latitudes.[1] Eleven arcs have been measured in Europe, one in the Andes of equatorial America, and two in the East Indies; but a comparison of no two gives the same result, which shows that the earth has a slightly irregular form. From a mean of ten of these arcs M. Bessel found that the equatorial radius of the earth is 3963·025 miles, and the polar radius 3949·8 miles nearly. Whence, assuming the earth to be a sphere, the length of a mean degree of the meridian is 69·05 British statute miles; therefore 360 degrees, or the whole circumference of the globe, is 24,858 miles; the diameter, which is something less than a third of the cirumference, is about 8286 statute miles; and the length of a geographical mile of 60 to a degree is 6086·76 feet. The breadth of the torrid zone is 2815 geographical miles, the breadth of each of the temperate zones is 2854 miles, and that of each of the spaces within the arctic and antartic circles 1140 miles nearly. The Astronomer Royal Mr. Airy's results, obtained ten years afterwards, only differ from those of M. Bessel by 127 feet in the equatorial, and 138 feet in the polar radius, quantities not greater than the length of a good sized ball-room. In consequence of the round form of the earth, the dip or depression of the horizon is a fathom for every three miles of dis-

[1] The theoretical investigation of the figure of the earth, the method employed for measuring arcs of the meridian, and that of finding the form of the earth from the oscillations of the pendulum, are given in the 'Connexion of the Physical Sciences,' by Mary Somerville, 7th Section, 7th edition

tance; that is to say, an object a fathom or six feet high would be hid by the curvature of the earth at the distance of three miles. Since the dip increases as the square, a hill 100 fathoms high would be hid at the distance of ten miles, and the top of Kunchinjunga, the most elevated point of the Himalaya, hitherto measured 28,178 feet high, would be seen to sink beneath the horizon by a person about 167 miles off; thus, when the height is known, an estimate can be formed of the distance of a mountain.

The oscillations of the pendulum have afforded another method of ascertaining the form of the earth. Like all heavy bodies, its descent and consequently its oscillations are accelerated in proportion to the force of gravitation, which increases from the equator to the poles. In order, therefore, that the oscillations may be everywhere performed in the same time, the length of the pendulum must be increased progressively in going from the equator to the poles, according to a known law,[1] from whence the compression or flattening at the poles may be deduced. Experiments for that purpose have been made in a great number of places, but, as in the measurement of the arcs, no two sets give exactly the same results; the mean of the whole, however, differs very little from that given by the degrees of the meridian and the perturbations of the moon; and as the three methods are so entirely independent of each other, the figure and dimensions of the earth may be considered to be known. The sea has little effect on these experiments, both because its mean density is less than that of the earth, and that its mean depth of perhaps four miles is inconsiderable when compared with 3956 miles, the mean terrestrial radius.[2]

[1] A pendulum which oscillates 86,400 times in a mean day at the equator, will do the same at every point of the earth's surface if its length be increased progressively to the pole as the square of the sine of the latitude. The sine of the latitude is a perpendicular line drawn from any point of a terrestrial meridian to the equatorial radius of the earth. That line expressed in feet or miles, and multiplied by itself, is the square of the sine of the latitude. Gravitation increases from the equator to the poles according to that law, and the length of the degrees augments very nearly in the same ratio.

[2] The compression deduced by M. Bessel from arcs of the meridian is $\frac{1}{299}$; that deduced by Colonel Sabine from his experiments with the pendulum is $\frac{1}{288 \cdot 7}$. Other pendulum experiments have also given a compression of $\frac{1}{298 \cdot 2}$ and $\frac{1}{266 \cdot 4}$. The protuberant matter at the earth's equator produces inequalities in the moon's motions, from whence the compression of the earth is found to be $\frac{1}{305 \cdot 05}$; and although the reciprocal action of the

The discrepancies in the results, from the comparison of the different sets of pendulum experiments, and also of degrees of the meridian, arise from local attraction, as well as from irregularities in the form of the earth. These attractions, arising from dense masses of rock in mountains, cause the plumb-line to deviate from the vertical, and when under ground they alter the oscillations of the pendulum. Colonel Sabine, who made experiments with the pendulum from the equator to within ten degrees of the north pole, discovered that the intensity is greatly augmented by volcanic islands. A variation to the amount of a tenth of a second in twenty-four hours can be perfectly ascertained in the rate of the pendulum; but from some of these local attractions a variation of nearly ten seconds has occurred during the same period. The islands of St. Helena, Ascension, St. Thomas, the Isle of France, are some of those noted by Colonel Sabine.

There are other remarkable instances of local disturbance, arising from the geological nature of the soil; for example, the intensity of gravitation is very small at Bordeaux, from whence it increases rapidly to Clermont-Ferrand, Milan, and Padua, where it attains a maximum (owing, probably, to dense masses of rock under ground), and from thence it extends to Parma. In consequence of this local attraction, the degrees of the meridian in that part of Italy seem to increase towards the equator through a small space, instead of decreasing, as if the earth were drawn out instead of flattened at the poles.

It appears from this, that the effect of the whole earth on a pendulum or torsion balance may be compared with the effect of a small part of it, and thus a comparison may be instituted between the mass of the earth and the mass of that part of it. Now a leaden ball was weighed against the earth by comparing the effects of each upon a balance of torsion; the nearness of the smaller mass making it produce a sensible effect as compared with that of the larger, for by the laws of attraction the whole earth must be considered as collected in its centre; in this manner a value of the mass of the earth was obtained, and, as its volume was known, its mean density was found to be 5·675 times greater than that of water at the temperature of 62° of Farenheit's thermometer. Now, as that mean density is double that of basalt, and more than double that of granite, rocks which undoubtedly emanate from very great depths beneath the surface of

---

moon on the protuberant matter at the earth's equator does not actually give the compression, it proves that it must be between $\frac{1}{279}$ and $\frac{1}{573}$. Coincidences so near and so remarkable, arising from such different methods, show how nearly the irregular figure of the earth has been determined. The inequalities in the motions of the moon and earth alluded to are explained in Sections 5 and 11, 'Connexion of Physical Sciences.'

the earth, it affords another proof of the increase in density towards the earth's centre. These experiments were first made by Cavendish and Mitchell, and latterly with much greater accuracy by the late Mr. Baily, who devoted four years of unremitted attention to the accomplishment of this important and difficult object.[1]

Although the earth increases in density regularly from the surface to the centre, as might naturally be expected, from the increasing pressure, yet the surface consists of a great variety of substances of different densities, some of which occur in amorphous masses; others are disposed in regular layers or strata, either horizontal or inclined at all angles to the horizon. By mining, man has penetrated only a very little way; but by reasoning from the dip or inclination of the strata at or near the surface, and from other circumstances, he has obtained a pretty accurate idea of the structure of our globe to the depth of about ten miles. All the substances of which we have any information are divided into four classes, distinguished by the manner in which they have been formed: namely,—plutonic and volcanic rocks, both of igneous origin, though produced under different circumstances; aqueous or stratified rocks, entirely due to the action of water, as the name implies; and metamorphic rocks, deposited by water, according to the opinion of many eminent geologists, and consequently stratified, but subsequently altered and crystallized by heat. The aqueous and volcanic rocks are formed at or near the surface of the earth, the plutonic and metamorphic at great depths; but all of them have originated simultaneously during every geological period, and are now in a state of slow and constant progress. The antagonist principles of fire and water have ever been and still are the cause of the perpetual vicissitudes to which the crust of the earth is liable.

It has been ascertained by observation that the plutonic rocks, consisting of the granites and some of the porphyries, were formed in the deep and fiery caverns of the earth, of melted matter, which crystallized as it slowly cooled under enormous pressure, and was then heaved up in unstratified masses, by the elastic force of the internal heat, even to the tops of the highest mountains, or forced in a semi-fluid state into fissures of the superincumbent strata, sometimes into the cracks of the previously formed granite; for that rock, which constitutes the base of so large a portion of the earth's crust, has not been all formed at once; some portions had been solid, while others were yet in a liquid state. This class of rocks is completely destitute of fossil remains.

[1] It is clear that the mean density of the earth may be found from the attraction of the plumb-line by mountains, or by the irregularity in the oscillations of the pendulum, but the torsion balance is a much more sensible instrument than either. The density determined by M. Reich differs from that found by Mr. Baily by only one twenty-eighth part.

Although granite and the volcanic rocks are both due to the action of fire, their nature and position are very different; granite, fused in the interior of the earth, has been cooled and consolidated before coming to the surface: besides, it generally consists of few ingredients, so that it has nearly the same character in all countries. But as the volcanic fire rises to the very surface of the earth, fusing whatever it meets with, volcanic rocks take various forms, not only from the different kinds of strata which are melted, but from the different conditions under which the liquid matter has been cooled, though most frequently on the surface—a circumstance that seems to have had the greatest effect on its appearance and structure. Sometimes it assumes a crystalline granitic structure, at other times it becomes glass; in short, all those massive, unstratified, and occasionally columnar rocks, as basalt, greenstone, porphyry, and serpentine, are due to volcanic fires, and are devoid of fossil remains.

There seems scarcely to have been any age of the world in which volcanic eruptions have not taken place in some part of the globe. Lava has pierced through every description of rocks, spread over the surface of those existing at the time, filled their crevices, and flowed between their strata. Ever changing its place of action, it has burst out at the bottom of the sea as well as on dry land. Enormous quantities of scoriæ and ashes have been ejected from numberless craters, and have formed extensive deposits in the sea, in lakes, and on the land, in which are embedded the remains of the animals and vegetables of the epoch. Some of these deposits have become hard rock, others remain in a crumbling state; and as they alternate with the aqueous strata of almost every period, they contain the fossils of all the geological epochs, chiefly fresh and salt-water testaceæ.[1]

According to a theory now generally adopted, which originated with Sir Charles Lyell, whose works are models of philosophical investigation, the metamorphic rocks, which consist of gneiss, mica-schist, clay-slate, statuary marble, &c., were formed of the sediment of water in regular layers, differing in kind and colour, but, having been deposited near the place where plutonic rocks were generated, they have been changed by the heat transmitted from the fused matter, and, in cooling under heavy pressure and at great depths, they have become as highly crystallized as the granite itself, without losing their stratified form. An earthy stratum has sometimes been changed into a highly crystallized rock, to the distance of a quarter of a mile from the point of contact, by transmitted heat; and there are instances of dark-coloured limestone, full of fossil shells, that has been changed into statuary marble from that cause. Such alterations may frequently be seen to a small extent on rocks adjacent to a stream of lava. There is seldom a trace of organic remains in the

[1] Testaceæ are shell-fish.

metamorphic rocks; their strata are sometimes horizontal, but they are usually tilted at all angles to the horizon, and form some of the highest mountains and most extensive table-lands on the face of the globe. Although there is the greatest similarity in the plutonic rocks in all parts of the world, they are by no means identical; they differ in colour, and even in ingredients, though these are few.

Aqueous rocks are all stratified, being the sedimentary deposits of water. They originate in the wear of the land by rain, streams, or the ocean. The débris carried by running water is deposited at the bottom of the seas and lakes, where it is consolidated, and then raised up by subterraneous forces, again to undergo the same process after a lapse of time. By the wasting away of the land the lower rocks are laid bare, and, as the materials are deposited in different places according to their weight, the strata are exceedingly varied, but consist chiefly of arenaceous or sandstone rocks, composed of sand, clay, and carbonate of lime. They constitute three great classes, which, in an ascending order, are the primary and secondary fossiliferous strata and the tertiary formations.

The primary fossiliferous or palæozoic strata, the most ancient of all the sedimentary rocks, consisting of limestone, sandstones, and shales, are entirely of marine origin, having been formed far from land at the bottom of a very deep ocean; consequently they contain the exuviæ of marine animals only, and after the lapse of unnumbered ages the ripple-marks of the waves are still distinctly visible on some of their strata. This series of rocks is subdivided into the Cambrian and the upper and lower Silurian and Carboniferous systems, each distinguished by the differences in their fossil remains.

In the Cambrian rocks, sometimes many thousand yards thick, organic remains are of comparatively rare occurrence, but the Silurian rocks abound in them more and more as the strata lie higher in the series. In the lower Silurian group are the remains of shell-fish, almost all of extinct genera, and the few that have any affinity to those alive are of extinct species; crinoidea, or stone lilies, which had been fixed to the rocks like tulips on their stems, are coëval with the earliest inhabitants of the deep; and the trilobite, a jointed creature of the crab kind, with prominent eyes, are almost exclusively confined to the Silurian strata, but the last traces of them are found in the coal-measures above. In the upper Silurian group are abundance of marine shells of almost every order, together with crinoidea, vast quantities of corals, and some sea-weeds: several sauroid fishes,[1] of extinct genera, but of a high organization, have been found in the highest beds—the only vertebrated animals that have yet been discovered among the countless profusion of the lower orders of animals that are entombed in the primary fossiliferous strata. The re-

[1] Sauroid fish have somewhat of the form and organization of the lizard tribe.

mains of one or more land plants, in a very imperfect state, have been found in the Silurian rocks of North America, which shows that there had been land with vegetation at that early period. The type of these plants, as well as the size of the shells and the quantity of the coral, indicate that a uniformly warm temperature had then prevailed over the globe. During the Silurian period an ocean covered the northern hemisphere, islands and lands of moderate size had just begun to rise, and earthquakes, with volcanic eruptions from insular and submarine volcanoes, were frequent towards its close.

The secondary fossiliferous strata, which comprise a great geological period, and constitute the principal part of the high land of Europe, were deposited at the bottom of an ocean, like the primary, from the débris of all the others, carried down by water, and still bear innumerable tokens of their marine origin, although they have for ages formed a part of the dry land. Calcareous rocks are more abundant in these strata than in the crystalline, probably because the carbonic acid was then, as it still is, driven off from the lower strata by the internal heat, and came to the surface as gas or in calcareous springs, which either rose in the sea and furnished materials for shell-fish and coral insects to build their habitations and form coral reefs, or deposited their calcareous matter on the land in the form of rocks.

The Devonian or old Red Sandstone group, in many places 10,000 feet thick, consisting of strata of dark red and other sandstones, marls, coralline limestones, conglomerates, &c., is the lowest of the secondary fossiliferous strata, and forms a link between them and the Silurian rocks, by an analogy in their fossil remains. It has fossils peculiarly its own, but it has also some shells and corals common to the strata both above and below it. There are various families of extinct sauroid fishes in this group, some of which were gigantic, others had strong bony shields on their heads, and one genus, covered with enamelled scales, had appendages like wings. The shark approaches nearer to some of these ancient fish than any other now living.[1]

During the long period of tranquillity that prevailed after the Devonian group was deposited, a very warm, moist, and extremely equable climate, which extended all over the globe, had clothed the islands and lands in the ocean then covering the northern hemisphere with exuberant tropical forests and jungles. Subsequent inroads of fresh water, or of the sea, or rather partial sinkings of the land, had submerged these forests and jungles, which being mixed with layers

[1] The old red sandstone of Scotland, where it is remarkably well developed, has been admirably illustrated in two recent works, by one of our most industrious and talented northern geologists, Mr. Hugh Millar. See 'Old Red Sandstone,' and the recently published work, 'Footprints of the Creator,' 1 vol. 12mo., 1850.

of sand and mud, had in time been consolidated into one mass, and were then either left dry by the retreat of the waters or gently raised above the surface.

These constitute the remarkable group of the carboniferous strata, which consist of numberless layers of various substances filled with a prodigious quantity of the remains of fossil land-plants intermixed with beds of coal, which is entirely composed of vegetable matter. In some cases the plants appear to have been carried down by floods, and deposited in estuaries; but in most instances the beauty, delicacy, and sharpness of the impressions show that they had grown on to the spot where the coal was formed. More than 300 fossil plants have been collected from the strata where they abound, frequently with their seeds and fruits, so that enough remains to show the peculiar nature of this flora, whose distinguishing feature is the preponderance of ferns; among these there were tree-ferns which must have been 40 or 50 feet high. There were also plants resembling the fox-fail tribe, of gigantic size, others like the tropical club mosses; an aquatic plant of an extinct family was very abundant, besides many others, to which we have nothing analogous. Forest-trees of great magnitude, of the pine and fir tribes, flourished at that period. The remains of an extinct araucaria, one of the largest of the pine family, have been found in the British coal-fields; the existing species now grow in countries in the southern hemisphere; a few rare instances occur of grasses, palms, and liliaceous plants. The botanical districts were very extensive when the coal-plants were growing, for some species are nearly identical throughout the coal-fields of Europe and America. From the extent of the ocean, the insular structure of the land, the profusion of ferns and fir-trees, and the warm, moist, and equable climate, the northern hemisphere, during the formation of the coal strata, is thought to have borne a strong resemblance to the South Pacific, with its fern and fir-clothed lands of New Zealand, Kerguelen-land, and others.

The animal remains of this period are in the mountain limestone, a rock occasionally 900 feet thick, which lies beneath the coal-measures, or sometimes alternates with the shale and sandstone. They consist of crinoidea and marine testaceæ, among which the size of the chambered shells, as well as that of the coral, shows that the ocean was very warm at that time, even in the high northern latitudes. The footsteps of a very large reptile of the frog tribe have been found on some of the carboniferous strata of North America.

The coal strata have been very much broken and deranged in many places by earthquakes and igneous eruptions, giving rise to faults or dykes, basaltic veins, which frequently occurred during the secondary fossiliferous period, and from time to time raised islands and land from the deep. However, these and all other changes that have taken place on the earth have been gradual and partial, whether brought about by fire or water. The older rocks are more shattered by the earthquakes than the newer, because the movement came from below; but these convulsions have never extended all over the earth at the same time—they have always been local: for example, the Silurian strata have been dislocated and tossed in Britain, while a vast area in the south of Sweden and Russia still retains a horizontal position. There is no proof that any mountain-chain has ever been raised at once; on the contrary, the elevation has always been produced by a long-continued and reiterated succession of internal convulsions with intervals of repose. In many instances the land has risen up or sunk down by an imperceptible equable motion continued for ages, while in other places the surface of the earth has remained stationary for long geological periods.

The magnesian limestone, or permian formation, comes immediately above the coal-measures, and consists of breccias or conglomerates, gypsum, sandstone, marl, &c.; but its distinguishing feature in England is a yellow limestone rock, containing carbonate of magnesia, which often takes a granular texture, and is then known as dolomite. The permian formation has a fossil flora and fauna peculiar to itself, mingled with those of the coal strata. Here the remnant of an earlier creation gradually tends to its final extinction, and a new one begins to appear. The flora is, in many instances, specifically the same with that in the coal strata below. Certain fish are also common to the two, which never appear again. They belong to a race universal in the early geological periods, forming a kind of passage from the first tribe to saurian reptiles, and therefore called Sauroid. A small number of existing genera only, such as the shark and sturgeon, make some approach to the structure of these ancient inhabitants of the waters. The new creation is marked by the introduction of two species of saurian reptiles;[1] the fossil remains of one have been found in the magnesian limestone in England, and those of the other in a corresponding formation in Germany. They are the earliest members of a family which was to have dominion on the land and in the water for ages.

A series of red marls, rock-salt, and sandstones, which have arisen from the disintegration of metamorphic slates and porphyritic trap, containing oxide of iron, and known as the trias or new red sandstone system, lies above the magnesian limestone. In England this formation is particularly rich in rock-salt, which, with layers of gypsum and marl, is sometimes 600 feet thick; but in this country the muschelkalk, a peculiar kind of shelly limestone, is wanting, which in Germany and on the southern declivity of the Alps, is so remarkable for the quantity of organic remains. At this time creatures like frogs, of enormous dimensions, had been frequent, as they have left their footsteps on what must then have been a soft shore.

---

[1] Saurian reptiles are crocodiles, lizards, iguanas, &c

Forty-seven genera of fossil remains have been found in the trias in Germany, consisting of shells, cartilaginous fish, encrinites, &c., all distinct in species, and many distinct in genera, from the organic fossils of the magnesian limestone below, and also from those entombed in the strata above.

During a long period of tranquillity the oolite or Jurassic group was next deposited in a sea of variable depth, and consists of sands, sandstones, marls, clays, and limestone. At this time there was a complete change in the aqueous deposits all over Europe. The red iron-stained arenaceous rocks, the black coal, and dark strata, were succeeded by light-blue clays, pale-yellow limestones, and, lastly, white chalk. The water that deposited the strata must have been highly charged with carbonate of lime, since few of the formations of that period are without calcareous matter, and calcareous rocks were formed to a prodigious extent throughout Europe: the Pyrenees, Alps, Apennines, and Balkan abound in them; and the Jura mountains, which have given their name to the series, are formed of them. The European ocean then teemed with animal life; whole beds consist almost entirely of marine shells and corals. Belemnites and ammonites, from an inch in diameter to the size of a cart-wheel, are entombed by myriads in the strata; whole forests of that beautiful encrinite the stone-lily flourished on the surface of the oolite, then under the waters; and the Pentacrinite, one of the same family, is embedded in millions in the enchorial shell-marble, which occupies such extensive tracts in Europe. Fossil fish are numerous in these strata, but different from those of the coal series, the permian formation, and trias; not one genus of the fish of this period is now in existence. The newly-raised islands and lands were clothed with vegetation like that of the large islands of the intertropical archipelagoes of the present day, which, no less rich than during the carboniferous period, still indicate a very moist and warm climate. Ferns were less abundant, as they were associated with various genera and species of the cycadeæ, which had grown on the southern coast of England, and in other parts of northern Europe, congeners of the present cycas and zamia of the tropics. These plants had been very numerous, and the pandanus, or screw-pine, the first tenant of the new lands in ancient and modern times, is a family found in a fossil state in the inferior oolite of England, which was but just rising from the deep at that time. The species now flourishing grows only on the coasts of such coral islands in the Pacific as have recently emerged from the waves. In the upper strata of this group, however, the confervæ and monocotyledonous plants[1] become more rare — an indication of a change of climate.

[1] Confervæ are plants with nearly imperceptible fructification, found in ponds, damp places, and in the sea.

Monocotyledonous plants are grasses, palms, and others, having onlv one seed-lobe.

The new lands that were scattered on the ocean of the oolitic period were drained by rivers, and inhabited by huge crocodiles and saurian reptiles of gigantic size, mostly of extinct genera. The crocodiles come nearest to modern reptiles; but the others, though bearing a remote similitude in general structure to living forms, were quite anomalous, combining in one the structure of various distinct creatures, and so monstrous that they must have been more like the visions of a troubled dream than things of real existence; yet in organization a few of them came nearer to the type of living mammalia than any existing reptiles do. Some of these had lived in rivers, others in the ocean—some were inhabitants of the land, others were amphibious; and the various species of one genus even had wings like a bat, and fed on insects. They were both herbivorous and predacious saurians; and from their size and strength they must have been formidable enemies. Besides, the numbers deposited are so great, especially in the lias, a marine stratum of clay and limestone, the lowest of the oolite series, that they must have swarmed for ages in the estuaries and shallow seas of the period. They gradually declined towards the end of the secondary fossiliferous epoch; but as a class they lived in all subsequent eras, and still exist in tropical countries, although the species are very different from their ancient congeners. Tortoises of various kinds were contemporary with the saurians, also a family that still exists. In the Stonefield slate, a stratum of the oolitic group, there are the remains of insects, and the bones of two small quadrupeds have been found there belonging to the marsupial tribe,[1] such as the opossum — a very remarkable circumstance, not only as the most ancient animal of the class of mammalia, but because that family of animals at the present time is confined to Australia, South America, and North America, as far north as Pennsylvania at least. The great changes in animal life during this period were indications of the successive alterations that had taken place on the earth's surface.

The cretaceous strata follow the oolite in ascending order, consisting of clay, green and iron sands, blue limestone, and chalk, probably formed of the decay of coral and shells, which predominates so much in England and other parts of Europe, that it has given the name and its peculiar feature to the whole group. It is, however, by no means universal; the chalk is wanting in many parts of the world where the other strata of this series prevail, and then their connexion with the group can only be ascertained by the identity of their fossil remains. With the exception of some beds of coal among the oolitic series, the Wealden clay, the lowest of the cretaceous group in England, is the only fresh-water formation, and the

[1] Marsupial animals have pouches in which their young take refuge and are nourished till they are matured. The opossum and kangaroo are marsupials.

tropical character of its flora shows that the climate was still very warm. Plants allied to the zamias and cycadeæ of our tropical regions, many ferns and pines of the genus araucaria, characterized its vegetation, and the upright stem of a fossil forest at Portland show that it had been covered with trees. It was inhabited by tortoises approaching to families now living in warm countries, and saurian reptiles of five different genera swarmed in the lakes and estuaries. This clay contains fresh-water shells and fish of the carp kind. The Wealden clay is one of the various instances of the subsidence of land which took place during this period.

The cretaceous strata above our Wealden clay are full of marine exuviæ. There are vast tracts of sand in Northern Europe, and many very extensive tracts of chalk; but in the southern part of the Continent the cretaceous rocks assume a different character. There and elsewhere extensive limestone rocks, filled with very peculiar shells, show that, when the cretaceous strata were forming, an ocean extended from the Atlantic into Asia, which covered the south of France, all Southern Europe, part of Syria, the isles of the Ægean Sea, the coasts of Thrace and the Troad. The remains of turtles have been found in the cretaceous group, quantities of coral, and abundance of shells of extinct species; some of the older kinds still existed, new ones were introduced, and some of the most minute species of microscopic shells, which constitute a large portion of the chalk, are supposed to be the same with creatures now alive, the first instances of identity of species in the ancient and modern creation. An approximation to recent times is to be observed also in the arrangement of organized nature, since at this early period, and even in the Silurian and oolitic epochs, the marine fauna was divided, as now, into distinct geographical provinces. The great saurians were on the decline, and many of them were found no more, but a gigantic creature, allied to the monitor and iguana,[1] lived at this period. From the permian group to the chalk inclusive only two instances of fossil birds recur, one in a chalk deposit in the Swiss Alps, and the other a kind of albatross in the chalk in England; in North America, however, the foot-marks of a variety of birds have been found in the strata between the coal and lias, some of which are larger than those of the ostrich.

An immense geological cycle elapsed between the termination of the secondary fossiliferous strata and the beginning of the tertiary. With the latter a new order of things commenced, approaching more closely to the actual state of the globe. During the tertiary formation the same causes under new circumstances produced an infinite variety in the order and kind of the strata, accompanied by a corresponding change in animal and vegetable life. The old creation,

---

[1] The monitor and iguana, creatures of the lizard tribe, still existing.

which had nothing in common with the existing order of things, had passed away, and given place to one more nearly approaching to that which now prevails. Among the myriads of beings that inhabited the earth and the ocean during the secondary fossiliferous epoch scarcely one species is to be found in the tertiary. Two planets could hardly differ more in their natural productions. This break in the law of continuity is the more remarkable, as hitherto some of the newly-created animals were always introduced before the older were extinguished. The circumstances and climate suited to the one became more and more unfit for the other, which consequently perished gradually, while their successors increased. It is possible that, as observations become more extended, this hiatus may be filled up.

The series of rocks, from the granite to the end of the secondary fossiliferous strata, taken as a whole, constitute the solid crust of the globe, and in that sense are universally diffused over the earth's surface. The tertiary strata occupy the hollows formed in this crust, whether by subterraneous movements, by lakes, or denudation by water as in the estuaries of rivers, and consequently occur in irregular tracts, often, however, of prodigious thickness and extent. Indeed, they seem to have been as widely developed as any other formation, though time has been wanting to bring them into view.

The innumerable basins and hollows with which the continents and larger islands had been indented for ages after the termination of the secondary fossiliferous series had sometimes been fresh-water lakes, and at other times inundated by the sea; consequently, the deposits which took place during these changes alternately contain the spoils of terrestrial and marine animals. The frequent intrusion of volcanic strata among the tertiary formations show that, in Europe, the earth had been in a very disturbed state, and that these repeated vicissitudes had been occasioned by elevations and depressions of the soil, as well as by the action of water.

There are three distinct groups in these strata: the lowest tertiary or Eocene group, so called by Sir Charles Lyell, because, among the myriads of fossil shell-fish which it contains, very few are identical with those now living; the Miocene, or middle group, has a greater number of the exuviæ of existing species of shells; and the Pliocene, or upper tertiary group, still more. Though frequently heaved up to great elevations on the flanks of the mountain-chains, as, for example, on the Alps and Apennines, by far the greater part of the tertiary strata maintain their original horizontal position in the very places where they are formed. Immense insulated deposits of this kind are to be met with all over the world; Europe abounds with them, London, Paris, and Vienna stand on such basins, and they cover immense tracts both in North and South America.

The monstrous reptiles had mostly disappeared, and the mammalia now took possession of the earth, of forms scarcely less anomalous than their predecessors, though approaching more nearly to those now living.

Numerous species of extinct animals that lived during the earliest or Eocene period have been found in various parts of the world, especially in the Paris basin, of the order of Pachydermata,[1] to the greater number of which we have nothing analogous; they were mostly herbivorous quadrupeds, which had frequented the borders of the rivers and lakes that covered the greater part of Europe at that time. This is the more extraordinary, as existing animals most similar to these, the tapirs for instance, are confined to the torrid zone. These creatures were widely diffused, and some of them were associated with genera still existing, though of totally different species; such as animals allied to the racoon and dormouse, the ox, bear, deer, the fox, the dog, and others. Although these quadrupeds differ so widely from those of the present day, the same proportion existed then as now between the carnivorous and herbivorous genera. The spoils of marine mammalia[2] of this period have also been found, sometimes at great elevations above the sea, all of extinct species, and some of these cetacea were of huge size. This marvellous change of the creative power was not confined to the earth and the ocean; the air was now occupied by many extinct races of birds allied to the owl, buzzard, quail, curlew, &c. The climate must still have been warmer than at present, from the remains of land and sea plants found in high latitudes. Even in England bones of the opossum, monkey, and boa have been discovered, all animals of warm countries, besides fossil sword and saw fish, both of genera foreign to the British seas.

During the Miocene period new amphibious quadrupeds were associated with the old, of which the deinotherium is the most characteristic and much the largest of the mammalia yet found, surpassing the largest elephant in size, and of a singular form.

The palæotherium was of this period, and also the mastodon, both of large dimensions. Various families, and even genera, of quadrupeds now existing were associated with these extraordinary creatures, though of extinct species, such as the elephant, rhinoceros, hippopotamus, tapir, horse, bear, wolf, hyæna, weasel, beaver, ox, buffalo, deer, &c.; and also marine mammalia, as dolphins, walruses, and lamantins. Indeed, in the constant increase of animal life manifested throughout the whole of the tertiary strata, the forms approach nearer to the living species as their remains lie high in the series.

---

[1] Pachydermata, thick-skinned animals, as the rhinoceros, elephant, and hippopotamus.

[2] Marine mammalia, which suckle their young like land animals, are seals, whales, porpoises, &c.

In the older Pliocene period some of the large amphibious quadrupeds, and other genera of mammalia of the earlier tertiary periods, appear no more; but there were the mastodon, and the Elephas primigenius or mammoth, some species of which, of prodigious size, were associated with numerous quadrupeds of existing genera, but lost species. Extinct species of almost all the quadrupeds now alive seem to have inhabited the earth at that time; their bones have been discovered in caverns; they were embedded in the breccias and in most of the strata of that epoch — as the hippopotamus, rhinoceros, elephant, horse, bear, wolf, water-rat, hyæna, tiger, and various birds. It is remarkable that in the caverns of Australia the fossil bones all belong to extinct species of gigantic kangaroos and wombats, animals belonging to the marsupial family, which are so peculiarly the inhabitants of that country at the present day, but of diminished size. The newer Pliocene strata show that the same analogy existed between the extinct and recent mammalia of South America, which, like their living congeners, as far as we know, belonged to that continent alone; for the fossil remains, quite different from those in the old world, are of animals of the same families with the sloths, ant-eaters, and armadilloes which now inhabit that country, but of vastly superior size and different species. In fact, there were giants in the land in those days. Were change of species possible, one might almost fancy that these countries had escaped the wreck of time, and that their inhabitants had pined and dwindled under the change of circumstances. The megatherium and Equus curvidens, or extinct horse, had so vast a range in America, that, while Sir Charles Lyell collected their bones in Georgia in 33° N. latitude, Mr. Darwin brought them from the corresponding latitude in South America. The Equus curvidens differed as much from the living horse as the quagga or zebra does, and the European fossil horse is also probably a distinct and lost species.

A comparison of the fossil remains with the living forms has shown the analogy between these beings of the ancient world and those that now people the earth; and the greatest triumph of the geologist is the certainty with which he can decide upon the nature of animals that have been extinct for thousands of years, from a few bones entombed on the earth's surface. Baron Cuvier will ever be celebrated as the founder of this branch of comparative anatomy, which Professor Owen, following in his steps, has brought to the highest perfection. Among many discoveries, the latter has found, by the most minute microscopic observation, that the structure of the tissue of which teeth are formed is different in different classes of animals, and that the species can in many instances be determined from the fragment of a tooth. A small portion of a bone enabled him to decide on the nature of an extinct race of birds, and the subsequent discovery of the whole skeleton confirmed the accuracy of this determination.

The greater part of the land in the northern hemisphere was elevated above the deep during the tertiary period, and such lands as already existed acquired additional height; consequently the climate, which had previously been tropical, became gradually colder, for an increase of land, which raises the temperature between the tropics, has exactly the contrary effect in higher latitudes. Hence excessive cold prevailed during the latter part of the Pliocene period, and a great part of the European continent was discovered by an ocean full of floating ice, not unlike that seen at this day off the north-eastern coast of America.[1]

During the latter part of the Pliocene period, however, the bed of that glacial ocean rose partially, and after many vicissitudes the European continent assumed nearly the form it now bears. There is every reason to believe that the glacial sea extended also over great portions of the arctic lands of Asia and America. Old forms of animal and vegetable life were destroyed by these alterations in the surface of the earth, and the consequent change of temperature; and when, in the progress of the Pliocene period, the mountain-tops appeared as islands above the water, they were clothed with the flora and peopled by the animals they still retain; and new forms were added as the land rose and became dry and fitted to receive and maintain the races of animals now alive, all of which had possession of the earth for ages prior to the historical or human period. Some of the extinct animals had long resisted the great vicissitudes of the times; of these the mammoth, or Elephas primigenius, whose fossil remains are found all over Europe, Asia, and America, but especially in the gelid soil of Siberia, alone outlived its associates, the last remnant of a former world. In two or three instances this animal has been discovered entire, entombed in frozen mud, with its hair and flesh so fresh that wolves and dogs feed upon it. The globe of the eye of one found by M. Middendorf at Tas, between the rivers Oby and Jenesei, was so perfect that it is now preserved in the museum at Moscow. It has been supposed that, as the Siberian rivers flow for hundreds of miles from the southern part of the country to the Arctic Ocean, these elephants might have been drowned by floods while browsing in the milder regions, and that their bodies were carried down by the rivers and embedded in mud,

---

[1] If a line be drawn from the north-eastern coast of North America within the limit of floating ice, and if it be continued across the southern half of Ireland and England, and prolonged eastward so as to strike against the Ural mountains, it will mark the boundary of the European portion of the Glacial Sea. It submerged part of Russia to the depth of 1000 feet.—Essay on the British Fauna and Flora, by Professor E. Forbes, in the 'Memoirs of the Geological Survey of Great Britain,' vol. i.

and frozen before they had time to decay. Mr. Darwin has suggested that, if the climate of Siberia has at any time been similar to that of the high latitudes of South America, where the line of perpetual snow in the Andes, and its sudden flexure in Southern Chile, come close to a nearly tropical vegetation, such a vegetation may have prevailed south of the frozen regions in Siberia. On the other hand, although the congeners of this animal are now inhabitants of the torrid zone, they may have been able to endure the cold of a Siberian winter; for Baron Cuvier found that this animal differed as much from the living elephant as the horse does from the ass. Mr. Darwin has shown that the supply of food in summer was probably sufficient, since the quantity requisite for the maintenance of the larger animals is by no means in proportion to their bulk; or these elephants may have migrated to a more genial climate in the colder months.

Shell-fish seem to have been more able to endure all the great geological changes than any of their organic associates, but they show a constant approximation to modern species during the progress of the tertiary period. The whole of these strata contain enormous quantities of shells of extinct species; in the oldest, three and a half per cent. of the shells are identical with species now existing, while in the uppermost strata of this geological period there are not less than from ninety to ninety-five in a hundred identical with those now alive.

Of all the fossil fishes, from the Silurian strata to the end of the tertiary, scarcely one is specifically the same with living forms: the Mallotus villosus, or caplan, of the salmon family, is an exception, and perhaps a few others of the most recent of these periods. In the Eiocene strata one-third belong to extinct genera.

Under the vegetable mould in every country there is a stratum of loose sand, gravel, and mud, lying upon the subjacent rocks, often of great thickness, called alluvium, which in the high latitudes of North America and Europe is mixed with enormous fragments of rock, sometimes angular, and sometimes rounded and water-worn, which have been transported hundreds of miles from their origin. It is there known as the Boulder formation, or Northern Drift, because, from the identity of the boulders with the rocks of the northern mountains, they evidently have come from them, and their size becomes less as the distance increases. In Russia there are blocks of great magnitude that have been carried 800 and even 1000 miles south-east from their origin in the Scandinavian range. There is much reason to believe that such masses, enormous as they are, have been transported by ice-bergs, and deposited when the northern parts of the continents were covered by the glacial sea, by which part of Russia was submerged to the depth of at least 1000 feet The same process is now in progress in the high southern latitudes, where icebergs have been met with covered with fragments of rock and boulders.[1]

The last manifestation of creative power, with few exceptions, differs specifically from all that preceded it; the recent strata contain only the exuviæ of animals now living, often mixed with the works of man.

The solid earth thus tells us of mountains washed down into the sea with their forests and inhabitants; of lands raised from the bottom of the ocean, loaded with the accumulated spoils of centuries; of torrents of water and torrents of fire. In the ordinances of the heavens no voice declares a beginning, no sign points to an end; in the bosom of the earth, however, the dawn of life appears, the time is obscurely marked when first living things moved in the waters, when the first plants clothed the land. There we see that during ages of tranquillity the solid rock was forming at the bottom of the ocean, that during ages it was tossed and riven by fire and earthquake. What years must have gone by since that ocean flowed which has left its ripple-marks on the sand, now a solid mass on the mountain—since those unknown creatures left their foot-prints on the shore, now fixed by time on the rock for ever! time, which man measures by days and years, nature measures by thousands of centuries.

The thickness of the fossiliferous strata up to the end of the tertiary formation has been estimated at about seven or eight miles; so that the time requisite for their deposition must have been immense. Every river carries down mud, sand, or gravel, to the sea: the Ganges brings more than 700,000 cubic feet of mud every hour, the Yellow River in China 2,000,000,[2] and the Mississippi still more; yet, notwithstanding these great deposits, the Italian hydrographer Manfredi has estimated that, if the sediment of all the rivers on the globe were spread equally over the bottom of the ocean, it would require 1000 years to raise its bed one foot; so that at that rate it would require 3,960,000 years to raise the bed of the ocean alone to a height nearly equal to the thickness of the fossiliferous strata, or seven miles and a half, not taking account of the waste of the coasts by the sea itself: but if the whole globe be considered, instead of the bottom of the sea only, the time would be nearly four

---

[1] Sir James Ross and Captain Wilkes met with icebergs covered with mud and stones in the antarctic seas, and even in 66° 5′ lat. One block seen by Sir James Ross was estimated to weigh many tons. — Antarctic Voyages. — [Narrative of the United States' Exploring Expedition. By Charles Wilkes, U. S. Navy.]

[2] Account of the Ganges and Brahmapootra, by Major Rennell. — 'Phil. Trans.,' 1781. Sir George Staunton's 'Embassy to China.' 'Élie de Beaumont, Leçons de Géologie,' 1 vol. 8vo. The latter work contains a very elaborate essay on alluvial deposits by rivers, &c.

times as great, even supposing as much alluvium to be deposited uniformly both with regard to time and place, which it never is. Besides, in various places the strata have been more than once carried to the bottom of the ocean, and again raised above its surface by subterranean fires after many ages, so that the whole period from the beginning of these primary fossiliferous strata to the present day must be great beyond calculation, and only bears comparison with the astronomical cycles, as might naturally be expected, the earth being without doubt of the same antiquity with the other bodies of the solar system. What then shall we say if the time be included which the granitic, metamorphic, and recent series occupied in forming? These great periods of time correspond wonderfully with the gradual increase of animal life and the successive creation and extinction of numberless orders of being, and with the incredible quantity of organic remains buried in the crust of the earth in every country on the face of the globe.

Every great geological change in the nature of the strata was accompanied by the introduction of a new race of beings, and the gradual extinction of those that had previously existed, their structure and habits being no longer fitted for the new circumstances in which these changes had placed them. The change, however, was never abrupt; and it may be observed that there is no proof of progressive development of species by generation from a low to a high organization, for animals and plants of high organization appeared among the earliest of their kind, yet throughout the whole the gradual approach to living and more perfect forms is undoubted, not by change of species, but by increasing similarity of type.

The geographical distribution of animated beings was much more extensive in the ancient seas and land than in later times. In very remote ages the same animal inhabited the most distant parts of the sea; the corallines built from the equator to within ten or fifteen degrees of the pole; and previous to the formation of the carboniferous strata there appears to have been even a greater uniformity in the vegetable than in the animal world, though New Holland had formed even then a peculiar district, supposing the coal in that country to be of the same age as in Europe and America; but as the strata became more varied, species were less widely diffused. Some of the saurians were inhabitants of both the Old and New World, while others lived in the latter only. During the tertiary period the animals of Australia and America differed nearly as much from those of Europe as they do at the present day. The world was then, as now, divided into great physical regions, each inhabited by a peculiar race of animals; and even the different species of mollusca of the same sea were confined to certain shores. Of 405 species of the latter which inhabited the Atlantic Ocean during the early and middle parts of the tertiary period, only 12 were common to the American and European coasts. In fact, the divisions of the animal and vegetable creation into geographical districts had been in the latter periods contemporaneous with the rise of the land, each portion of which, as it rose above the deep, had been clothed with a vegetation and peopled with creatures suited to its position with regard to the equator, and to the existing circumstances of the globe; and the marine creatures had, no doubt, been divided into districts at the same periods, because the bed of the ocean had been subject to similar changes.

The quantity of fossil remains is so great that, with the exception of the metals and some of the primary rocks, probably not a particle of matter exists on the surface of the earth that has not at some time formed part of a living creature. Since the commencement of animated existence, zoophytes have built coral reefs extending hundreds of miles, and mountains of limestone are full of their remains all over the globe. Mines of shells are worked to make lime; ranges of hills and rocks, many hundred feet thick, are almost entirely composed of them, and they abound in every mountain-chain throughout the earth. The prodigious quantity of microscopic shells discovered by M. Ehrenberg is still more astonishing; shells not larger than a grain of sand form entire mountains; a great portion of the hills of San Casciano, in Tuscany, consist of chambered shells so minute that Padre Soldani collected 10,454 of them from one ounce of stone. Chalk is often almost entirely composed of them. Tripoli, a fine powder long in use for polishing metals, is also almost wholly composed of shells which owe their polishing property to their silicious coats; and there are even hills of great extent consisting of this substance, the débris of an infinite variety of microscopic insects.

The facility with which many clays and slates are split is owing, in some instances, to layers of minute shells. Fossil fish are found in all parts of the world, and in all the fossiliferous strata with the exception of some of the lowest, but each great geological period had species peculiar to itself.

The remains of the great saurians are innumerable; those of extinct quadrupeds are very numerous; but there is no circumstance in the whole science of fossil geology more remarkable than the inexhaustible multitudes of fossil elephants that are found in Siberia. Their tusks have been an object of traffic in ivory for centuries, and in some places they have been in such prodigious quantities, that the ground is tainted with the smell of animal matter. Their huge skeletons are found from the frontier of Europe through all Northern Asia to its extreme eastern point, and from the foot of the Altaï Mountains to the shores of the Frozen Ocean, a surface equal in extent to the whole of Europe. Some islands in the Arctic Sea, as, for instance, the first of the Läcbow group, are chiefly composed of their remains, mixed with the bones of various other animals of living genera, but extinct species.[1]

Equally wonderful is the quantity of fossil plants that still remain, if it be considered that, from the frail nature of many vegetable substances, multitudes must have perished without leaving a trace behind. The vegetation that covered the terrestrial part of the globe previous to the formation of the carboniferous strata had far surpassed in exuberance the rankest tropical jungles. There are many coal-fields of great extent in various parts of the earth, especially in North America, where that of Pittsburg occupies an area of about 14,000 square miles, and that in the Illinois is not much inferior to the area of all England.[2]

As coal is entirely a vegetable substance, some idea may be formed of the richness of the ancient flora; in later times it was less exuberant, and never has again been so luxuriant, probably on account of the decrease of temperature during the deposition of the tertiary strata, and in the glacial period which immediately preceded the creation of the present tribes of plants and animals. Even after their introduction the temperature must have been very low, but by subsequent changes in the distribution of the sea and land the cold was gradually mitigated, till at last the climate of the northern hemisphere became what it now is.

Such is the marvellous history laid open to us on the earth's surface. Surely it is not the heavens only that declare the glory of God—the earth also proclaims His handiwork![3]

---

[1] Lieut. Anjou's Polar Voyage.

[2] [See 'Statistics of Coal,' by Richard Cowling Taylor, Philadelphia, 1848.]

[3] The author's geological information rests on the authority of those distinguished authors whose works are in the hands of every one, namely, Baron Cuvier, Sir Charles Lyell, Sir Roderick Murchison, Sir Henry de la Beche, Professor Owen, M. Elie de Beaumont, and the Memoirs of the Geological Society.

CHAPTER II.

Direction of the Forces that raised the Continents — Proportion of Land and Water—Size of the Continents and Islands—Outline of the Land—Extent of Coasts, and proportion they bear to the Areas of the Continents — Elevation of the Continents — Forms of Mountains — Forms of Rocks — Connexion between Physical Geography of Countries and their Geological Structure — Contemporaneous Upheaval of parallel Mountain Chains — Parallelism of Mineral Veins or Fissures — Mr. Hopkins's Theory of Fissures — Parallel Chains similar in Structure — Interruptions in Continents and Mountain Chains—Form of the Great Continent —The High Lands of the Great Continent—The Atlas, Spanish, French, and German Mountains—The Alps, Balkan, and Apennines—Glaciers—Geological Notice.

At the end of the tertiary period the earth was much in the same state as it is at present with regard to the distribution of land and water. The preponderance of land in the northern hemisphere indicates a prodigious accumulation of internal energy under these latitudes at a very remote geological period. The forces that raised the two great continents above the deep, when viewed on a wide scale, must evidently have acted at right angles to one another, nearly parallel to the equator in the old continent, and in the direction of the meridian in the new; yet the structure of the opposite coasts of the Atlantic points at some connexion between the two.

The mountains, from their rude and shattered condition, bear testimony to repeated violent convulsions similar to modern earthquakes; while the high table-lands, and that succession of terraces by which the continents sink down from their mountain-ranges to the plains, to the ocean, and even below it, show also that the land must have been heaved up occasionally by slow and gentle pressure, such as appears now to be gradually elevating the coast of Scandinavia and many other parts of the earth. The periods in which these majestic operations were effected must have been incalculable, since the dry land occupies an area of nearly 38,000,000 of square miles.

The ocean covers nearly three-fourths of the surface of the globe, but the distribution is very unequal, whether it be considered with regard to the northern and southern hemispheres, or the eastern and western. Independently of Victoria Land, whose extent is unknown, the quantity of land in the northern hemisphere is three times greater than in the southern. In the latter it occupies only one-sixteenth of the space between the Antarctic Circle and the thirteenth parallel of south latitude, while between the corresponding parallels in the northern hemisphere the extent of land and water is nearly equal. If the globe be divided into two hemispheres by a meridian passing through the island of Teneriffe, the land will be found to predominate greatly on the eastern side of that line, and the water on the western. In consequence of the very unequal arrangement of the solid and liquid portions of the surface of the earth, England is nearly in the centre of the greatest mass of land, and its antipode, the island of New Zealand, is in the centre of the greatest mass of water; so that a person raised above Falmouth, which is almost the central point, till he could perceive a complete hemisphere, would see the greatest possible expanse of land, while, were he elevated to the same height above New Zealand, he would see the greatest possible extent of ocean.[1] In fact, only one-twenty-seventh of the land has land directly opposite to it in the opposite hemisphere, and under the equator five-sixths of the circumference of the globe is water. It must however be observed that there is still an unexplored region within the Antarctic Circle more than twice the size of Europe, and of the north polar basin we know nothing. With regard to the land alone, the great continent has an area of about 24,000,000 square miles, while the extent of America is 11,000,000, and that of Australia with its islands scarcely 3,000,000. Africa is more than three times the size of Europe, and Asia is more than four times as large. The extent of the continents is twenty-three times greater than that of all the islands taken together.[2]

Of the polar lands little is known. Greenland probably is part of a continent, the domain of perpetual snow; and the recent discovery of so extensive a mass of high volcanic land near the south pole is an important event in the history of physical science, though the stern severity of the climate must for ever render it unfit for the abode of animated beings, or even for the support of vegetable life. It seems to form a counterpoise to the preponderance of dry land in the northern hemisphere. There is something sublime in the contemplation of these lofty and unapproachable regions — the awful realm of ever-during ice and perpetual fire, whose year consists of

[1] M. Gay Lussac, at the height of four miles and a quarter, must have seen 10,857 square miles of the earth's surface from his balloon. Mr. Green, who ascended to the height of five miles, must have seen 13,154 square miles of the globe, the greatest extent viewed by man.

[2] The proportion of land to water referred to in the text was estimated by Mr. Gardner. According to his computation, the extent of land is about 37,673,000 square British miles, independently of Victoria Continent [discovered by Charles Wilkes, U. S. N.] and the sea occupies 110,849,000. Hence the land is to the sea as 1 to 4 nearly. The unexplored region within the Arctic Circle is about 7,620,000 square miles.

one day and one night. The strange and terrible symmetry in the nature of the lands within the polar circles, whose limits are to us a blank, where the antagonist principles of cold and heat meet in their utmost intensity, fills the mind with that awe which arises from the idea of the unknown and the indefinite.

The tendency of the land to assume a peninsular form is very remarkable, and it is still more so that almost all the peninsulas tend to the south — circumstances that depend on some unknown cause which seems to have acted very extensively. The continents of South America, Africa, and Greenland are peninsulas on a gigantic scale, all tending to the south; the Asiatic peninsula of India, the Indo-Chinese peninsula, those of Corea, Kamtschatka, of Florida, California, and Aliaska, in North America, as well as the European peninsulas of Norway and Sweden, Spain and Portugal, Italy and Greece, take the same direction. All the latter have a rounded form except Italy, whereas most of the others terminate sharply, especially the continents of South America and Africa, India and Greenland, which have the pointed form of wedges; while some are long and narrow, as California, Aliaska, and Malacca. Many of the peninsulas have an island or group of islands at their extremity, as South America which terminates with the group of Tierra del Fuego: India has Ceylon; Malacca has Sumatra and Banca; the southern extremity of Australia ends in Van Diemen's Land; a chain of islands run from the end of the peninsula of Aliaska; Greenland has a group of islands at its extremity; and Sicily lies close to the termination of Italy. It has been observed, as another peculiarity in the structure of peninsulas, that they generally terminate boldly, in bluffs, promontories, or mountains, which are often the last portions of the continental chains. South America terminates in Cape Horn, a high promontory, which is the visible termination of the Andes; Africa with the Cape of Good Hope; India with Cape Comorin, the last of the Ghauts; New Holland ends with South-East Cape in Van Diemen's Land: and Greenland's farthest point is the elevated bluff of Cape Farewell.[1]

There is a strong analogy between South America and Africa in form and the unbroken mass which their surface presents, while North America resembles Europe, in being much indented by inland seas, gulfs, and bays. Eastern Asia is evidently continued in a subaqueous continent from the Indian Ocean across the Pacific nearly to the west coast of America, of which New Holland, the Indian Archipelago, the islands of the Asiatic coast and of Oceanica are

---

[1] This very general view of the structure of the globe originated chiefly with the celebrated German geologist Von Buch, and has been much extended and developed by M. Elie de Beaumont, one of the most philosophical of modern geologists.

the great table-lands and summits of its mountain-chains. With the exception of a vast peninsula in Siberia between the mouths of the rivers Yenesei and Khatanga and the unknown regions of Greenland, the two great continents terminate in a very broken line to the north; and as they sink beneath the Icy Ocean, the tops of their high lands and mountains rise above the waves and stud the coast with innumerable snow-clad rocks and islands. The 70th parallel is the average latitude of these northern shores, which have a great similarity on each side of Behring's Straits in form, direction, and in the adjacent islands.

The peninsular form of the continents adds greatly to the extent of their coasts, of such importance to civilization and commerce. All the shores of Europe are deeply indented and penetrated by the Atlantic Ocean, which has formed a number of inland seas of great magnitude, so that it has a greater line of maritime coast, compared with its size, than any other quarter of the world. The extent of coast from the Straits of Waigatz, in the Polar Ocean, to the Strait of Caffa, at the entrance of the Sea of Azoff, is about 17,000 miles. The coast of Asia has been much worn by currents, and possibly also by the action of the ocean occasioned by the rotation of the earth from west to east. On the south and east especially it is indented by large seas, bays, and gulfs; and the eastern shores are rugged and encompassed by chains of islands which render navigation dangerous. Its maritime coast is about 33,000 miles in length.

The coast of Africa, 16,000 miles long, is very entire, except perhaps at the Gulf of Guinea and in the Mediterranean. The shores of North America have probably been much altered by the equatorial current and the Gulf-stream. There is little doubt that these currents, combined with volcanic action, have hollowed out the Gulf of Mexico, and separated the Antilles and Bahama Islands from the continent. The coast is less broken on the west, but in the Icy Ocean there is a labyrinth of gulfs, bays, and creeks. The shores of South America on both sides are very entire, except towards Southern Chile and Cape Horn, where the tremendous surge and currents of the Ocean in those high latitudes have eaten into the mountains, and produced endless sounds and fiords which run far into the land. The whole continent of America has a sea-coast of 31,000 miles. Thus it appears that the ratio of the number of linear miles in the coast-line to the number of square miles in the extent of surface, in each of these great portions of the globe, is 164 for Europe, 376 for Asia, 530 for Africa, and 359 for America. Hence the proportion is most favourable to Europe, with regard to civilization and commerce; America comes next, then Asia, and last of all Africa, which has every natural obstacle to contend with, from the extent and nature of its coasts, the desert character of the country, and the insalubrity of its climate, on the Atlantic coast at least.

The continents had been raised from the deep by a powerful effort of the internal forces acting under widely extended regions, and the stratified crust of the earth either remained level, rose in undulations, or sank into cavities, according to its intensity. Some thinner portion of the earth's surface, giving way to the internal forces, had been rent into deep fissures, and the mountain masses had been raised by violent concussions, perceptible in the convulsed state of their strata. The centres of maximum energy are marked by the plutonic rocks,[1] which generally form the nucleus or axis of the mountain masses, on whose flanks the stratified rocks are tilted at all angles to the horizon, whence, declining on every side, they sink to various depths, or stretch to various distances in the plains. Enormous as the mountain-chains and table-lands are, and prodigious as the forces that elevated them, they bear a very small proportion to the mass of the level continents and to the vast power which raised them even to their inferior altitude. Both the high and the low lands have been elevated at successive periods; some of the very highest mountain-chains are but of recent geological date, and some chains that are now far inland once stood up as islands above the ocean, while marine strata filled their cavities and formed round their bases. The influence of mountain-chains on the extent and form of the continents is beyond a doubt.

Notwithstanding the various circumstances of their elevation, there is everywhere a certain regularity of form in mountain masses, however unsymmetrical they may appear at first, and rocks of the same kind have identical characters in every quarter of the globe. Plants and animals vary with climate, but a granite mountain has the same peculiarities in the southern as in the northern hemisphere —at the equator as near the poles. Single mountains, insulated on plains are rare, except when they are volcanic; they generally appear in groups intersected by valleys in every direction, and more frequently in extensive chains symmetrically arranged in a series of parallel ridges, separated by narrow longitudinal valleys, the highest and most rugged of which occupy the centre:[2] when the chain

---

[1] Plutonic rocks are granite and others owing their origin to fire.

[2] According to M. Elie de Beaumont, every system of mountains occupies a portion of a great circle of the globe, the cleft being more easily made in that, than in any other direction, and he shows that the mountain chains are parallel to one another, even when in opposite hemispheres; thus the Central Alps and Carpathians, the Caucasus and Himalaya, lie nearly in the same direction. The great circle of the sphere, that would pass through that part of the Apennines lying between Genoa and the sources of the Tiber, is parallel to the mountains in Achaia, to the Pyrenees, to the Alleghanies in North America, and to the Ghauts in Malabar. The Western Alps are parallel to the Spanish mountains from Cape San Maritimo to

is broad and of the first order in point of magnitude, peak after peak arises in endless succession. The lateral ridges and valleys are constantly of less elevation, and are less bold, in proportion to their distance from the central mass, till at last the most remote ridges sink down into gentle undulations. Extensive and lofty branches diverge from the principal chains at various angles, and stretch far into the plains. They are often as high as the chains from which they spring, and it happens not unfrequently that these branches are united by transverse ridges, so that the country is often widely covered by a net-work of mountains, and, at the point where these offsets diverge, there is frequently a knot of mountains spreading over hundreds of square miles.

One side of a mountain-range is usually more precipitous than the other, but there is nothing in which the imagination misleads the judgment more than in estimating the steepness of a declivity. In the whole range of the Alps there is not a single rock which has 1600 feet of perpendicular height, or a vertical slope of 90°. The declivity of Mont Blanc towards the Allée Blanche, precipitous as it seems, does not amount to 45°; and the mean inclination of the Peak of Teneriffe, according to Baron Humboldt, is only 12° 38′. The Silla of Caraccas, which rises precipitously from the Caribbean Sea, at an angle of 53° 28′, to the height of between 6000 and 7000 feet, is a majestic instance of perhaps the nearest approach to perpendicularity of any great height yet known.

The circumstances of elevation are not the only causes of that variety observed in the summits of mountains. A difference in the composition and internal structure of a rock has a great influence upon its general form, and on the degree and manner in which it is worn by the weather. Thus dolomite [magnesian limestone] assumes generally the form of high insulated peaks; crystalline schists and gneiss assume the form of needles, as in the Alps; slates and quartziferous schists take the form of triangular pyramids; calcareous rocks a rounded shape; serpentine and trachyte are often of a dome form; phonolites assume a pyramidal form; dark walls, like those in Greenland, are of trap and basalt; and volcanoes are indicated by blunt

---

Cape de Gatte; they are parallel to the African mountains along the coast of the Atlantic, to the chain of Brazil between St. Roque and Monte Video, and to the Scandinavian chain; the range of Monte Viso in the Piedmontese Alps is parallel to the Apennines of the Roman and Neapolitan States, to Pindus, and to the chain of Taigetus as far as Cape Matapan. — The Southern part of the Ural is parallel to the system of Corsica and Sardinia; another part is parallel to the Tanare. Monte Laputa, on the coast of South Africa, is parallel to the mountains of Madagascar, those of Egypt and the Red Sea are parallel to the Thuringerwald; and many of the Chinese chains observe a parallelism with the Andes, in running from East to West.

cones and craters. Thus the mountain-peaks often indicate by their form their geological nature.

Viewing things on a broad scale, it appears that there is also a very striking connexion between the physical geography or external aspect of different countries and their geological structure. By a minute comparison of the different parts of the land, M. Boué has shown that similarity of outward forms, while indicating similarity in the producing causes, must also to a large extent indicate identity of structure, and therefore from the external appearance of an unexplored country its geological structure may be inferred, at least to a certain extent. This he illustrates by pointing out a correspondence, even in their most minute details, between the leading features of Asia and Europe, and the identity of their geological structure. It has been justly observed, that when the windings of our continents and seas are narrowly examined, and the more essential peculiarities of their contours contemplated, it is evident that Nature has not wrought after an indefinite number of types or models, but that, on the contrary, her fundamental types are very few, and derived from the action of definite constructive forces on a primary base.[1] The whole of our land and sea, in fact, may be decomposed into a less or greater number of masses, either exhibiting all these fundamental forms or merely a portion of them. The peninsular structure of the continents with their accompanying islands is a striking illustration of the truth of this remark, and many more might be adduced. It follows, as a consequence of that law in Nature's operations, that analogy of form and contour throws the greatest light on the constitution of countries far removed from each other. Even the picturesque descriptions of a traveller often afford information of which he may be little aware.[2]

The determination of the contemporaneous upheaval of parallel mountain-chains, by a comparison of the ages of the inclined and horizontal strata resting on them, is one of the highest steps of generalization which has been attempted by geologists, and is due to M. Elie de Beaumont. It was first observed by the miners of the Freyberg school, and established as a law by Werner, that veins of the same nature in mines occur in parallel fissures opened at the same time, and probably filled with metal, also simultaneously at a subse-

[1] M. Boué.

[2] The author avails herself with much pleasure of an opportunity of expressing her admiration of the accuracy, extent, and execution of Mr. Keith Johnston's Physical Atlas, and of the valuable information contained in the letterpress which accompanies it, which has afforded her the greatest assistance. As Mr. Johnston has published a smaller and cheap edition of his Atlas, well fitted to illustrate these volumes, the necessity of inserting in them any similar maps, which was at one time contemplated, is no longer necessary.—Physical Atlas, 1 vol., folio edition, 1848; Physical Atlas, 1 vol., quarto edition, 1850.

quent period; and that fissures differing in direction differ also in age. As these veins and fissures are rents through the solid strata, often of unfathomable depth and immense length, there is the strongest analogy between them and those enormous fissures in the solid mass of the globe through which the mountain-chains have been heaved up. Were the analogy perfect, it ought to follow that parallel mountain-chains have been raised simultaneously, that is, by forces acting during the same geological periods. By a careful examination of the relative ages of the strata resting on the flanks of many of the mountain systems, M. Elie de Beaumont has shown that all strata elevated simultaneously assume a parallel direction, or, that parallel chains of mountains are contemporaneous. Should this be confirmed, parallel chains in the most distant regions will no longer be regarded as insulated masses. They will indicate the course of enormous fissures that have simultaneously rent the solid globe and passed through the bed of the ocean from continent to continent, from island to island. M. Von Buch has found that four systems of mountains in Germany accord with this theory, and Mr. Sedgwick has observed the same in the Westmoreland system of mountains, believed to be the most ancient of which the globe can now furnish any traces. This theory of elevation of mountain-chains, which originated with M. Elie de Beaumont, has already led to the discovery of twenty different periods of fracture and elevation in the European continent alone.[1]

Mr. Hopkins, of Cambridge, has taken a purely mathematical

---

[1] Mountain Systems of Europe, according to M. Elie de Beaumont:—

1. System of the Hundsruck, and of the Eifel in Rhenish Prussia, and of Westmoreland in England — direction..............E. 25° N.
2. . . of the Vosges, and of the Bocages in Western France..E. 15° S.
3. . . of the N. of England ....N. 5° W.
4. . . of the Low Countries ....E. 5° S.
5. . . of the Rhine..N. 21° E.
6. . . of the Morvan, and of the Mountains of Central Germany..E. 40°. S.
7. . . of Mount Pilat, and of the Côte d'Or......................E. 40° N.
8. . . of Monte Viso...N. 22° W.
9. . . of the Pyrenees and Northern Apennines.................E. 8° S.
10. . . of Corsica and Sardinia..N. & S.
11. . . of the Western Alps..N. 26° E.
12. . . of the principal chain of the Alps ....E. 16° N.
13. . . from Cape Tenare to the S. extremity of the Morea..N. 10° W.
14. . . of La Vendée ....N. 22° 30′ W.
15. . . of the Finisterre ....E. 21° 45′ N.
16. . . of Longmynd, 25° E. at Church Stretton, and N. 31° 15′ E. at Bingenloch, owing to the difference of longitude.
17. . . of Morbithan ....W. 38° 15′ N
18. . . of the Forez..N. 15° 3′ W
19. . . of Mount Tatra..W. 4° 50′ N.
20. . . of the Sancerrois.......................................E. 26° 0′ N.

view of the subject, and has proved that, when an internal expansive force acts upwards upon a single point in the earth's crust, the splits or cracks must all diverge from that point like radii in a circle, which is exactly the case in many volcanic districts; that when the expansive force acts uniformly from below on a wide surface or area, it tends to stretch the surface, so that it would split or crack where the tension is greatest, that is, either in the direction of the length or breadth; and if the area yields in more places than one, he found that the fissures would necessarily be parallel to one another, which agrees with the law of arrangement of veins in mines. These results are greatly modified by the shape of the area, but the modification is according to a fixed law, which, instead of interfering with that of the parallelism of the fissures, actually arises from the same action which produces it. This investigation agrees in all its details with the fractures in the districts in England to which they were applied, so that theory comes to the aid of observation in this still unsettled question.[1]

It seems to bear on the subject, that parallel mountain-chains are similar in geological age, even when separated by seas. For instance, the mountains of Sweden and Finland are of the same structure, though the Gulf of Bothnia is between them; those of Cornwall, Brittany, and the north-west of Spain are similar; the Atlas and the Spanish mountains, the chains in California and those on the adjacent coast of America, and, lastly, those of New Guinea and the north-east of Australia, furnish examples. The same correspondence in geological epoch prevails in chains that are not parallel, but that are convergent from the form of the earth. This observation is also extensively exemplified in those that run east and west, as the Alps, the Balkan, Taurus, Paropamisus with its prolongation, the Hindoo Coosh, the Himalaya, and in America the mountains of Parima and the great chain of Venezuela.

Continents and mountain chains are often interrupted by posterior geological changes, such as clefts and cavities formed by erosion, as evidently appears from the correspondence of the strata. The chalk cliffs on the opposite sides of the British Channel show that Britain once formed part of the continent; the formation of the Orkney Islands and Ireland is the same with that of the Highlands of Scotland; the formation is the same on each side of the Straits of Gibraltar: that of Turkey in Europe passes into Asia Minor, the Crimea into the Caucasus, a volcanic region bounds the Straits of Babelmandel, and Behring's Straits divide the ancient strata of a similar age. This is particularly the case with coast islands.[2]

Immediately connected with the mountains are the high table-

[1] 'On the Parallel Lines of Simultaneous Elevation in the Weald of Kent and Sussex,' by —— Hopkins, Esq

[2] M. Boué.

lands which form so conspicuous a feature in the Asiatic and American continents. These perpetual storehouses of the waters send their streams to refresh the plains, and to afford a highway between the nations. Table-lands of less elevation, sinking in terraces of lower and lower level, constitute the links between the high ground and the low, the mountains and the plains, and thus maintain the continuity of the land. They frequently are of the richest soil, and enjoy the most genial climate, affording a delightful and picturesque abode to man, though the plains are his principal dwelling. Sloping imperceptibly from the base of the inferior table-lands, or from the last undulations of the mountains, to the ocean, the plains carry off the superfluous waters. Fruitfulness and sterility vary their aspect; immense tracts of the richest soil are favoured by climate and hardly require culture; a greater portion is only rendered productive by hard labour, compelling man to fulfil his destiny; while vast regions are doomed to perpetual barrenness, never gladdened by a shower.

The form of the great continent has been determined by an immense zone of mountains and table-lands, lying between the 30th and 40th or 45th parallels of north latitude, which stretches across it from W.S.W. to E.N.E. from the coasts of Barbary and Portugal, on the Atlantic Ocean, to the farthest extremity of Asia, at Behring's Straits, in the North Pacific. North of this lies a vast plain, extending almost from the Pyrenees to the extremity of Asia, the greater portion of which is a dead level, or low undulations, uninterrupted except by the Scandinavian and British system on the north, and the Ural chain, which is of small elevation. The low lands south of the mountainous zone are much indented by the ocean, and of the most diversified aspect. The greater part of the flat country lying between the China Sea and the river Indus is of the most exuberant fertility, while that between the Persian Gulf and the foot of the Atlas is, with some happy exceptions, one of the most desolate tracts on the earth. The southern lowlands, too, are broken by a few mountain systems of considerable extent and height.

The Atlas and Spanish mountains form the western extremity of that great zone of high land that girds the old continent almost throughout its extent: these two mountain systems were certainly at one time united, and from their geological formation, and also the parallelism of their mountain-chains, they must have been elevated by forces acting in the same direction; now, indeed, the Strait of Gibraltar, a sea-filled chasm 960 fathoms deep, divides them.[1]

[1] By the soundings of Captain Smyth, R. N., the Strait is 960 fathoms deep between Gibraltar and Ceuta, and varying from 160 to 500 in the narrowest part.

A very elevated and continuous mountain region extends in a broad belt along the north-west of Africa, from the promontory of Gher, on the Atlantic, to the Gulf of Sidra, in the Mediterranean, enclosing all the high lands of Morocco, Algiers, and Tunis. It is bounded by the Atlantic and Mediterranean, and insulated from the rest of Africa by the desert of Sahara.

This mountain system consists of three parts. The chain of the Greater Atlas, which is farthest inland, extends from Cape Gher, on the Atlantic, to the Lesser Syrtis; and, in Morocco, forms a knot of mountains 15,000 feet high, covered with perpetual snow.

The Lesser Atlas begins at Cape Spartel (the ancient Cape Cotes) opposite to Gibraltar, and keeps parallel to the Mediterranean till it attains the Gharian range in Tripoli, the last and lowest of the Little Atlas, which runs due east in a uniformly diminishing line till it vanishes in the plain of the Great Syrtis. That long, rugged, but lower chain of parallel ridges and groups which forms the bold coasts of the Straits of Gibraltar and the Mediterranean, is only a portion of the Lesser Atlas, which rises above it majestically, covered with snow. The flanks of the mountains are generally clothed with forests, but their summit is one uninterrupted line of bare inaccessible rocks, and they are rent by fissures frequently not more than a few feet wide—a peculiar feature of the whole system.

The Middle Atlas, lying between the two great chains, consists of a table-land, rich in valleys and rivers, which rises in successive terraces to the foot of the Greater Atlas, separated by ridges of hills parallel to it. This wide and extensive region has a delightful climate, abounds in magnificent forests, and valleys full of vitality. The Greater Atlas is calcareous in its central portion, and composed of granite and schistose rocks near the sea-coast.

The Spanish peninsula consists chiefly of a table-land traversed by parallel ranges of mountains, and is surrounded by the sea, except where it is separated from France by the Pyrenees, which extend from the Mediterranean to the Bay of Biscay, but are continued by the Cantabrian chain to Cape Finisterre on the Atlantic.

The Pyrenean chain is of moderate height at its extremities, but its summit maintains a waving line whose mean altitude is 7990 feet; it rises to a greater height on the east; its highest point is the Malahite or Nethou, 11,170 feet above the sea. The snow lies deep on these mountains during the greater part of the year, and is perpetual on the highest parts; but the glaciers, which are chiefly on the northern side, are neither so numerous nor so large as in the Alps.

The greatest breadth of this range is about 60 miles, and its length 270. It is so steep on the French side, so rugged and so notched, that from the plains below its summits look like the teeth of a saw, whence the term Sierra has been appropriated to mountains of this form. On the Spanish side, gigantic sloping offsets, separated by deep precipitous valleys, extend to the banks of the Ebro. All the Spanish mountains are torn by deep crevices, the beds of torrents and rivers.

The interior of Spain is a table-land with an area of 93,000 square miles, nearly equal to half of the peninsula. It dips to the Atlantic from its western side, where its altitude is about 2300 feet. There it is bounded by the Iberian mountains, which begin at the point where the Pyrenees take the name of the Cantabrian chain, and run in a tortuous south-easterly direction through all Spain, constituting the eastern boundary of Valencia and Murcia, and sending many branches through these provinces to the Mediterranean; its most elevated point is the Sierra Urbino.

Four nearly parallel ranges of mountains originate in this limiting chain, running from E.N.E. to W.S.W. diagonally across the peninsula to the Atlantic. Of these the high Castilian chain of the Gaudarama and the Sierra de Toledo cross the table-land, the Sierra Morena, so called from the dingy colour of its forests of Hermes oak, on the southern edge; and lastly, the Sierra Nevada, though only 100 miles long and 50 broad, the finest range of mountains in Europe after the Alps, traverses the plains of Andalusia and Granada. The table-land is monotonous and bare of trees; the plains of Old Castile are as naked as the Steppes of Siberia, and uncultivated, except along the banks of the rivers. Corn and wine are produced in abundance on the wide plains of New Castile and Estremadura: other places serve for pasture. The table-land becomes more fertile as it extends towards Portugal, which is altogether more productive than Spain, though the maritime provinces of the latter on the Mediterranean are luxuriant and beautiful, with a semi-tropical vegetation.

Granite, crystalline, and paleozoic rocks prevail chiefly in the Spanish mountains, and give them their peculiar, bold, serrated aspect. Some of the valleys between the parallel ranges, through which the great Spanish rivers flow to the Atlantic, appear to have been at one time the basins of lakes.

The mass of high land is continued through the south of France, at a much lower elevation, by chains of hills and table-lands, the most remarkable of which are the Montagnes Noires, and the great plateau of Auvergne, once the theatre of violent volcanic action, which continued from the beginning to the middle of the tertiary period, presenting cones and craters very perfect: some of the highest, as the Puy de Dôme, are trachytic domes. The trachytic group of Mont Dore, the highest peak of which, the Puy de Sancy, rises to the height of 6188 feet, and includes an immense crater of elevation.[1] The volcanic mountains of Auvergne, and the Cévennes, which are a little lower, are the most remarkable of the French system; the eastern offsets from the latter reach the right bank of the Rhone. In fact, the French mountains are the link between the more elevated masses of Western and Eastern Europe.

The eastern and highest part of the European portion of the mountain-zone begins to rise above the low lands about the 52nd parallel of north latitude, ascending by terraces, groups, and chains of mountains, through six or seven degrees of latitude, till it reaches its highest point in the great range of the Alps and Balkan. The descent on the south side of this lofty mass is much more rapid and abrupt, and the immediate offsets from the Alps shorter; but, taking a very general view, the Apennines and mountains of Northern Sicily, those of Greece and the southern part of Turkey in Europe, with all the islands of the adjacent coasts, are but outlying members of the general protuberance.

The principal chain of the Hyrcanian mountains, the Sudetes, and the Carpathian mountains, form the northern boundary of these high lands: the first, consisting of three parallel ridges, extends from the right bank of the Rhine to the centre of Germany, about 51° or 52° of N. lat., with a mean breadth of about 100 miles, and terminates in the knot of the Fichtelberge, covering an area of 9000 square miles, on the confines of Bavaria and Bohemia. The Sudetes begin on the east of this group, and, after a circuit of 300 miles round Bohemia, terminate at the small elevated plain of the Upper Oder, which connects them with the Carpathian mountains. No part of these limiting ranges attains the height of 5000 feet, except the Carpathians, some of which are very high. The latter consist of mountain-groups united by elevated plains, rather than of a single chain: the Tatra mountains, bisected by the 20th meridian, is their loftiest point. This range is high also in Transylvania, before it reaches the Danube, which divides it from a secondary branch of the Balkan. Spurs decline in undulations from these limiting chains on the great northern plain, and the country to the south, intervening between them and the Alps, is covered with an intricate network of mountains and plains of moderate elevation.

The higher Alps, which form the western crest of the elevated zone, may be said to begin at Cape della Melle on the Gulf of Genoa, and bend round by the west and north to Mont Blanc; then turning E. N. E. they run through the Grisons and Tyrol to the Great Glockner, in 40° 7′ N. lat., and 12° 43′ E. long., where the

[1] A crater of elevation is a mountain, generally dome-shaped, whose top has sunk into a crater or hollow, after the internal force which raised it was withdrawn, but from which no lava had issued. Dome-shaped mountains owe their form to internal pressure, probably from lava, but which have not sunk into a crater.

higher Alps terminate a course 420 miles long. All this chain is lofty; much of it is above the line of perpetual congelation; the most elevated part lies between the Col de la Seigne, on the western shoulder of Mont Blanc, and the Simplon. The highest mountains in Europe are comprised within this space, not more than 60 miles long, where Mont Blanc, the highest of all, has an absolute elevation of 15,759 feet. The central ridge of the higher Alps is jagged with peaks, pyramids, and needles of bare and almost perpendicular rock, rising from fields of perpetual snow and rivers of ice to an elevation of 14,000 feet. Many parallel chains and groups, alike rugged and snowy, press on the principal crest, and send their flanks far into the lower grounds. Innumerable secondary branches, hardly lower than the main crest, diverge from it in various directions: of these the chain of the Bernese Alps is the highest and most extensive. It separates at the St. Gothard, in a line parallel from the principal chain, separates the Valais from the Canton of Berne, and with its ramifications forms one of the most remarkable groups of mountain scenery in Europe. Its endless maze of sharp ridges and bare peaks, mixed with gigantic masses of pure snow, fading coldly serene into the blue horizon, present a scene of sublime quiet and repose, unbroken but by the avalanche or the thunder.

At the Great Glockner the chain of the Alps, hitherto undivided, splits into two branches, the Noric and Carnic Alps: the latter is the continuation of the chief stem. Never rising to the height of perpetual snow, it separates the Tyrol and Upper Carinthia from the Venetian States, and, taking the name of the Julian Alps at Mont Terglou, runs east till it joins the Eastern Alps, or Balkan, under the 18th meridian. Offsets from this chain cover all the neighbouring countries.

It is difficult to estimate the width of the Alpine chain: that of the higher Alps is about 100 miles; it increases to 150 east of the Grisons, and amounts to 200 between the 15th and 16th meridians, but is not more than 80 at its junction with the Balkan.

The Stelvio, 9177 feet above the sea, is the highest carriage-pass in these mountains. That of St. Gothard (6808) goes directly over the crest of the Alps. Passes very rarely go over the summit of a mountain; they generally cross the watershed, ascending by the valley of a torrent, and descending by a similar path on the other side.

The frequent occurrence of extensive deep lakes is a peculiar feature in European mountains, rarely to be met with in the Asiatic system, except in the Altaï and on the elevated plains.

With the exception of the Jura, whose pastoral summit is about 3000 feet above the sea, there are no elevated table-lands in the Alps; the tabular form, so eminently characteristic of the Asiatic high lands, begins in the Balkan. The Oriental peninsula rises by degrees from the Danube to Bosnia and Upper Macedonia, which are some hundred feet above the sea; and the Balkan extends 600 miles along this elevated mass, from the Julian Alps to Cape Eminec on the Black Sea. It begins by a table-land 70 miles long, traversed by low hills ending, towards Albania and Myritida, in precipitous limestone rocks from 6000 to 7000 feet high. Rugged mountains, all but impassable, succeed to this, in which the domes and needles of the Schandach, or ancient Scamus, are covered with perpetual snow. Another table-land follows, whose marshy surface is bounded by mural precipices at Mount Arbelus, near the town of Sophia. There the Hemus, or Balkan properly so called, begins, and runs in parallel ridges, separated by longitudinal valleys, to the Black Sea, dividing the plains between the Lower Danube and the Propontis into nearly equal parts. The central ridge is passable in few places, and where there is no lateral ridge the precipices descend at once to the plains.

The Balkan is everywhere rent by terrific fissures across the chains and table-lands, so deep and narrow that daylight is almost excluded. These chasms afford the safest passes across the range; the others along the faces of the precipices are frightful.

The Mediterranean is the southern boundary of the elevated zone of Eastern Europe, whose last offsets rise in rocky islands along the coasts. The crystalline mountains of Sardinia and Corsica are outlying members of the Maritime Alps, while shorter offsets end in the plains of Lombardy, forming the magnificent scenery of the Italian lakes. Even the Apennines, whose elevation has given its form to the peninsula of Italy, are but secondary on a greater scale to the broad central band, as well as the mountains and high land in the north of Sicily, which form the continuation of the Calabrian chain.

The Apennines, beginning at the Maritime Alps, enclose the Gulf of Genoa, and run through the centre of Italy in parallel ranges to the middle of Calabria, where they split into two branches, one of which goes to Capo de Leuca, on the Gulf of Tarento, the other to Cape Spartivento, in the Straits of Messina. The whole length is about 800 miles. None of the Apennines come within the line of perpetual snow, though it lies nine months in the year on the Monte Corno or Gran Sasso d'Italia, 9521 feet high in Abruzzo Ulteriore.

Offsets from the Julian and Eastern Alps render Dalmatia and Albania perhaps the most rugged tract in Europe; and the Pindus, which forms the water-shed of Greece, diverges from the latter chain, and, running south 200 miles, separates Albania from Macedonia and Thessaly.

Greece is a country of mountains, and, although none are perpetually covered with snow, it lies nine months on several of their summits. The chains terminate in strongly projecting headlands, which reach far into the sea, and reappear in the numerous islands and rocks which stud that deeply indented coast. The Grecian mountains, like the Balkan, are torn by transverse fractures. The defile of Blatamana and the Gulf of Salonica are examples. The Adriatic, the Dardanelles, and the Sea of Marmora limit the secondaries of the southern part of the Balkan.

The valleys of the Alps are deep, long, and narrow; those among the mountains of Turkey in Europe and Greece are mostly caldron-shaped hollows, often enclosed by mural rocks. Many of these cavities of great size lie along the foot of the Balkan. In the Morea they are so encompassed by mountains that the water has no escape but through the porous soil, consisting of tertiary strata, some of which have formed the bottom of lakes. Caldron-shaped valleys occur also in most volcanic countries, as Italy, Sicily, and central France.

The table-lands which constitute the tops of mountains or of mountain-chains are of a different character from those terraces by which the high lands slope to the low. The former are on a small scale in Europe, and of a forbidding aspect, with the exception of the Jura, which is pastoral, whereas the latter are almost always habitable and cultivated. The mass of high land in south-eastern Europe shelves on the north to the great plain of Bavaria, 3000 feet high; Bohemia, which slopes from 1500 to 900 feet; and Hungary, from 4000 above the sea to 300. The descent on the south of the Alps is six or seven times more rapid, because the distance from the axis of the chain is shorter.

It is scarcely possible to estimate the quantity of ice on the Alps: it is said, however, that, independent of the glaciers in the Grisons, there are 1500 square miles of ice in the Alpine range, from 80 to 600 feet thick. There are no glaciers east of the Great Glockner, except on the small group of Hallstadt. Thirty-four bound the snowy regions of Mont Blanc, and 95 square miles of snow and ice clothe that mountain. Some glaciers have been permanent and stationary in the Alps time immemorial, while others now occupy ground formerly bearing corn or covered with trees, which the irresistible force of the ice has swept away. These ice-rivers, formed on the snow-clad summits of the mountains, fill the hollows and high valleys, hang on the declivities, or descend by their gravity through the transverse valleys to the plains, where they are cut short by the increased temperature, and deposit those accumulations of rocks and rubbish, which had fallen upon them from the heights above, forming those accumulations called moraines; but their motion is so slow that generations may pass before a stone fallen on the upper end of a long glacier can reach the moraine. In the Alps the glaciers move at the rate of from 12 to 25 feet annually, and, as in rivers, the motion is most rapid in the centre, and slower at the sides and bottom on account of friction. It is slower in winter, yet it does not cease, because the winter's cold penetrates the ice, as it does the ground, only to a limited depth. Glaciers are not of solid ice; they consist of a mixture of ice, snow, and water, so that they are in some degree flexible and viscous, but acquire more solidity as they descend to lower levels: evaporation goes on at their surface, but they are not consumed by it. The front is perpetually melting, but maintains a permanent form; it is steep and inaccessible, owing to the figure of the ground over which it tumbles in its icy cascade, sometimes 1000 feet high. The middle course is rather level, the higher part very steep, and the surface is convex and uneven, and rent by crevices into which the purest blue streams fall in rushing cascades while the sun is up; but they freeze at his setting, and then a death-like silence prevails. The rocks and large stones that fall on them from the surrounding heights protect the ice below from the sun which melts it all around, so that at last they rest on elevated pinnacles, till they fall off by their weight, and in this manner those numerous pyramids are formed with which the surface is bristled. Small stones, on the contrary, absorb the sun's heat, and melt the ice under them into holes in which they are buried. Throughout much of the length of a glacier the winter's snow melts from its surface as completely as it does from the ground: it is fed from above, for in the upper part the snow never melts, but accumulates in a stratified form, and is consolidated. In some of the largest glaciers, where there is a difference of 4000 feet in height between the origin and termination, the pressure is enormous and irresistible, carrying all before it; even the thickest forest is overwhelmed and crushed.

Glaciers advance or retreat according to the severity or mildness of the season: they have been advancing in Switzerland of late years, but they are subject to cycles of unknown duration. From the moraines, as well as the striæ engraven on the rocks over which they have passed, M. Agassiz has ascertained that the valley of Chamouni was at one time occupied by a glacier that had moved towards the Col de Balme. A moraine 2000 feet above the Rhone at St. Maurice would appear to indicate that, at a remote period, glaciers had covered Switzerland to the height of 2155 feet above the Lake of Geneva.

Their increase is now limited by various circumstances — as the mean temperature of the earth, which is always above the freezing-point in those latitudes; excessive evaporation; and blasts of hot air, which occur at all heights, in the night as well as in the day, from some unknown cause. They are not peculiar to the Alps, but have been observed also in the glaciers of the Andes. From the heat of the valley thawing the ice, the natural springs that rise under the glacier as they do elsewhere, the heat of the earth, the melting of the glacier itself, the rain that falls on its surface, which rushes down its crevices, a stream of turbid water is formed which works out an icy cavern at the termination of the glacier, and flows through it into the lower ground. Thus a glacier "begins in the clouds, is formed by the mountains, and ends in the ocean."[1]

Granite no doubt forms the base of the mountain system of Eastern Europe, though it more rarely comes into view than might have been expected. Crystalline schists of various kinds are enormously developed, and generally form the most elevated pinnacles of the Alpine crest and its offsets, and also the principal chains in Greece and Turkey in Europe; but the secondary fossiliferous strata constitute the chief mass, and often rise to the highest summits; indeed, secondary limestones occupy a great portion of the high land of Eastern Europe. Calcareous rocks form two great mountain-zones on each side of the central chain of the Alps, and rise occasionally to altitudes of 10,000 or 12,000 feet. They constitute a great portion of the central range of the Apennines, and fill the greater part of Sicily. They are extensively developed in Turkey in Europe, where the plateau of Bosnia, with its high lands on the south, part of Macedonia, and Albania with its islands, are principally composed of them.[2] Tertiary strata of great thickness rest on the flanks of the Alps, and rise in some places to a height of 5000 feet; zones of the older Pliocene period flank the Apennines on each side, filled with organic remains, and half of Sicily is covered with the Pliocene strata. It appears that the Atlas, the Sierra Morena and most of the Spanish mountains, the central chain of the Caucasus, and the Balkan, were raised before the period of the erratic blocks.

From numerous dislocations in the strata, the Alps appear to have been heaved up by many violent and repeated convulsions, separated by intervals of repose, and different parts of the chain have been raised at different times; for example, the Maritime Alps and the south-western part of the Jura mountains were raised previously to the formation of the chalk: but the tertiary period appears to have been that of the greatest commotions; for nearly two-thirds of the lands of Europe have risen since the beginning of that epoch, and those that existed acquired additional height, though some sank below their original level. During that time the Alps acquired an additional elevation of between 2000 and 3000 feet; Mont Blanc then attained its present altitude, the Apennines rose 1000 or 2000

[1] The reader who may wish for a more detailed view on this subject is referred to Professor James Forbes' volume on Glaciers, a work which is a model of exact observation, combined with such accurate physical and mechanical deductions as could only be arrived at by one conversant with the highest principles of physics and mathematical investigations.

[2] Dr. Boué.

feet higher, and the Carpathians seemed to have gained an accession of height since the seas were inhabited by the existing species of animals.[1]

CHAPTER III.

The High Lands of the Great Continent, *continued*—The Caucasus—The Western Asiatic Table-Land and its Mountains.

THE Dardanelles and the Sea of Marmora form but a small break in the mighty girdle of the old continent, which again appears in immense table-lands, passing through the centre of Asia, of such magnitude that they occupy nearly two-fifths of the continent. Here everything is on a much grander scale than in Europe; the table-lands rise above the mean height of the European mountains, and the mountains themselves that gird and traverse them surpass those of every other country in altitude. The most barren deserts are here to be met with, as well as the most luxuriant productions of animal and vegetable life. The earliest records of the human race are found in this cradle of civilization, and monuments still remain which show the skill and power of those nations which have passed away, but whose moral influence is still visible in their descendants. Customs, manners, and even prejudices, carry us back to times beyond the record of history or even of tradition, while the magnitude with which the natural world is here developed evinces the tremendous forces that must have been in action at epochs immeasurably anterior to the existence of man.

The gigantic mass of high land which extends for 6000 miles between the Mediterranean and the Pacific is 2000 miles broad at its eastern extremity, 700 to 1000 in the middle, and somewhat less at its termination. Colossal mountains and elevated terraces form the edges of the lofty plains.

Between the 47th and 68th eastern meridians, where the low plains of Hindostan and Bokhara press upon the table-land and reduce its width to 700 or 1000 miles, it is divided into two parts by an enormous knot of mountains formed by the meeting of the Hindoo Coosh, the Himalaya, the Tsung-lin, and the transverse ranges of the Beloot Tagh, or Cloudy Mountains: these two parts differ in height, form, and magnitude.

The western portion, which is the table-land of Persia or plateau of Iran, is oblong, extending from the shores of Asia Minor to the Hindoo Coosh and the Solimaun range, which skirts the right bank

[1] Sir Charles Lyell.

of the Indus. It occupies an area of 1,700,000 square miles, generally about 4000 feet above the sea, and in some places 7000. The Oriental plateau or table-land of Tibet, much the largest, has an area of 7,600,000 square miles, a mean altitude of 14,000 feet, and in some parts of Tibet an absolute altitude of 17,000 feet.

As the table-lands extend from S.W. to N.E., so also do the principal mountain-chains, as well those which bound the high lands as those which traverse them. Remarkable exceptions to this equatorial direction of the Asiatic mass, however, occur in a series of meridional chains, whose axes extend from S.S.E. to N.N.W., between Cape Comorin, opposite to Ceylon and the Arctic Ocean, under the names of the Western Ghauts, the Solimaun range, (which forms the eastern boundary of the table-land of Persia,) the Beloot Tagh, or Bolor (which is the western limit of the Oriental plateau), and the Ural Mountains. These chains, rich in gold, lie in different longitudes, and so alternate among themselves that each begins only in that latitude which has not yet been attained by the preceding one. The Khinghan, in China, also extends from south to north along the eastern slopes of the table-land, and forms its boundary at that end.[1]

The lofty range of the Caucasus, which extends 700 miles between the Black and Caspian Seas, is an outlying member of the Asiatic high lands. Offsets diverge like ribs from each side of the central crest, which penetrate the Russian Steppes on one hand and on the other cross the plains of Kara, or valley of the Kour and Rioni, and unite the Caucasus to the table-land. Some parts of these mountains are very high; the Elbruz, on the western border of Georgia, is 17,796 feet. The central part of the chain is full of glaciers, and the limit of perpetual snow is at the altitude of 11,000 feet, which is higher than in any other chain of the old continent, except the Himalaya.

Anatolia, the most western part of the table-land of Iran, 3000 feet above the sea, is traversed by short chains and broken groups of mountains, separated by fertile valleys, which sink rapidly towards the Archipelago and end in promontories and islands along the shores of Asia Minor, which is a country abounding in vast, luxuriant, but solitary plains, watered by broad rivers — in Alpine platforms and mountain-ridges broken up by great valleys, opening seawards, with meandering streams. Single mountains of volcanic formation are conspicuous objects on the table-land of Anatolia, which is rich in pasture, though much of the soil is saline and covered with lakes and marshes. A triple range of limestone mountains, 6000 or 7000 feet high, divided by narrow but beautiful valleys, is the limit of the Anatolian table-land along the shores of the Black Sea. Two-thirds of their height are covered with forests,

[1] Johnston's Physical Atlas.

and broken by wooded glens, leaving a narrow coast, except near Trebizond, where it is broad and picturesque. The high land is bounded on the south by the serrated snowy range of the Taurus, which, beginning in Rhodes, Cos, and other islands, in the Mediterranean, fills the south-western parts of Asia Minor with ramifications, and, after following the sinuosities of the iron-bound coast of Karamania in a single lofty range, extends to Samisat, where the Euphrates has pierced a way through this stony girdle.

About the 50th meridian the table-land is compressed to nearly half its width, and there the lofty mountainous regions of Armenia, Kourdistan, and Azerbijan tower higher and higher between the Black Sea, the Caspian, and the Gulf of Scanderoon in the Mediterranean. Here the cold treeless plains of Armenia, the earliest abode of man, 7000 feet above the sea, bear no traces of the Garden of Eden; Mount Ararat, on which the Ark is said to have rested, stands a solitary majestic volcanic cone, 17,112 feet above the sea, shrouded in perpetual snow. Though high and cold, the soil of Armenia is richer than that of Anatolia, and is better cultivated. It shelves on the north in luxuriant and beautiful declivities to the low and undulating valley of Kara, south of the Caucasus; and on the other hand, the broad and lofty belt of the Kourdistan mountains, rising abruptly in many parallel ranges from the plains of Mesopotamia, form its southern limit, and spread their ramifications wide over its surface. They are rent by deep ravines, and in many places are so rugged that communication between the villages is always difficult, and in winter impracticable from the depth of snow. The line of perpetual snow is decided and even along their sides; their flanks are wooded, and their valleys populous and fertile.

A thousand square miles of Kourdistan is occupied by the brackish lake Van, which is seldom frozen, though 566 feet above the sea, and surrounded by lofty mountains.

The Persian mountains, of which Elbruz is the principal chain, extend along the northern brink of the Plateau, from Armenia, almost parallel to the shores of the Caspian Sea, maintaining a considerable elevation up to the volcanic peak of Demavend, near Tehrân, their culminating point, which, though 90 miles inland, is a landmark to sailors on the Caspian. Elevated offsets of these mountains cover the volcanic table-land of Azerbijan, the fire-country of Zoroaster, and one of the most fertile provinces of Persia; there the Koh Salavan elevates its volcanic cone. Beautiful plains, pure streams, and peaceful glades, interspersed with villages, lie among the mountains, and the Vale of Khosran Shah, a picture of sylvan beauty, is celebrated as one of the five paradises of Persian poëtry. The vegetation at the foot of these mountains on the shores of the Caspian has all the exuberance of a tropical jungle. The Elbruz loses its height to the east of Demavend, and then joins the mountains of Khorasan and the Paropamisan range, which appear to be chains of mountains when viewed from the low plains of Khorasan and Balkh, but on the table-land of Persia they merely form a broad hilly country of rich soil, till they join the Hindoo Coosh.

The table-land of Iran is bounded for 1000 miles along the Persian Gulf and Indian Ocean by a mountainous belt of from three to seven parallel ranges, having an average width of 200 miles, and extending from the extremity of the Kourdistan Mountains to the mouth of the Indus. The Lasistán Mountains, which form the northern part of this belt, and bound the vast level plain of the Tigris, rise from it in a succession of high table-lands divided by very rugged mountains, the last ridge of which, mostly covered with snow, abuts on the table-land of Persia. Oaks clothe their flanks; the valleys are of generous soil, verdant, and cultivated; and many rivers flow through them to swell the stream of the Tigris. Insulated hill-forts, from 2000 to 5000 feet high, occur in this country, with flat cultivated tops some miles in extent, accessible only by ladders, or holes cut in their precipitous sides. These countries are full of ancient inscriptions and remains of antiquity. The moisture decreases more and more south from Shiraz, and then the parallel ridges, repulsive in aspect and difficult to pass, are separated by arid longitudinal valleys, which ascend like steps from the narrow shores of the Persian Gulf to the table-land. The coasts of the gulf are burning hot sandy solitudes, so completely barren, that the country from Bassora to the Indus, a distance of 1200 miles, is nearly a sterile waste. In the few favoured spots on the terraces where water occurs, there is vegetation, and the beauty of these valleys is enhanced by surrounding sterility.[1]

With the exception of Mazanderan and the other provinces bordering upon the Caspian, and in the Paropamisan range, Persia is arid, possessing few perennial springs, and not one great river; in fact, three-tenths of the country is a desert, and the table-land is nearly a wide scene of desolation. A great salt-desert occupies 27,000 square miles between Irak and Khorasan, of which the soil is a stiff clay, covered with efflorescence of common salt and nitre, often an inch thick, varied only by a few saline plants and patches of verdure in the hollows. This dreary waste joins the large sandy and equally dreary desert of Kerman. Kelat, the capital of Belochistan, is 7000 feet above the level of the sea: round it there is cultivation, but the greater part of that country is a lifeless plain, over which the brick-red sand is drifted by the north wind into ridges like the waves of the sea, often 12 feet high, without a vestige of vegetation. The blast of the desert, whose hot and pestilential breath is fatal to man and animals, renders these dismal sands impassible at certain seasons.

---

[1] Sir John Malcolm on Persia, and Mr. Morier's Travels.

Barren lands or bleak downs prevail at the foot of the Lukee and Solimaun ranges, formed of bare porphyry and sandstone, which skirt the eastern edge of the table-land, and dip to the plains of the Indus. In Afghanistan there is little cultivation except on the banks of the streams that flow into the Lake Zerrah, but vitality returns towards the north-east. The plains and valleys among the offsets from the Hindoo Coosh are of surpassing loveliness, and combine the richest peaceful beauty with the majesty of the snow-capped mountains by which they are encircled.

CHAPTER IV.

The High Lands of the Great Continent, *continued.*—The Oriental Table-Land and its Mountains.

THE Oriental plateau, or table-land of Tibet, is an irregular four-sided mass stretching from S.W. to N.E., enclosed and traversed by the highest mountains in the world. It is separated from the table-land of Persia by the Hindoo Coosh, which may be considered as the western prolongation of the Himalaya, occupying the terrestrial isthmus between the low lands of Hindostan and Bucharia.

The cold dreary plateau of Tibet is separated on the south from the glowing luxuriant plains of Hindostan by the Himalaya, which extends from the eastern extremity of the Hindoo Coosh in Cabulistan to about the 95th meridian, where it joins the immense mountain-knot which renders the south-western corner of the table-land and the Chinese province of Yun-nan one of the most elevated regions on the earth. On the north the table-land is bounded by the Altaï chain which separates it from the Siberian plains, and on the west, it has its limits in the chain of the Bolor or Beloot Tagh, the "Cloudy Mountains," the Tartash Tagh of the natives, a transverse range which detaches itself from the Hindoo Coosh nearly at a right angle about the 72d degree of E. longitude, and, pursuing a northerly direction forms magnificent mountain-knots with the diagonal chains of the table-land, and is the watershed between the valley of the Oxus and Chinese Tartary. It descends in a succession of tiers or terraces through the countries of Bokhara and Balkh to the deep cavity in which the Caspian Sea and the Sea of Azoff lie, and forms, with the Western Ghauts, the Solimaun range, and the Ural, a singular exception to the general parallelism of the Asiatic mountains. Two narrow difficult passes lead over the Beloot Tagh from the low plains of Bucharia and Independent Tourkistan to Kashgar and Yarkund, on the table-land in Chinese Tartary. The north-eastern edge of the table-land is bounded by the Khing-han Mountains, a serrated granitic chain running from south to north, which separates the plateau of Mongolia from the country of Mantchouria, and joins the Yablonoi branch of the Altaï at right angles about the 55th degree of north latitude. Little more is known of the south-eastern boundary of the table-land than that it is a mass of exceedingly high mountains. In fact, between the sources of the Brahmapootra and the Altaï chain, nearly 1,000,000 of square miles of the Chinese empire is covered with mountains.

The table-land itself is traversed from west to east by two great chains. The Kuenlun, or Chinese range, begins about 35° 30′ N. lat. at the mountain-knot of Tsung-lin, formed by the Hindoo Coosh and Himalaya, and, running eastward, it terminates about the 110th meridian, but probably covers a great part of the western provinces of China with its branches. The Thian-shan, or "Celestial Mountains," lie more to the north; they begin at the Bolor or Beloot Tagh, and, running along the 42d parallel, sink to the desert of the Great Gobi about the centre of the plateau, but, rising again, they are continued under the name of Shan-Garjan, which runs to the north-east and ends on the shores of the Japan Sea. The Thian-shan is exceedingly volcanic, and, though so far inland, some of its peaks pour forth lava, and exhibit all the other phenomena of volcanic districts.

Tibet is a mountain valley, enclosed between the chains of the Himalaya on the south, and the Kuenlun on the north; Tungut, or Chinese Tartary, lies between the latter chain and the Thian-shan, or Celestial Mountains; and Zungary, or Mongolia, between the Celestial range and the Altaï. The meridional chain of the Bolor encloses Chinese Tartary on the west; and Mongolia, which is entirely open on the west, is shut in on the east by the Khinghan range, also running from south to north. The Himalaya and Altaï ranges diverge in their easterly courses, so that the table-land, which is only from 700 to 1000 miles wide at its western extremity, is 2000 between the Chinese province of Yunnan and the country of the Mantchou Tonguses.[1]

Of all these vast chains of mountains the Himalaya, and its principal branch the Hindoo Coosh, are best known; though even of these a great part has never been explored, on account of their enormous height and the depth of snow, which make it impossible to approach the central ridge, except in a very few places.

The range consists of three parts: the Hindoo Coosh, or Indian Caucasus, which extends from the Paropamisan range in Afghanistan to Cashmere; the Himalaya, or Imaus of the ancients, which stretches from the valley of Cashmere to Bhotan; and, lastly, the

[1] Johnson's Physical Atlas and Humboldt's Asie Centrale

mountains of Bhotan and Assam—the three making one magnificent unbroken chain.

The Hindoo Coosh, which has its name from a mountain of great height (20,232 feet), north of the city of Cabul, is very broad to the west, extending over many degrees of latitude, and, together with the offsets of the Beloot Tagh, fills the countries of Kafferistan, Kooduz, and Budakshan. From the plains to the south it seems to consist of four distinct ranges running one above another, the last of which abuts on the table-land, and is so high that its snowy summits are visible at the distance of 150 miles. A ridge of stupendous height encloses the beautiful valley of Cashmere, to the east of which the chain takes the name of Himalaya, "the dwelling of snow." From the great mountain-knot of Tsung-lin, the Himalaya no longer maintains its direct easterly course, but takes an E. S. E. direction, extending to the Brahmapootra, varying in breadth from 250 to 350 miles, and occupying an area of 600,000 square miles.[1] The general structure of the Himalaya is very regular: the first range of hills that rises above the plains of Hindostan is alluvial, north of which lies the Tariyani, a tract from 10 to 30 miles wide, 1000 feet above the sea, covered with dense pestilential jungle, and extending along the foot of the range. North of this region are rocky ridges 5000 or 6000 feet high. Between these and the higher ranges lie the peaceful and well-cultivated valleys of Nepaul, Sikim, Bhotan, and Assam, interspersed with picturesque and populous towns and villages. Behind these are mountains from 10,000 to 12,000 feet high, flanked by magnificent forests; and lastly, the snowy ranges rise in succession to the table-land.

The mean height of the Himalaya is stupendous. Captain Gerard and his brother estimated that it could not be less than from 16,000 to 20,000 feet; but, from the average elevation of the passes over these mountains, Baron Humboldt thinks it must be under 15,700 feet. Colonel Sabine estimates it to be only 11,510 feet, though the peaks exceeding that elevation are not to be numbered, especially near the sources of the Sutlej and the Ganges; indeed, from that river to the Kalee, the chain exhibits an endless succession of the loftiest mountains on earth; forty of them surpass the height of Chimborazo, one of the highest of the Andes, and several reach the height of 25,000 feet at least. So rugged is this part of the magnificent chain, that the military parade at Sabathoo, half a mile long and a quarter of a mile broad, is said to be the only level ground between it and the Tartar frontier on the north, or the valley of Nepaul on the east. Towards the fruitful valleys of Nepaul and Sikim the Himalaya is more lofty still, some of the mountains exceeding 28,000 feet in height; but it is narrower, and the descent

[1] Johnson's Physical Atlas.

to the plains excessively rapid, especially in the territory of Bhotan, where the dip from the table-land is more than 10,000 feet in ten miles. The valleys are crevices so deep and narrow, and the mountains that hang over them in menacing cliffs are so lofty, that these abysses are shrouded in perpetual gloom, except where the rays of a vertical sun penetrate their depths. From the steepness of the descent the rivers shoot down with the swiftness of an arrow, filling the caverns with foam and the air with mist. At the very base of this wild region lies the elevated and peaceful valley of Bhotan, vividly green, and shaded by magnificent forests. Another rapid descent of 1000 feet leads to the plain of the Ganges.

The Himalaya still maintains great height along the north of Assam; and where the Brahmapootra cuts through it, the parent stem and its branches extend in breadth over two degrees of latitude, forming a vast mountain-knot of great elevation. Beyond this point nothing certain is known of the range, but it or some of its branches are supposed to cross the southern provinces of the Chinese empire and to end in the volcanic island of Formosa. Little more is known of the northern side of the mountains than that the passes are about 5000 feet above the plains of Tibet.

The passes over the Hindoo Coosh, though not the highest, are very formidable: there are six from Cabul to the plains of Turkistan; and so deep and so much enclosed are the defiles, that Sir Alexander Burnes never could obtain an observation of the pole-star in the whole journey from Bameean till within thirty miles of Turkistan.

Most of the passes over the Himalaya are but little lower than the top of Mont Blanc; many are higher, especially near the Sutlej, where they are from 18,000 to 19,000 feet high; and that north-east of Khoonawur is 20,000 feet above the level of the sea—the highest that has been attempted. All are terrific, and the fatigue and suffering from the rarity of the air in the last 500 feet is not to be described. Animals are as much distressed as human beings, and many die; thousands of birds perish from the violence of the wind, the drifting snow is often fatal to travellers, and violent thunder-storms add to the horror of the journey. The Niti Pass, by which Mr. Moocroft ascended to the sacred lake of Manasarowar, in Tibet, is tremendous; he and his guide had not only to walk bare-footed, from the risk of slipping, but they were obliged to creep along the most frightful chasms, holding by twigs and tufts of grass, and sometimes they crossed deep and awful crevices on a branch of a tree, or on loose stones thrown across. Yet these are the thoroughfares for commerce in the Himalaya, never repaired nor susceptible of improvement from frequent land-slips and torrents.

The loftiest peaks being bare of snow gives great variety of colour and beauty to the scenery, which in these passes is at all times magnificent. During the day, the stupendous size of the mountains, their interminable extent, the variety and sharpness of their forms, and, above all, the tender clearness of their distant outline melting into the pale blue sky contrasted with the deep azure above, is described as a scene of wild and wonderful beauty. At midnight, when myriads of stars sparkle in the black sky, and the pure blue of the mountains looks deeper still below the pale white gleam of the earth and snow-light, the effect is of unparalleled solemnity, and no language can describe the splendour of the sunbeams at daybreak streaming between the high peaks, and throwing their gigantic shadows on the mountains below. There, far above the habitation of man, no living thing exists, no sound is heard: the very echo of the traveller's footsteps startles him in the awful solitude and silence that reigns in these august dwellings of everlasting snow.

Nature has in mercy mitigated the intense rigour of the cold in these high lands in a degree unexampled in other mountainous regions. The climate is mild, the valleys are verdant and inhabited, corn and fruit ripen at elevations which in other countries—even under the equator—would be buried in permanent snow.

It is also a peculiarity in these mountains that the higher the range the higher likewise is the limit of snow and vegetation. On the southern slopes of the first range Mr. Gerard found cultivation 10,000 feet above the sea, though it was often necessary to reap the corn still green and unripe; while in Chinese Tartary good crops are raised 16,000 feet above the sea. Captain Gerard saw pasture and low bushes up to 17,009 feet; and corn as high as even 18,544 feet, which is 2805 feet higher than the top of Mont Blanc, and 1279 feet above the snow-line in the province of Quito under the equator. Birch-trees with tall stems grow at the elevation of 14,068 feet, and the vine and other fruits thrive in the valleys of these high plains. The temperature of the earth has probably some influence on the vegetation; as many hot springs exist in the Himalaya at great heights, there must be a source of heat beneath these mountains, which in some places comes near the surface, and possibly may be connected with the volcanic fires in the central chains of the table-land. Hot springs abound in the valley of Jumnotra; and as it is well known that many plants thrive in very cold air if their roots are well protected, it may be the cause of pine-trees thriving at great elevations in that valley, and of the splendid forests of the Deodar, a species of cypress that grows to a gigantic size even to the snow.

According to Captain and Mr. Gerard, the line of perpetual congelation is at an elevation of only 12,981 feet on the southern slopes of the Himalaya, while on the northern side the limit is 16,620 feet; but although the main fact of the great difference in the height of the snow-line and of vegetation is beyond a doubt, the mean height of the table-land of Tibet, and the relative elevation of the line of perpetual snow on the two declivities of the Himalaya, require to be further investigated. The greater height of the snow-line on the northern side is the joint result of the serenity of the sky, the less frequent formation of snow in very cold, dry, and elevated atmospheres, and the radiation of heat from the neighbouring plains, which, being so near, have much greater effect on the temperature than the warmer but more distant plains on the south. There are fewer glaciers in the Asiatic mountains than might have been expected from the great mass of snow: they are chiefly on the Thibetian side of the Himalaya and on the Kuenlun. There is a very large one at the source of the Indus, and another at the source of the Ganges, on the southern face of the Himalaya.

Various secondary chains of great length detach themselves from the eastern extremity of the Himalaya, or rather the vast knot of mountains, near the sources of the Brahmapootra in the Chinese province of Yun-nan, which is a terra incognita; their origin therefore is unknown. But in Upper Assam they run cross to the equatorial system of Asiatic mountains, and, extending in a southerly but diverging direction, they spread like the spokes of a fan through the countries east of the Ganges and the Indo-Chinese peninsula, leaving large and fertile kingdoms between them. The Birmano-Siamese chain is the most extensive, reaching to Cape Romania, at the southern extremity of the Malay peninsula, the most southerly point of the Asiatic continent; it may be traced through the island of Sumatra parallel to the coast, and also in the islands of Banca and Biliton, where it ends.

Another range, called the Laos-Siamese chain, forms the eastern boundary of the kingdom of Siam, and the Annamatic chain, from the same origin, separates the empire of Annam from Tonquin and Cochin China.

These slightly diverging lines of mountains yield gold, ores of silver and tin, and precious stones, as rubies and sapphires. Mountains in low latitudes have nothing of the severe character of those in less favoured climes. Magnificent forests reach their summit; trees yielding spices, dyes of brilliant tints, medicinal and odoriferous plants clothe their declivities; and in the low grounds the fruits of India and China grow in perfection, in a soil which yields three crops of grain in the year.

The crest of the Himalaya is of stratified crystalline rocks, especially gneiss, with large granitic veins, and beds of quartz of huge magnitude. The zone between 15,000 and 18,000 feet above the level of the sea is formed chiefly of Palæozoic strata; granite is most frequent at the base, and probably forms the foundation of the chain. Strata of the comparatively modern age of the British oolites, occur at great elevations. These sedimentary formations, prevailing also on the acclivities of the Alps and Apennines, show that the epochs of elevation in parts of the earth widely remote from one another, if not simultaneous, were at least not very different. There can be no doubt that very great geological changes have taken place at a comparatively recent period, in the Himalaya, and through an extensive part of the Asiatic continent.

The Altaï mountains, which form the northern margin of the table-land, are unconnected with the Ural chain; they are separated from it by 400 miles of a low marshy country, part of the steppe of the Kirghiz, and by the Dalaï mountains, a low range never above 2000 feet high, which runs between the 64th meridian and the left bank of the Irtish. The Altaï chain rises on the right bank of that river, at the north-west angle of the table-land, and extends in a serpentine line to the Pacific, south of the Gulf of Okhotzk, dividing the high lands of Tartary and China from the wastes of Asiatic Siberia. Under various names, its branches skirt the northwest side of the Sea of Okhotzk, and thence stretching to Behring's Straits, it ends at Eastern Cape, the most eastern extremity of the old continent, the whole length of the chain being 4500 miles. The breadth of this chain varies from 400 to 1000 miles, but towards the 105th meridian it is contracted to about 150 by a projection of the desert of the Great Gobi. Its height bears no proportion to its length and breadth. The Altaï, the only part of the chain properly so called, can only be regarded as a succession of terraces of a swelling outline, descending by steps from the table-land, and ending in the promontories on the Siberian plains. There are numerous large lakes on these terraces and in the valleys, as in the mountain systems of Europe. The general form of this part of the chain is monotonous from the prevalence of straight lines and smooth rounded outlines — long ridges with flattened summits or small table-lands not more than 6000 feet high, which rarely attain the line of perpetual congelation; snow, however, is permanent on the Korgon table-land, 9900 feet above the sea, supposed to be the culminating point of this part of the chain. These table-lands bear a strong resemblance to those in the Scandinavian mountains in baldness and sterility, but their flanks are clothed with forests, verdant meadows, and pastoral valleys.

East of the 86th meridian this region of low mountains splits into three branches, enclosing longitudinal valleys for 450 miles. The Sayansk and Zongnou mountains, which are the northern and central branches, form a mountain-knot nearly as large as England, which projects like a huge promontory on the Siberian plains[1] west of Lake Baikal, and is celebrated for the richness of its mines. The

[1] Johnston's Physical Atlas.

third branch, which is the Ulangomula, lies south of Lake Oubsa. The principal part of the Baikal group is 500 miles long, from 10 to 60 wide, high, and snow-capped, and said to be without glaciers. It flanks Lake Baikal on the north, the largest of Alpine lakes, so embedded in a knot of mountains, partly granitic, partly volcanic, that rocks and pillars of granite rise from its bed. The mountains south of the lake are but the face of the table-land; a traveller ascending them finds himself at once in the desert of Gobi, which stretches in unbroken sadness to the great wall of China.

The Daouria mountains, a volcanic portion of the Altaï, which borders the table-land on the north-east, follow the Baikal chain; and farther east, at the sources of the Aldan, the Altaï range takes the name of the Yablonnoi Khrebet, and stretches south of the Gulf of Okhotzk to the coast of the Pacific opposite to the island of Saghalian; while another part, 1000 miles broad, fills the space between the Gulf of Okhotzk and the river Lena, and then, bending to the north-east, ends in the peninsula of Kamtchatka. Between the western end of Lake Baikal and the Yablonnoi Khrebet the mountain-chains are parallel, and extend from the W.S.W. to the E.N.E., which is the general direction of the high lands in the most easterly regions of Asia.

A great part of the Altaï chain is unknown to Europeans; the innumerable branches that penetrate the Chinese empire are completely so; those belonging to Russia abound in a great variety of precious and rare metals and minerals—silver, copper, and iron. In the Yablonnoi range and other parts there are whole mountains of porphyry, with red and green jasper; coal is also found; and in a branch of the Altaï between the rivers Obi and Yenissei there are mines of coal which, having been set on fire by lightning, have continued to burn for more than a century. The Siberian mountains far surpass the Andes in the richness of their gold mines. The eastern flank of the Ural chain, and some of the northern spurs of the Altaï, have furnished a vast quantity of gold; but a region as large as France has lately been discovered in Siberia covered with the richest gold alluvium, lying above rocks abounding in that metal. The precious metals of the Ural and Altaï are situated principally in metamorphic rocks, adjacent to the greenstones, syenites, and serpentines that have caused their change; and as the same formation prevails throughout the greater part of the Altaï and Aldan chains almost to Kamtchatka, there is every reason to believe that the whole of that vast region is auriferous: besides, as many of the northern offsets of the Altaï are particularly rich, it may be concluded that the southern branches in the Chinese empire are equally so. Thus Southern Siberia and Chinese Tartary form an auriferous district, probably greater in area than all Europe, which extends even to our dominions in Hindostan, where the formations containing gold are unexplored.[1]

The sedimentary deposits in this extensive mountain-range are more ancient than the granite, syenite, and porphyries; consequently these igneous rocks have not here formed part of the original crust of the globe. Rocks of the Palæozoic series occupy the greater part of the Altaï, and probably there are none more modern. There are no volcanic rocks properly speaking, ancient or modern, west of the Yenissei, but they abound to the east of that river, even to Kamtchatka, which is full of them.

The physical characters and the fossil remains of this extensive mountain system have little relation with the geological formations of Europe and America. Eastern Siberia seems even to form an insulated district by itself, and that part between the town of Yakoutzk and the mouth of the Lena appears to have been raised at a later period than the part of Siberia stretching westward to the Sayansk mountains; moreover, the elevation of the western part of the Altaï was probably contemporaneous with that of the Ural mountains.[2] On the whole, the chains in the direction of parallels of latitude in the Old Continent are much more numerous and extensive than those in the direction of the meridian; and as they lie chiefly towards the equator, the internal forces that raised them were probably modified by the rotation of the earth.

The table-land of Tibet is only 4000 feet above the sea towards the north, but it rises in Little Tibet to between 11,000 and 12,000 feet. The Kuen-lun, the most southerly of the two diagonal mountain-chains that cross the table-land, begins at the Hindoo Coosh, in latitude 35° 30′, being, in fact, a branch of that chain, and extends eastward in two branches, which surround the lake Tengri-Nor, and again unite in the K'han of eastern Tibet. The most southerly of the two branches known as the Ice Mountains, and which is crossed by the Kara-Korum Pass, 18,600 feet above the sea, maintains a curved course parallel to the Himalaya, and then bends north towards the Kuen-lun, which pursues a more direct line across the table-land. Chains more or less connected with these form an elevated mountain plain round Lake Koko-Nor, nearly in the centre of the table-land, from whence those immense mountain-ranges diverge which render the south-western provinces of China the most elevated region on earth. The country of Tibet lying between the Himalaya and the

[1] Sir Roderick I. Murchison.

[2] From the observations of Sir Roderick Murchison, M. Middendorf, M. de Verneuil, and Count Keyserling, it appears also that the low land of Siberia has been extended since the existing species of shell-fish inhabited the northern seas; a circumstance that may have rendered the Siberian climate still more severe, and materially affected that of all the northern parts of Europe and Asia.

Kuen-lun consists of rocky mountainous ridges, extending from N.W. to S.E., separated by long valleys, in which flow the upper courses of the Brahmapootra, Sutlej, and Indus. According to Lieut. Strachey, the sacred lakes of Manasa, in Great Tibet, and the surrounding country, are 15,250 feet above the sea—higher than Monte Rosa, the second giant of the Alps. In this elevated region wheat and barley grow, and many of the fruits of southern Europe ripen. The city of H'Lassa in eastern Tibet, the residence of the Grand Lama, is surrounded by vineyards, and is called by the Chinese the "Realm of Pleasure." There are some trees in this country; but the ground in cultivation bears a small proportion to the grassy steppes, which extend in endless monotony, grazed by thousands of the shawl-wool goats, sheep, and cattle. There are many lakes in the table-land: some about Ladak contain borax[1]—a salt very useful in the arts, found only here.

In summer the sun is powerful at mid-day, the air is of the purest transparency, and the azure of the sky so deep that it seems black as in the darkest night. The rising moon does not enlighten the atmosphere, no warning radiance announces her approach, till her limb touches the horizon, and the stars shine with the distinctness and brilliancy of suns. In southern Tibet the verdure is confined to favoured spots; the bleak mountains and high plains are sternly gloomy—a scene of barrenness not to be conceived. Solitude reigns in these dreary wastes, where there is not a tree, nor even a shrub to be seen of more than a few inches high. The scanty, short-lived verdure vanishes in October; the country then looks as if fire had passed over it, and cutting dry winds blow with irresistible fury, howling in the bare mountains, whirling the snow through the air, and freezing to death the unfortunate traveller benighted in their defiles.

Yarkand and Khotan, provinces of Chinese Tartary, which lie beyond the two diagonal chains, are less elevated and more fertile than Tibet; yet it is so cold in winter that the river Yarkiang is frozen for three months. They are watered by five rivers, and contain several large cities; Yarkand, the most considerable of these, is the emporium of commerce between Tibet, China, Turkistan, Bookahra, and Persia and Russia. Gold, rubies, silk, and other productions are exported.

The Tartar range of the Thian-Shan is very high; the Bogda

[1] Borax (borate of soda), for a long time exclusively brought from Tibet, is now manufactured in large quantities, by combining boracic acid with soda. Boracic acid exists in abundance in the hot springs of Monte Cerboli and Castel Nuovo in Tuscany, and in an extinct crater of the Island of Volcano, one of the Lipari group; but Tibet appears to be the only place where native Borax, or Tincal, is found. Borax and Boracic acid are extensively used as fluxes in the making of glass, &c.

Oola, or "Holy Mountain," near Lake Lop, its highest point, is always covered with snow, and has two active volcanoes, one on each side. This range runs along the 42nd parallel of north latitude, forming at its western extremity a mountain-knot with the Beloot Tagh, in the centre of which lies the small table-land of Pamir, 15,630 feet high, called by the natives the Bami Dunya, or "Terraced Roof of the World." Its remarkable elevation was first described by the celebrated Venetian traveller Marco Polo, six centuries ago. The Amu or Oxus issues from the western extremity of the small Alpine lake Sir-i-Kol,[1] situated on this elevated plateau; and the rivers of Yarkund and Kohan also rise towards the eastern side of the same plain, which is intensely cold in winter, and in summer is alive with flocks of sheep and goats. Snow lies deep on the Thian-Shan range in winter, yet little falls on the plains on account of the dryness of the air. There are only two or three showers of rain annually on these mountains, for a very short time, and the drops are so minute as scarcely to wet the ground, yet the streams from them suffice for irrigation.

Zungary, or Mongolia, the country between the Thian-Shan and the Altaï, is hardly known, further than that its grassy steppes, intersected by many lakes and offsets from the Altaï, are the pasture-grounds of the wandering Kirghiz.

The remarkable feature of the table-land is the desert of the Great Gobi, which occupies an area of 300,000 square miles in its eastern extremity, interrupted only by a few spots of pasture and low bushes. Wide tracts are flat and covered with small stones or sand, and widely separated from one another are low hills destitute of wood and water; its general elevation is 4220 feet above the sea, but it is intersected from west to east by a depressed valley, aptly named Shamo, or the "Sea of Sand," which is also mixed with salt. West from it lies the Han-Hai, the "Dry Sea," a barren plain of shifting sand blown into high ridges. Here, as in all deserts, the summer sun is scorching, no rain falls, and when thick fog occurs it is only the precursor of fierce winds. All the plains of Mongolia are intensely cold in winter, because the hills to the north are too low to screen them from the polar blast, and being higher than the Siberian deserts, they are bitterly cold; no month in the year is free from frost and snow, yet it is not deep enough to prevent cattle from finding pasture. Sandy deserts like that of the Great Gobi occupy much of the country south of the Chinese branches of the Altaï.

Such is the stupendous zone of high land that girds the old continent throughout its whole length. In the extensive plains on each side of it several independent mountain systems rise, though much inferior to it in extent and height.

---

[1] Lieut. Wood, Voyage to the Source of the River Oxus, 1 vol., 8vo.

CHAPTER V.

Secondary Mountain Systems of the Great Continent — That of Scandinavia — Great Britain and Ireland — The Ural Mountains — The Great Northern Plain.

THE Great Northern Plain is broken by two masses of high land, in every respect inferior to those described; they are the Scandinavian system and the Ural mountains, the arbitrary limit between Europe and Asia.

The range of primary mountains which has given its form to the Scandinavian peninsula begins at Cape Lindesnaes, the most southerly point of Norway, and, after running along its western coast 1000 miles in a north-easterly direction, ends at Cape Nord Kyn, on the Polar Ocean, the extremity of Europe. The highest elevation of this chain is not more than 8412 feet. It has been compared to a great wave or billow, rising gradually from the east, which, after having formed a crest, falls perpendicularly into the sea in the west. There are 3696 square miles of this peninsula above the line of perpetual snow.

The southern portion of the chain consists of ridges following the general direction of the range, 150 miles broad. At the distance of 360 miles from Cape Lindesnaes the mountains form a single elevated mass, terminated by a table-land which maintains an altitude of 4500 feet for 100 miles. It slopes towards the east, and plunges at once in high precipices into a deep sea on the west.

The surface is barren, marshy, and bristled with peaks; besides an area of 600 square leagues is occupied by the Snae Braen, the greatest mass of perpetual snow and glaciers on the continent of Europe. A prominent cluster of mountains follows, from whence a single chain, 25 miles broad, maintains an uninterrupted line to the island of Megaree, where it terminates its visible career in North Cape, a huge barren rock perpetually lashed by the surge of the Polar Ocean, but from the correspondence in geological structure it must be continued under the sea to where it reappears, according to M. Boué, in the schistose rocks of Spitzbergen. Offsets from these mountains cover Finland and the low rocky table-land of Lapland; the valleys and countries along the eastern side of the chain abound in forests and Alpine lakes.

The iron-bound coast of Norway is a continued series of rocky islands, capes, promontories, and precipitous cliffs, rent into chasms which penetrate miles into the heart of the mountains. These chasms, or fiords, are either partly or entirely filled by arms of the sea; in the former case the shores are fertile and inhabited, and the whole country abounds in the most picturesque scenery. Fiords are not peculiar to the coast of Norway; they are even more extensive in Greenland and iceland, and of a more stern character, overhung by snow-clad rocks and glaciers.

As the Scandinavian mountains, those of Feroe, Britain, Ireland, and the north-eastern parts of Iceland have a similar character, and follow the same general directions, they must have been elevated by forces acting in parallel lines, and therefore may be regarded as belonging to the same system.

The Feroe islands, due west from Norway, rise at once in a table-land 2000 feet high, bounded by precipitous cliffs, which dip into the ocean.

The rocky islands of Zetland and those of Orkney form part of the mountain system of Scotland; the Orkney islands have evidently been separated from the mainland by the Pentland Firth, where the currents run with prodigious violence. The north-western part of Scotland is a table-land from 1000 to 2000 feet high, which ends abruptly in the sea, covered with heath, peat-mosses, and pasture. The general direction of the Scottish mountains, like those of Scandinavia, is from north-east to south-west, divided by a long line of lakes in the same direction, extending from the Moray Firth completely across the island to south of the island of Mull. Lakes of the most picturesque beauty abound among the Scottish mountains. The Grampian hills, with their offsets and some low ranges, fill the greater part of Scotland north of the Clyde and Forth. Ben Nevis, only 4368 feet above the sea, is the highest mountain in the British islands.

The east coast of Scotland is generally bleak, though in many parts it is extremely fertile, and may be cited as a model of good cultivation; and the midland and southern counties are not inferior either in the quality of the soil or the excellence of the husbandry. To the west the country is wildly picturesque; the coast of the Atlantic, penetrated by the sea, which is covered with islands, bears a strong resemblance to that of Norway.

There cannot be a doubt that the Hebrides formed part of the mainland at some remote geological period, since they follow the direction of the mountain system in two parallel lines of islands, of rugged and imposing aspect, never exceeding the height of 3200 feet. The undulating country on the borders of Scotland becomes higher in the west of England and North Wales, where the hills are wild, but the valleys are cultivated like gardens, and the English lake scenery is of the most gentle beauty.

Evergreen Ireland is mostly a mountainous country, and opposes to the Atlantic storms an iron-bound coast of the wildest aspect;

but it is rich in arable land and pasture, and possesses the most picturesque lake scenery: indeed, freshwater lakes in the mountain valleys, so peculiarly characteristic of the European system, are the great ornaments of the high lands of Britain.

Various parts of the British islands were dry land while most of the continent of Europe was yet below the ancient ocean. The high land of Lammermuir and the Grampian hills in Scotland, and those of Cumberland in England, were raised before the Alps had begun to appear above the waves. In general all the highest parts of the British mountains are of granite and stratified crystalline rocks. The earliest fossiliferous strata are of immense thickness in Cumberland and in the north of Wales, and the old red sandstone, many hundred feet thick, stretches from sea to sea along the flanks of the Grampians. The coal strata are developed on a great scale in the south of Scotland and the north of England; and examples of every formation, with the exception of the muschelkalk, are to be found in these islands. Volcanic fires had been very active in early times, and nowhere is the columnar structure more beautifully exhibited than in Fingal's Cave and the Storr of Skye, in the Hebrides: and in the North of Ireland a base of 800 square miles of mica-slate is covered with volcanic rocks, which end on the coast in the magnificent columnar cliffs of the Giant's Causeway.

The Ural chain, the boundary between Europe and Asia, is the only interruption to the level of the great northern plain, and is altogether unconnected with and far separated from the Altaï mountains by salt lakes, marshes, and deserts. The central ridge may be traced from between the Lake of Aral and the Caspian Sea to the northern extremity of Nova Zembla, a distance of more than 1700 miles; but as a chain it really begins on the right bank of the Ural river, at the steppes of the Kirghiz, about the 51st degree of north latitude, and runs due north in a long narrow ridge to the Karskaïa Gulf, in the Polar Ocean, though it may be said to terminate in dreary rocks on the west side of Nova Zembla. The Ural range is about the height of the mountains in the Black Forest or the Vosges; and, with few exceptions, it is wooded to the top, chiefly by the Pinus cembra. The immense mineral riches of these mountains—gold, platina, magnetic iron, and copper—lie on the Siberian side, and mostly between the 54th and 60th degrees of north latitude: the only part that is colonized, and one of the most industrious and civilized regions of the Russian empire. To the south the chain is pastoral, about 100 miles broad, consisting of longitudinal ridges, the highest of which does not exceed 3498 feet: in this part diamonds are found. To the north of the mining district the narrow mural mass is covered with impenetrable forests and deep morasses, altogether uninhabitable and unexplored. Throughout the Ural mountains there are neither precipices, transverse gorges, nor any of the characteristics of a high chain; the descent on both sides is so gentle that in many places it is difficult to know where the plain begins; and the road over the chain from Russia by Ekaterinburg is so low that it hardly seems to be a mountain-pass. The gentle descent and sluggishness of the streams produce extensive marshes along the Siberian base of the range. To the arduous and enterprising researches of Sir Roderick Murchison we are indebted for almost all we know of these mountains: he found them on the western side to be composed of Silurian, Devonian, and carboniferous rocks, more or less altered and crystallized; on the eastern declivity the mines are in metaphoric strata, mixed with rocks of igneous origin; and the central axis is of quartzose and chloritic rocks.

The great zone of high land which extends along the old continent from the Atlantic to the shores of the Pacific Ocean divides the low lands into two very unequal parts. That to the north, only broken by the Ural range and the Valdai table-land of still less elevation, stretches from the Thames or the British hills and the eastern bank of the Seine to Behring's Straits, including more than 190° of longitude, and occupying an area of at least 4,500,000 square geographical miles, which is a third more than all Europe. The greater part of it is perfectly level, with a few elevations and low hills, and in many places a dead level extends hundreds of miles. The country between the Carpathian and Ural mountains is a flat, on which there is scarcely a rise in 1500 miles; and in the steppes of southern Russia and Siberia the extent of level ground is immense. The mean absolute height of the flat provinces of France is 480 feet. Moscow, the highest point of the European plain, is also 480 feet high, from whence the land slopes imperceptibly to the sea, both on the north and south, till it absolutely dips below its level. Holland, on one side, would be overflowed, were it not for its dykes, and towards Astrakan the plain sinks still lower. With the exception of the plateau of Ust-Urt, of no great elevation, situated between the Caspian and Aral, and which is the extreme southern ridge of the Ural chain, the whole of that extensive country north and east of the Caspian Sea and around the Lake of Aral forms a vast cavity of 18,000 square leagues, all considerably below the level of the ocean; and the surface of the Caspian Sea itself, the lowest point, has a depression of rather more than 82 feet.

The European part of the plain is highly cultivated and very productive in the more civilized countries in its western and middle regions and along the Baltic. The greatest amount of cultivated land lies to the north of the watershed which stretches from the Carpathians to the centre of the Ural chain, yet there are large heaths which extend from the extremity of Jutland through Luneburg and Westphalia to Belgium. The land is of excellent quality to the south of it. Round Polkova and Moscow there is an extent of the finest vegetable mould, equal in size to France and the Spanish peninsula together, which forms part of the High Steppe, and is mostly in a state of nature.

A large portion of the great plain is pasture-land, and wide tracts are covered with natural forests, especially in Poland and Russia, where there are millions of acres of pine, fir, and deciduous trees.

The quantity of waste land in Europe is very great, and there are also many swamps. A morass as long as England extends from the 52d parallel of latitude, following the course of the river Prepit, a branch of the Dnieper, which runs through its centre. There are swamps at the mouths of many of the sluggish rivers in Central Europe. They cover 1970 miles in Denmark, and mossy quagmires occur frequently in the more northerly parts.

Towards the eastern extremity of Europe the great plain assumes the peculiar character of desert called a *steppe*, a word supposed to be of Tartar origin, signifying a level waste destitute of trees: hence the steppes may vary according to the nature of the soil. They commence at the river Dnieper, and extend along the shores of the Black Sea. They include all the country north and east of the Caspian Lake and Independent Tartary; and passing between the Ural and Altaï mountains, they may be said to occupy all the low lands of Siberia. Hundreds of leagues may be traversed east from the Dnieper without variation of scene. A dead level of thin but luxuriant pasture, bounded only by the horizon, day after day the same unbroken monotony fatigues the eye. Sometimes there is the appearance of a lake, which vanishes on approach, the phantom of atmospheric refraction. Horses and cattle beyond number give some animation to the scene, so long as the steppes are green; but winter comes in October, and then they become a trackless field of spotless snow. Fearful storms rage, and the dry snow is driven by the gale with a violence which neither man nor animal can resist, while the sky is clear and the sun shines cold and bright above the earthly turmoil. The contest between spring and winter is long and severe, for

> "Winter oft at once resumes the breeze,
> Chills the pale morn, and bids his driving sleets
> Deform the day, delightless."

Yet when gentler gales succeed, and the waters run off in torrents through the channels which they cut in the soft ground, the earth is again verdant. The scorching summer's sun is as severe in its consequences in these wild regions as the winter's cold. In June the steppes are parched, no shower falls, nor does a drop of dew refresh the thirsty and rent earth. The sun rises and sets like a globe of fire, and during the day he is obscured by a thick mist from the evaporation. In some seasons the drought is excessive: the air is filled with dust in impalpable powder, the springs become dry, and cattle perish in thousands. Death triumphs over animal and vegetable nature, and desolation tracks the scene to the utmost verge of the horizon, a hideous wreck.

Much of this country is covered by an excellent but thin soil, fit for corn, which grows luxuriantly wherever it has been tried; but a stiff cold clay at a small distance below the surface kills every herb that has deep roots, and no plants thrive but those which can resist the extreme vicissitudes of climate. A very wide range is hopelessly barren. The country from the Caucasus, along the shores of the Black and Caspian Seas — a dead flat, twice the size of the British islands — is a desert destitute of fresh water. Saline efflorescences cover the surface like hoar-frost. Even the atmosphere and the dew are saline, and many salt lakes in the neighbourhood of Astrakan furnish great quantities of common salt and nitre. Saline plants, with patches of verdure few and far between, are the only signs of vegetable life, but about Astrakan there is soil and cultivation. Some low hills occur in the country between the Caspian and the Lake of Aral, but it is mostly an ocean of shifting sand, often driven by appalling whirlwinds.

Turkistan is a sandy desert, except on the banks of the Oxus and the Jaxartes, and as far on each side of them as canals convey the fertilizing waters. To the north, barrenness gives place to verdure between the river Ural and the terraces and mountains of Central Asia, where the steppes of the Kirghiz afford pasture to thousands of camels and cattle belonging to these wandering hordes.

Siberia is either a dead level or undulating surface of more than 7,000,000 of square miles, between the North Pacific and the Ural mountains, the Polar Sea and the Altaï range, whose terraces and offsets end in those plains, like headlands and promontories in the ocean. M. Middendorf, indeed, met with a chain of most desolate mountains on the shores of the Polar Ocean, in the country of the Samoides; and the almost inapproachable coast far to the east is unexplored. The mineral riches of the mountains have brought together a population who inhabit towns of considerable importance along the base of the Ural and Altaï chains, where the ground yields good crops and pasture; and there are forests on the undulations of the mountains and on the plains. There are many hundred square miles of rich black mould covered with trees and grass, uninhabited, between the river Tobol and the upper course of the Obi, within the limit where corn would grow; but even this valuable soil is studded with small lakes of salt and fresh water, a chain of which, 300 miles long, skirts the base of the Ural mountains.

North of the 62nd parallel of latitude corn does not ripen on account of the biting blasts from the Icy Ocean which sweep supreme over these unprotected wastes. In a higher latitude, even the interminable forests of gloomy fir are seen no more: all is a wide-spreading desolation of salt steppes, boundless swamps, and lakes of salt and fresh water. The cold is so intense there that the spongy soil is perpetually frozen to the depth of some hundred feet below the surface; and the surface itself, not thawed before the end of June, is again ice-bound by the middle of September, and deep snow covers the ground nine or ten months in the year. Happily gales of wind are not frequent during winter, but when they do occur no living thing ventures to face them. The Russian Admiral Wrangel, who travelled during the most intense cold from the mouth of the river Kolyma to Behring's Strait, gives an appalling account of these deserts. "Here endless snows and ice-covered rocks bound the horizon, nature lies shrouded in all but perpetual winter, life is a constant conflict with privation and with the terrors of cold and hunger—the grave of nature, which contains only the bones of another world. The people, and even the snow smoke, and this evaporation is instantly changed into millions of needles of ice, which make a noise in the air like the sound of torn satin or thick silk. The reindeer take to the forest, or crowd together for heat, and the raven alone, the dark bird of winter, still cleaves the icy air with slow and heavy wing, leaving behind him a long line of thin vapour, marking the track of his solitary flight. The trunks of the thickest trees are rent with a loud noise, masses of rock are torn from their sites, the ground in the valleys is rent into yawning fissures, from which the waters that are underneath rise, giving off a cloud of vapour, and immediately become ice. The atmosphere becomes dense, and the glistening stars are dimmed. The dogs outside the huts of the Siberians burrow in the snow, and their howling, at intervals of six or eight hours, interrupts the general silence of winter."[1] In many parts of Siberia, however, the sun, though long absent from these dismal regions, does not leave them to utter darkness. The extraordinary brilliancy of the stars, and the gleaming snowlight, produce a kind of twilight, which is augmented by the splendid coruscations of the aurora borealis.

The scorching heat of the summer's sun produces a change like magic on the southern provinces of the Siberian wilderness. The snow is scarcely gone before the ground is covered with verdure, and flowers of various hues blossom, bear their seed, and die in a few months, when Winter resumes his empire. A still shorter-lived

---

[1] In the year 1820, Admiral (then Lieutenant) Wrangel travelled from the mouth of the Kolyma to Behring's Straits on sledges drawn by dogs, and made a bold but vain attempt to reach the North Pole. Lieutenant Anjou, at the same time, sailed from the mouth of the Jana river, reached 76½ degrees of north latitude, and passed round the group of the New Siberian Islands.

vegetation scarcely covers the plains in the far north, and, on the shores of the Icy Ocean, even reindeer-moss grows scantily.

The abundance of fur-bearing animals in the less rigorous parts of the Siberian deserts has tempted the Russians to colonize and build towns on these frozen plains. Yakutsk, on the river Lena, in 62° 1′ 30″ N. lat., is probably the coldest town on the earth. The ground is perpetually frozen to the depth of more than 400 feet, of which three feet only are thawed in summer, when Fahrenheit's thermometer is frequently 77° in the shade; and as there is in some seasons no frost for four months, larch forests cover the ground, and wheat and rye produce from fifteen to forty fold. In winter the cold is so intense that mercury is constantly frozen two months, and occasionally even three.

In the northern parts of Europe the Silurian, Devonian, and carboniferous strata are widely developed, and more to the south they are followed in ascending order by immense tracts of the higher series of secondary rocks, abounding in the huge monsters of a former world. Very large and interesting tertiary basins fill the ancient hollows in many parts of the plain, which are crowded with the remains of animals that no longer exist. Of these the most important are the London, Paris, Vienna, and Moscow basins, with many others in the north of Germany and Russia; and alluvial soil covers the greater part of the plain. In the east Sir Roderick Murchison has determined the boundary of a region twice as large as France, extending from the Polar Ocean to the southern steppes, and from beyond the Volga to the flanks of the Ural chain, which consists of a red deposit of sand and marl, full of copper in grains, belonging to the Permian system. This and the immense tract of black loam already mentioned are among the principal features of Eastern Europe.

CHAPTER VI.

The Southern Low Lands of the Great Continent, with their Secondary Table-Lands and Mountains.

THE low lands to the south of the great mountain girdle of the old continent are much broken by its offsets, by separate groups of mountains, and still more by the deep indentation of bays and large seas. Situate in lower latitudes, and sheltered by mountains from the cutting Siberian winds, these plains are of a more tropical character than those to the north; but they are strikingly contrasted in their different parts — either rich in all the exuberance that heat, moisture, and soil can produce, or covered by wastes of bare sand—in the most advanced state of cultivation, or in the wildest garb of nature.

The barren parts of the low lands lying between the eastern shores of China and the Indus bear a small proportion to the riches of a soil vivified by tropical warmth and watered by the periodical inundations of the mighty rivers that burst from the icy caverns of Tibet and the Himalaya. On the contrary, the favoured regions in that part of the low lands lying between the Persian Gulf, the Euphrates, and the Atlas mountains, are small when compared with the immense expanse of the Arabian and African deserts, scorched and calcined by an equatorial sun. The blessing of a mountain-zone, pouring out its everlasting treasures of moisture, the life-blood of the soil, is nowhere more strikingly exhibited than in the contrast formed by these two regions of the globe.

The Tartar country of Manchouria, watered by the river Amour, but little known to Europeans, lies immediately south of the Yablonnoi branch of the Altaï chain, and consequently partakes of the desert aspect of Siberia, and, in its northern parts, even of the Great Gobi. It is partly intersected by mountains, and covered by dense forests; nevertheless, oats grow in the plains, and even wheat in sheltered places. Towards Corea the country is more fertile; in that peninsula there are cultivated plains at the base of its central mountain-range.

China is the most productive country on the face of the earth; an alluvial plain of 210,000 square miles, formed by one of the most extensive river systems in the old world, occupies its eastern part. This plain, seven times the size of Lombardy, is no less fertile, and perfectly irrigated by canals. The Great Canal traverses the eastern part of the plain for 700 miles, of which 500 are in a straight line of considerable breadth, with a current in the greater part of it. Most of the plain is in rice and garden grounds, the whole cultivated with the spade. The tea-plant grows on a low range of hills between the 30th and 32nd parallels of north latitude, an offset from the Pe-ling chain. The cold in winter is much greater than in the corresponding European latitudes, and the heat in summer is proportionally excessive.

The Indo-Chinese peninsula, lying between China and the river Brahmapootra, has an area of 77,700 square miles, and projects 1500 miles into the ocean. The plains lying between the offsets descending from the east end of the Himalaya, and which divide it longitudinally, as before mentioned, are very extensive. The Birman empire alone, which occupies the valley of the Irrawaddy, is said to be as large as France, and not less fertile, especially its southern part, which is the granary of the empire. Magnificent rivers intersect the alluvial plains, whose soil they have brought down from the table-land of Tibet, and still continue to deposit in great quantities in the deltas at their mouths.

The plains of Hindostan extend 2000 miles along the southern slopes of the Himalaya, between the Brahmapootra and the Indus, and terminate on the south in the Bay of Bengal, the table-land of the Deccan, and the Indian Ocean — a country embracing in its range every variety of climate from tropical heat and moisture to the genial temperature of southern Europe.

The valley of the Ganges is one of the richest on the globe, and contains a greater extent of vegetable mould, and of land under cultivation, than any other country in this continent, except perhaps the Chinese empire. In its upper part, Sirhind and Delhi, the seat of the ancient Mogul empire, still rich in splendid specimens of Indian art, are partly arid, although in the latter there is fertile soil. The country is beautiful where the Jumna and other streams unite to form the Ganges. These rivers are often hemmed in by rocks and high banks, which in a great measure prevent the periodical overflow of the waters; this, however, is compensated by the coolness and moisture of the climate. The land gradually improves towards the east, as it becomes more flat, till at last there is not a stone to be seen for hundreds of miles down to the Gulf of Bengal. Wheat and other European grain are produced in the upper part of this magnificent valley, while in the south every variety of Indian fruit, rice, cotton, indigo, opium, and sugar, are the staple commodities. The ascent of the plain of the Ganges from the Bay of Bengal is so gradual that Saharampore, nearly at the foot of the Himalaya, is only 1100 feet above the level of Calcutta; the consequence of which is that the Ganges and Brahmapootra, with their branches, in the rainy season between June and September, lay Bengal under water for hundreds of miles in every direction, like a great sea. When the water subsides, the plains are verdant with rice and other grain; but when harvest is over, and the heat is intense, the scene is changed — the country, divested of its beauty, becomes parched and dusty everywhere, except in the extensive jungles. It has been estimated that one-third of the British territory in India is covered with these rank marshy tracts.[1]

The peninsula of Hindostan is occupied by the triangular-shaped table-land of the Deccan, which is much lower, and totally unconnected with the table-land of Tibet. It has the primary ranges of the Ghauts on the east and west, and the Vendhya mountains on the north, sloping by successive levels to the plains of Hindostan Proper. A trace of the general equatorial direction of the Asiatic high land is still perceptible in the Vendhya mountains, sometimes

[1] The estimate was made by Lord Cornwallis, and confirmed by Mr. Colebrooke

called the central chain of India, and in the Saulpoora range to the south, both being nearly parallel to the Himalaya.[1] The surface of the Deccan, between 3000 and 4000 feet above the sea, is a combination of plains, ridges of rocks, and insulated flat-topped hills, which are numerous, especially in its north-eastern parts. These solitary and almost inaccessible heights rise abruptly from the plains, with all but perpendicular sides, which can only be scaled by steps cut in the rock, or by very dangerous paths. Many are fortified, and were the strongholds of the natives, but they never have withstood the determined intrepidity of British soldiers.

The peninsula terminates with the table-land of the Mysore, 7000 feet above the sea, surrounded by the Nilgherry or Blue Mountains, which rise 2941 feet higher.

The base of this plateau, and indeed of all the Deccan, is granite, and there are also many syenitic and trap rocks, with abundance of primary and secondary fossiliferous strata. Though possessing the diamond mines of Golconda, the true riches of the country consist in its vegetable mould, which in the Mysore is 100 feet thick, an inexhaustible source of fertility. The sea-coasts on the two sides of the peninsula are essentially different; that of Malabar on the western side is rocky, but in many parts well cultivated, and its mountains covered with forests form a continuous wall of very simple structure, 510 miles long, and rather more than 5000 feet high. On the coast of Coromandel the mountains are bare, lower, frequently interrupted, and the wide maritime plains are generally parched.

The island of Ceylon, nearly equal in extent to Ireland, is almost joined to the southern extremity of the peninsula by sandbanks and small islands, between which the water is only six feet deep in spring-tides. The Sanscrit name of the "Resplendent" may convey some idea of this island, rich and fertile in soil, adorned by lofty mountains, numerous streams, and primeval forests; in addition to which it is rich in precious stones, and has the pearl oyster on its coast.

The Asiatic low lands are continued westward from the Indian peninsula by the Punjab and the great Indian desert. "The Punjab, or country of the five rivers," lies at the base of the Western Himalaya. Its most northern part consists of fertile terraces highly cultivated, and valleys at the foot of the mountains. It is very productive in the plain within the limits of the periodical inundations of the rivers, and where it is watered by canals; in other parts it is pastoral. The kingdom of Lahore occupies the chief part of the Punjab, and the city of that name near the Ravee, the ancient Hydraotes, once the rival of Delhi, lies on the high road from Persia to India, and was made the capital of the kingdom by Runjeet Sing.

[1] Johnston's Physical Atlas.

The lower valley of the Indus throughout partakes of the character of the Punjab; it is fertile only where it is within reach of water; much of it is delta, which is occupied by rice-grounds; the rest is pasture, or sterile salt marshes.

South of the Punjab, and between the fertile plains of Hindostan and the left bank of the Indus, lies the great Indian desert, which is about 400 miles broad, and becomes more and more arid as it approaches the river. It consists of a hard clay, covered with shifting sand, driven into high waves by the wind, with some parts that are verdant after the rains. In the province of Cutch, south of the desert, a space of 7000 square miles, known as the Run of Cutch, is alternately a sandy desert and an inland sea. In April the waves of the sea are driven over it by the prevailing winds, leaving only a few grassy eminences, the resort of wild asses. The desert of Mekran, an equally barren tract, extends along the Gulf of Oman from the mouths of the Indus to the Persian Gulf: in some places, however, it produces the Indian palm, and the aromatic shrubs of Arabia Felix. It was the line followed by Alexander the Great returning with his army from India.

The scathed shores of the Arabian Gulf, where not a blade of grass freshens the arid sands, and the uncultivated valleys of the Euphrates and Tigris, separate Asia from Arabia and Africa, the most desert regions in the old world.

The peninsula of Arabia, divided into two parts by the Tropic of Cancer, is about four times the size of France. No rivers, and few streams or springs nourish the thirsty land, whose barren sands are scorched by a fierce sun. The central part is a table-land of moderate height, which however is said to have an elevation of 8000 feet in the province of Haudramaut. To the south of the tropic it is an almost interminable ocean of drifting sand, wafted in clouds by the gale, and dreaded even by the wandering Bedouin. At wide intervals, long narrow depressions cheer the eye with brushwood and verdure. More to the north, mountains and hills cross the peninsula from S.E. to N.W., enclosing cultivated and fine pastoral valleys adorned by groves of the date-palm and aromatic shrubs. Desolation once more resumes its domain where the table-land sinks into the Syrian desert, and throughout the rest of its circumference it descends in terraces or parallel ranges of mountains and hills to a flat sandy coast from 30 to 100 miles wide, which surrounds the greater part of the peninsula, from the mouth of the Euphrates to the Isthmus of Suez. The hills come close to the beach in the province of Oman, which is traversed by chains, and broken into piles of arid mountains not more than 3500 feet high, with the exception of the Jebel Okkdar, which is 6000 feet above the sea, and is cleft by temporary streams and fertile valleys. Here the ground is cultivated and covered with verdure, and still farther south there is a line of oases fed by subterraneous springs, where the fruits common to Persia, India, and Arabia are produced.

The south-eastern coast is scarcely known, except towards the provinces of Haudramaut and Yemen, or Arabia Felix, where ranges of mountains, some above 5000 feet high, line the coast, and in many places project into the ocean, sometimes forming excellent harbours, as that of Aden, which is protected by projecting rocks. In the intervals there are towns and villages, cotton-plantations, date-groves, and cultivated ground.

On the northern side of these granite ranges, where the table-land is 8000 feet above the sea, and along the edge of the desert of El Aklaj in Haudramaut, there is a tract of sand so loose and so very fine, that a plummet was sunk in it by Baron Wrede to the depth of 360 feet without reaching the bottom. There is a tradition in the country that the Sabæan army of King Suffi perished in attempting to cross this desert. Arabia Felix, which merits its name, is the only part of that country with permanent streams, though they are small. Here also the mountains and fertile ground run far inland, producing grain, pasture, coffee, odoriferous plants, and gums. High cliffs line the shores of the Indian Ocean and the Strait of Bab-el-man-deb—"the Gate of Tears." The fertile country is continued a considerable way along the coast of the Red Sea, but the character of barrenness is resumed by degrees, till at length the hills and intervening terraces, on which Mecca and Medina, the holy cities of the Mahomedans, stand, are sterile wastes wherever springs do not water them. The blast of the desert, loaded with burning sand, sweeps over these parched regions. Mountains skirt the table-land to the north; and the peninsula, between the Gulfs of Akabah and Suez on the Red Sea, the Eliath of Scripture, is filled by the mountain-group of Sinai and Horeb. Jebel Houra, Mount Horeb, on which Moses received the Ten Commandments, is 8593 feet high, surrounded by higher mountains, which are covered with snow in winter. The group of Sinai abounds in springs and verdure. At its northern extremity lies the desert of El-Teh, 70 miles long and 30 broad, in which the Israelites wandered forty years. It is covered with long ranges of high rocks, of most repulsive aspect, rent into deep clefts only a few feet wide, hemmed in by walls of rock sometimes 1000 feet high, like the deserted streets of a Cyclopean town. The journey from Sinai to Akabah, by the Wadee-el-Ain, or Valley of the Spring, is perfectly magnificent, and the sight of Petra itself is a tremendous confusion of black and brown mountains. It is a considerable basin closed in by rocks, with chasms and defiles in the precipices. The main street is two miles long, and not more than from 10 to 30 feet wide, enclosed between perpendicular rocks from 100 to 700 feet high, which so nearly meet as to leave only a strip of sky. A stream runs through the street which must once have been a considerable torrent, and the precipitous rocks are excavated into thousands of caverns once inhabited — into conduits, cisterns, flights of steps, theatres, and temples, forming altogether one of the most wonderful remains of antiquity. The whole of Arabia Petrea, Edom of the sacred writers, presents a scene of appalling desolation, completely fulfilling the denunciation of prophecy.[1]

A sandy desert, crossed by low limestone ridges, separates the table-land of Arabia from the habitable part of Syria, which the mountains of Lebanon divide into two narrow plains. These mountains may almost be considered offsets from the Taurus chain; at least they are joined to it by the wooded range of Gawoor, the ancient Amanus, impassable except by two defiles, celebrated in history as the Amanic and Syrian Gates. The group of Lebanon begins with the Jebel Okrab (Mount Casius), which rises abruptly from the sea in a single peak to the height of 7000 feet, near the mouth of the Orontes. From thence the chain runs south, at a distance of about twenty miles from the shores of the Mediterranean, in a continuous line of peaks to the sources of the Jordan, where it splits into two nearly parallel naked branches, enclosing the wide and fertile plain of Beka or Ghor, the ancient Cœlo-Syria, in which are the ruins of Balbec.

The Lebanon branch terminates at the sea near the mouth of the river Leontes, a few miles north of the city of Old Tyre; while the Anti-Libanus, which begins at Mount Hermon, 9000 feet high, runs west of the Jordan through Palestine in a winding line, till its last spurs, south of the Dead Sea, sink into rocky ridges on the desert of Sinai.

The tops of all these mountains, from Scanderoon to Jerusalem, are covered with snow in winter; it is permanent on Lebanon only, whose absolute elevation is 9517 feet. The precipices are terrific, the springs abundant, and the spurs of the mountains are studded with villages and convents; there are forests in the higher grounds, and lower down vineyards and gardens. Many offsets from the Anti-Libanus end in precipices on the coast between Tripoli and Beyrout, among which the scenery is superb.

The valleys and plains of Syria are full of rich vegetable mould, particularly the plain of Damascus, which is brilliantly verdant, though surrounded by deserts, the barren uniformity of which is relieved on the east by the broken columns and ruined temples of Palmyra (Tadmor). The Assyrian wilderness, however, is not everywhere absolutely barren. In the spring-time it is covered with a thin but vivid verdure, mixed with fragrant aromatic herbs of very

---

[1] From Miss Martineau's spirited and picturesque account of her journey to Egypt and Syria.

short duration. When these are burnt up, the unbounded plains resume their wonted dreariness. The country, high and low, becomes more barren towards the Holy Land, yet even here some of the mountains—as Carmel, Bashan, and Tabor—are luxuriantly wooded, and many of the valleys are fertile, especially the valley of the Jordan, which has the appearance of pleasure-grounds with groves of wood and aromatic plants, but almost in a state of nature. One side of the Lake of Tiberias in Galilee is savage; on the other there are gentle hills and wild romantic vales, adorned with palm-trees, olives, and sycamores—a scene of calm solitude and pastoral beauty. Jerusalem stands on a declivity encompassed by severe stony mountains, wild and desolate. The greater part of Syria is a desert compared with what it formerly was. Mussulman rule has blighted this fair region, once flowing with milk and honey—the Land of Promise.

Farther south desolation increases; the valleys become narrower, the hills more denuded and rugged, till, south of the Dead Sea, their dreary aspect announces the approach of the desert.

The valley of the Jordan affords the most remarkable instance known of the depression of the land below the general level of the ocean. This hollow, which extends from the Gulf of Akabah on the Red Sea to the bifurcation of Lebanon, is 620 feet below the Mediterranean at the Lake of Tiberias, and the acrid waters of the Dead Sea have a depression of 1300 feet.[1] The lowness of the valley had been observed by the ancients, who gave it the descriptive name of Cœlo-Syria, "Hollow Syria." It is absolutely walled in by mountains between the Dead Sea and Lebanon, where it is from ten to fifteen miles wide.

A shrinking of the strata must have taken place along this coast of the Mediterranean, from a sudden change of temperature in the earth's crust, or perhaps in consequence of some of the internal props giving way, for the valley of the Jordan is not the only instance of a depression of the soil below the sea-level: the small bitter lakes on the Isthmus of Suez are cavities of the same kind, as well as the Natron lakes on the Libyan desert, west from the delta of the Nile, and perhaps a part of the date-bearing district of Beskra in the regency of Tunis.

[1] By the trigonometrical measurement of Lieutenant Anthony Symonds, confirmed by French authorities, and adopted by Baron Humboldt, the depression of the Dead Sea is, as stated in the text, 1300 feet; out MM. Bertou and Russiger made it out to be 1388 by the barometer. See Lieut. Molyneux's paper in the Journal of the Royal Geographical Society, 1848. Subsequently the American expedition, under Lieut. Lynch, found "the depression of the Dead Sea below the Mediterranean a little over 1300 feet."

CHAPTER VII.

Africa — Table Land — Cape of Good Hope and Eastern Coast — Western Coast — Abyssinia — Senegambia — Low Lands and Deserts.

THE continent of Africa is 4330 geographical miles long from Cape Lagullus, east of the Cape of Good Hope, to Cape Bianco, near Bizerta, its northern extremity, and 4000 between Cape Guardafuï, on the Indian Ocean, and Cape Verd, on the Atlantic; but from the irregularity of its figure it has an area of only 12,000,000 of square miles. It is divided in two by the equator, consequently the greater part of it lies under a tropical sun. The high and low lands of this portion of the old continent are so distinctly separated from the Mountains of the Moon, or rather of Komri, that, with the exception of the mountainous territory of the Atlas, and the small table-land of Barca, it may be said to consist of two parts only, a high country and a low.

An extensive, though not very elevated, table-land occupies all Southern Africa, and even reaches to six or seven degrees north of the equator. On three sides it shelves down in tiers of narrow parallel terraces to the ocean, separated by mountain-chains which rise in height as they recede from the coast; and there is reason to believe that the structure of the northern declivity is similar, though its extremities only are known—namely, Abyssinia on the east, and the high land of Senegambia on the west; both of which project farther to the north than the central part.

The borders of the table-land are very little known to Europeans, and still less its surface, which no white man has crossed north of the Tropic of Capricorn. A comparatively small part, north from the Cape of Good Hope, has been explored by European travellers. Mr. Truter and Mr. Somerville were the first white men whom the inhabitants of Litakoo had seen. Of an expedition that followed their track, a few years after, no one returned.

North of the Cape the land rises to 600 feet above the sea; and the Orange River, or Gareep, with its tributaries, may be more aptly said to drain than to irrigate the arid country through which they flow; many of the tributaries, indeed, are only the channels through which torrents, from the periodical rains, are carried to the Orange River, and are destitute of water many months in the year. The "Dry River," the name of one of these periodical streams, is in that country no misnomer. Their margins are adorned with mimosas, and the sandy plains have furnished treasures to the botanist; zoology is no less indebted to the whole continent of Africa for the various animals it produces.

Dr. Smith crossed the Tropic of Capricorn in a journey from the Cape of Good Hope, where the country had still the same arid character. North from that there is a vast tract unexplored. In 1802 two native travelling merchants crossed the continent, which is 1590 miles wide, from Loando on the Atlantic to Zanzibar on the Mozambique Channel. They found various mercantile nations, considerably advanced in civilization, who raise abundance of maize and millet, though the greater part of the country is in a state of nature. Ridges of low hills yielding copper, the staple commodity of this country, run from S. E. to N. W. to the west of the dominions of the Cambeze, a country full of rivers, morasses, and extensive salt marshes which supply this part of the continent with salt. The travellers crossed 102 rivers, most of them fordable. The leading feature of this country is Lake N'yassi, of great but unknown length, and comparatively narrow. It begins 200 miles north from the town of Tete, on the Zambeze, and extends from S. E. to N. W., flanked on the east by a range of mountains of the same name running in the same direction, at the distance of 350 miles from the Mozambique Channel. This is all we know from actual observation of the table-land of South Africa, till about the 8th northern parallel of latitude, where M. d'Abbadie's Abyssinian journey terminated. It is probable, however, that there can be no very high mountains covered with perpetual snow in the interior of the table-land, for, if there were, Southern Africa would not be destitute of great rivers; nevertheless the height of the table-land on its northern edge must be considerable to supply the perennial sources of the Nile, the Senegal, and the Niger.

The edges of the table-land are better known. At the Cape of Good Hope the African continent is about 700 miles broad, and ends in three narrow parallel ridges of mountains, the last of which is the highest and abuts on the table-land. All are cleft by precipitous deep ravines, through which winter torrents flow to the ocean. The longitudinal valleys, or karroos, that separate them, are tiers, or steps, by which the plateau dips to the maritime plains. The descent is rapid, as both these plains and the mountain ranges are very narrow. On the western side the mountains form a high group and end in steep promontories on the coast, where Table Mountain, at Cape Town, 3582 feet high, forms a conspicuous landmark for mariners.

Granite, which is the base of Southern Africa, rises to a considerable height in many places, and is generally surmounted by vast horizontal beds of sandstone, which give that character of flatness peculiar to the summits of many of the Cape mountains.

The karroos are arid deserts in the dry season, but soon after the rains they are covered with verdure and a splendid flora. The maritime plains partake of the same temporary aridity, though a large portion is rich in cereal productions, vineyards, fruits, and pasture.

The most inland of the parallel ranges, about the 20th meridian east, is 10,000 feet high, and, though it sinks to some groups of hills at its eastern extremity, it rises again about the 37th meridian, in a truly alpine and continuous chain—the Quotlamba mountains, which follow the northerly direction of Natal, and are continued in the Lupata range of hills, 89 miles inland, through Zanguebar.

At Natal the coast is grassy, with clumps of trees, like an English park. The Zambeze and other streams from the table-land refresh the plains on the Mozambique Channel and Zanguebar, where, though some parts are marshy and covered with mangroves, groves of palm-trees adorn the plains, which yield prodigious quantities of grain, and noble forests cover the mountains; bnt from 4° N. latitude to Cape Guardafuï is a continued desert. There is also a barren tract at the southern end of the Lupata chain, where gold is found in masses and grains on the surface and in the water-courses, which tempted the Portuguese to make settlements on these unwholesome coasts.

The island of Madagascar, with its magnificent range of mountains, full of tremendous precipices, and covered with primeval forests, is parallel to the African coast, and only separated from it by the Mozambique Channel, 300 miles broad, so it may be presumed that it rose from the deep at the same time as the Lupata chain.

The contrast between the eastern and western coasts of South Africa is very great. The escarped bold mountains round the Cape of Good Hope, and its rocky coast, which extends a short way along the Atlantic to the north, are succeeded by ranges of sandstone of small elevation, which separate the internal sandy desert from the equally parched sandy shore. The terraced dip of the Atlantic coast for 900 miles between the Orange River and Cape Negro has not a drop of fresh water.

At Cape Negro, ranges of mountains separated by long level tracts begin and make a semicircular bend into the interior, leaving plains along the coast 140 miles broad. In Benguela these plains are healthy and cultivated; farther north there are monotonous grassy savannahs, and forests of gigantic trees. The ground, in many places saturated with water, bears a tangled crop of mangroves and tall reeds which even cover the shoals along the coast; hot pestilential vapours hang over them, never dissipated by a breeze.

The country of Calbongos is the highest land on the coast, where a magnificent group of mountains, covered almost to their tops with large timber, lie not far inland. The low plains of Biafra and Benin, west of them, and especially the delta of the Niger, consist entirely of swamps loaded with rank vegetation.

The angel of Death brooding over these regions in noisome exhalations, guards the interior of that country from the aggressions of the European, and has hitherto baffled his attempts to form settlements on the banks of this magnificent river.

Many portions of North Guinea are so fertile that they might vie with the valley of the Nile in cereal riches, besides various other productions; and though the temperature is very high, the climate is not very unhealthy.

The chain of mountains bordering the great African table-land on the east, or towards the Indian Ocean, attains a great height between the third and fourth degrees of south latitude. It is in this space that, according to M. Rebman, the giant mountain of Africa, the snow-capped Kilimanjaro, rises to an elevation of 20,000 feet in 3° 40′ south, in the country of Mono Moezi; it is supposed by some authors that the highest branch of the Nile rises in this remote part of the continent, and as Moezi in the language of the country signifies Moon, the origin of the Nile in the Mountains of the Moon, as described by Ptolemy, would be confirmed. As to the chain of Komri, made to stretch in an equatorial direction across the African continent from the Arabic Gulf to the Bay of Benin on the Atlantic, it has probably no other existence than the imagination of map-makers, the vast extent of country it is made to traverse being entirely unknown to modern geographers and travellers.

The vast alpine promontory of Abyssinia or Ethiopia,[1] 700 miles wide, projects from the table-land for 300 miles into the low lands of North Africa. It dips to a low swampy region on the north, to the plains of Senaar and Kordofan on the west, and on the east sinks abruptly to the coast at a short distance from the Red Sea. It is there from 8000 to 9000 feet high on the plateau of Tigraj, but declines to the westward, so that in the 15th parallel of N. latitude the eastern slope of the table-land towards the Red Sea is nearly twenty times greater than the counter-slope towards the Nile; the edge of the latter, however, is from 3000 to 4000 feet above the plains.[2] The character of Abyssinia is in that respect like the Deccan, or Southern India, where the Ghauts rise abruptly near the

---

[1] The name of Ethiopia is still used by the Abyssinians, as stated by M A. d'Abbadie, the talented traveller, who has resided so many years among them, as including Abyssinia proper, the Bija country as far as Sawakin, the Afar (Aidal of our maps), the Somaly, Gurage, and Galla countries. The word Abyssinia is better employed in the Arab sense, for those populations, chiefly Christian, which have lost all idea of tribe, according to the same traveller.

[2] Estimated from N.E. to S.W., the proportion of the two slopes of the Abyssinian table-land is as 12.6 to 1.

coast of Malabar, and the surface falls gradually towards that of Coromandel. The table-land of Abyssinia is a succession of undulating plains, broken by higher insulated mountain-masses, which in Simën,[1] Gojjam, and in Kaffa more to the south, attain an absolute altitude of from 11,000 to 13,500 feet. The plains are intersected by numerous streams which form the Nile and its tributaries on the one hand, and the Hawásh and its affluents, which flow towards the Indian Ocean, to be lost in a swamp, on the other. The edge of the table-land towards the Nile is steep; the streams run to the low lands through valleys from 3000 to 4000 feet deep, so that a traveller in ascending them might imagine that he is crossing a mountain range, whereas, on coming to the top, he finds himself on a high plain. This elevated country has lakes, swamps, verdant meadows, and cultivated land, producing various grains, and occasionally coffee. The plain of Dambia, the granary of the country, enjoys perpetual spring. M. A. d'Abbadie and Dr. Beke, to whom we are indebted for so much valuable information with regard to this part of Africa, travelled to within eight degrees of the equator, and, from the accounts given by them, the country south of Abyssinia appears to be similar to those of Shoa and Gojjam — extensive undulating plains, with occasional mountain-masses, and traversed by numerous streams; wide tracts must be 7000 or 8000 feet high, as they only produce barley: the country towards Kaffa and the sources of the Gojeb is still higher, and in some parts desert; but the caravan-road between Wallega and Kaffa passes through a vast forest impervious to the rays of the sun, which, according to the accounts of the merchants, is not seen for four or five days successively; and west of the Did-ësa there are immense grassy plains, the elephant-hunting grounds of the Galla tribes.

The geological structure of Ethiopia is somewhat similar to that of the Cape of Good Hope, the base being granite and the superstructure sandstone, occasionally limestone, schist, and breccia. The granite comes to the surface in the lower parts of Abyssinia, but sandstone predominates in the upper parts and assumes a tabular form, often lying on the tops of the mountains in enormous flat masses, only accessible by steps cut in the rocks or by ladders: such insulated spots are used as state prisons. Large tracts of ancient volcanic rocks occur, especially in Shoa. Trap rocks also abound in Simën. A great part of Gojjam and Gudru is formed of prismatic basalt lying under red clay; it is likewise found in Inarya. Many of the hill forts in Abyssinia are basaltic.

Senegambia, the appendage to the western extremity of the table-land, also projects far into the low lands, and is the watershed

[1] The highest inhabited village visited by M. d'Abbadie was that of Arquiaze, in the province of Simën, 12,450 feet above the sea.

whence the streams flow on one side to the plains of Soudan, where they join the Joliba or Niger; and from the other side, the Gambia, Senegal, and other rivers run into the Atlantic over a rich cultivated plain, but unhealthy from the rankness of the vegetation.

The moisture that descends from the northern edge of the table-land of South Africa, under the fiery radiance of a tropical sun, fertilizes a tract of country stretching from sea to sea across the continent, the commencement of the African low lands. A great part of this region, which contains many kingdoms and commercial cities, is a very productive country. The abundance of water, the industry of the natives in irrigating the ground, the periodical rains, and the tropical heat, leave the soil no repose. Agriculture is in a rude state, but nature is so bountiful that rice and millet are raised in sufficient quantity to supply the wants of a numerous population. Gold is found in the river courses, and there are elephants in the forests; but man is the staple of their commerce—a disgrace to the savage who sells his fellow-creature, but a far greater disgrace to the more savage purchaser who dares to assume the sacred name of Christian.

This long belt of never-failing vitality, which has its large lakes, poisonous swamps, deep forests of gigantic trees, and vast solitudes in which no white men ever trode, is of small width compared with its length. In receding from the mountains, the moisture becomes less and the soil gradually worse, sufficing only to produce grass for the flocks of the wandering Bedouin. At last a hideous barren waste begins, which extends northwards 800 miles in unvaried desolation to the grassy steppes at the foot of the Atlas; and for 1000 miles between the Atlantic and the Red Sea the nakedness of this blighted land is unbroken but by the valley of the Nile and a few oases.

In the west about 760,000 miles, an area equal to that of the Mediterranean Sea, and, in some parts, of a lower level, is covered by the trackless sands of the Sahara desert, which is even prolonged for miles into the Atlantic Ocean in the form of sand-banks. The desert is alternately scorched by heat and pinched by cold. The wind blows from the east nine months in the year; and at the equinoxes it rushes in a hurricane, driving the sand in clouds before it, producing the darkness of night at midday, and overwhelming caravans of men and animals in common destruction. Then the sand is heaped up in waves ever varying with the blast; even the atmosphere is sand. The desolation of this dreary waste, boundless to the eye as the ocean, is terrific and sublime; the dry heated air is like a red vapour, the setting sun seems to be a volcanic fire, and at times the burning wind of the desert is the blast of death. There are many salt lakes to the north, and even the springs are of brine, thick incrustations of dazzling salt cover the ground, and the particles, carried aloft by whirlwinds, flash in the sun like diamonds

Sand is not the only character of the desert; tracts of gravel and low bare rocks occur at times, not less barren and dreary; but on the eastern and northern borders of the Sahara fresh water rises near the surface, and produces an occasional oasis where barrenness and vitality meet. The oases are generally depressed below the level of the desert, with an arenaceous or calcareous border enclosing their emerald verdure like a frame. The smaller oases produce herbage, ferns, acacias, and some shrubs; forests of date-palms grow in the larger, which are the resort of lions, panthers, gazelles, reptiles, and a variety of birds.

In the Nubian and Libyan deserts, to the east of the Sahara, the continent shelves down towards the Mediterranean in a series of terraces, consisting of vast level sandy or gravelly deserts, lying east and west, separated by low rocky ridges. This shelving country, which is only 540 feet above the sea at the distance of 750 miles inland, is cut transversely by the Nile, and by a deep furrow parallel to it, in which there is a long line of oases. This furrow, the Nile, and the Red Sea, nearly parallel to both, are flanked by rocky eminences which run north from the table-land.

On the interminable sands and rocks of these deserts no animal —no insect—breaks the dread silence; not a tree nor a shrub is to be seen in this land without a shadow. In the glare of noon the air quivers with the heat reflected from the red sand, and in the night it is chilled under a clear sky sparkling with its host of stars. Strangely but beautifully contrasted with these scorched solitudes is the narrow valley of the Nile, threading the desert for 1000 miles in emerald green, with its blue waters foaming with rapids among wild rocks, or quietly spreading in a calm stream amidst fields of corn and the august monuments of past ages.

At the distance of a few days' journey west from the Nile, over a hideous flinty plain, lies the furrow already mentioned, trending to the north, and containing the oases of Darfour, Selime, the Great and Little Oases, and the parallel valleys of the Natron Lakes, and Bahr-Belama or the "Dry River." The Great Oasis, or Oasis of Thebes, is 120 miles long and 4 or 5 broad; the lesser Oasis, separated from it by 40 miles of desert, is of the same form. Both are rich in verdure and cultivation, with villages amid palm-groves and date-plantations, mixed with the ruins of remote antiquity, offering scenes of peaceful and soft beauty contrasted with the surrounding gloom. The Natron Lakes are in the northern part of the Valley of Nitrùn, 35 miles west of the Nile; the southern part is a beautiful quiet spot, that became the retreat of Christian monks in the middle of the second century, and at one time contained 360 convents, of which 4 only remain; from these some very valuable manuscripts of old date have recently been obtained.

Another line of oases runs along the latitude of Cairo, with fresh-water lakes—consequently no less fertile than the preceding. The ruins of the Temple of Jupiter Ammon are in one of them.

Hundreds of miles on the northern edge of the desert, from the Atlantic along the southern foot of the Atlas to the Great Syrtis, are pasture-lands without a tree—an ocean of verdure. At the Great Syrtis the Sahara comes to the shores of the Mediterranean; and, indeed, for 1100 miles between the termination of the Atlas and the little table-land of Barca, the ground is so unprofitable that the population only amounts to about 30,000, and these are mostly wandering tribes who feed their flocks on the grassy steppes. Magnificent countries lie along the Mediterranean coast north of the Atlas, susceptible of cultivation. History, and the ruins of many great cities, attest their former splendour; even now there are many populous commercial cities, and much grain is raised, though a great part of these valuable kingdoms is badly cultivated or not cultivated at all.

The base of the sandy parts of North Africa is stiff clay; in Lower Nubia, between the parallels of Assouan and Esneh, red and white granite prevail, followed by argillaceous sandstone; Middle Egypt is calcareous; and lower down the alluvium of the Nile covers the surface.

It would appear that Southern Africa, though similar in its unbroken surface and peninsular shape to South America, bears no resemblance to it in other respects, but has a great analogy to the Deccan in its triangular form, its elevated platform, and in the position of its encompassing mountain-chains, if, as there is reason to believe, from the fertile region to the north, either that South Africa descends in a succession of terraces to the low lands, or that the Komri mountains have a real existence, and run directly across the continent. From the connexion already mentioned between external appearance and internal structure, as well as from partial information, it is surmised that the mountains surrounding the two triangles in question are of corresponding constitution; that, if any secondary strata do exist in this part of Africa, they must be exterior to these chains, and neither on the summits of the high mountains nor in the interior; and that any tertiary strata on the table-land must, as in the Deccan, have been formed in the basins of fresh-water lakes.[1]

The prodigious extent of desert is one of the most extraordinary circumstances in the structure of the old continent. A zone of almost irretrievable desolation prevails from the Atlantic Ocean across Africa and through Central Asia almost to the Pacific Ocean, at least 120 degrees of longitude. There are also many long districts of the same sterile nature in Europe; and if to these sandy

[1] Johnston's Physical Atlas.

plains the deserts of Siberia be added, together with all the barren and rocky mountain tracts, the unproductive land in the Old World is prodigious. The quantity of salt on the sandy plains is enormous, and proves that they have been part of the bed of the ocean or of inland seas at no very remote geological period. The low lands round the Black Sea and Caspian, and the Lake of Aral, seem to have been the most recently reclaimed, from the great proportion of shells in them identical with those now existing in these seas. The same may be said of the Sahara desert, where salt and recent shells are plentiful.

CHAPTER VIII.

American Continent—The Mountains of South America—The Andes—The Mountains of the Parima and Brazil.

Some thinner portion of the crust of the globe under the meridians that traverse the continent of America from Cape Horn to the Arctic Ocean must have yielded to the expansive forces of the subterranean fires, or been rent by contraction of the strata in cooling. Through this the Andes had arisen, producing the greatest influence on the form of the continent, and the peculiar simplicity that prevails in its principal mountain systems, which, with very few exceptions, have a general tendency from north to south. The continent is 9000 miles long, and, its form being two great peninsulas joined by a long narrow isthmus, it is divided by nature into three parts, of South, Central, and North America; yet these three are connected by the mighty chain of the Andes, but little inferior in height to the Himalaya, running along the coast of the Pacific from within the Arctic nearly to the Antarctic circle. In this course every variety of clime is to be met with, from the rigour of polar congelation to the scorching heat of the torrid zone; while the mountains are so high that the same extremes of heat and cold may be experienced in the journey of a few hours from the burning plains of Peru to the snow-clad peaks above. In this long chain there are three distinct varieties of character, nearly, though not entirely, corresponding to the three natural divisions of the continent. The Andes of South America differ materially from those of Central America and Mexico, while both are dissimilar to the North American prolongation of the chain, generally known as the Chippewayan or Rocky Mountains.

The greatest length of South America from Cape Horn to the Isthmus of Panamá is about 4020 geographical miles. It is very narrow at its southern extremity, but increases in width northwards to the latitude of Cape Roque on the Atlantic, between which and Cape Blanco on the Pacific it attains its greatest breadth of nearly 2750 miles. It consists of three mountain systems, separated by the basins of the three greatest rivers in the world. The Andes run along the western coast from Cape Horn to the Isthmus of Panamá, in a single chain of no inconsiderable width but majestic height, dipping rapidly to the narrow maritime plains of the Pacific, but descending on the east in high valleys and occasional offsets to plains of vast extent, whose dead level is for hundreds of miles as unbroken as that of the ocean by which they are bounded. Nevertheless two detached mountain systems rise on these plains, one in Brazil between the Rio de la Plata and the river of the Amazons; the other is that of Parima and Guiana, lying between the river of the Amazons and the Orinoco.

The great chain of the Andes first raises its crest above the waves of the Antarctic Ocean in the majestic dark mass of Cape Horn, the southernmost point of the archipelago of Tierra del Fuego. This group of mountainous islands, equal in size to Britain, is cut off from the main land by the Straits of Magellan. The islands are penetrated in every direction by bays and narrow inlets of the sea, or fiords, ending often in glaciers fed by the snow on the summits of mountains 6000 feet high. Peat-mosses cover the higher declivities of these mountains, and their flanks are beset with densely entangled forests of brown beech, which never lose their dusky leaves, producing altogether a savage, dismal scene. The mountains which occupy the western side of this cluster of islands sink down to wide level plains to the east, like the continent itself, of which the archipelago is but the southern extremity.[1]

The Pacific washes the very base of the Patagonian Andes for about 1000 miles, from Cape Horn to the 40th parallel of south latitude. The whole coast is lined by a succession of archipelagoes and islands, separated from the iron-bound shores by narrow arms of the sea, which, in the more southern part, are in fact profound longitudinal valleys of the Andes filled by the ocean, so that the chain of islands running parallel to the axes of the mountains is but the summits of an exterior range rising above the sea.

The coast itself for 650 miles is begirt by walls of rock, which sink into unfathomable depths, torn by long crevices or fiords, similar to those on the Norwegian shore, ending in tremendous glaciers, whose masses, falling with a crash like thunder, drive the sea in

---

[1] The Voyages of Captains King and Fitzroy, R.N., Mr. Darwin's 'Journal of a Naturalist,' Dr. Pœppig's 'Travels in South America,' are the authorities for the account of Tierra del Fuego, Patagonia, and Chile; Baron Humboldt, Mr. Pentland, Drs. Pœppig and Meyer of Berlin, for Peru and the Andean Chain to the Isthmus of Panamá.

sweeping breakers through these chasms. The islands and the mainland are thickly clothed with forests, which are of a less sombre aspect as the latitude decreases.

Between the Pass of Chacabuco north of Santiago, the capital of Chile, and the archipelago of Chiloe, a chain of hills, composed in general of crystalline rocks, borders the coast; between which and the Andes exists a longitudinal valley, well watered by the rivers descending from the central chain, and which constitutes the most fertile portion, nay the garden of the Chilian republic — the rich provinces of Santiago, Talca, Cauquenes, and Concepcion. This longitudinal depression may be considered as a prolongation of the strait that separates Chiloe from the mainland. Many peaks of the Andes enter within the limits of perpetual snow, between the 40th and 31st parallels; some of which are active volcanoes. In lat. 32° 39′ rises the giant of the American Andes, the Nevado of Aconcagua, which towers over the Chilian village of the same name, and is so clearly visible from Valparaiso. Although designated as a volcano, a term generally applied in Chile to every elevated and snowy peak, it offers no trace of modern igneous origin. It appears to be composed of a species of porphyry generally found in the centre of the Chilian chain. Its height, according to Captain Beechey's very accurate observations, exceeds 23,910 feet.[1]

About the latitude of Concepcion the dense forests of Araucarias and of other semi-tropical plants cease with the humid equable climate; and as no rain falls in central Chile for nine months in the year, the brown, purple, and tile-red hills and mountains are only dotted here and there with low trees and bushes; very soon, however, after the heavy showers have moistened the cracked ground, it is covered with a beautiful but transient flora. In some valleys it is more permanent and of a tropical character, mixed with alpine plants.[2] In Northern Chile rain falls only once in two or three years, the consequence of which is sterility on the western precipitous and unbroken descent of the Andes; but on the east, two secondary branches leave the central Cordillera, which extend 300 or 400 miles into the plains, wooded to a great height. The Sierra de Cordova, the most southern of these, begins between the 33d and 31st parallels, and extends in the direction of the Pampas; more to the north, Sierra di Salti and Jujuy stretches from the valley of Catamarca and Tucuman towards the Rio Vermejo, one of the tributaries of the Rio de la Plata.

The chain takes the name of the Peruvian Andes about the 24th degree of south latitude, and is separated from the Pacific by a range

[1] This great height has been deduced, adopting the position of the Peak as fixed by Captain Fitz Roy, and employing the angles of elevation observed near Valparaiso by Captain Beechey.

[2] Dr. Pœppig's Travels.

of hills composed of crystalline rocks, parallel to the sea coast, and by an intervening sandy desert, seldom above 60 miles broad, on which rain scarcely ever falls, where bare rocks pierce through the moving sand. The width of the coast is nearly the same to the Isthmus of Panamá, but damp luxuriant forests, full of orchideæ, begin about the latitude of Payta, and continue northwards through the provinces of Guayaquil, Las Esmeraldas, and Darien.

From its southern extremity to the Nevado of Chorolque, in 21° 30′ S. lat., the Andes are merely one grand and continuous range of mountains; but north of that the chain divides into longitudinal ridges, which enclose a series of valleys or table-lands, forming so many basins, enclosed at various points by transverse groups or mountain-knots, or by single ranges crossing between them like dykes, a structure that prevails to Pasto, 1° 13′ 6″ N. lat.

Unlike the table-lands of Asia of the same elevation, where cultivation is confined to the more sheltered spots, or those still lower in Europe, which are only fit for pasture, these lofty regions of the Andes yield exuberant crops of every European grain, and have many populous cities enjoying the luxuries of life, with universities, libraries, civil and religious establishments, at altitudes equal to that of the Peak of Teneriffe, which is 12,170 feet above the sea-level. Villages are placed and mines are worked at heights as great or even greater than the top of Mont Blanc.[1] This state is not limited to the present times, since these table-lands were made the centre of civilization by a race of mankind which "bear the same relation to the Incas and the present inhabitants that the Etruscans bear to the ancient Romans and to the Italians of our own days."

The table-land or valley of Desaguadero, one of the most remarkable of these, has an absolute altitude of 12,900 feet, and a breadth varying from 30 to 60 miles: it stretches 400 miles between the two parallel chains of the Andes, and between the transverse mountain-groups of Lipez, in 20° S. lat., and the great mountain knot of Vilcañota, which, extending from east to west, shuts in the valley on the north-west, and occupies an area three times as large as Switzerland, some of the snowy peaks rising 8300 feet above the surface of the table-land, from which an idea may be formed of the gigantic scale of the Andes. This table-land or valley is bounded on each side by the two grand chains of the Bolivian Andes: that on the west is the Cordillera of the coast; the range on the east is the Bolivian Cordillera, properly speaking; and on its north-west prolongation the Cordillera Real.[2] These two rows of mountains lie so near to each other that the whole breadth of the

[1] The celebrated silver-mines of Potosi were, until the last fifty years, worked to the very summit of that metalliferous mountain, 16,150 feet above the sea-level.

[2] Baron Humboldt and Mr. Pentland.

table-land, including both, is only 226 miles. All the snowy peaks of the Cordilleras of the coast are either active volcanoes or of igneous origin, and are all situate near the maritime declivity of the chain; consequently, the descent to the shores of the Pacific is every where very abrupt. The eastern Cordillera, which begins near the metalliferous mountain of Potosi, is below the level of perpetual snow to the south, but its northern portion contains the three peaked mountains of Ancohuma or Nevado of Sorata, of Supäíwasi, and Illimani, and is one of the most magnificent portions of the Andes.[1] The snowy part begins with the gigantic mass of Illimani, whose serrated ridges are elongated in the direction of the axis of the chain. The lowest glacier on its southern slope does not descend below 16,500 feet, and the valley of Tortoral, a mere gulf in which Vesuvius might stand, comes between Illimani and the Nevado of La Mesada, from whence the eastern Cordillera runs to the north-west in a continuous line of snow-clad peaks to the group of Vilcañota, where it unites with the Western Cordillera.

The valley of the Desaguadero, occupying 150,000 square miles, has a considerable variety of surface; in the south, throughout the mining districts, it is poor and cold. Potosi, the highest city in the world, stands at an absolute elevation of 13,330 feet, at the foot of a mountain celebrated for its silver-mines. Chiquisaca, the capital of Bolivia, containing 13,000 inhabitants, lies to the north-east of Potosi, in the midst of cultivated fields. The northern part of the valley is populous, and produces wheat, barley, and other grain; and the Lake of Titicaca, twenty times as large as the Lake of Geneva, fills the north-western portion of this great basin. The islands and shores of this lake still exhibit ruins of gigantic magnitude, monuments of a people more ancient than the Incas. The modern city of La Paz with 40,000 inhabitants, a few leagues from its southern shores, stands in the most sublime situation that can be imagined, having in full view the vast Nevado of Illimani to the east-south-east, at a distance of seven leagues.

Many offsets leave the eastern side of the Bolivian Cordillera, which terminates in the great plain of Chiquitos and Paraguay; the most important is the Cordillera of Yuracaraës, which bounds the rich valley of Cochabamba on the north, and ends near the town of Santa Cruz de la Sierra.

There are some fertile valleys in the snow-capped group of Vilcañota and Cusco. The city of Cusco, which contains nearly 50,000 inhabitants, was the capital of the empire of the Incas: it still contains numerous ruins of that dynasty, among which the remains of

[1] The breadth of the table-land, and the two Cordilleras of the Bolivian Andes given in the text, was measured by Mr. Pentland; he also determined the heights of Illimani to be 21,150 feet; of Supäíwasi, or Huayna Potosi, 20,260 feet; and of Ancohuma or the Nevado of Sorata, 21,290 feet.

the Temple of the Sun, and the Cyclopean Fortress that towers over it on the north, still mark its former splendour. Four ancient Peruvian roads led from Cusco to the different parts of the empire, little inferior in many respects to the old Roman ways: all crossing mountain-passes higher than the Peak of Teneriffe. On the northern prolongation of the chain, in lat. 11° S., encircled by the Andes, is the elevated plain of Bonbon, near to the celebrated silver-mines of Pasco, at a height of 14,000 feet above the sea. In it is situated the Lake of Lauricocha, which may be considered, from its remoteness, as one of the sources of the Amazon. There are many small lakes on the table-lands and high valleys of the Andes, some even within the range of perpetual snow. They are very cold and deep, often of the purest sea-green colour; some of them may have been craters of extinct volcanoes.

The chain of the Andes is divided into three ranges of mountains running from south to north in the transverse group or mountain-knot of Pasco and Huanuco, which shuts in the valley of Bonbon between the 11th and 10th parallels of south latitude: that in the centre separates the wide fertile valley of the Upper Marañon from the still richer valley of the Huallaga, whilst the more eastern forms the barrier between the latter and the tropical valley of the Yucayali. The western chain alone reaches the limit of perpetual snow, and, if we except the Nevado of Huaylillas, in 7° 50′, no mountain north of this for nearly 400 miles to the Andes of Quito arrives at the snow-line.

In lat. 4° 50′ S. the Andes form the mountain-knot of Loxa, once celebrated for its forests, in which the Cinchona or Peruvian bark was first discovered. From this knot the chain divides into two great longitudinal ridges or cordilleras, in an extent of 350 miles passing through the republic of the Equator to the mountain-group of Los Pastos in that of New Grenada. These ridges enclose a vast longitudinal valley, which, divided by the cross ridges of Assuay, and Chisinche, into three basins, form the valleys of Cuenca, La Tapia, and Quito. The plain of Cuenca offers little interest; that of La Tapia is magnificent; whilst the valley of Quito is one of extraordinary beauty: on either side rise a series of snow-capped peaks, celebrated in every way in the history of science, as the valley itself is in that of the aboriginal races of the New World Here the energies of volcanic action have been studied with the greatest advantage; here, more than one hundred years ago, took place that measurement of an arc of the meridian which afforded the most accurate data at the time towards the determination of the mass and form of our planet, and which has conferred eternal honour on the body with which it originated, the French Academy of Sciences; and celebrity on the names of Bouguier, La Condamine,

Juan, Ulloa, and Godin, who conducted it on the part of the crowns of France and of Spain.

The Cordillera or ridge which hems in the valley of Quito on the east contains the snow-capped peaks of Antisana, Cotopaxi, one of the most beautiful of active volcanoes, whose dazzling cone rises to a height of 18,775 feet, of Tungaragua, and el-Altar, the latter once equal to Chimborazo in height, and Sangay. The western range includes the gigantic Chimborazo, which may be seen from the coast of the Pacific, the pyramidal peak of Illinissa, the wreck of an ancient volcano. The height of Illinissa was measured by the French Academicians, by a very careful direct operation—above the level of the ocean, the latter being visible from it; and by its means the absolute elevation of the valley of Quito, and of the other peaks that encircle it, was deduced, as well as the first approximate value of the barometrical coefficient. North of Chimborazo and near it is the Carguäirazo, and close to the city of Quito rises the scarcely less celebrated volcano of Pichincha, whilst the Nevado of Cayambè, whose summit, elevated 19,535 feet, is traversed by the terrestrial equator, perhaps the greatest and most remarkable landmark on the earth's surface, closes the north-east extremity of the valley.

The valley of Quito, one of the finest in the Andes, is 200 miles long and 30 wide, with a mean altitude of 10,000 feet, bounded by the most magnificent series of volcanoes and mountains in the New World. A peculiar interest is attached to two of the many volcanoes in the parallel Cordilleras that flank it on each side. The beautiful snow-clad cone of Cayambè Urcu, as already stated, traversed by the equator, the most remarkable division of the globe, closes it on the north; and in the western cordillera the cross still stands on the summit of Pichincha, 15,924 feet above the Pacific, which served for a signal to Bouguier and La Condamine in their memorable measurement of the Equatorial arc of the meridian.[1]

Some parts of the plain of Quito to the south are sterile, but the soil generally is good, and perpetual spring clothes it with exuberant vegetation. The city of Quito, containing 70,000 inhabitants, on the side of Pichincha has an absolute height of 9540 feet. The city is well built and handsome; the churches are splendid; it possesses universities, the comforts and luxuries of civilized life, in a situation of unrivalled grandeur and beauty. Thus, on the very summit of the Andes there is a world by itself, with its mountains and its valleys, its lakes and rivers, populous towns and cultivated fields. Many monuments of the Incas are still found in good preservation in these plains, where the scenery is most noble—eleven

[1] Baron Humboldt

volcanic cones are visible from one spot. Although the Andes are inferior in height to the Himalaya, yet the domes of trachyte, the truncated cones of the active volcanoes, and the serrated ruins of those that are extinct, mixed with the bold features of primary mountains, give an infinitely greater variety to the scene, while the smoke, and very often the flame, issuing from these regions of perpetual snow, increase its sublimity. Stupendous as these mountains appear even when viewed from the plains of the table-land, they are merely the inequalities of the tops of the Andes, the serrated summit of that mighty chain.

Between the large group of Los Pastos, containing several active volcanoes, and the group of Las Papas, in the second degree of north latitude, the bottom of the valley is only 6920 feet above the sea; and north of the latter mountain-knot the crest of the Andes splits into three Cordilleras, which diverge not again to unite. The most westerly of these, the chain of Choco, which may be considered the continuation of the great chain, divides the valley of the river Cauca from the Pacific; it is only 5000 feet high, and the lowest of the three. Though but 20 miles broad, it is so steep and so difficult of access, that travellers cannot cross it on mules, but are carried on men's shoulders: it is rich in gold and platina. The central branch, or Cordillera of Quindiü, runs due north between the Magdalena and Cauca, rising to a great height in the volcanic Peak of Tolima. The two last chains are united by the mountain-knot of Antioquia, of which little more is known than that it forms two great masses, which, after separating the streams of the Magdalena, Cauca, and Atrato, trends to the N. W., greatly reduced in height, and with the chain of Choco forms the low mountains of the Isthmus of Panamá. The most easterly of the three Cordilleras, called the Sierra de la Summa Paz, spreads out on its western declivity into the table-lands of Bogota, Tunja, and others, the ancient Cundinamarca, which have an elevation of about 9000 feet; whilst on its eastern slope rise the rivers of Guaviari and Meta, which form the head waters of the Orinoco. The tremendous crevice of Icononzo occurs in the path leading from the city of Santa Fé de Bogota to the banks of the Magdalena. It probably was formed by an earthquake, and is like an empty mineral vein, across which are two natural bridges: the lowest is composed of stones that have been jammed between the rocks in their fall.[1] This Cordillera comprises the Andes of Cundinamarca and Merida, and goes north-east through New Grenada to the 10th northern parallel, where it joins the coast-chain of Venezuela or Caraccas, which runs due east, and ends at Cape Paria in the Caribbean Sea, or rather at the eastern extremity of the island of Trinidad. This coast-chain is so majestic

[1] Baron Humboldt.

and beautiful that Baron Humboldt says it is like the Alps rising out of the sea without their snow. The insulated group of Santa Martha, 19,000 feet high, deeply covered with snow, stands on an extensive plain between the delta of the Magdalena and the sea-lake of Maracaybo, and is a landmark to mariners far off in the Caribbean Sea.

The passes over the Chilian Andes are numerous: that of the Portillo, leading from Santiago, the capital of Chile, to Mendoza, is the highest; it crosses two ridges, offering a valley between, a diminutive representation of the great Peru-Bolivian depression and of the valley of Quito; the most elevated is so high that vegetation ceases far below its summit. Those in Peru are higher, though very few reach the snow-line. In Bolivia the mean elevation of the passes in the western and eastern Cordillera is 14,892 and 14,422 feet respectively. That leading from Sorata to the auriferous valley of Tipuani is perhaps the highest in Bolivia. From the total absence of vegetation and the intense cold, it is supposed to exceed 16,000 feet above the Pacific; those to the north are but little lower. The pass of Quindiü in Colombia, though not so high, is the most difficult of all across the Andes; but those crossing the mountain-knots from one table-land to another are the most dangerous; for example, that over the Paramo del Assuay, in the plain of Quito, where the road is nearly as high as Mont Blanc, and travellers not unfrequently perish from cold winds in attempting it.[1]

On the western side of the Andes little or no rain falls, except at their most southern extremity, and scanty vegetation appears only on spots or in small valleys, watered by streams from the Andes. Excessive heat and moisture combine to cover the eastern side and its offsets with tangled forests of large trees and dense brushwood. This exuberance diminishes as the height increases, till at last the barren rocks are covered only by snow and glaciers. In the Andes near the equator glaciers descending below the snow-line are unknown. The steepness of the declivities and the dryness of the air, at such great elevations, prevent any accumulation of infiltrated water: the annual changes of temperature besides are small. Nothing can surpass the desolation of these regions, where nature has

[1] It appears by the measurements of Mr. Pentland in the Peru-Bolivian Andes, that many of their passes are higher than in the equatorial portion of the chain. The passes of Rumihuasi, on the high road from Cusco to Arequipa, of Toledo (between Arequipa and Puno), of Gualillas and Chullunquiani (between Arica and La Paz), all in the Western Cordillera, attain the respective elevations of 16,160, 15,790, 14,750, and 15,160 feet; whilst in the Eastern or Bolivian Cordillera the passes of Challa (between Oruro and Cochabamba), of Pecuani (between La Paz and Coröico), of Pumapacheta (between the lake of Titicaca and the affluents to the Amazon), of Vilcañota (between the valley of the Collao and that of the river Yucay), rise to the heights of 13,600, 15,350, 13,600, and 14,520 English feet.

been shaken by terrific convulsions. The dazzling snow fatigues the eye; the huge masses of bold rock, the mural precipices, and the chasms yawning into dark unknown depths, strike the imagination; while the crash of the avalanche or the rolling thunder of the volcano startles the ear. In the dead of night, when the sky is clear and the wind hushed, the hollow moaning of the volcanic fire fills the Indian with superstitious dread in the deathlike stillness of these solitudes.

In the very elevated plains in the transverse groups, such as that of Bonbon, however pure the sky, the landscape is lurid and colourless: the dark-blue shadows are sharply defined, and from the thinness of the air it is hardly possible to make a just estimate of distance. Changes of weather are sudden and violent; clouds of black vapour arise and are carried by fierce winds over the barren plains; snow and hail are driven with irresistible impetuosity; and thunder-storms come on, loud and awful, without warning. Notwithstanding the thinness of the air, the crash of the peals is quite appalling: while the lightning runs along the scorched grass, and, sometimes issuing from the ground, destroys a team of mules or a flock of sheep at one flash.[1]

Currents of warm air are occasionally met with on the crest of the Andes—an extraordinary phenomenon on such gelid heights, which is not yet accounted for: they generally occur two hours after sunset, are local and narrow, not exceeding a few fathoms in width, similar to the equally partial blasts of hot air in the Alps. A singular instance, probably, of earth-light occurs in crossing the Andes from Chile to Mendoza. On this rocky scene a peculiar brightness occasionally rests, a kind of undescribable reddish light, which vanishes during the winter rains, and is not perceptible on sunny days. Dr. Pœppig ascribes the phenomenon to the dryness of the air: he was confirmed in his opinion from afterwards observing a similar brightness on the coast of Peru, and it has also been seen in Egypt.

The Andes descend to the eastern plains by a series of cultivated levels, as those of Tucuman, Sata, and Jujuy, in the republic of La Plata, with many others. That of Tucuman is 2500 feet above the sea—the garden of the republic.

The low lands to the east of the Andes are divided by the table-lands and mountains of Parima and Brazil into three parts of very different aspect—the deserts and pampas of Patagonia and Buenos Ayres, the Silvas or woody basin of the Amazons, and the Llanos or grassy steppes of the Orinoco. The eastern table-lands nowhere exceed 2500 feet of absolute height; the plains are so low and flat, especially at the foot of the Andes, that a rise of 1000 feet in the Atlantic Ocean would submerge more than half the continent of South America.

[1] Dr. Pœppig.

The system of Parima is a group of mountains scattered over a table-land not more than 2000 feet above the sea, which extends 600 or 700 miles from east to west, between the river Orinoco, the Rio Negro, the Amazons, and the Altantic Ocean. It is quite unconnected with the Añdes, being 80 leagues east from the mountains of New Grenada. It begins 60 or 70 miles from the coast of Venezuela, and ascends by four successive terraces to undulating plains, which come within one or two degrees of the equator, and is twice as long as it is broad.

Seven chains, besides groups of mountains, cross the table-land from west to east, of which the chief is the Sierra del Parima. Beginning at the mouth of the Meta, it crosses the plains of Esmeralda to the frontier of Brazil. This chain is not more than 600 feet high, is everywhere abrupt, and forms the watershed between the tributaries of the Amazons and those of the Orinoco. The Orinoco rises on the northern side of the Sierra del Parima, and in its circuitous course over the plains of Esmeralda it breaks through that chain and the parallel chain of the Maypures 36 miles to the south; dashing with violence against the transverse shelving rocks and dykes, it forms the magnificent series of rapids and cataracts of Maypures and Atures, from whence the Parima mountains have got the name of the Cordillera of the cataracts of the Orinoco. The chain is of granite, which forms the banks and fills the bed of the river, covered with luxuriant tropical vegetation, especially palm-forests. In the district of the Upper Orinoco, near Charichana, there is a granite rock which emits musical sounds at sunrise, like the notes of an organ, occasioned by the difference of temperature of the external air and that which fills the deep narrow crevices with which the rock is everywhere torn. Something of the same kind occurs at Mount Sinai.[1]

The other parallel chains that extend over the table-land in Venezuela and Guiana, though not of great height, are very rugged and often crowned with mural ridges; they are separated by flat savannahs, generally barren in the dry season, but after the rains covered with a carpet of emerald-green grass, often six feet high, mixed with flowers. The vegetation in these countries is beautiful beyond imagination: the regions of the Upper Orinoco and Rio Negro, and of almost all the mountains and banks of rivers in Guiana, are clothed with majestic and impenetrable forests, whose moist and hot recesses are the abode of the singular and beautiful race of the Orchideæ and tangled creepers of many kinds.

Although all the mountains of the system of Parima are wild and rugged, they are not high; the inaccessible peak of the Cerro Duida, which rises insulated 7155 feet above the plain of Esmeralda,

[1] Baron Humboldt.

is the culminating point, and one of the highest mountains in South America east of the Andes. The fine savannahs of the Rupununi were the country of romance in the days of Queen Elizabeth. South of Pacaraime, near an inlet of the river, the far-famed city of Manoa was supposed to stand, the object of the unfortunate expedition of Sir Walter Raleigh; about 11 miles south-west of which is situate the lake Amucu, "the Great Lake with golden banks,"—great only during the periodical floods.[1]

On the southern side of the basin of the river Amazon lies the table-land of Brazil, nowhere more than 2500 feet high, which occupies half that empire, together with part of the republics of the Rio de la Plata and Uruguay. Its form is a triangle, whose apex is at the confluence of the rivers Mamore and Beni, and its base extends, near the shores of the Atlantic, from the mouth of the Rio de la Plata to within three degrees of the equator. It is difficult to define the limits of this vast territory, but some idea may be formed of it by following the direction of the rapids and cataracts of the rivers descending from it to the plains around. Thus a line drawn from the fall of the river of the Tocantins, in 3° 30′ S. latitude, to the cataracts of the Madeira, in the eighth degree of south latitude, will nearly mark its northern boundary; from thence the line would run S. W. to the junction of the Mamore and Beni; then turning to the S. E. along the ridges of mountains called the Cordillera Geral, and Serra Pareicis, it would proceed south to the cataract of the Paraná, called the Sete Quedas, in 24° 30′ S. lat.; and lastly from thence, by the great falls of the river Iguassu, to the Morro de Santa Martha, in lat. 28° 40′, south of the island of St. Catherine.

Chains of mountains, nearly parallel, extend from south-west to north-east, 700 miles along the base of the triangle, with a breadth of about 400 miles. Of these the Serra do Mar, on the "coast-chain," reaches from the river Uruguay to Cape San Roque, never more distant than 20 miles from the Atlantic, except to the south of the bay of Santos, where it is 80. Offsets diverge to the right and left; the granitic peaks of the Corcovado and Tejuco, [2000 feet above the sea level], which form such picturesque objects in that most magnificent of panoramas, the bay of Rio de Janeiro, are the ends of one. The parallel chain of Espenhaço, beginning near the town of San Paulo, and the continuous chains of the Serra Frio, and forming the western boundary of the basin of the Rio San Francisco, is the highest in Brazil, one of its mountains, Itambe, being 8426 feet above the sea. All the mountains in Brazil have a general tendency from S.W. to N.E., except the transverse chain of the Serra dos Vertentes, which begins 60 miles south of Villa Rica

[1] Baron Humboldt's Personal Narrative.

and runs in a tortuous line to its termination near the junction of the Mamore and Beni. It forms the watershed of the tributaries of the San Francisco and Amazon on the north, and those of the Rio de la Plata on the south; its greatest height is 3500 feet above the sea: its western part, the Serra Pareicis, is merely a succession of detached hills. This chain, the coast-chain of Venezuela, and the mountains of Parima, are the only ranges in the continent of America that do not entirely, or in some degree, lie in the direction of the meridians.

Magnificent forests of tall trees, bound together by tangled creeping and parasitical plants, clothe the declivities of the mountains and line the borders of the Brazilian rivers, where the soil is rich and the verdure brilliant. Many of the plains on the table-land bear a coarse nutritious grass after the rains only, others forests of dwarf trees; but vast undulating tracts are always verdant with excellent pasture intermixed with fields of corn: some parts are bare sand and rolled quartz; and the Campos Pareicis, north of the Serra dos Vertentes, in the province of Matto Grosso, is a sandy desert of unknown extent, similar to the Great Gobi on the table-land of Tibet. [Matto Grosso is the most western of the provinces of Brazil. Its surface, estimated at 540,000 square miles, is abundantly watered by lakes and rivers. It includes the famous diamond district. Among its mineral productions are gold, iron, clays of different colours, salt, nitre and various gems. Valuable timber abounds; plants yielding gum-elastic, dragons'-blood, gums, balsams, jalap, indigo and vanilla grow spontaneously, and in quantities almost without limit.]

CHAPTER IX.

The Low Lands of South America — Desert of Patagonia — The Pampas of Buenos Ayres — The Silvas of the Amazon — The Llanos of the Orinoco and Venezuela — Geological Notice.

THE southern plains are the most barren of the three great tracts of American low lands; they stretch from Tierra del Fuego over 27 degrees of latitude, or 1900 miles, nearly to Tucuman and the mountains of Brazil. Palms grow at one end, deep snow covers the other many months in the year. This enormous plain, of 1,620,000 square miles, begins on the eastern side of Tierra del Fuego, which is a flat covered with trees, and therefore superior to its continuation on the continent through eastern Patagonia, which, for 800 miles from the land's end to beyond the Rio Colorado, is a desert of shingle.[1] It is occasionally diversified by huge boulders, tufts of brown grass, low bushes armed with spines, brine-lakes, incrustations of salt, white as snow, and by black basaltic platforms, like plains of iron, at the foot of the Andes, barren as the rest. Eastern Patagonia, however, is not one universal flat, but a succession of shingly horizontal plains at higher and higher levels, separated by long lines of cliffs or escarpments, the gable-ends of the tiers or plains. The ascent is small, for even at the foot of the Andes the highest of these platforms is only 3000 feet above the ocean. The plains are here and there intersected by a ravine or a stream, the waters of which do not fertilize the blighted soil. The transition from intense heat to intense cold is rapid, and piercing winds often rush in hurricanes over these deserts, shunned even by the Indian, except when he crosses them to visit the tombs of his fathers. The shingle ends a few miles to the north of the Rio Colorado: there the red calcareous earth of the Pampas begins, monotonously covered with coarse tufted grass, without a tree or bush. This country, nearly as level as the sea and without a stone, extends almost to the table-land of Brazil, and for 1003 miles between the Atlantic and the Andes, interrupted only at vast distances by a solitary umbú, the only tree of this soil, rising like a great landmark. This wide space, though almost destitute of water, is not all of the same description. In the Pampas of Buenos Ayres there are four distinct regions. For 180 miles west from Buenos Ayres they are covered with thistles and lucern of the most vivid green so long as the moisture from the rain lasts. In spring the verdure fades, and a month afterwards the thistles shoot up 10 feet high, so dense and so protected by spines that they are impenetrable. During the summer the dried stalks are broken by the wind, and the lucern again spreads freshness over the ground. The Pampas for 430 miles west of this region is a thicket of long tufted luxuriant grass, intermixed with gaudy flowers, affording inexhaustible pasture to thousands of horses and cattle; this is followed by a tract of swamps and bogs, to which succeeds a region of ravines and stones, and, lastly, a zone, reaching to the Andes, of thorny bushes and dwarf trees in one dense thicket. The flat plains in Entre Rios in Uruguay, those of Santa Fé, and a great part of Cordova and Tucuman, are of sward, with cattle farms. The banks of the Paraná, and other tributaries of the La Plata, are adorned with an infinite variety of tropical productions, especially the graceful tribe of palms; and the river islands are bright with orange-groves. A desert of sand, called El Gran Chaco,[2] exists

[1] Captain King, R.N., and Mr. Darwin.

[2] [See, Noticias Historicas y Descriptivas sobre el gran pais del Chaco y Rio Bermejo. Por José Arenales, Buenos Aires, 1833.]

west of the Paraguay, the vegetable produce of which is confined to a variety of the aloe and cactus tribes. Adjoining this desert are the Bolivian provinces of Chiquitos and Moxos, covered with forests and jungle, the scene of the most laborious and beneficent exertions of the Jesuit Missionaries towards the civilization of the aborigines of South America in the last century.

The Pampas of Buenos Ayres, 1000 feet above the sea, sink to a low level along the foot of the Andes, where the streams from the mountains collect in large lakes, swamps, lagoons of prodigious size, and wide-spreading salines. The swamp or lagoon of Ybera, of 1000 square miles, is entirely covered with aquatic plants. These swamps are swollen to thousands of square miles by the annual floods of the rivers, which also inundate the Pampas, leaving a fertilizing coat of mud. Multitudes of animals perish in the floods, and the drought that sometimes succeeds is more fatal. Between the years 1830 and 1832, two millions of cattle died from want of food. Millions of animals are sometimes destroyed by casual and dreadful conflagrations in these countries when covered with dry grass and thistles.[1]

The Silvas of the river of the Amazon, lying in the centre of the continent, form the second division of the South American low lands. This country is more uneven than the Pampas, and the vegetation is so dense that it can only be penetrated by sailing up the river or its tributaries. The forests not only cover the basin of the Amazon from the Cordillera of Chiquitos to the mountains of Parima, but also its limiting mountain-chains, the Serra dos Vertentes and Parima, so that the whole forms an area of woodland more than six times the size of France, lying between the 18th parallel of south latitude and the 7th of north; consequently intertropical and traversed by the equator. There are some marshy savannahs between the 3d and 4th degrees of north latitude, and some grassy steppes south of the Pacaraimo chain; but they are insignificant compared with the Silvas, which extend 1500 miles along the river, varying in breadth from 350 to 800 miles, and probably more. According to Baron Humboldt, the soil, enriched for ages by the spoils of the forest, consists of the richest mould. The heat is suffocating in the deep and dark recesses of these primeval woods, where not a breath of air penetrates, and where, after being drenched by the periodical rains, the damp is so excessive that a blue mist rises in the early morning among the huge stems of the trees, and envelops the entangled creepers stretching from bough to bough. A death-like stillness prevails from sunrise to sunset, then the thousands of animals that inhabit these forests join in one loud discord-

[1] Sir Woodbine Parish on Buenos Ayres, and Sir Francis Head's Journey over the Pampas.

ant roar, not continuous, but in bursts. The beasts seem to be periodically and unanimously roused by some unknown impulse, till the forest rings in universal uproar. Profound silence prevails at midnight, which is broken at the dawn of morning by another general roar of wild chorus. Nightingales too have their fits of silence and song; after a pause, they

> "—— all burst forth in choral minstrelsy,
> As if some sudden gale had swept at once
> A hundred airy harps." *Coleridge.*

The whole forest often resounds when the animals, startled from their sleep, scream in terror at the noise made by bands of its inhabitants flying from some night-prowling foe. Their anxiety and terror before a thunder-storm is excessive, and all nature seems to partake in the dread. The tops of the lofty trees rustle ominously, though not a breath of air agitates them; a hollow whistling in the high regions of the atmosphere comes as a warning from the black floating vapour; midnight darkness envelops the ancient forests, which soon after groan and creak with the blast of the hurricane. The gloom is rendered still more hideous by the vivid lightning and the stunning crash of thunder. Even fishes are affected with the general consternation; for in a few minutes the Amazon rages in waves like a stormy sea.

The Llanos of the Orinoco and Venezuela, covered with long grass, form the third department of South American low lands, and occupy 153,000 square miles between the deltas of the Orinoco and the river Coqueta, flat as the surface of the sea. It is possible to travel over these flat plains for 1100 miles from the delta of the Orinoco to the foot of the Andes of Pasto; frequently there is not an eminence a foot high in 270 square miles. They are twice as long as they are broad; and as the wind blows constantly from the east, the climate is the more ardent the farther west. These steppes for the most part are destitute of trees or bushes, yet in some places they are dotted with the maurita and other palm-trees. Flat as these plains are, there are in some places two kinds of inequalities; one consists of banks or shoals of grit or compact limestone, five or six feet high, perfectly level for several leagues, and imperceptible except on their edges: the other inequality can only be detected by the barometer or levelling instruments; it is called a Mesa, and is an eminence rising imperceptibly to the height of some fathoms. Small as the elevation is, a mesa forms the watershed from S.W. to N.E., between the affluents of the Orinoco and the streams flowing to the northern coast of Terra Firma. In the wet season, from April to the end of October, the tropical rains pour down in torrents, and hundreds of square miles of the Llanos are inundated by the floods of the rivers. The water is sometimes 12 feet deep in the hollows, in which so many horses and other animals perish, that the ground smells of musk, an odour peculiar to many South American quadrupeds. From the flatness of the country too, the waters of some affluents of the Orinoco are driven backwards by the floods of that river, especially when aided by the wind, and form temporary lakes. When the waters subside, these steppes, manured by the sediment, are mantled with verdure, and produce ananas, with occasional groups of fan palm-trees, and mimosas skirt the rivers. When the dry weather returns, the grass is burnt to powder; the air is filled with dust raised by currents occasioned by difference of temperature, even where there is no wind. If by any accident a spark of fire falls on the scorched plains, a conflagration spreads from river to river, destroying every animal, and leaves the clayey soil sterile for years, till vicissitudes of weather crumble the brick-like surface into earth.

The Llanos lie between the equator and the Tropic of Cancer; the mean annual temperature is about 84° of Fahrenheit The heat is most intense during the rainy season, when tremendous thunder-storms are of common occurrence.

GEOLOGY OF SOUTH AMERICA.

THE most remarkable circumstance in the geological features of the South American continent is the vast development of volcanic force, which is confined to the chain of the Andes, where it has acquired a considerable breadth, as in the Peru-Bolivian portion, to the part nearest the sea-coast. It would be wrong, however, to say that there are no traces of modern volcanic action at a great distance from the sea:[1] it is one of those theories which recent discoveries in both continents have proved the fallacy of. The volcanic vents occur in the Andes in linear groups: the most southern of these is that of Chile, extending from the latitude of Chiloe to that of Santiago, 42° to 33° S.: in this space exist five well-authenticated

[1] Mr. Pentland found a very perfect volcanic crater, with well-marked currents of lava issuing from it—a rare occurrence in the higher craters of the Andes—not far from San Pedro de Cacha in the valley of the Yucay (lat. 14° 12′, long. 71° 15′ W., and at an elevation of 12,000 feet), near to the ruins of the Temple of the Inga Viracocha, a monument and a locality celebrated in Peruvian legend, the nearest point of the sea-coast being 175 miles distant. It is probable that some of the most celebrated mining districts of Alto Peru—Potosi, for instance, situated in a porphyry—have been upheaved at a very recent period. Modern volcanic rocks are not wanting in the valley of the Desaguadero; volcanic conglomerates exist in the deep ravines round the city of La Paz, lat. 16° 39′; and the mountain of Litanias, which furnishes the building-stone for that Bolivian city (lat 16° 42′, long. 68° 19½′), is composed of a most perfect trachyte, and rises to a height of 14,500 feet above, and at a distance of 160 miles from the Pacific.

craters in ignition—the most southern is the volcano of Llanquihue or Osorno, observed by M. Gaye, and the most northern that of Maypu,[1] the fires of which are sometimes seen from the capital of Chile. Between the 33rd parallel and the Bolivian frontier there does not appear to be a single volcanic vent, but in the province of Atacama rises the volcano of San Pedro of Atacama. The mountain of Isluga, in the province of Tarapaca, is said to be an active volcano; but the great centre of volcanic action in this part of the Western Cordillera extends from 18° 10′ to 16° 20′, where the Andes have changed their direction from being parallel to the meridian to one inclined nearly 45 degrees to that line. The trachytic giant domes of the Andes, Sahama, and the Nevado of Chuquibamba mark the N. and S. limits of this line of vents: the former, one of the most perfect trachytic pyramids in the Andes, rises to a height of 22,350 feet, in lat. 18° 7′ and long. 68° 54′ W.; near to it are the twin Nevados of Pomarape and Parinacota, one of which appears to emit smoke. The group of snowy peaks seen from Arica, the centre of which, the Nevado of Tacora, is in lat. 17° 43′, offers a broken-down crater, and an active solfaterra, on one of its sides. Between this point and the volcano of Arequipa no active volcano has been observed. It is well known that the latter has vomited flames and ashes, and spread desolation around, at a comparatively recent period;[2] the crater of Uvinas, active in the 16th century, is now filled up and completely extinct. Between the latitude of Arequipa (16° 24′) and the Equatorial group of volcanoes, the Andes do not present a single active crater. This Equatorial group extends over a meridional line of 3¼ degrees—between the Peak of Sangay and the volcano of Los Pastos. The most remarkable of these volcanic vents are the Sangay, Tunguragua, and Cotopaxi, all situated in the Cordillera most remote from the ocean. Pichincha burned as recently as 1831; and north of the Equator, Imbaburu, the volcanoes of Chiles, of Cumbal, of Tuqueres or Los Pastos, of Sotara and Purace, mark the extension of actual volcanic action into our hemisphere.

Granite, which seems to be the base of the whole continent, is widely spread to the east and south: it appears in Tierra del Fuego and in the Patagonian Andes abundantly, and at great elevations, and in Chile and southern Peru forms the line of hills parallel to

---

[1] Between the volcanoes of Osorno and Maypu are situated those of Villarica, Antuco, and Chillan. The volcano of Antuco was, in 1845, when visited by M. Domeyko, in great activity; its height, as determined by that naturalist, 8918 feet only, and the snow-line on its sides 7996 feet above the sea; the volcano of Villarica is 120 miles S., and that of Chillan 80 miles N. of the volcano of Antuco.

[2] Dr. Weddell, in 1847, visited the crater of this volcano, which at that period only emitted masses of aqueous vapour from its fumaroles.

the Pacific, wherein are situated the mineral riches of the former republic; but it comes into view so rarely in the northern parts of the chain, that Baron Humboldt says a person might travel years in the Andes of Peru and Quito without falling in with it. He never saw it at a greater height above the sea than 11,500 feet. Gneiss is here and there associated with the granite, but mica schist is by much the most common of the crystalline rocks. Quartz rock, probably of the Devonian period, is much developed, generally mixed with mica, and rich in gold and specular iron. It sometimes extends several leagues in the western declivities of Peru, 6000 feet thick. Red sandstone, with its gypseous and saliferous marls, of the age of our English red marl, of vast dimensions, occurs in the Andes, and on the table-land east of them, where in some places, as in Colombia, it spreads over thousands of square miles to the shores of the Atlantic. It is widely extended at altitudes of 10,000 and 12,000 feet—for example, on the plains of Tarqui and in the valley of Cuença. Coal is sometimes associated with it, and is found in the Andes of Pasco, in Peru, 17,450 feet above the sea.

Porphyry abounds all over the Andes, from Patagonia to Colombia, at every elevation, on the slopes and summits of the mountains, rising to the greatest elevation, but of very different ages and mineralogical characters. One variety which frequently occurs is rich in metals, and hence has been designated as *metalliferous:* in it are situated some of the most celebrated silver-mines of Peru, those of Potosi, Oruro, Puno. The bare and precipitous porphyry-rocks give great variety to the colouring of the Andes, especially in Chile, where purple, tile-red, and brown are contrasted with the snow on the summit of the chain.[1]

Trachyte, often so difficult to distinguish from porphyry, is perhaps still more abundant than it is in the Andes; many of the loftiest rocks, and all the great dome-shaped mountains, are formed of it. Masses of this rock, from 14,000 to 18,000 feet thick, form the Chimborazo and the Pichincha. Prodigious quantities of volcanic products, lava, tufa, and obsidian, occur on the western face of the Andes, where volcanoes are active. On the eastern side there are none. This is especially the case in that part of the chain lying between the equator and Chile. The Bolivian Cordilleras, which encircle the valley of Desaguadero, furnish a striking example. The Cordillera of the coast is composed of crystalline and stratified rocks at its base, and of trachytes, obsidian, and trachytic conglomerates at greater elevations, while the eastern Cordillera consists of stratified rocks of the Silurian system, with granites, quartziferous porphyries, and syenites, injected, and of secondary rocks of the triassic period, red marls, containing gypsum, oolitic limestone, and

[1] Dr. Pœppig.

rock-salt of the most beautiful colours. Towards Chile, and throughout the Chilian range, the case is different, because active volcanoes are there in the centre of the chain.

Sea-shells of different geological periods are found at various elevations, which shows that many upheavings and subsidences have taken place in the chain of the Andes.[1] The whole range, after twice subsiding some thousand feet, was brought up by a slow movement in mass during the Eocene period, after which it sank down once more several hundred feet, to be again uplifted to its present level by a slow and often interrupted motion. These vicissitudes are very perceptible, especially at its southern extremity. Stems of large trees, which Mr. Darwin found in a fossil state in the Uspallata range, on the eastern declivity of the Chilian Andes, now 700 miles distant from the Atlantic, exhibit a remarkable example of such vicissitudes. These trees, with the volcanic soil on which they had grown, had sunk from the beach to the bottom of a deep ocean, from which, after five alternations of sedimentary deposits and deluges of submarine lava of prodigious thickness, the whole mass was raised up, and now forms the Uspallata chain. Subsequently, by the wearing of streams, the embedded trunks have been brought into view in a silicified state, projecting from the soil in which they grew — now solid rock.

"Vast and scarcely comprehensible as such changes must ever appear, yet they have all occurred within a period recent when compared with the history of the Cordillera; and the Cordillera itself is absolutely modern compared with many of the fossiliferous strata of Europe and America."[2]

From the quantity of shingle and sand in the valleys in the lower ridges, as well as at altitudes from 7000 to 9000 feet above the present level of the sea, it appears that the whole area of the Chilian Andes has been rising for centuries by a gradual motion; and the coast is now rising by the same imperceptible degrees, though it is sometimes suddenly elevated by a succession of small upheavings of a few feet by earthquakes similar to that which shook the continent for 1000 miles on the 20th of February, 1835.

On the eastern side of the Andes the land from Tierra del Fuego to the Rio de la Plata has been raised *en masse* by one great elevating force, acting equally and imperceptibly for 2000 miles, within the period of the shell-fish now existing, which, in many parts of these plains, even still retain their colours. The gradual upward movement was interrupted by at least eight long periods of rest,

[1] Mr. Pentland found fossil shells of the Silurian period at a height of 17,500 feet, on the Bolivian Nevado of Antakäua, lat. 16° 21′, and those of the carboniferous limestone as high as 14,200 in several parts of Upper Peru.

[2] Mr. Darwin's Journal of Travels in South America.

marked by the edges of the successive plains, which, extending from south to north, had formed so many lines of sea-coast, as they rose higher and higher between the Atlantic and the Andes. It appears, from the shingle and fossil shells found on both sides of the Cordillera, that the whole south-western extremity of the continent has been rising slowly for a long time, and indeed the whole Andean chain. The rise on the coast of Chile has been at the rate of several feet in a century; but it has diminished eastward, till, in the Patagonian plains and Pampas, it has been only a few inches in the same line.

The instability of the southern part of the continent is less astonishing, if it be considered that at the time of the earthquake of 1835 the volcanoes in the Chilian Andes were in eruption contemporaneously for 720 miles in one direction and 400 in another, so that in all probability there was a subterranean lake of burning lava below this end of the continent twice as large as the Black Sea.[1]

The terraced plains of Patagonia, which extend hundreds of miles along the coast, are tertiary strata, not in basins, but in one great deposit, above which lies a thick stratum of a white pumaceous substance, extending at least five hundred miles, a tenth part of which consists of marine infusoria. Over the whole lies the shingle already mentioned, spread over the coast for 700 miles in length, with a mean breadth of 200 miles, and 50 feet thick. These myriads of pebbles, chiefly of porphyry, have been torn from the rocks of the Andes, and water-worn, at a period subsequent to the deposition of the tertiary strata—a period of incalculable duration. All the plains of Tierra del Fuego and Patagonia, on both sides of the Andes, are strewed with huge boulders, which are supposed to have been transported by icebergs which had descended to lower latitudes in ancient times than they do now — observations of great interest which we owe to Mr. Darwin.

The stunted vegetation of these sterile plains was sufficient to nourish large animals, now extinct, of the family of the Pachydermata, even at a period when the present shell-fish of the Patagonian seas existed.

The Pampas of Buenos Ayres are partly alluvial, the deposit of the Rio de la Plata. Granite prevails, to the extent of 2000 miles along the coast of Brazil, and with syenite forms the base of the table-land. The superstructure of the latter consists of metamorphic and old igneous rocks, sandstone, clay-slate, limestone; in the latter are large caverns containing bones of several species of extinct animals. Gold is found in the alluvial soil on the banks of the rivers, and in the slate rocks of the Palæozoic period, from the destruction of which this alluvium has been derived, and diamonds,

---

[1] Mr. Darwin's Journal of Travels in South America

so abundant in that country, in a ferruginous conglomerate of a very recent period.

The fertile soil of the Silvas has travelled from afar: washed down from the more elevated regions, it has been gradually deposited and manured by the decay of a thousand forests. Granite again appears, in more than its usual ruggedness, in the table-land and mountains of the Parima system. The sandstone of the Andes is found there also; and on the plains of Esmeralda it caps the granite of the solitary prism-shaped Duida, the culminating mountain of the Parima system. Limestone appears in the Brigantine or Cocollar, the most southern of the three ranges of the coast-chain of Venezuela; the other two are of granite, metamorphic rocks, and crystalline schists, torn by earthquakes and worn by the sea, which has deeply indented that coast. The chain of islands in the Spanish main is merely the wreck of a more northern ridge, broken up into detached masses by these irresistible powers.

CHAPTER X.

Central America — West Indian Islands — Geological Notice.

TAKING the natural divisions of the continent alone into consideration, Central America may be regarded as lying between the Isthmus of Panamá and Darien and the Isthmus of Tehuantepec, and consequently in a tropical climate. This narrow tortuous strip of land, which unites the continents of North and South America, stretches, from S.E. to N.W. about 1200 miles, varying in breadth from 20 to 300 or 400 miles.

As a regular chain, the Andes descend suddenly at the Isthmus of Panamá,[1] but as a mass of high land they continue through Central America and Mexico, in an irregular mixture of table-lands and mountains. The mass of high land which forms the central ridge of the country, and the watershed between the two oceans, is very steep on its western side, and runs near the coast of the Pacific, where Central America is narrow; but to the north, where it becomes wider, it recedes to a greater distance from the shore than the Andes do in any other part between Cape Horn and Mexico.

This country consists of three distinct groups, divided by valleys

[1] From the American survey of the Isthmus of Panamá by Col. Hughes, the highest point to be traversed by the projected railroad is Baldwin's Summit, 299 feet above the sea — which the road is to cut through by a tunnel 254 feet above the same level.

which run from sea to sea, namely, Costarica, the group of Honduras and Nicaragua, and the group of Guatemala.[1]

The plains of Panamá, very little raised above the sea, and in some parts studded with hills, follow the direction of the isthmus for 280 miles, and end at the Bay of Parita. From thence the forest-covered Cordillera of Veragua, supposed to be 9000 feet high, extends to the small but elevated table-land of Costarica, surrounded by volcanoes, and terminates at the plain of Nicaragua, which, together with its lake, occupies an area of 30,000 square miles, and forms the second break in the great Andean chain. The lake is only 128 feet above the Pacific, from which it is separated by a line of active volcanoes. The river of San Juan flows from its eastern end into the Caribbean Sea, and at its northern extremity it is connected with the smaller lake of Managua or Leon by the river Penaloya or Tipitapa. By this water-line it has been projected to unite the two seas. The high land begins again, after an interval of 170 miles, with the Mosquito country and Honduras, which mostly consist of table-lands and high mountains, some of which are volcanoes.

Guatemala is a table-land intersected by deep valleys, which lies between the plain of Comayagua and the Isthmus of Tehuantepec. It spreads to the east in the peninsula of Yucatan, which terminates at Cape Catoche, and encompasses the Bay of Honduras with terraces of high mountains. The table-land of Guatemala consists of undulating verdant plains of great extent, of the absolute height of 5000 feet, fragrant with flowers. In the southern part the cities of Old and New Guatemala are situate, 12 miles apart. The portion of the plain on which the new city stands is bounded on the west by the three volcanoes of Pacaya, Fuego, and Agua; these, rising from 7000 to 10,000 feet above the plain, lie close to the new city on the west, and form a scene of wonderful boldness and beauty. The Volcano de Agua, at the foot of which Old Guatemala stands, is a perfect cone, verdant to its summit, which occasionally pours forth torrents of boiling water and stones. The old city has been twice destroyed by it, and is now nearly deserted on account of earthquakes. The Volcano del Fuego generally emits smoke from one of its peaks; and the Volcano de Pacaya is only occasionally active. The wide grassy plains are cut by deep valleys to the north, where the high land of Guatemala ends in parallel ridges of mountains, called the Cerro Pelado, which run from east to west along the 94th meridian, filling half the Isthmus of Tehuantepec, which is 140 miles broad, and unites the table-land of Guatemala with that of Mexico.

Though there are large savannahs on the high plains of Guatemala, there are also magnificent primeval forests, as the name of the

[1] Johnston's Physical Atlas.

country implies, Guatemala signifying, in the Mexican language, a place covered with trees. The banks of the Rio de la Papian, or Usumasinta, a tributary of which rises in the alpine lake of Peten, and flows over the table-land to the Gulf of Mexico, are beautiful beyond description.

The coasts of Central America are generally narrow, and in some places the mountains and high lands come close to the water's edge. The sugar-cane is indigenous, and on the low lands of the eastern coast all the ordinary produce of the West Indian islands is raised, besides much that is peculiar to the country.

As the climate is cool on the high lands, the vegetation of the temperate zone is there in perfection. On the low lands, as in other countries where heat and moisture are in excess, and where nature is for the most part undisturbed, vegetation is vigorous to rankness: forests of gigantic timber seek the pure air above an impenetrable undergrowth, and the mouths of the rivers are dense masses of jungle with mangroves and reeds 100 feet high, yet delightful savannahs vary the scene, and wooded mountains dip into the water.

Nearly all the coast of the Pacific is skirted by an alluvial plain, of small width, and generally very different in character from that on the Atlantic side. In a line along the western side of the table-land and the mountains there is a continued succession of volcanoes, at various distances from the shore, and at various heights, on the declivity of the table-land. It seems as if a great crack or fissure had been produced in the earth's surface, along the junction of the mountains and the shore, through which the internal fire had found a vent. There are more than 20 active volcanoes in succession between the 10th and 20th parallels of north latitude; some higher than the mountains of the central ridge, and several subject to violent eruptions. Altogether there are 39 in Central America, 17 of which are in Guatemala—a greater number than in any other country, Java excepted.

The Colombian Archipelago, or West Indian Islands, which may be regarded as the wreck of a submerged part of the continent of South and Central America, consists of three distinct groups, namely, the Lesser Antilles or Caribbean Islands, the Greater Antilles, and the Bahama or Lucay Islands. Some of the Lesser Antilles are flat, but their general character is bold, with a single mountain or group of mountains in the centre, which slopes to the sea all around, more precipitously on the eastern side, which is exposed to the force of the Atlantic current. Trinidad is the most southerly of a line of magnificent islands, which form a semi-circle, enclosing the Caribbean Sea, with its convexity facing the east. The range is single to the island of Guadaloupe, where it splits into two chains, known as the Windward and Leeward Islands. Trinidad, Tobago, St. Lucia, and Dominica are particularly mountainous, and the mountains are cut by deep narrow ravines, or gullies, covered by ancient forests. The volcanic islands, which are mostly in the single part of the chain, have conical mountains bristled with rocks of a still more rugged form; but almost all the islands of the Lesser Antilles have a large portion of excellent vegetable soil in a high state of cultivation. Most of them are surrounded by coral reefs, which render navigation dangerous, and there is little intercourse between these islands, and still less with the Greater Antilles, on account of the prevailing winds and currents, which make it difficult to return. The Lesser Antilles terminate with the group of the Virgin Islands, which are small and flat, some only a few feet above the sea, and most of them are mere coral rocks.

The four islands which form the group of the Greater Antilles are the largest and finest in the Archipelago. Porto Rico, Haiti or San Domingo, and Jamaica, separated from the Virgin Islands by a narrow channel, lie in a line parallel to the coast-chain of Venezuela, from east to west; while Cuba, by a serpentine bend, separates the Caribbean Sea, or Sea of the Antilles, from the Gulf of Mexico. Porto Rico is 90 miles long and 36 broad, with wooded mountains passing through its centre nearly from east to west, which furnish abundance of water. There are extensive savannahs in the interior, and very rich soil on the northern coast, but the climate near the sea is unhealthy.

Haiti or San Domingo, 340 miles long and 132 broad, has a chain of mountains in its centre, extending from east to west like all the mountains in the Greater Antilles, the highest point of which is 9000 feet above the sea. A branch diverges from the main stem to Cape Tiburon, so that Haiti contains a great proportion of high land. The mountains are susceptible of cultivation nearly to the summit, and are clothed with undisturbed tropical forests. The extensive plains are well watered, and the soil, though not deep, is productive.

Jamaica, the most valuable of the British possessions in the West Indies, has an area of 4256 square miles, of which 110,000 acres are cultivated, chiefly as sugar-plantation. The principal chain of the Blue Mountains lies in the centre of the island, from east to west, with so sharp a crest that in some places it is only four yards across. The offsets from it cover all the eastern part of the island; some of them are very high. The more elevated ridges are flanked by lower ranges, descending to verdant savannahs. The escarpments are wild, the declivities steep, and mingled with stately forests. The valleys are very narrow, and not more than a twentieth part of the island is level ground. There are many small rivers, and the coast-line is 500 miles long, with at least 30 good harbours. The mean summer-heat is 80° of Fahrenheit, and that of winter is 75°.

The plains are often unhealthy, but the air in the mountains is salubrious; fever has never prevailed at the elevation of 2500 feet.

Cuba, the largest island in the Colombian Archipelago, has an area of 3615 square leagues, and 200 miles of coast, but so beset with coral reefs, sandbanks, and rocks, that only a third of it is accessible. Its mountains, which attain the height of 8000 feet, occupy the centre, and fill the eastern part of the island, in a great longitudinal line. No island in these seas is more important with regard to situation and natural productions; and although much of the low ground is swampy and unhealthy, there are vast savannahs, and about a seventh part of the island is cultivated.

The Bahama Islands are the least valuable and least interesting part of the Archipelago. The group consists of about 500 islands, many of them mere rocks, lying east of Cuba and the coast of Florida. Twelve are rather large, and are cultivated; and, though arid, they produce Log-wood and Mahogany. The most intricate labyrinth of shoals and reefs, chiefly of corals, madrepores, and sand, encompass these islands; some of them rise to the surface, and are adorned with groves of palm-trees. The Great Bahama is the first part of the New World on which Columbus landed—the next was Haiti, where his ashes rest.

The geology of Central America is little known; nevertheless it appears, from the confused mixture of table-lands and mountain-chains in all directions, that the subterraneous forces must have acted more partially and irregularly than either in South or North America. Granite, gneiss, and mica-slate form the substrata of the country; but the abundance of igneous rocks bears witness to strong volcanic action, both in ancient and in modern times, which still maintains its activity in the volcanic groups of Guatemala and Mexico.

From the identity of the fossil remains of extinct quadrupeds, there is every reason to believe that the West Indian Archipelago was once part of South America, and that the rugged and tortuous isthmus of Central America, and the serpentine chain of islands winding from Cumana to the peninsula of Florida, are but the shattered remains of an unbroken continent. The powerful volcanic action in Central America and Mexico, the volcanic nature of many of the West Indian Islands, and the still-existing fire in St. Vincent's, together with the tremendous earthquakes to which the whole region is subject, render it more than probable that the Caribbean Sea and the Gulf of Mexico are one great area of subsidence, which possibly has been increased by the erosion of the Gulf-stream and ground-swell—a temporary current of great impetuosity, common among the West Indian Islands from October to May.

The subsidence of this extensive area must have been very great, since the water is of considerable depth between the islands. It must have taken place after the destruction of the great quadrupeds, and consequently at a very recent geological period. The elevation of the table-land of Mexico may have been a contemporaneous event. In the Colombian Archipelago, volcanic action is confined to the smaller islands, which, forming a line in a meridional direction, extend from 12° to 18° N., and may be designated as the Caribbean range: it begins with Grenada and ends with St. Eustatius. St. Vincent, St. Lucia, Martinique, a great portion of Guadaloupe, Montserrat, Nevis, and St. Kitts are volcanic; most of them possess craters recently extinct, which have vomited ashes and lava within historical periods; whilst the less elevated of the Leeward and Windward Islands, Tobago, Barbadoes, Deseada, Antigua, Barbuda, and St. Bartholomew's, with the Virgin Islands and Bahamas, are composed either of calcareous or coral rocks.

CHAPTER XI.

North America — Table-Land and Mountains of New Mexico — The Rocky Mountains — The Maritime Chain and Mountains of Russian America.

According to the natural division of the continent, North America begins about the 20th degree of north latitude, and terminates in the Arctic Ocean. It is longer than South America, but the irregularity of its outline renders it impossible to estimate its area. Its greatest length is about 3100 miles, and its breadth at the widest part is 3500 miles.

The general structure of North America is still more simple than that of the southern part of the continent. The table-land of Mexico and the Rocky Mountains, which are the continuation of the high land of the Andes, run along the western side, but at so great a distance from the Pacific as to admit of another system of mountains along the coast. The immense plains to the east are divided longitudinally by the Alleghany Mountains, which stretch from the Carolinas to the Gulf of St. Lawrence, parallel to the Atlantic, and at no great distance from it. Although the general direction of the mountains is from south to north, yet, as they maintain a degree of parallelism to the two coasts, they diverge towards the north — one inclining towards the north-west and the other towards the north-east. The long narrow plain between the Atlantic and the Alleghanies is divided, throughout its length, by a line of cliffs not more than 200 or 300 feet above the Atlantic plain, the outcropping edge of the Second Terrace, or Atlantic slope, whose rolling surface goes west to the foot of the mountains.

An enormous table-land occupies the greater part of Mexico or Anahuac. It begins at the Isthmus of Tehuantepec, and extends north-west to the 42nd parallel of north latitude — a distance of 1600 miles, which is nearly equal to the distance from the northern extremity of Scotland to Gibraltar. It is narrow towards the south, and expands towards the north-west till about the latitude of the city of Mexico, where it attains its greatest breadth of 360 miles, and there also it is highest. The most easterly part in that parallel is about 5000 feet above the sea, from whence it rises towards the west to the height of 7482 feet at the city of Mexico, and then gradually diminishes to 4000 feet towards the Pacific.

Its height in California is not known, but it still bears the character of a table-land, and maintains an elevation of 6000 feet along the east side of the Serra Madre, even to the 32nd degree of north latitude, where it sinks to a lower level before joining the Rocky Mountains. The descent from this plateau to the low lands is very steep on all sides; on the east, especially, it is so precipitous that, seen from a distance, it is like a range of high mountains. There are only two carriage-roads to it from the Mexican Gulf, by passes 500 miles asunder — one at Xalapa, near Vera Cruz, the other at Santilla, west of Monterey. The descent to the shores of the Pacific is almost equally rapid, and that to the south no less so, where, for 300 miles between the plains of Tehuantepec and the Rio Yopez, it presses on the shores of the Pacific, and terminates in high mountains, leaving only a narrow margin of hilly maritime coast. Where the surface of the table-land is not traversed by mountains, it is as level as the ocean. There is a carriage-road over it for 1500 miles, without hills, from the city of Mexico to Santa Fé.

The southern part of the plateau is divided into four parts or distinct plains, surrounded by hills from 500 to 1000 feet high. In one of these, the plain of Tenochtitlan, surrounded by a wall of porphyritic mountains, stands the city of Mexico, once the capital of the empire of Montezuma, which must have far surpassed the modern city in extent and splendour, as many remains of its ancient glory testify. It is 7482 feet above the sea.

One of the singular crevices through which the internal fire finds a vent, stretches from the Gulf of Mexico to the Pacific, directly across the table-land, in a line about 16 miles south of the city of Mexico. A very remarkable row of active volcanoes occurs along this parallel; Tuxtla, the most eastern of them, is in the 95th degree of west longitude, near the Mexican Gulf, in a low range of wooded hills. More to the west stands the snow-shrouded cone of Orizaba, with its ever-fiery crater, seen like a star in the darkness of the night, which has obtained it the name of Citlaltepetl — the "Mountain of the Star." Popocatepetl, the loftiest mountain in Mexico, 17,720 feet above the sea, lies still farther west, and is in a state of constant eruption, which, with the peaks of Iztacihuatl and of Toluca, form a kind of volcanic circus, in the midst of which the city of Mexico and its lake are situated. A chain of smaller volcanoes unites the three. On a plain on the western slope of the table-land, and about 70 miles in a straight line from the Pacific, is the volcanic cone of Jorullo.[1] It suddenly appeared, and rose 1683 feet above the plain, on the night of the 29th of September, 1759, and is the highest of six mountains which have been thrown up on this part of the table-land since the middle of last century. The great cone of Colima, the last of this volcanic series, stands insulated in the plain of that name, between the western declivity of the table-land and the Pacific.

A high range of mountains extends along the eastern margin of the table-land to the Real de Catorce, and the surface of the plain is divided into two parts by the Serra Madre, which begins at 21° of N. lat.; and, after running north about 60 miles, its continuity is broken into the insulated ridges of the Serra Altamina, and the group containing the celebrated silver-mines of Fresnillo and Zacatecas: it soon after resumes its character of a regular chain, and, with a breadth of 100 miles, proceeds in parallel ridges and longitudinal valleys to New Mexico, where it skirts both banks of the Rio Bravo del Norte, and joins the Serra Verde, the most southern part of the Rocky Mountains, in 40° of N. lat.

To the south, some points of the Sierra Madre are said to be 10,000 feet high and 4000 feet above their base; and between the parallels of 36° and 42°, where the chain is the watershed between the Rio Colorado and the Rio Bravo del Norte, they are still higher, and perpetually covered with snow. The mountains on the left bank of the last-mentioned river are the eastern ridges of the Serra Madre, and contain the sources of the innumerable affluents of the Missouri and other rivers that flow into the Mississippi and Mexican Gulf.

Deep cavities, called Barancas, are a characteristic feature of the table-lands of Mexico: they are long rents, two or three miles in breadth, and many more in length, often descending 1000 feet below the surface of the plain, with a brook or the tributary of some river flowing through them. Their sides are precipitous and rugged, with overhanging rocks covered with large trees. The intense heat adds to the contrast between these hollows and the bare plains, where the air is more cool.

Vegetation varies with the elevation; consequently the splendour which adorns the low lands vanishes on the high plains, which, though producing much grain and pasture, are often saline, sterile,

---

[1] Baron Humboldt.

and treeless, except in some places where oaks grow to an enormous size, free of underwood.

The Rocky Mountains run 1800 miles in two parallel chains from the mountains of Anahuac in latitude 40° N. to the mouth of the Mackenzie river in the Arctic Ocean, sometimes united by a transverse ridge. In some places the eastern range rises to the snow-line, and even far above it, as in Mounts Hooker and Brown, 15,700 and 15,990 feet high; but the general elevation is only above the line of trees. The western range is not so high till north of the 55th parallel, where both ranges are of the same height, and frequently higher than the snow-line. They are generally barren, though the transverse valleys have fertile spots with grass, and sometimes trees. Their only offset in the south is the Saba and Ozark mountains, which run through Texas to the Mississippi. The long valley between the two rows of the Rocky Mountains, which is 100 miles wide, must have considerable elevation in the south, since the tributaries of the Colombia river descend from it in a series of rapids and cataracts for nearly 100 miles; and it is probably still higher towards the sources of the Peace river, where the mountains, only 1500 feet above it, are perpetually covered with snow. The Serra Verde is 490 miles from the Pacific, but as the coast trends due north to the Sound of Juan de Fuca, the western range of the Rocky Mountains maintains a distance of 380 miles from the ocean, from that point to the latitude of Behring's Sea, in 60° of N. lat.

The mountains on the west coast consist of two chains, one of which, beginning in Mexico, about the same latitude with the Sierra Madre, skirts the Gulf of California on the east, and maintains rather an inland course till north of the Oregon river, where it forms the Sea Alps of the coast; and then, increasing in breadth as it passes through Russian America, it ends at Nootka Sound.

The other chain, known as the Sea Alps of California, begins at the extremity of the peninsula, and, running northward with increasing height close to the Pacific, it passes through the island of Quadra at Vancouver, and after joining the Alps of the north-west coast, it terminates at Mount St. Elias, which is 17,860 feet high. A range of very high snowy mountains, which begins at Cape Mendocino, goes directly across both of these coast-chains, and unites them to the Rocky Mountains. It forms the watershed between the Colorado, which goes to the Gulf of California, and the affluents of the Oregon or Columbia river, which flows into the Pacific, and is continued to the east of the Rocky Mountains, at a less elevation, under the name of the Black Mountains, which stretch to the Missouri. Prairies extend between this coast-chain and the Rocky Mountains from California to the north of the Oregon river. The Oregon coast, for 200 miles, is a mass of undisturbed forest-thickets and marshes; and north from it, with few exceptions, is a mountainous region of bold aspect, often reaching above the snow-line. A branch of the Sea Alps, which runs westward to Bristol Bay, has many active volcanoes, and so has that which fills the promontory of Aliaska.

Archipelagoes and islands along the coast, from California to the promontory of Aliaska, have the same bold character as the mainland, and may be regarded as the tops of a submarine chain of table-lands and mountains which constitute the most westerly ridge of the maritime chains. Prince of Wales's Archipelago contains seven active volcanoes.

The mountains on the coasts of the Pacific and the islands are in many places covered with colossal forests, but wide tracts in the south are sandy deserts.

CHAPTER XII.

North America, continued — The Great Central Plains, or Valley of the Mississippi—The Alleghany Mountains—The Atlantic Slope—The Atlantic Plain — Geological Notice — The Mean Height of the Continents.

The great central plain of North America, lying between the Rocky and Alleghany Mountains, and reaching from the Gulf of Mexico to the Arctic Ocean, includes the valleys of the Mississippi, St. Lawrence, Nelson, Churchill, and most of those of the Missouri, Mackenzie, and Coppermine rivers. It has an area of 3,245,000 square miles, which is 245,000 square miles more than the central plain of South America, and about half the size of the great plain of the old continent, which is less fertile; for although the whole of America is not more than half the size of the old continent, it contains at least as much productive soil.

The plain, 5000 miles long, becomes wider towards the north, and has no elevations except a low table-land which crosses it at the line of the Canadian lakes and the sources of the Mississippi, and is nowhere above 1500 feet high, and rarely more than 700: it is the watershed between the streams that go to the Arctic Ocean and those that flow to the Misssissippi. The character of the plain is that of perfect uniformity, rising by a gentle regular ascent from the Gulf of Mexico to the sources of the Mississippi, which river is the great feature of the North American low lands. The ground rises in the same equable manner from the right bank of the Mississippi to the foot of the Rocky Mountains, but its ascent from the left bank to the Alleghanies is broken into hill and dale, containing the most fertile territory in the United States. Under so wide a range of latitude the plain embraces a great variety of soil, climate, and productions; but, being almost in a state of nature, it is characterized in its middle and southern parts by interminable grassy savannahs, or prairies, and enormous forests, and in the far north by deserts which rival those of Siberia in dreariness.

In the south a sandy desert, 400 or 500 miles wide, stretches along the base of the Rocky Mountains to the 41st degree of N. lat. The dry plains of Texas and the upper region of the Arkansas have all the characteristics of Asiatic table-lands; more to the north the bare treeless steppes on the high grounds of the far west are burned up in summer, and frozen in winter by biting blasts from the Rocky Mountains; but the soil improves toward the Mississippi. At its mouth, indeed, there are marshes which cover 35,000 square miles, bearing a rank vegetation, and its delta is a labyrinth of streams and lakes, with dense brushwood. There are also large tracts of forest and saline ground, especially the Grand Saline between the rivers Arkansas and Nesuketonga, which is often covered two or three inches deep with salt like a fall of snow. All the cultivation on the right bank of the river is along the Gulf of Mexico and in the adjacent provinces, and is entirely tropical, consisting of sugar-cane, cotton, and indigo. The prairies, so characteristic of North America, then begin.

To the right of the Mississippi these savannahs are sometimes rolling, but oftener level, and interminable as the ocean, covered with long rank grass of tender green, blended with flowers chiefly of the liliaceous kind, which fill the air with their fragrance. In the southern districts they are sometimes interspersed with groups of magnolia, tulip, and cotton trees, and in the north with oak and black walnut. These are rare occurrences, as the prairies may be traversed for many days without finding a shrub, except on the banks of the streams, which are beautifully fringed with myrtles, azaleas, kalmias, andromedas, and rhododendrons. On the wide plains the only objects to be seen are countless herds of wild horses, bisons, and deer. The country assumes a more severe aspect in higher latitudes. It is still capable of producing rye and barley in the territories of the Assiniboine Indians, and round Lake Winnepeg there are great forests; a low vegetation with grass follows, and towards the Icy Ocean the land is barren and covered with numerous lakes.

East of the Mississippi there is a magnificent undulating country about 300 miles broad, extending 1000 miles from south to north between that great river and the Alleghany Mountains, mostly covered with trees. When America was discovered, one uninterrupted forest spread over the country, from the Gulf of St. Lawrence and

11 *

the Canadian lakes to the Gulf of Mexico, and from the Atlantic Ocean it crossed the Alleghany Mountains, descended into the valley of the Mississippi on the north, but on the south it crossed the main stream of that river altogether, forming an ocean of vegetation of more than 1,000,000 of square miles, of which the greater part still remains. Although forests occupy so much of the country, there are immense prairies on the east side of the river also. Pine barrens, stretching far into the interior, occupy the whole coast of the Mexican Gulf eastward from the Pearl river, through Alabama and a great part of Florida.

These vast monotonous tracts of sand, covered with forests of gigantic pine-trees, are as peculiarly a distinctive feature of the continent of North America as the prairies, and are not confined to this part of the United States; they occur to a great extent in North Carolina, Virginia, and elsewhere. Tennessee and Kentucky, though much cleared, still possess large forests, and the Ohio flows for hundreds of miles among magnificent trees, with an undergrowth of azaleas, rhododendrons, and other beautiful shrubs, matted together by creeping plants. There the American forests appear in all their glory: the gigantic deciduous cypress, and the tall tulip-tree, overtopping the forest by half its height, a variety of noble oaks, black walnuts, American plane, hickory, sugar-maple, and the lireodendron, the most splendid of the magnolia tribe, the pride of the forest.

The Illinois waters a country of prairies ever fresh and green, and five new states are rising round the great lakes, whose territory of 280,000 square miles contains 180,000,000 of acres of land of excellent quality. These states, still mostly covered with wood, lie between the Lakes and the Ohio, and they reach from the United States to the Upper Mississippi—a country twice as large as France, and six times the size of England.

The quantity of water in the north-eastern part of the central plain greatly preponderates over that of the land; the five principal lakes, Huron, Superior, Michigan, Erie, and Ontario, cover an area equal to Great Britain [and Ireland], without reckoning small lakes and rivers innumerable.

[The north-west country, or Upper Mississippi valley, comprehends about ten degrees of latitude, from 39° to 49° north, and about fourteen degrees of longitude, from 87° to 101° (from 10° to 24° from the meridian of Washington), and contains about 300,000 square miles. A large part of this tract, consisting of the northern portion, is still held by the aborigines.

This country has some very peculiar natural features. The most remarkable of these is the numberless lakes which spangle its northern surface, the remains, no doubt, of a vast sea that once covered the whole country, extending north from the Gulf of Mexico, possibly to Hudson's Bay.

The country from the outlets of the Illinois and Missouri rivers to St. Peter's, and from Lake Michigan to Council Bluffs, and beyond that point westerly, is a vast gently-inclined plane, ascending to the north and to the west. Between the Mississippi and the lake elevation above the Atlantic, has been found to be a little more than 500 feet; and west of the river on the same parallel, towards the Missouri, something more than 700 feet. At St. Peter's it is about 700 feet. Nicollet states that Council Bluffs is 1037 feet above the Gulf; and the elevation of Rock Island, in the same latitude on the Mississippi, he says, is 528; and the height of Fort Pierre Chouteau, on the Missouri, on the same authority, is 1456; the lower end of Lake Pepin, in the same latitude (44° 24′ north), is 710 feet, and the mouth of the St. Peter's, in about latitude 45°, is 744 feet. There are few elevations above the general range, called mounds; but with the exception of these, the surface is marked only by ravines running down to the beds of the streams, which are usually from one to two hundred feet lower.

There are large tracts of this north-west country wholly destitute of tree or shrub, and covered only with a luxuriant growth of wild grass, and beautifully interspersed with flowers of every hue, each successively making the prairie to look gay with their presence from April till October. This beautiful natural meadow yields bountiful returns for culture and toil bestowed upon it. It consists of a very dark-brown vegetable mould, and is mellow beyond the conception of those who are acquainted only with the hard, stiff soils of the Atlantic slope. This mould is from one and a half to two feet deep, and entirely free from gravel. The subsoil is yellow light clay or clay loams, which resembles the soil of timbered lands. The country is a limestone formation. Timber is found only along the streams: it consists of elm, ash, black walnut, butternut, maple, mulberry, and iron-wood, on the bottoms; and on the upland, white, red, black and burr-oaks, shell-bark and common hickory, with, occasionally, linden, birch, wild plum and cherry, locust, and some other trees. On the Wisconsin and St. Croix Rivers are heavy growths of pine, from which supplies of lumber are carried down the Mississippi River.[1]

The mighty rivers of this region must be measured by travel, the prairies must be crossed, and the lakes seen, before the mind fully comprehends a description of them. "To look at a prairie up or down," says Nicollet, "to ascend one of its undulations; to reach a small plateau, or, (as the voyageurs call it, a *prairie planché*), moving from wave to wave over alternate swells and depressions; and, finally, to reach the vast interminable low prairie that extends

---

[1] [Notes on the North-West, or Valley of the Upper Mississippi. By W. J. A. Bradford, New York, 1846.]

itself in front, — be it for hours, days, or weeks, one never tires; pleasurable and exhilarating sensations are all the time felt; *ennui* is never experienced. Doubtless there are moments when excessive heat, a want of fresh water, and other privations, remind one that life is toil; but these drawbacks are of short duration. There is almost always a breeze over them. The security that one feels in knowing that there are no concealed dangers — so vast is the extent the eye takes in — no difficulties of road; a far-spreading verdure, relieved by a profusion of variously-coloured flowers; the azure of the sky above, or the tempest that can be seen from its beginning to its end; the beautiful modifications of the changing clouds; the curious looming of objects between the earth and sky, tasking the ingenuity every moment to rectify; — all, every thing, is calculated to excite the perceptions and keep alive the imagination. In the summer season, especially, every thing upon the prairies is cheerful, graceful, and animated. The Indians, with herds of deer, antelope and buffalo, give life and motion to them. It is then they should be visited; and I pity the man whose soul could remain unmoved under such a scene of excitement."]

The Canadas contain millions of acres of good soil, covered with immense forests. Upper Canada is the most fertile, and in many respects is one of the most valuable of the British colonies in the West: every European grain, and every plant that requires a hot summer and can endure a cold winter, thrives there. The forest consists chiefly of black and white spruce, the Weymouth and other pines—trees which do not admit of undergrowth: they grow to great height, like bare spars, with a tufted crown, casting a deep gloom below. The fall of large trees from age is a common occurrence, and not without danger, as it often causes the destruction of those adjacent; and an ice-storm is awful.

After a heavy fall of snow, succeeded by rain and a partial thaw, a strong frost coats the trees and all their branches with transparent ice often an inch thick; the noblest trees bend under the load, icicles hang from every bough, which come down in showers with the least breath of wind. The hemlock-spruce especially, with its long drooping branches, is then like a solid mass. If the wind freshens, the smaller trees become like corn beaten down by the tempest, while the large ones swing heavily in the breeze. The forest at last gives way under its load, tree comes down after tree with sudden and terrific violence, crushing all before them, till the whole is one wild uproar, heard from afar like successive discharges of artillery. Nothing, however, can be imagined more brilliant and beautiful than the effect of sunshine in a calm day on the frozen boughs, where every particle of the icy crystals sparkles, and nature seems decked in diamonds.[1]

---

[1] Mr. Taylor.

Although the subsoil is perpetually frozen at the depth of a few feet below the surface beyond the 56th degree of north latitude, yet trees grow in some places up to the 64th parallel. Farther north the gloomy and majestic forests cease, and are succeeded by a bleak, barren waste, which becomes progressively more dreary as it approaches the Arctic Ocean. Four-fifths of it are like the wilds of Siberia in surface and climate, covered many months in the year with deep snow. During the summer it is the resort of herds of rein-deer and bisons, which come from the south to browse on the tender short grass which then springs up along the streams and lakes.

The Alleghany or Appalachian chain, which constitutes the second or subordinate system of North American mountains, separates the great central plain from that which lies along the Atlantic Ocean. Its base is a strip of table-land, from 1000 to 3000 feet high, lying between the sources of the rivers Alabama and Yazoo, in the southern states of the Union, and New Brunswick, at the mouth of the river St. Lawrence. This high land is traversed throughout 1000 miles, between Alabama and Vermont, by from three to five parallel ridges of low mountains, rarely more than 3000 or 4000 feet high, and separated by fertile longitudinal valleys, which occupy more than two-thirds of its breadth of 100 miles. In Virginia and Pennsylvania, the only part of the chain to which the name of the Alleghany mountains properly belongs, it is 150 miles broad, and the whole is computed to have an area of 20,000 square miles. The parallelism of the ridges, and the uniform level of their summits, are the characteristics of this chain, which is lower and less wild than the Rocky Mountains. The uniformity of outline in the southern and middle parts of the chain is very remarkable, and results from their peculiar structure.[1] These mountains have no central axis, but consist of a series of convex and concave flexures, forming alternate hills and longitudinal valleys, running nearly parallel throughout their length, and cut transversely by the rivers that flow to the Atlantic on one hand, and to the Mississippi on the other. The watershed nearly follows the windings of the coast from the point of Florida to the north-western extremity of the State of Maine.

The picturesque and peaceful scenery of the Appalachian mountains is well known; they are generally clothed with a luxuriant vegetation, and their western slope is considered one of the finest countries in the United States. To the south they maintain a distance of 200 miles from the Atlantic, but approach close to the coast in the south-eastern part of the State of New York, from whence their general course is northerly to the river St. Lawrence. But

[1] Sir Charles Lyell's Travels in North America.

the Blue Mountains, which form the most easterly ridge, are continued in the double range of the Green Mountains to Gaspé Point in the Gulf of St. Lawrence. They fill the Canadas, Maine, New Brunswick, and Nova Scotia with branches as high as the mean elevation of the principal chain, and extend even to the dreary regions of Baffin's Bay. The chief Canadian branches are parallel to the river St. Lawrence. One goes N.E. from Quebec; and the Mealy Mountains, which are of much greater length, extend from Ottawa River to Sandwich Bay, and, though low, are always covered with snow. Little is known of the high lands within the Arctic circle, except that they probably extend from S.E. to N.W.

The country between Hudson's Bay, the mouths of the Churchill and that of the Mackenzie River, is also an unknown region; on the east it descends steeply to the coast, but the western part, known as the Barren Ground, is low and destitute of wood, except on the banks of the streams. The whole is covered with low precipitous hills. Not only the deep forests, but vegetation in general, diminish as the latitude increases, till on the arctic shores the soil becomes incapable of culture, and the majestic forest is superseded by the arctic birch, which creeps on the ground. Many of the islands along the north-eastern coasts, though little favoured by nature, produce flax and timber; and Newfoundland, as large as England and Wales, maintains a population of 70,000 souls by its fisheries: it is nearer to Britain than any part of America—the distance from the port of St. John to the harbour of Valentia in Ireland is only 1626 geographical miles.

The long aud comparatively narrow plain which lies between the Appalachian mountains and the Atlantic extends from the Gulf of Mexico to the eastern coast of Massachusetts. At its southern extremity it joins the plain of the Mississippi, and gradually becomes narrower in its northern course to New England, where it merely includes the coast islands. It is divided throughout its length by a line of cliffs from 200 to 300 feet high, which begins in Alabama and ends on the coast of Massachusetts. This escarpment is the eastern edge of the terrace known as the Atlantic Slope, which rises above the Maritime or Atlantic Plain, and undulates westward to the foot of the Blue Mountains, the most eastern ridge of the Appalachian chain. It is narrow at its extremities in Alabama and New York, but in Virginia and the Carolinas it is 200 miles wide. The surface of the slope is of great uniformity; ridges of hills and long valleys run along it parallel to the mountains, close to which it is 600 feet high. It is rich in soil and cultivation, and has an immense water-power in the streams and rivers flowing from the mountains across it, which are precipitated over its rocky edge to the plains on the west. More than twenty-three rivers of considerable size fall in cascades down this ledge between New York and the Mississippi, affording scenes of great beauty.[1]

Both land and water assume a new aspect on the Atlantic Plain. The rivers, after dashing over the rocky barrier, run in tranquil streams to the ocean, and the plain itself is a monotonous level, not more than 100 feet above the surface of the sea. Along the coast it is scooped into valleys and ravines, with innumerable creeks.

The greater part of the magnificent countries east of the Alleghanies is in a high state of cultivation and commercial prosperity, with natural advantages not surpassed in any country. Nature, however, still maintains her sway in some parts, especially where pine-barrens and swamps prevail. [The area of the thirty-one states which now form the Union (1853) is 1,485,870 square miles, with an average population of 15·48 to the square mile. The total area of the territory of the United States is 3,220,595 square miles, with an average population of 7·219 to the square mile. The areas of the great lakes which lie on the north, and the bays which indent the Atlantic and Pacific coasts, are not included in this statement. The total population on the 1st of June 1850, according to the late census, was 23,246,301; and of this number 19,619,366 are white. The rate of decennial increase of the white population is 37·14 per cent., and the rate of annual increase of the total population is 3½ per cent.] The territory of the United States is capable of producing everything that is useful to man, but not more than a twenty-sixth part of it has been cleared. [According to recent statements, 1,400,000,000 acres of the public lands remain to be sold.] The climate is generally healthy, the soil fertile, abounding in mineral treasures, and it possesses every advantage from navigable rivers and excellent harbours. The outposts of Anglo-Saxon civilization have already reached the Pacific, and the tide of white men is continually and irresistibly pressing onwards to the ultimate extinction of the original proprietors of the soil—a melancholy, but not a solitary, instance of the rapid extinction of a whole race.

Crystalline and Palæzoic rocks, rich in precious and other metals, form the substratum of Mexico, for the most part covered with plutonic and volcanic formations and secondary limestone; granite comes to the surface on the coast of Acapulco, and occasionally on the plains and mountains of the table-land. The Rocky Mountains are mostly Silurian, except the eastern ridge, which is of stratified crystalline rocks, amygdaloid and ancient volcanic productions. The coast-chain has the same character, with immense tracts of volcanic

[1] The author is indebted to the 'Physical Geography of North America,' by H. D. Rogers, Esq., and to the very interesting 'Travels' of Sir Charles Lyell in the United States, for the greater part of what she has said on the Physical Geography and Geology of that portion of the New World.

rocks, both ancient and modern, especially obsidian, which is nowhere developed on a greater scale, except in Mexico and the Andes.

In North America, as in the southern part of the continent, volcanic action is entirely confined to the coast and high land along the Pacific. The numerous vents in Mexico and California [in which there are five] are often in great activity, and hot springs abound. Though a considerable interval occurs north of them, where the fire is dormant, the country is full of igneous productions, and it again finds vent in Prince of Wales's Archipelago, which has seven active volcanoes. From Mount St. Elias westward through the whole southern coast of the peninsula of Russian America and the Aleutian Islands, which form a semicircle between Cape Aliaska, in America, and the peninsula of Kamtchatka, volcanic vents occur, and in the latter peninsula there are three of great height.

From the similar nature of the coasts and the identity of the fossil mammalia on each side of Behring's Strait, it is more than probable that the two continents were united, even since the sea was inhabited by the existing species of animals. Some of the gigantic quadrupeds of the old continent are supposed to have crossed, either over the land or over the ice, to America; and to have wandered southward through the longitudinal valleys of the Rocky Mountains, Mexico, and Central America, and to have spread over the vast plains of both continents, even to their utmost extremity.[1] An extinct species of horse, the mastodon, a species of elephant, three gigantic edentata, and a hollow-horned ruminating animal roamed over the prairies of North America—certainly since the sea was peopled by its present inhabitants, probably even since the existence of the Indians. The skeletons of these creatures are found in great numbers in the saline marshes on the prairies called the Licks, which are still the resort of the existing races.[2]

There were, however, various animals peculiar to America, as well as to each part of that continent, at least so far as yet known. South America still retains in many cases the type of its ancient inhabitants, though on a very reduced scale. But on the Patagonian plains, and on the Pampas, skeletons of creatures of gigantic size and anomalous forms have been found, one a quadruped of great magnitude, covered with a prodigious coat of mail similar to that of the armadillo; others like rats or mice, as large as the hippopotamus—all of which had lived on vegetables, and had existed at the same time with those already mentioned. These animals were not destroyed by the agency of man, since creatures not larger than a rat disappeared from Brazil within the same period.

---

[1] Dr. Richardson on the Fauna of the High Latitudes of North America.

[2] Sir Charles Lyell.

The geological outline of the United States, the Canadas, and all the country of the Polar Ocean, though highly interesting in itself, becomes infinitely more so when viewed in connection with that of northern and middle Europe. A remarkable analogy exists in the structure of the land on each side of the North Atlantic basin. Gneiss, mica-schist, and occasional granite, prevail over wide areas in the Alleghanies, on the Atlantic Slope, and still more in the northern latitudes of the American continent; and they range also through the greater part of Scandinavia, Finland, and Lapland. In the latter countries, and in the more northern parts of America, Sir Charles Lyell has observed that the fossiliferous rocks belong either to the most ancient or to the newest formations,[1] to the Silurian strata, or to such as contain shells of recent species only, no intermediate formation appearing through immense regions. Palæozoic strata extend over 2000 miles in the middle and high latitudes of North America; they occupy a tract nearly as great between the most westerly headlands of Norway and those that separate the White Sea from the Polar Ocean; Sir Roderick Murchison has traced them through central and eastern Europe, and the Ural Mountains, even to Siberia; Messrs. Abich and Thiatcheff through the Caucasus and Altaï. They have been seen also by Messrs. Dorbigny and Pentland to constitute the most elevated pinnacles of the Peru-Bolivian Andes, and Lieut. Strachey has recently discovered them at great elevations in the Himalaya, where they form the summits of the gigantic Junnotri, and with fossils analogous to those found in the Ural, and the Andes. Throughout these vast regions, both in America and in the old continent, the Silurian strata are followed in ascending order by the Devonian and carboniferous formations, which are developed on a stupendous scale in the United States, chiefly in the Alleghany mountains and on the Atlantic Slope. The Devonian and carboniferous strata together are a mile and a half thick in the State of New York, [in which there is no coal,] and three times as much in Pennsylvania, where one [continuous] coal-field, [extending] between the northern limits of that State and [the northern sections of] Alabama, occupies 63,000 square miles. [What is termed the Pittsburg seam, a part of the great Appalachian coal-field, according to the surveys of the Professors Rogers, measures 14,000 square miles; but the anthracite coal-bed does not probably exceed 2000 square miles.] There are many others of great magnitude, both in the United States and to the north of them, so that most valuable of all minerals is inexhaustible, which is not the least of the many advantages enjoyed

[1] This remarkable analogy between the fossil remains of the Palæozoic systems in the Old and New World has been more particularly shown by the researches of Messrs. de Verneuil and Sharpe.

by that flourishing country. The coal formation is also developed in New Brunswick, and traces of it are found on the shores and in the islands of the Polar Ocean, on the east coast of Greenland, and even in Spitzbergen.

A vast carboniferous basin exists in Belgium, above the Silurian strata; two or three of less importance in France; and a great portion of Britain is perfectly similar in structure to North America. The Silurian rocks in many instances are the same, and the coal-fields of New England are precisely similar to those in Wales, 3000 miles off.

In all the more northern countries that have been mentioned, so very distant from one another, the general range of the rocks is from north-east to south-west; and in northern Europe, the British isles, and North America, great lakes are formed along the junction of the strata, the whole analogy affording a proof of the wide diffusion of the same geological conditions in the northern regions at a very remote period. At a later time those erratic blocks, which are now scattered over the higher latitudes of both continents, were, most likely, brought from the north by drift ice or currents, while the land was still covered by the deep. Volcanic agency has not been wanting to complete the analogy. The Silurian and overlying strata have been pierced in many places by trappean rocks on both continents, and they appear also in the islands of the North Atlantic and Polar Seas. Even now the volcanic fires are in great activity in the very centre of that basin in Iceland, and in the very distant and less-known island of Jan Meyen.

The average height of the continents above the level of the sea is the mean between the height of all the high lands and all the low. Baron Humboldt, by whom the computation was effected, found that the table-lands, with their slopes, on account of their great extent and mass, have a much greater influence upon the result than mountain-chains. For example, if the range of the Pyrenees were pulverized, and strewed equally over the whole of Europe, it would only raise the soil 6 feet; the Alps, which occupy an area four times as great as that on which the Pyrenees stand, would only raise it 22 feet; whereas the compact plateau of the Spanish peninsula, which has only 1920 feet of mean height, would elevate the soil of Europe 76 feet; so that the table-land of the Spanish peninsula would produce an effect four times as great as the whole system of the Alps.[1]

[1] A chain of mountains is assumed to be a three-sided horizontal prism, whose height is the mean elevation of the chain, and the base the mean length and breadth of the same, or the area on which the chain stands, and thus its mass may be computed approximately. It is evident that a table-land must have a greater effect on the mean height of a continent than a chain of mountains, for, supposing both to be of the same base and altitude, one would be exactly double the other; and even if the mountains be the higher of the two, their upper parts contain much less solid matter than their lower, on account of the intervals and deep valleys between the peaks.

A great extent of low land necessarily compensates for the high —at least it diminishes its effect. The mean elevation of France, including the Pyrenees, Jura, Vosges, and all the other French mountains, is 870 feet, while the mean height of the whole European continent, of 1,720,000 square miles, is only 670 feet, because the vast European plain, which is nine times as large as France, has a mean altitude of but 380 feet, although it has a few intumescences, which, however, are not much above 1000 feet high, so that it is 200 feet lower than the mean height of France.[1]

The great table-land of Eastern Asia, with its colossal mountain-chains, has a much less effect on the mean height of Asia than might have been expected, on account of the depression round the Caspian Sea; and still more from the very low level and the enormous extent of Siberia, which is a third larger than all Europe. The intumescences in these vast plains are insignificant in comparison with their vast area, for Tobolsk is only 115 feet above the level of the sea; and even on the Upper Angora, at a point nearer the Indian than the Arctic Sea, the elevation is only 830 feet, which is not half the height of the city of Munich, and the third part of Asia has a mean height of only 255 feet. The effect of the Great Gobi, that part of the table-land lying between Lake Baikal and the wall of China, is diminished by a vast hollow 2560 feet deep, the dry basin of an ancient sea of considerable extent near Ergé, so that this great desert has a mean height of but 4220 feet, and consequently it only raises the general level of the Asiatic continent 128 feet, though it is twice as large as Germany. The table-land of Tibet, whose mean elevation, according to Baron Humboldt, is 11,600 feet, together with the chains of the Himalaya and Kuenlun, which enclose it, only produces an effect of 358 feet. On the whole, the mean level of Asia above the sea is 1150 feet.[2]

Notwithstanding the height and length of the Andes, their mass has little effect on the continent of South America on account of the extent of the eastern plains, which are one-third larger than Europe. For if these mountains were reduced to powder and strewn equally over them, it would not raise them above 518 feet; but when the minor mountain systems and the table-land of Brazil are added to

[1] According to Mr. Charpentier, the area of the base of the Pyrenees is 1720 square English miles. As the mean elevation of the passes gives the mean height of the mountains, Baron Humboldt estimated from the height of 23 passes over the Pyrenees that the mean crest of that chain is 7990 feet high, which is 300 feet higher than the mean height of the Alps, though the peaks in the Alps have a greater elevation than those of the Pyrenees in the ratio 1·4 to 1.

[2] The Russian Academicians MM. Fuss and Bunge found by barometrical measurement the mean height of that part of the Eastern Asiatic table-land lying between Lake Baikal and the Great Wall of China to be only about 6960 feet. The smallness of this mean is owing to hollows in the table-land, especially in the desert of the Great Gobi.

the Andes, the mean height of the whole of South America is 1130 feet. North America, whose mountain-chains are far inferior to those in the southern part of the continent, has its mean elevation increased by the table-land of Mexico, so that it has 750 feet of mean height.

The mean elevation of the whole of the New World is 930 feet, and of the continental masses of Europe and Asia above the level of the sea, 1010 feet. Thus it appears that the internal action in ancient times has been most powerful under Asia, somewhat less under South America, considerably less under North America, and least of all under Europe. In the course of ages changes will take place in these results, on account both of the sudden and gradual rise of the land in some parts of the earth, and its depression in others. The continental masses of the north are the lowest portions of our hemisphere, since the mean heights of Europe and North America are 670 and 750 feet.[1]

So little is known of the bed of the ocean that no inference can be drawn with regard to its heights and hollows, and what relation its mean depth bears to the mean height of the land. From its small influence on the gravitating force, La Place assumed it to be about four miles.[2] As the mean height of the continents is about 1000 feet, and their extent only about a fourth of that of the sea, they might be easily submerged, were it not that, in consequence of the sea being only one-fifth of the mean density of the earth, and the earth itself increasing in density towards its centre, La Place has proved that the stability of the equilibrium of the ocean can

[1] By the mensuration and computation of Baron Humboldt and Mr. Pentland, the elevation of the highest peaks, and the mean heights of the Himalaya, of the Equatorial and Bolivian Andes and the Alps, are as follows:—

		Peaks.	Mean Height.
Himalaya		28,178	15,670
Andes between 5° N. and 2° S. lat.		21,424	11,380
Eastern Cordillera	Between 18° and 15° S. lat.	21,300	15,250
Western Cordillera		22,350	14,900
Alps		15,739	7,353

The Peak of Dhawalaghiri is 26,862 feet high, and the Kunchinginga in Sikim 28,178. Captain Gerard gives 18,000 or 19,000 feet as the height of the snow-line on the mountains in the middle of the Asiatic table-land, and 30,000 feet as the absolute elevation of the Kuenlun, but Colonel Sabine observes that the latter figures require confirmation, no direct measures of the peaks of the Kuenlun having been ever executed.

[2] The greatest depth hitherto attained by soundings was six statute miles, or about 10,500 yards, in the North Atlantic, by the American expedition lately sent to ascertain the existence of the false Bermudas. [The deepest soundings ever made prior to the experiment referred to, was 4000 fathoms or 24,000 feet, by an officer of the British navy, but it was not considered to be very satisfactory.] See official despatch of Lieut. Maury, in a Washington paper of November 8, 1850.

never be subverted by any physical cause: a general inundation from the mere instability of the ocean is therefore impossible.

[On the 15th of November 1849, in latitude 31° 59′ north, longitude 58° 43′ west, Lieut. J. C. Walsh, U.S. Navy, (under the instructions of Lieut. M. F. Maury, Superintendent of the National Observatory at Washington), sounded with 5700 fathoms of wire without reaching the bottom. The inference from this experiment is that the depth of the ocean exceeds, at that place, 34,200 feet, or more than six statute miles.

Commander Barron, on board the U. S. Ship John Adams, reports, in May 1851, deep sea soundings made while crossing the Atlantic. They are as follows:—

Date.	North Latitude.	West Longitude.	Fathoms.	Feet.	
May 3	33° 50	52° 34′	2,500	= 15,000	
9	32° 06′	44° 47	5,500	= 33,000	got bottom
10	31° 01	44° 31	2,300	= 13,800	got bottom
17	Peak of Pico in sight		670	= 4,020	got bottom
21	35° 07′	25° 43′	1,040	= 6,240	got bottom

These soundings indicate that the great basin which holds the waters of the Atlantic Ocean, has a surface broken into irregular depressions and elevations analogous to the deep valleys and cloud-capped mountains of the dry land. The persevering efforts of Lieut. Maury to ascertain the laws which regulate the motions of the sea and of the air will result in adding largely to the knowledge of many phenomena which have hitherto been deemed mysteries, and eluded the satisfactory investigation of physicists. And amongst the inquiries not the least interesting are those relating to the depth of the ocean, and the form of its bottom.[1]]

CHAPTER XIII.

The Continent of Australia—Tasmania, or Van Diemen's Land—Islands—Continental Islands—Pelasgic Islands—New Zealand—New Guinea—Borneo—Atolls—Encircling Reefs—Coral Reefs—Barrier Keefs—Volcanic Islands—Areas of Subsidence and Elevation in the Bed of the Pacific—Active Volcanoes—Earthquakes—Secular Changes in the Level of the Land.

THE continent of Australia, situate in the Eastern Pacific Ocean is so destitute of large navigable rivers that probably no very high

[[1] See Maury's Explanations and Sailing Directions to accompany the Wind and Current Charts. Washington, 1851.]

land exists in its interior, which, as far as it has been explored seems to be singularly flat and low, but it is still so little known that no idea can be formed of its mean elevation. It is 2400 miles from east to west, and 1700 from north to south, and is divided into two unequal parts by the Tropic of Capricorn; consequently it has both a temperate and a tropical climate. New Guinea, separated from Australia by Torres Straits, and traversed by the same chain of mountains with New Holland and Van Diemen's Land, is so perfectly similar in structure, that it forms but a detached member of the adjacent continent.

The coasts of Australia are indented by very large bays, and by harbours that might give shelter to all the navies in Europe. The most distinguishing feature of the eastern side, which is chiefly occupied by the British colony of New South Wales, is a long chain of mountains which never retires far from the coast, and, with the exception of some short deviations in its southern part, maintains a meridional direction through 35 degrees of latitude. It is continued at one extremity from Torres Straits, at the north of the Gulf of Carpentaria, far into the interior of New Guinea; and at the other it traverses the whole of Van Diemen's Land. It is low in the northern parts of New Holland, being in some places merely a high land; but about the 30th degree of south latitude it assumes the form of a regular mountain-chain, and running in a very tortuous line from N.E. to S.W., terminates its visible course at Wilson's Promontory, the southern extremity of the continent. It is continued, however, by a chain of mountainous islands across Bass's Straits to Cape Portland, in Van Diemen's Land; from thence the range proceeds in a zigzag line of high and picturesque mountains to South Cape, where it ends, having, in its course of 1500 miles, separated the drainage of both countries into eastern and western waters.

The distance of the chain from the sea in New South Wales is from 50 to 100 miles, but at the 32nd parallel, it recedes to 150, yet soon returns, and forms the wild group of the Corecudgy Peaks, from whence, under the names of the Blue Mountains and Australian Alps, its highest part, it proceeds in a general westerly direction to the land's end.

The average height of these mountains is only from 2400 to 4700 feet above the level of the sea, and even Mount Kosciusko, the loftiest of the Australian Alps, is not more than 6500 feet high; yet its position is so favourable, that the view from its snowy and craggy top sweeps over an area of 7000 square miles. The rugged and savage character of these mountains far exceeds what might be expected from their height: in some places, it is true, their tops are rounded and covered with forests: but by far the greater part of the chain, though wooded along the flanks, is crowned by naked needles, tooth-formed peaks, and flat crests of granite or porphyry, mingled with patches of snow. The spurs give a terrific character to these mountains, and in many places render them altogether inaccessible, both in New South Wales and Van Diemen's Land. These shoot right and left from the ridgy axis of the main range, equal to it in height, and separated from it and from one another, by dark and almost subterraneous gullies, like rents in the bosom of the earth, iron-bound by impracticable precipices, and streams flowing through them in black silent eddies, or foaming torrents. The intricate character of these ravines, the danger of descending into them, and the difficulty of getting out again, render this mountain-chain, in New South Wales at least, almost a complete barrier between the country on the coast and that in the interior — a circumstance very unfavourable to the latter.[1]

In New South Wales the country slopes westward from these mountains to a low, flat, unbroken plain. On the east side, darkly verdant and round-topped hills and ridges are promiscuously grouped together, leading to a richly-wooded undulating country, which gradually descends to the coast, and forms the valuable lands of the British colony. Discovered by Cook in the year 1770, it was not colonized till 1788. It has become a prosperous country; and although new settlers in the more remote parts suffer the privations and difficulties incident to their position, yet there is educated society in the towns, with the comforts and luxuries of civilized life.

The coast-belt on the western side of Australia is generally of inferior land, with richer tracts interspersed near the rivers, and bounded on the east by a range of primary mountains from 3000 to 4000 feet high, in which granite occasionally appears. Beyond this the country is level, and the land better, though nowhere very productive except in grass.

None of the rivers of Australia are navigable to any great distance from their mouths. The want of water is severely felt in the interior, which, as far as it is known, is a treeless desert of sand, swamps, and jungle; yet a belief prevails that there is a large sea or fresh-water lake in its centre; and this opinion is founded partly on the nature of the soil, and also because all the rivers that flow into the sea on the northern coast, between the Gulfs of Van Diemen and Carpentaria, converge towards their sources, as if they served for drains to some large body of water.

However unpropitious the centre of the continent may be — and the shores generally have the same barren character—there is abundance of fine country inland from the coast. On the north all tropical productions might be raised, and in so large a continent there must be extensive tracts of arable land, though its peculiar character

[1] Memoirs of Count Strzelecki.

is pastoral. There are large forests on the mountains and elsewhere, yet that moisture is wanting which clothes other countries in the same latitude with rank vegetation. In the colonies, the clearing of a great extent of land has modified in some degree the mean annual temperature, so that the climate has become hotter and drier, and not thereby improved.

Van Diemen's Land, of triangular form, has an area of 27,200 square miles, and is very mountainous. No country has a greater number of deep, commodious harbours; and as most of the rivers, though not navigable to any distance, end in arms of the sea, they afford secure anchorage for ships of any size. The mountain chain that traverses the colony of New South Wales and the islands in Bass's Straits, rises again from Cape Portland, and, winding through Van Diemen's Land in the form of the letter Z, separates it into two nearly equal parts, with a mean height of 3750 feet, and at an average distance of 40 miles from the sea. It encloses the basins of the Derwent and Heron rivers, and, after sending a branch between them to Hobart Town, ends at South Cape. The offsets which shoot in all directions are as savage and full of impassable chasms as it is itself. There are cultivable plains and valleys along the numerous rivers and large lakes by which the country is well watered; so that Van Diemen's Land is more agricultural and fertile than the adjacent continent, but its climate is wet and cold. The uncleared soil of both countries, however, is far inferior to that in the greater part of North or South America.[1]

Granite constitutes the entire floor of the western portion of New South Wales, and extends far into the interior of the continent, bearing a striking resemblance in character to a similar portion of the Altaï chain described by Baron Humboldt. The central axis of the mountain-range, in New South Wales and in Van Diemen's Land, is of granite, syenite, and quartz rock; but in early times there had been great invasions of volcanic substances, as many parts of the main chain, and most of its offsets, are of the older igneous rocks. The fossiliferous strata of the two colonies are mostly of the Palæozoic period, but their fossil fauna is poor in species. Some are identical with, and others are representatives of, the species of other countries, even of England. It appears from their coal-measures that the flora of these countries was as distinct in appearance from that of the northern hemisphere, previous to the carboniferous period, as it is at the present day.

[" Geological researches into the structure of the globe show that a succession of physical changes have modified its surface from the earliest period up to the present time; and that these changes have been accompanied with variations, not only in the phases of animal

[1] Count Strzelecki.

and vegetable life, but often in the development also of organization; and as these changes cannot be supposed to have been operating uniformly over the entire surface of the globe in the same periods of time, we should naturally be prepared for finding the now-existing fauna of some regions exhibiting a higher state of development than that of others: accordingly, if we contrast the old continents of geographers with the zoology of Australia and New Zealand, we find a wide difference in the degree of organization which creation has reached in these respective regions. In New Zealand, with the exception of a *vespertilis* (bat), and a *mus* (mouse,) which latter is said to exist there, but which has not yet been sent to England, the most highly organized animal hitherto discovered, either fossil or recent, is a bird. In Australia, if compared with New Zealand, creation appears to have considerably advanced, but even here the order of *Rodentia* (the gnawing animals), is the highest in the scale of its indigenous animal productions; the great majority of its quadrupeds being the *Marsupiata* (kangaroos, &c.,) and the *Monotremata*, (*Echidna*, and *Ornithorynchus*), which are the very lowest of the mammalia; and its ornitho logy being characterized by the presence of certain peculiar genera — *Talegalla*, *Leipoa*, *Megapodius*,— birds which do not incubate their own eggs, and which are perhaps the lowest representatives of their own class; while the low organization of its botany is indicated by the remarkable absence of its fruit-bearing trees, the *Cerealia*, (wheat, rye, barley, &c.,)"][1]

Though the innumerable islands that are scattered through the ocean and seas differ much in size, form, and character, they have been grouped by M. Von Buch into the two distinct classes of Continental and Pelasgic islands, most of the latter being either of volcanic or coral formation. Continental islands are long in proportion to their breadth, and follow each other in succession along the margin of the continents, as if they had been formed during the elevation of the mainland, or had subsequently been separated from it by the action of the sea, and still mark its ancient boundary. These islands, which follow one another in their elongated dimensions, generally run parallel to the maritime chains of mountains, and are mostly of the same structure, so that they suggest the idea of a submarine portion of the maritime range that has not yet completely emerged from the deep — or, if sinking, has not yet disappeared below the waves.

America offers numerous examples of this kind of island. On the north-western coast there is a long chain of them, beginning with the New Norfolk group, and ending with Vancouver's Island, all similar and parallel to the maritime chain. Another range of

[1] Gould.— Birds of Australia.

Continental islands occurs at the southern extremity of America, extending from Chiloe to Cape Horn, evidently an exterior range of the Patagonian Andes, and the southern prolongation of the granitic or coast chain of Chile; in the Gulf of Mexico, the ancient margin of the mainland is marked by the curved group of Porto Rico, San Domingo, Jamaica, and Cuba, which nearly joins the peninsula of Yucatan. The various islands along the American coast of the Polar Ocean are the shattered fragments of the continent.

The old continent also affords innumerable examples; along the whole coast of Norway, from North Cape southwards, there is a continuous chain of rocky islands similar and parallel to the great range of the Scandinavian Alps; Great Britain itself, with the Hebrides, Orkney, and Zetland islands, are remarkable instances of Continental islands. It would be superfluous to mention the various instances which occur in the Mediterranean, where many of the islands are merely the prolongations of the mountain-chains of the mainland rising above the sea, as Corsica and Sardinia, which are a continuation of the Maritime Alps.

The great central chain of Madagascar and its elongated form, parallel to the Lupata Mountains and south-eastern margin of the great African table-land, show that the island once formed part of the continent. Asia, also, abounds in instances, as Sumatra, Java, and the Moluccas, and another vast chain extends along the western coast of Asia from Formosa to Kamtchatka.

Pelasgic islands have risen from the bed of the ocean, independently of the continents, and generally far from land. They are mostly volcanic, altogether or in part; often very lofty; sometimes single, and frequently in groups, and each group has, or formerly has had, a centre of volcanic action in one or more of the islands, round which the others have been formed. Many have craters of elevation, that is to say, they have been raised up in great hollow domes by the internal elastic vapours, and have either remained so, have become rent at the surface into gigantic fissures, or have collapsed into hollow cups, in which craters have formed, by the eruption of loose incoherent matter, or of lava currents, when the pressure from below was removed:[1] a considerable number have active vents.

The small islands and groups scattered at enormous distances from one another, within the Antarctic Circle, are all of volcanic formation, though none are active. In the Atlantic, Tristan da Cunha, St. Helena, Ascension, and Madeira are volcanic, though not now actively so; whereas the Cape de Verde, Canaries, and Azores have each volcanic vents:[2] the peak of Teyde, in Teneriffe, is one of the most magnificent volcanic cones in the world.

---

[1] M. Von Buch.

[2] These two last groups of islands have been admirably illustrated, since the publication of the first edition of this work, by the beautiful charts by

The labyrinth of islands scattered over the Pacific Ocean for more than 30 degrees on each side of the equator, and from the 130th eastern meridian to Sumatra, which all but unites this enormous archipelago to the continent of Asia, has the group of New Zealand or Tasmania, and the continent of Australia, with its appendage, Van Diemen's Land, on the south, and altogether forms a region which, from the unstable nature of the surface of the earth, is partly the wreck of a continent that has been engulfed by the ocean, and partly the summits of a new one rising above the waves. This extensive portion of the globe is in many parts terra incognita; the Indian Archipelago has been little explored, and, with the exception of our colonies in New Holland and New Zealand, is little known.

M. Von Buch conceives that the enormous circuit, beginning with New Zealand and extending through Norfolk Island, New Caledonia, New Hebrides, Solomon's Island, New Britain, New Hanover, New Ireland, Louisiade, and New Guinea, once formed the western and northern boundary of the Australian continent.

New Zealand, divided into three islands, by rocky and dangerous channels, is superior to Australia in richness of soil, fertility, and beauty; it abounds in a variety of vegetable and mineral productions. High mountains, of volcanic origin, run through the islands, which, in the most northerly, rise to nearly 10,000 feet[1] above the stormy ocean around, buried two-thirds of their height in permanent snow and glaciers, exhibiting on the grandest scale all the alpine characters, with the addition of active volcanoes on the eastern and western coasts: that of Tangariro pours forth deluges of boiling water, which deposit vast quantities of siliceous sinter like the Geysers in Iceland; and such is the vitality of the vegetation that plants grow richly on the banks, and even in water too hot to be endured.[2] The coast is a broken country, overspread with a most luxuriant but dark and gloomy vegetation. There are undulating tracts and table-lands of great extent without a tree, overrun by ferns and a low kind of myrtle; but the mountain-ridges are clothed with dense and gigantic forests. There is much good land and many lakes, with navigable rivers, the best of harbours, and a mild climate; so that no country is better suited for a prosperous and flourishing colony. It may be considered, even at this early period of its colonial existence, as the Great Britain of the southern hemis phere.

Captains Arlett and Vidal, published at the Admiralty under Sir Francis Beaufort's directions. They are equally interesting to the geologist and to the navigator.

[1] The highest peaks hitherto measured are Mount Egmont, 8840, and Mount Edgecumbe, a very perfect cone, near the settlement of New Plymouth, 9630 feet above the sea.

[2] — Mansel, Esq.

A very different scene from the stormy seas of New Zealand presents itself to the north of Australia. There, vivified by the glowing sun of the equator, the islands of the Indian Archipelago are of matchless beauty, crowned by lofty mountains, loaded with aromatic verdure, that shelve to the shore, or dip into a transparent glassy sea. Their coasts are cut by deep inlets, and watered by the purest streams, which descend in cascades rushing through wild crevices. The whole is so densely covered with palms and other beautiful forms of tropical vegetation that they seem to realize a terrestrial paradise.

Papua or New Guinea, the largest island in the Pacific after New Holland, is 1100 miles long and 400 in width, with mountains rising above mountains, till in the west they attain the height of 16,000 feet, capped with snow. From its position so near the equator it is probable that New Guinea has the same vegetation with the Spice Islands to the east, and, from the little that is known of it, must be one of the finest countries in existence. Storms are frequent; rain falls in torrents; earthquakes are rare and not violent.[1]

Borneo, next in size to New Guinea, is a noble island, divided into two nearly equal parts by the equator, and traversed through its whole length by magnificent chains of mountains, which end in three branches at the Java Sea. Beautiful rivers flow from them to the plains, and several of these spring from a spacious lake on the table-land in the interior, among the peaks of Keni-balu, the highest point of the island. Diamonds, gold, and antimony are among its minerals; gums, gutta percha, precious woods, and all kinds of spices and tropical fruits are among its vegetable productions.

Situate in the centre of a vast archipelago, and in the direct line of an extensive and valuable commerce, it will in the course of time become the seat of a great nation, whose civilization and prosperity will hand down to posterity the name of the enterprising, philanthropic Sir James Brooke, Rajah of Sarāwak, with the highest honour to which man can aspire. The climate is healthy, tempered by sea-breezes, and in some parts even European; the small island of Labuan and the adjacent coasts of Borneo being rich in coal, situated in the route of steam-vessels between India and China, are likely to exercise a very great influence on the trade between Europe and the Celestial empire, and on the civilization of the barbarous and piratical tribes of the Eastern Archipelago.

A volume might be written on the beauty and riches of the Indian Archipelago. Many of the islands are hardly known; the interior of the greater number has never been explored, so that they offer a wide field of discovery to the enterprising traveller, as they are now

[1] Moniteur des Indes Orientales, ii. p. 45.

of easier access since the seas have been cleared of pirates by the exertions of Sir James Brooke and the officers of Her Majesty's Navy.

They have become of much importance since our relations with China have been extended, on which account surveys of their coasts have been already made, and are going on, under the able direction of the Hydrographer of the Navy, Sir F. Beaufort. The great intertropical islands of the Pacific, likewise other large islands, as Ceylon and Madagascar, in the Indian Seas, which by the way do not differ in character from the preceding, are really continents in miniature, with their mountains and plains, their lakes and rivers; and in the climate they vary, like the main land, with the latitude, only that continental climates are more extreme both as to heat and cold.

It is a singular circumstance, arising from the instability of the crust of the earth, that all the smaller tropical pelasgic islands in the Pacific and Indian Oceans are either volcanic or coralline, except New Caledonia and the Seychelles; and it is a startling fact, that in most cases where there are volcanoes the land is rising by slow and almost imperceptible degrees above the ocean, whereas there is every reason to believe that those vast spaces, studded with coral islands or atolls, are actually sinking below it, and have been for ages.[1]

There are four different kinds of coral formations in the Pacific and Indian Oceans, all entirely produced by the growth of organic being, and their detritus, namely, lagoon islands or atolls, encircling reefs, barrier reefs, and coral fringes. They are all nearly confined to the tropical regions; the atolls to the Pacific and Indian Oceans alone.

An atoll or lagoon island consists of a chaplet or ring of coral, enclosing a lagoon or portion of the ocean in its centre. The average breadth of the part of the ring above the surface of the sea is about a quarter of a mile, oftener less, and it seldom rises higher than from 6 to 10 or 12 feet above the waves. Hence the lagoon islands are not discernible, even at a very small distance, unless when they are covered with cocoa-nut, palm, or the pandanus, which is frequently the case. On the outer side this ring or circlet shelves down to the distance of 100 or 200 yards from its edge, so that the sea gradually deepens to 25 fathoms, beyond which the sides plunge at once into the unfathomable depths of the ocean, with a more rapid descent than the cone of any volcano. Even at the small distance of some hundred yards no bottom has been found with a sounding-line a mile and a half long. All the coral at a moderate depth below water is alive — all above is dead, being the detritus of

[1] Mr. Darwin on Coral Reefs. Dana on Corals and Corallines.

the living part, washed up by the surf, which is so tremendous on the windward side of the tropical islands of the Pacific and Indian Oceans, that it is often heard miles off, and is frequently the first warning to seamen of their approach to an atoll.

On the lagoon side, where the water is calm, the bounding ring or reef shelves into it by a succession of ledges, also of living coral, though not of the same species with those which build the exterior wall and the foundations of the whole ring. The perpetual change of water brought into contact with the external coral by the breakers probably supplies them with more food than they could obtain in a quieter sea, which may account for their more luxuriant growth. At the same time, they deprive the whole of the coral in the interior of the most nourishing part of their food, because the still water in the lagoon, being supplied from the exterior by openings in the ring, ceases to produce the hardier corals; and species of more delicate forms, and of much slower growth, take their place.[1] The depth of the lagoon varies, in different atolls, from 20 to 50 fathoms, the bottom being partly detritus and partly live coral. By the growth of the coral, some few of the lagoons have been filled up; but the process is very slow from the causes assigned, and also because there are marine animals that feed on the living coral, and prevent its indefinite growth. In all departments of nature, the exuberant increase of any one class is checked and limited by others. The coral is of the most varied and delicate structure, and of the most beautiful tints: dark brown, vivid green, rich purple, pink, deep blue, peach colour, yellow, with dazzling white, contrasted with deep shadows, shine through the limpid water; while fish of the most gorgeous hues swim among the branching coral, which are of many different kinds, though all combine in the structure of these singular islands. Lagoon islands are sometimes circular, but more frequently oval or irregular in their form. Sometimes they are solitary or in groups, but they occur most frequently in elongated archipelagoes, with the atolls elongated in the same direction. The grouping of atolls bears a perfect analogy to the grouping of the archipelagoes of ordinary islands.

The size of these fairy-rings of the ocean varies from 2 to 90 miles in diameter, and islets are frequently formed on the coral rings by the washing up of the detritus, for they are so low that the waves break over them in high tides or storms. They have openings or channels in their circuit, generally on the leeward side, where the tide enters, and by these ships may sail into the lagoons, which are excellent harbours, and even on the surface of the circlet or reef itself there are occasionally boat-channels between islets.

[1] Supplement to the Observations on the Temple of Serapis, by Charles Babbage, Esq.

Dangerous Archipelago, lying east of the Society Islands, is one of the most remarkable assemblages of atolls in the Pacific Ocean. There are 80 of them, generally in a circular form, surrounding very deep lagoons, and separated from each other by profound depths. The reefs or rings are about half a mile wide, and seldom rise more than 10 feet above the edge of the surf, which beats upon them with such violence that it may be heard at the distance of 8 miles; and yet on that side the coral insects build more vigorously, and vegetation thrives better, than on the other. Many of the islets are inhabited.

The Caroline Archipelago, the largest of all, lies north of the equator, and extends its atolls in 60 groups over 1000 miles. Many are of great size, and all are beat by a tempestuous sea and occasional hurricanes. The atolls in the Pacific Ocean and China Sea are beyond enumeration. Though less frequent in the Indian Ocean, none are more interesting, or afford more perfect specimens of this peculiar formation, than the Maldive and Laccadive archipelagoes, both nearly parallel to the coast of Malabar, and elongated in that direction. The former is 470 miles long and about 50 miles broad, with atolls arranged in a double row, separated by an unfathomable sea, into which their sides descend with more than ordinary rapidity. The largest atoll is 88 miles long, and somewhat less than 20 broad; Suadiva, the next in size, is 44 miles by 23, with a large lagoon in its centre, to which there is access by 42 openings. There are inhabited islets on most of the chaplets or rings not higher than 20 feet, while the reefs themselves are nowhere more than 6 feet above the surge.

The Laccadives run to the north of this archipelago in a double line of nearly circular atolls, on which are low inhabited islets.

Encircling reefs differ in no respect from atoll-reefs, except that they have one or more islands in their lagoon. They commonly form a ring round mountainous islands, at a distance of two or three miles from the shore, rising on the outside from a very deep ocean, and separated from the land by a lagoon or channel 200 or 300 feet deep. These reefs surround the submarine base of the island, and, rising by a steep ascent to the surface, they encircle the island itself. The Caroline Archipelago exhibits good examples of this structure in the encircled islands of Hogoleu and Siniavin; the narrow ring or encircling reef of the former is 135 miles in its very irregular circuit, on which are a vast number of islets: six or eight islands rise to a considerable height from its lagoon, which is so deep, and the opening to it so large, that a frigate might sail into it. The encircling reef of Siniavin is narrow and irregular, and its lagoon is so nearly filled by a lofty island, that it leaves only a strip of water round it from 2 to 5 miles wide and 30 fathoms deep.

Tahiti, the largest of the Society group, is another instance of an encircled island of the most beautiful kind: it rises in mountains 7000 feet high, with only a narrow plain along the shore, and, except where cleared for cultivation, it is covered with forests of cocoa-nut, palms, bananas, bread-fruit, and other productions of a tropical climate. The lagoon, which encompasses it like an enormous moat, is 30 fathoms deep, and is hemmed in from the ocean by a coral band of the usual kind, at a distance varying from half a mile to three miles.

Barrier-reefs are of precisely the same structure as the two preceding classes, from which they only differ in their position with regard to the land. A barrier-reef off the north-east coast of the continent of Australia is the grandest coral formation existing. Rising at once from an unfathomable ocean, it extends 1000 miles along the coast, with a breadth varying from 200 yards to a mile, and at an average distance of from 20 to 30 miles from the shore, increasing in some places to 60 and even 70 miles. The great arm of the sea included between it and the land is nowhere less than 10, occasionally 60 fathoms deep, and is safely navigable throughout its whole length, with a few transverse openings by which ships can enter. The reef is really 1200 miles long, because it stretches nearly across Torres Straits. It is interrupted off the southern coast of New Guinea by muddy water, which destroys the coral animals, probably from some great river on that island. There are also extensive barrier-reefs on the islands of Louisiade and New Caledonia, which are exactly opposite to the great Australian reef; and as atolls stud that part of the Pacific which lies between them, it is called the Coralline Sea. The rolling of the billows along the great Australian reef has been admirably described. "The long ocean-swell, being suddenly impeded by this barrier, lifted itself in one great continuous ridge of deep blue water, which, curling over, fell on the edge of the reef in an unbroken cataract of dazzling white foam. Each line of breaker ran often one or two miles in length with not a perceptible gap in its continuity. There was a simple grandeur and display of power and beauty in this scene that rose even to sublimity. The unbroken roar of the surf, with its regular pulsation of thunder, as each succeeding swell fell first on the outer edge of the reef, was almost deafening, yet so deep-toned as not to interfere with the slightest nearer and sharper sound. . . . . . Both the sound and sight were such as to impress the spectator with the consciousness of standing in the presence of an overwhelming majesty and power."[1]

Coral-reefs are distinct from all the foregoing; they are merely fringes of coral along the margin of a shore, and, as they line the

---

[1] By Mr. Jukes, Naturalist to the Surveying Voyage of Captain Blackwood, R.N., in Torres Straits.

shore itself, they have no lagoons. A vast extent of coast, both on the continents and islands, is fringed by these reefs, and, as they frequently surround shoals, they are very dangerous.

Lagoon islands are the work of various species of coral animals; but those particular zoophytes which build the external wall, the foundation and support of the whole ring or reef, are most vigorous when most exposed to the breakers; they cannot exist at a greater depth than 25 or 30 fathoms at most, and die immediately when left dry; yet the coral wall descends precipitously to unfathomable depths; and although the whole of it is not the work of these animals, yet the perpendicular thickness of the coral is known to be very great, extending hundreds of feet below the depth at which these polypi cease to live. From an extensive survey of the Coralline Seas of the tropics, Mr. Darwin has found an explanation of these singular phenomena in the instability of the crust of the earth.

Since there are certain proofs that large areas of the dry land are gradually rising, and others sinking down, so the bottom of the ocean is not exempt from the general change that is slowly bringing about a new state of things; and as there is evidence, on multitudes of the volcanic islands in the Pacific, of a rise in certain parts of the basis of the ocean, so the lagoon islands indicate a subsidence in others—changes arising from the expansion and contraction of the strata under the bed of the ocean.

There are strong reasons for believing that a continent once occupied a great part of the tropical Pacific, some part of which subsided by slow and imperceptible degrees. As portions of it gradually sank down below the surface of the deep, the tops of mountains and table-lands would remain as islands of different magnitude and elevation, and would form archipelagoes elongated in the direction of the mountain-chains. Now, the coral-animal, which constructs the outward wall and mass of the reefs, never builds laterally, and cannot exist at a greater depth than 25 or 30 fathoms. [It is asserted that the coral animal cannot exist at any depth of water which is beyond the reach or penetration of the light of the sun.] Hence, if it began to lay the foundation of its reef on the submerged flanks of an island, it would be obliged to build its wall upwards in proportion as the island sank down, so that at length a lagoon would be formed between it and the land. As the subsidence continued, the lagoon would increase, the island would diminish, and the base of the coral-reef would sink deeper and deeper, while the animal would always keep its top just below the surface of the ocean, till at length the island would entirely disappear, and a perfect atoll would be left. If the island were mountainous, each peak would form a separate island in the lagoon, and the encircled islands would have different forms, which the reefs would follow continuously. This theory, perfectly explains the appearances of the lagoon islands and barrier-reefs, the continuity of the reef, the islands in the middle of the lagoons, the different distances of the reefs from them, and the forms of the archipelago, so exactly similar to the archipelagoes of ordinary islands, all of which are but the tops of submerged mountain-chains, and generally partake of their elongated forms.[1]

Every intermediate form between an atoll and an encircling reef exists: New Caledonia is a link between them. A reef runs along the north-western coast of that island 400 miles, and for many leagues never approaches within 8 miles of its shore, and the distance increases to 16 miles near the southern extremity. At the other end the reefs are continued on each side 150 miles beyond the sub-marine prolongation of the land, marking the former extent of the island. In the lagoon of Keeling Atoll, situate in the Indian Ocean, 600 miles south of Sumatra, many fallen trees and a ruined storehouse show that it has subsided: these movements take place during the earthquakes at Sumatra, which are also felt in this atoll. Violent earthquakes have lately been felt at Vanikora (celebrated for the wreck of La Pérouse), a lofty island of the Queen Charlotte group, with an encircling reef in the western part of the South Pacific, and on which there are marks of recent subsidence. Other proofs are not wanting of this great movement in the beds of the Pacific and Indian Oceans.

The extent of the atoll formations, including under this name the encircling reefs, is enormous. In the Pacific, from the southern end of Low Archipelago to the northern extremity of Marshall or Radick Archipelago, a distance of 4500 miles, and many degrees of latitude in breadth, atolls alone rise above the ocean. The same may be said of the space in the Indian Ocean between Saya de Matha and the end of the Laccadives, which include 25 degrees of latitude — such are the enormous areas that have been, and probably still are, slowly subsiding. Other spaces of great extent may also be mentioned, as the large archipelago of the Carolinas, that in the

[1] Another theory relative to the formation of the lagoon islands is that the coral circuit is but the edge of a submarine elevation crater, on which the coral animals have raised their edifice. This view, which has been adopted by Von Buch and Captain Beechey, to whom we are indebted more than to any other navigator for positive information and admirable surveys of the coral islands of the Pacific, receives corroboration from the perfect conformity in shape between many of the lagoon islands of the Gambier group and the known elevation craters, and from the circumstance of a lagoon island having been seen to rise in 1825, in lat. 30° 14′, accompanied with smoke, and communicating so high a temperature to the surrounding sea as rendered it impossible to land.—See Beechey's Voyages, and Pœppig's Reise.

Coralline Sea off the north-west coast of Australia, and an extensive one in the China Sea.

Though the volcanic islands in the Pacific are so numerous, there is not one within the areas mentioned, and there is not an active volcano within several hundred miles of an archipelago, or even group of atolls. This is the more interesting, as recent shells and fringes of dead coral, found at various heights on their surfaces, show that the volcanic islands have been rising more and more above the surface of the ocean for a very long time.

The volcanic islands also occupy particular zones in the Pacific, and it is found from extensive observation that all the points of eruption fall on the areas of elevation.[1]

One of the most terribly active of these zones begins with the Banda group of islands, and extends through the Sunda group of Timor, Sumbawa, Bali, Java, and Sumatra, separated only by narrow channels, and altogether forming a gently curved line 2000 miles long; but as the volcanic zone is continued through Barren Island and Narcondam in the Bay of Bengal, northward through the islands along the coast of Aracan, the entire length of this volcanic range is a great deal more. During the last hundred years all the islands and rocks for 100 miles along the coast of Aracan have been gradually rising. The greatest elevation of 22 feet has taken place about the centre of the line of upheaval, in the north-west end of the island of Cheduba, containing two mud volcanoes, and is continued through Foul Island and the Terribles.[2]

The little island of Gonung-Api, belonging to the Banda group, contains a volcano of great activity; and such is the elevating pressure of the submarine fire in that part of the ocean, that a mass of black basalt rose up, of such magnitude as to fill a bay 60 fathoms deep, so quietly that the inhabitants were not aware of what was going on till it was nearly done. Timor and the other adjacent islands also bear marks of recent elevation.

There is not a spot of its size on the face of the earth that contains so many volcanoes as the island of Java.[3] A range of volcanic mountains, from 5000 to 14,000 feet high, forms the central crest of the island, and ends to the east in a series of 38 separate volcanoes with broad bases, rising gradually into cones. They all stand on a plain but little elevated above the sea, and each individual mountain seems to have been formed independently of the rest.

[1] Few books have more interest than Mr. Darwin's on Coral Reefs and Volcanic Islands, to which the author is much indebted. Consult also Captain Beechey's Voyages, and his beautiful charts of the Coral Islands in the Pacific. [Also, the United States' Exploring Expedition under Commander Charles Wilkes.]

[2] By the Nautical Survey in 1848.

[3] Sir Stamford Raffles on Java.

Most of them are of great antiquity, and are covered with thick vegetation. Some are extinct, or only emit smoke; from others sulphurous vapours issue with prodigious violence; one has a large crater filled with boiling water; and a few have had fierce eruptions of late years. The island is covered with volcanic spurs from the main ridge, united by cross chains, together with other chains of less magnitude, but not less active.

In 1772 the greater part of one of the largest volcanic mountains was swallowed up after a short but severe combustion; a luminous cloud enveloped the mountain on the 11th of August, and soon after the huge mass actually disappeared under the earth with tremendous noise, carrying with it about 90 square miles of the surrounding country, 40 villages, and 2957 of their inhabitants.

The northern coast of Java is flat and swampy, but the southern provinces are beautiful and romantic; yet in the lovely peaceful valleys the stillness of night is disturbed by the deep roaring of the volcanoes, many of which are perpetually burning with slow but terrific action.

Separated by narrow channels of the sea, Bali and Sumbawa are but a continuation of Java, the same in nature and structure, but on a smaller scale, their mountains being litttle more than 8000 feet high.

The intensity of the volcanic force under this part of the Pacific may be imagined from the eruption of Tomboro in Sumbawa in 1815, which continued from the 5th of April till July. The explosions were heard at the distance of 970 miles; and in Java, at the distance of 300 miles, the darkness during the day was like that of deep midnight, from the quantity of ashes that filled the air: they were carried to Bencoolen, a distance of 1100 miles, which, with regard to distance, is as if the ashes of Vesuvius had fallen at Birmingham. The country round was ruined, and the town of Tomboro was submerged by heavy rollers from the ocean.

In Sumatra the extensive granitic formations of Eastern Asia join the volcanic series which occupies so large a portion of the Pacific. This most beautiful of islands presents the boldest aspect; it is indented by arms of the most transparent sea, and watered by innumerable streams; it displays in its vegetation all the bright colouring of the tropics. Here the submarine fire finds vent in three volcanoes on the southern, and one on the northern side of the island. A few atolls, many hundreds of miles to the south, show that this volcanic zone alternates with an area of subsidence.

More to the north, and nearly parallel to the preceding zone, another line of volcanic islands begins to the north of New Guinea, and passes through New Britain, New Ireland, Solomon Islands, and the New Hebrides, containing many open vents. This range or area of elevation separates the Coralline sea from the great chain of atolls on the north between Ellice's group and the Caroline Islands, so that it lies between two areas of subsidence.

The third and greatest of all the zones of volcanic islands includes Gilolo, one of the Molucca group, which is bristled with volcanic cones; and from thence it may be traced northwards through the Philippine Islands and Formosa: bending thence to the north-east, it passes through Loo-Choo, the Japan Archipelago, and is continued by the Kurile Islands to the peninsula of Kamtchatka, where there are several volcanoes of great elevation.

The Philippine Islands and Formosa form the volcanic separation between the atoll region in the China Sea, and that of the Caroline and Pellew groups.

There are six islands east of Jephoon in the Japan Archipelago which are subject to eruptions, and the internal fire breaks through the Kurile Islands in 18 vents, besides having raised two new islands in the beginning of this century, one 4 miles round, and the other 3000 feet high, though the sea there is so deep that the bottom has not been reached with a line 200 fathoms long.

Thus some long rent in the earth had extended from the tropics to the gelid seas of Ochotsk, probably connected with the peninsula of Kamtchatka: a new one begins to the east of the latter in the Aleutian Islands, which are of the most barren and desolate aspect, perpetually beaten by the surge of a restless ocean, and bristled by the cones of 24 volcanoes; they sweep in a half-moon round Behring's Sea, till they join the volcanic peninsula of Russian America.

The line of volcanic agency has been followed far beyond the limits of the coral-working animals, which extend but a short way on each side of the tropics; but it has been shown that in the equatorial regions immense areas of elevation alternate with as great areas of subsidence: north of New Holland they are so mixed that it indicates a point of convergence.[1]

On the other side of the Pacific the whole chain of the Andes, and the adjacent islands of Juan Fernandez and the Galapagos, form a vast volcanic area, which is actually now rising; and though there are few volcanic islands north of the zone of atolls, yet those that be indicate great internal activity, especially in the Sandwich Islands, where the volcanoes of Hawaii or Owhyhee are inferior to none in awful sublimity. That of Kirawah, a lateral crater of eruption of the great volcano of Mauna Loa, was seen in high activity by Mr. Douglas in 1834; and subsequently by Mr. Dana. The former traveller describes it as a deep sunken pit, occupying five square miles, covered with masses of lava which had been in a state of recent fusion. In the midst of these were two lakes of liquid lava: in both there was a vast caldron in furious ebullition, occasionally spouting to the height of from 20 to 70 feet, whence streams of

[1] Mr. Darwin on Volcanic Islands.

lava, hurrying along in fiery waves, were finally precipitated down an ignited arch, where the force of the lava was partly arrested by the escape of gases, which threw back huge blocks, and literally spun them into threads of glass, which were carried by the wind like the refuse of a flax-mill. He says the noise could hardly be described — that of all the steam-engines in the world would be a whisper to it; and the heat was so overpowering, and the dryness of the air so intense, that the very eyelids felt scorched and dried up.[1]

It may be observed that, where there are coral fringes, the land is either rising or stationary; for, were it subsiding, lagoons would be formed. On the contrary, there are many fringing reefs on the shores of volcanic islands along the coasts of the Red Sea, the Persian Gulf, and the West Indian Islands, all of which are rising. Indeed, this occurrence, in numberless instances, coincides with the existence of upraised organic remains on the land.

As the only coral formations of the Atlantic are fringing reefs, and as there is not one in its central expanse, except in Bermuda, it may be concluded that the bed of the ocean is not sinking; and with the exception of the Leeward Islands, the Canaries, the Azores, and the Cape de Verd groups, there are no active volcanoes in the islands or on the coasts of that ocean.

At present the great continent has few centres of volcanic action in comparison with what it once had. The Mediterranean is still undermined by fire, which occasionally finds vent in Vesuvius and the stately cone of Etna. Though Stromboli constantly pours forth inexhaustible showers of incandescent matter, and a temporary island now and then starts up from the sea, the volcanic action is diminished, and Italy has become comparatively more tranquil.

The table-land of Western Asia, especially Azerbijan, had once been the seat of intense commotion, now spent, as evidenced by the volcanic peaks of the Seiban Dagh, Ararat, and by the still smoking cone of Demavend. The table-land of Eastern Asia furnishes the solitary instance of igneous explosion at a distance of 1500 miles from the sea, in the volcanic chain of the Thean-Tchan.

Besides the two active volcanoes of the Pe-shan and Ho-tcheou in the chain itself, at the distance of 670 miles from each other, with a solfatara between them, it is the centre of a most extensive volcanic district, extending northward to the Altaï Mountains, in which there are many points of connexion between the interior of the earth and the atmosphere, not by volcanoes, but by solfataras

[1] Mr. Douglas's Voyage to the Sandwich Islands in 1833–4.—Journal of the Royal Geographical Society of London. [Commander Charles Wilkes describes the crater of Mauna Loa, p. 111, Vol. IV., Narrative of the United States Exploring Expedition: he says he was surprised not to hear more noise.]

hot springs, and vapours. In the range of Targatabai, in the country of the Kirghiz, there is a mount said to emit smoke and even flame, which produces sulphur and sal-ammoniac in abundance. It is not ascertained that there are many mountains in China that eject lava, but there are many fire-hills and fire-springs; the latter are real Artesian wells, five or six inches wide, and from 1500 to 3000 feet deep: from some of these water rises containing a great quantity of common salt; from others gases issue: and when a flame is applied, fire rushes out with great violence, rising 20 or 30 feet high, with a noise like thunder. The gas, conducted in tubes of bamboo cane, is used in the evaporation of salt water from the neighbouring springs.

There are altogether about 270 active volcanoes, of which 190 are on the shores and islands of the Pacific. They are generally disposed in lines or groups. The chain of the Andes furnishes a magnificent example of linear volcanoes. The peak of Teneriffe, encompassed by the volcanic islands of Palma and Lancerote, is an equally good specimen of a central group. Eruptions are much more frequent in low than in high volcanoes; that in the island of Stromboli is in constant activity; whereas Cotopaxi, 18,875 feet high, and Tungaragua, 16,424, in the Andes, have only been active once in a hundred years. On account of the force requisite to raise lava to such great elevations, it rarely flows from very elevated cones. Antisana is the only instance to the contrary among all the lofty volcanoes of Equatorial America. In Etna also the pressure is so great that the lava forces its way through the sides of the mountain, or at the base of the cone.

An explosion begins by a dense volume of smoke issuing from the crater, mixed with aqueous vapour and gases, then masses of rock and molten matter in a half-fluid state are ejected with tremendous explosion and violence; after which lava begins to flow, and the whole terminates by a shower of ashes from the crater — often the most formidable part of the phenomenon, as was experienced at the destruction of Pompeii. There are several volcanoes which eject only streams of boiling water, as the Volcano de Agua in Guatemala; others pour forth boiling mud, as in the islands of Trinidad, Java, and Cheduba in the Bay of Bengal. A more feeble effort of the volcanic force appears in the numerous solfataras. Hot springs show that the volcanic fire is not extinguished, though not otherwise apparent. To these may be added acidulous springs, those of naphtha, petroleum, and various kinds of gas, as carbonic acid gas, the food of plants—and, when breathed, the destruction of animals, as is fearfully seen in the Guero Upas, or "Valley of Death," in Java: it is half a mile in circumference and about 35 feet deep, with a few large stones, and not a vestige of vegetation on the bottom, which is covered with the skeletons of human beings and the bones of animals and birds blanched white as ivory. On approaching the edge of the valley, which is situate on the top of a hill, a nauseous sickening sensation is felt; and nothing that has life can enter its precincts without being immediately suffocated.[1]

The seat of activity has been perpetually changing, but there always has been volcanic action, possibly more intense in former times, but even at present it extends from pole to pole.

Notwithstanding the numerous volcanic vents in the globe, many places are subject to violent earthquakes, which ruin the works of man, and often change the configuration of the country. The most extensive district of earthquakes comprises the Mediterranean and the adjacent countries, Asia Minor, the Caspian Sea, Caucasus, and the Persian mountains. It joins a vast volcanic district in Central Asia, whose chief focus seems to be the Thean-Tchan, which includes Lake Baikal and the neighbouring regions. A great part of the continent of Asia is more or less subject to shocks; but, with the exception of the shores of the Red Sea and the northern parts of Barbary, Africa is entirely free from these tremendous scourges; and it is singular that, notwithstanding the terrible earthquakes which shake the countries west of the Andes, the Andean chain itself, and all the countries round the Gulf of Mexico and the Caribbean Sea, they are extremely rare in the great eastern plains of South America. For the most part the shocks are transmitted in the line of the primary mountain-chains, and seem often to be limited by them in the other direction.

There must be some singular volcanic action underneath part of Great Britain, which has occasioned 255 slight shocks of earthquake, of which 139 took place in Scotland: the most violent of them have been felt at Comrie, in Perthshire, in 1839; of the rest, 14 took place on the borders of Yorkshire and Derbyshire, 30 in Wales, and 31 on the south coast of England: they were preceded by a sudden fall of the barometer, fogs, and unusual sultriness; the two latter phenomena are said to indicate these convulsions about Sienna, and in the Maremma of Tuscany, where they have of late years been attended with very disastrous effects.

Earthquakes are probably produced by fractures and sudden heavings and subsidences in the elastic crust of the globe, from the pressure of the liquid fire, vapour, and gases in its interior, which there find vent, relieve the tension which the strata acquire during their slow refrigeration, and restore equilibrium. But whether the initial impulse be eruptive, or a sudden pressure upwards, the shock originating in that point is propagated through the elastic surface of the earth in a series of circular or oval undulations, similar to those

[1] Letter from Alex. Loudon, Esq., in the Journal of the Geographical Society of London.

produced by dropping a stone into a pool, and like them they become broader and lower as the distance increases, till they gradually subside; in this manner the shock travels through the land, becoming weaker and weaker till it terminates. When the impulse begins in the interior of a continent, the elastic wave is propagated through the solid crust of the earth, as it is in sound through the air, and is transmitted from the former to the ocean, where it is finally spent and lost, or, if very powerful, is continued in the opposite land. Almost all the great earthquakes, however, have their origin in the bed of the ocean, far from land, whence the shocks travel in undulations to the surrounding shores.

No doubt many of small intensity are imperceptible: it is only the violent efforts of the internal forces that can overcome the pressure of the ocean's bed, and that of the superincumbent water. The internal pressure is supposed to find relief most readily in a belt of great breadth that surrounds the land at a considerable distance from the coast, and, being formed of the débris, the internal temperature is in a perpetual state of fluctuation, which would seem to give rise to sudden flexures and submarine eruptions.

When the original impulse is a fracture or eruption of lava in the bed of the deep ocean, two kinds of waves or undulations are produced and propagated simultaneously—one through the bed of the ocean, which is the true earthquake shock, and coincident with this a wave is formed and propagated on the surface of the ocean, which rolls to the shore, and reaches it in time to complete the destruction long after the shock or wave through the solid ocean-bed has arrived and spent itself on the land. The sea rose 50 feet at Lisbon and 60 at Cadiz after the great earthquake; it rose and fell 18 times at Tangier on the coast of Africa, and 15 times at Funchal in Madeira. At Kinsale a body of water rushed into the harbour, and the water in Loch Lomond in Scotland rose two feet four inches—so extensive was the oceanic wave.[1] The height to which the surface of the ground is elevated, or the vertical height of the shock-wave, varies from one inch to two or three feet. This earth-wave, on passing under deep water, is imperceptible; but when it comes to soundings, it carries with it to the land a long, flat, aqueous wave; on arriving at the beach, the water drops in arrear from the superior velocity of the shock, so that at that moment the sea seems to recede before the great ocean-wave arrives.

It is the small forced wave that gives the shock to ships, and not the great wave; but when ships are struck in very deep water, the centre of disturbance is either immediately under, or very nearly under, the vessel.

[1] Mitchell on the Causes of Earthquakes, in Philosophical Transactions for 1760.

Three other series of undulations are formed simultaneously with the preceding, by which the sound of the explosion is conveyed through the earth, the ocean, and the air, with different velocities. That through the earth travels at the rate of from 7000 to 10,000 feet in a second in hard rock, somewhat less in looser materials, and arrives at the coast a short time before, or at the same moment with, the shock, and produces the hollow sounds that are the harbingers of ruin; then follows a continuous succession of sounds, like the rolling of distant thunder, formed first, by the noise propagated in undulations through the water of the sea, which travels at the rate of 4700 feet in a second, and, lastly, by that passing through the air, which only takes place when the origin of the earthquake is a submarine explosion, and travels with the velocity of 1123 feet in a second. The rolling sounds precede the arrival of the great oceanic wave on the coasts, and are continued after the terrific catastrophe when the eruption is extensive.[1]

When there is a succession of shocks all the phenomena are repeated. Sounds sometimes occur when there is no earthquake: they were heard on the plains of the Apure, in Venezuela, at the moment the volcano in St. Vincent's, 700 miles off, discharged a stream of lava. The bellowings of Guanaxuato afford a singular instance: these subterraneous noises have been heard for a month uninterruptedly when there was no earthquake felt on the table-land of Mexico, nor in the rich silver-mines 1600 feet below its surface.

The velocity of the great oceanic wave varies as the square root of the depth; it consequently has a rapid progress through deep water, and less when it comes to soundings. That raised during the earthquake at Lisbon travelled to Barbadoes at the rate of 7·8 miles in a minute, and to Portsmouth at the rate of a little more than two miles in a minute. The velocity of the shock varies with the elasticity of the strata it passes through. The undulations of the earth are subject to the same laws as those of light and sound; hence, when the shock or earth-wave passes through strata of different elasticity, it will partly be reflected, and a wave will be sent back, producing a shock in a contrary direction, and partly refracted, or its course changed so that shocks will occur both upwards and downwards, to the right or to the left of the original line of transit. Hence most damage is done at the junction of deep alluvial plains

---

[1] Thus when an earthquake begins under the ocean, it occasions five distinct series of waves or undulations, all of which are subject to the same laws of motion, namely, the earth-wave, the water-wave, and three other series of waves arising from the passage of the sound, of the explosion through the air, the earth, the water. For the laws of Sound, see Connexion of the Physical Sciences, 8th edition. [Also, "Hand-Books of Natural Philosophy," by Dionysius Lardner, in which all that relates to the subject is very clearly explained.]

with the hard strata of the mountains, as in the great earthquake in Calabria in the year 1783.

When the height of the undulations is small, the earthquake will be a horizontal motion, which is the least destructive; when the height is great, the central and horizontal motions are combined, and the effect is terrible. The concussion was upwards in the earthquake which took place at Riobamba in 1797. Baron Humboldt mentions that some of the inhabitants were thrown across a river, several hundred feet in height, on a neighbouring mountain. The worst of all is a verticose or twisting motion, which nothing can resist; it is occasioned by the crossing of two waves of horizontal vibration, which unite at their point of intersection and form a rotatory movement. This, and the interferences of shocks arriving at the same point from different origins or routes of different lengths, account for the repose in some places, and those extraordinary phenomena that took place during the earthquake of 1783 in Calabria, where the shock diverged on all sides from a centre through a highly elastic base covered with alluvial soil, which was tossed about in every direction. The dynamics of earthquakes have been ably discussed by Mr. Mallet in a very interesting paper in the 'Transactions of the Royal Irish Academy.'

There are few places where the earth is long at rest, for, independently of those secular elevations and subsidences that are in progress over such extensive tracts of country, small earthquake shocks must be much more frequent than we imagine, though imperceptible to our senses, and only to be detected by means of instruments. The shock of an earthquake at Lyons in February, 1822, was not generally perceptible at Paris, yet the wave reached and passed under that city, and was detected by the swinging of the large declination needle at the Observatory, which had previously been at rest.

The undulations of some of the great earthquakes have spread to an enormous extent. The earthquake that happened in 1842 in Gaudaloupe was felt over an extent of 3000 miles in length; and that which destroyed Lisbon had its origin in the bed of the Atlantic, from whence the shock extended over an area of about 700,000 square miles, or a twelfth part of the circumference of the globe; the West Indian islands, and the lakes in Scotland, Norway, and Sweden, were agitated by it. In linear distance the effects of that earthquake extended through 300 miles, the shocks were felt through a line of 2700 miles, and the vibrations or tremors were perceptible in water through 4000 miles. It began without warning, and in five minutes the city was a heap of ruins.

The earthquake of 1783, in Calabria, which completely changed the face of the country, only lasted two minutes; but it was not very extensive, yet all the towns and villages for 22 miles round the small town of Oppido were utterly ruined. The destruction is generally accomplished in a fearfully short time; the earthquake at Caraccas, in March, 1812, consisted of three shocks, which lasted three or four seconds, separated by such short intervals that in 50 seconds 10,000 people perished. Baron Humboldt's works are full of interesting details on this subject, especially with regard to the tremendous convulsions in South America.

Sometimes a shock has been perceived under-ground which was not felt at the surface, as in the year 1802, in the silver-mine of Marienberg, in the Hartz. In some instances miners have been insensible to shocks felt on the surface above, which happened at Fahlun, in Sweden, in 1823—circumstances in both instances depending on the elasticity of the strata, the depth of the impulses, or obstacles that may have changed the course of the terrestrial undulation. During earthquakes, dislocations of strata take place, the course of rivers is changed, and in some instances they have been permanently dried up, rocks are hurled down, masses raised up, and the configuration of the country altered; but if there be no fracture at the point of original impulse, there will be no noise.

The power of the earthquake in raising and depressing the land has long been well known, but the gradual and almost imperceptible change of level through immense tracts of the globe is altogether a recent discovery; it has been ascribed to the expansion of rocks by heat, and subsequent contraction by the retreat of the melted matter from below them. It is not at all improbable that there may be motions, like tides, ebbing and flowing in the internal lava, for the changes are by no means confined to those enormous elevations and subsidences that appear to be in progress in the basin of the Pacific and its coasts, nor to the Andes and the great plains east of them —countries for the most part subject to earthquakes; they take place, to a vast extent, in regions where these convulsions are unknown. There seems to be an extraordinary flexibility in the crust of the globe from the 54th or 55th parallel of north latitude to the Arctic Ocean. There is a line crossing Sweden from east to west in the parallel of 56° 3′ N. lat., along which the ground is perfectly stable, and has been so for centuries. To the north of it for 1000 miles, between the Gottenburg and North Cape, the ground is rising, the maximum elevation, which takes place at North Cape, being at the rate of five feet in a century, from whence it gradually diminishes to three inches in a century at Stockholm. South of the line of stability, on the contrary, the land is sinking through part of Christianstad and Malmo, for the village of Stassten in Scania is now 380 feet nearer to the Baltic than it was in the time of Linnæus, by whom it was measured now 100 years ago. The coast of Denmark on the Sound, the island of Saltholm, opposite to Copenhagen, and that of Bornholm are rising, the latter at the rate of a foot in a century. The coast of Memel on the Baltic has actually risen a foot and four inches within the last 30 years, while the coast of Pillau has sunk down an inch and a half in the same period. The west coast of Denmark, part of the Feroe Islands, and the west coast of Greenland are all being depressed below their former level. In Greenland, the encroachment of the sea, in consequence of the change of level, has submerged ancient buildings on the low rocky islands, and on the main land. The Greenlander never builds near the sea on that account, and the Moravian settlers have had to move inland the poles to which they moor their boats. It has been in progress for four centuries, and extends through 600 miles from Igalito Firth to Disco Bay.[1] Mr. Robert Chambers has shown that in our own country the land has been for ages on the rise, and that the parallel roads in Glen Roy, which have so long afforded matter of discussion, are merely margins left by the retreat of the water, as the land alternately rose and remained stationary. In the present day the elevation is going on in many places, especially on the Moray Firth and in the Channel islands. The notice of this curious subject of the gradual changes of level on the land has been chiefly revived by Sir Charles Lyell, in whose admirable works on geology all the details will be found.[2]

CHAPTER XIV.

Arctic Lands—Greenland—Spitzbergen—Iceland—Its Volcanic Phenomena and Geysers — Jan Mayen's Land — New Siberian Islands — Antarctic Lands — Victoria Continent.

GREENLAND, the most extensive of the Arctic lands, begins with the lofty promontory of Cape Farewell, the southern extremity of a group of rocky islands, which are separated by a channel five miles wide from a table-land of appalling aspect, narrow to the south, but increasing in breadth northward to a distance of which only 1300 miles are known. This table-land is bounded by mountains rising from the deep in mural precipices, which terminate in needles and

---

[1] Captain Graah's Survey in 1823–4, and Dr. Pingel, 1830–2.

[2] Lyell's Principles of Geology, 8th edit., in 8vo., 1850. See also Mr. Darwin's observations on the same subject, in the Voyage of the Adventure and Beagle; M. Domeyko's paper 'Sur les Lignes d'ancien Niveau de l'Océan du Sud aux environs de Coquimbo,' Annales des Mines, 1848; and for an illustration of the whole of this chapter, the maps of active volcanoes, of volcanic phenomena, and earthquakes, in Johnston's Physical Atlases.

pyramids, or in parallel terraces, of alternate snow and bare rock, occasionally leaving a narrow shore. The coating of ice is so continuous and thick that the surface of the table-land may be regarded as one enormous glacier, which overlaps the rocky edges and dips between the mountain-peaks into the sea.

The coasts are beset with rocky islands, and cloven by fiords, which in some instances wind like rivers for an hundred miles into the interior. These deep inlets of the sea, now sparkling in sunshine, now shaded in gloom, are hemmed in by walls of rock often 2000 feet high, whose summits are hid in the clouds. They generally terminate in glaciers, which are sometimes forced on by the pressure of the upper ice-plains till they fill the fiord, and even project far into the sea like bold headlands, when, undermined by the surge, huge masses of ice fall from them with a crash like thunder, making the sea boil. These icebergs, carried by currents, are stranded on the Arctic coast, or are drawn into lower latitudes. The ice is very transparent and compact in the Arctic regions; its prevailing tints are blue, green, and orange, which, contrasted with the dazzling whiteness of the snow and the gloomy hue of the rocks, produce a striking effect.

A great fiord in the 68th parallel of latitude is supposed to extend completely across the table-land, dividing the country into south and north Greenland, which last extends indefinitely towards the pole; but it is altogether inaccessible from the frozen sea and the iron-bound shore, so that, excepting a very small portion of the coast, it is an unknown region.

In some sheltered spots in south Greenland, especially along the borders of the fiords, there are meadows where the service-tree bears fruit, beech and willow trees grow by the streams, but not taller than a man; still farther north the willow and juniper scarcely rise above the surface; yet this country has a flora peculiar to itself. South of the island of Disco on the west coast, Danish colonies and missionaries have formed settlements on some of the islands and at the mouths of fiords; the Esquimaux inhabit the coasts even to the extremity of Baffin's Bay.

The pelasgic islands in the Arctic Ocean are highly volcanic, with the exception of Spitzbergen. In the island of Spitzbergen the mountains spring sharp and grand from the margin of the sea in dark gloomy masses, mixed with pure snow and enormous glaciers, presenting a sublime spectacle. Seven valleys filled by glaciers ending at the sea, form a remarkable object on the east coast. One of the largest masses of ice seen by Captain Scoresby on the island was north of Horn Sound: it extended 11 miles along the shore, with a sea-face in one part more than 2000 feet high, from which he saw a huge fragment hurled into the sea, which it lashed into vapour, as it broke into a thousand pieces. The sun is not seen for several months in the year, and the cold is consequently intense. Many have perished in the attempt to winter in this island, yet a colony of Russian hunters and fishermen lead a miserable existence there, within 10° of the pole, the most northern inhabited spot on the globe.

Although the direct rays of the sun are powerful in sheltered spots within the Arctic Circle, the thermometer does not rise above 45° of Fahrenheit. July is the only month in which snow does not fall, and in the end of August the sea at night is covered with a thin coating of ice, and a summer often passes without one day that can be called warm. The snow-blink, the aurora, the stars, and the moon, which, when in her northern declination, appears above the horizon for ten or twelve days without intermission, furnish the principal light the inhabitants enjoy in their long and dreary winter.

Iceland is 200 miles east from Greenland, and lies south of the Arctic Circle, which its most northern part touches. Though a fifth part larger than Ireland, not more than 4000 square miles are habitable, all beside being a chaos of volcanoes and ice.

The peculiar feature of Iceland lies in a trachytic region which seems to rest on an ocean of fire. It consists of two vast parallel table-lands covered with ice-clad mountains, stretching from N.E. to S.W. through the very centre of the island, separated by a longitudinal valley nearly 100 miles wide, which reaches from sea to sea. These mountains assume rounded forms, with long level summits or domes with sloping declivities, as in the trachytic mountains of the Andes and elsewhere; but such huge masses of tufa and conglomerate project from their sides in perpendicular or overhanging precipices, separated by deep ravines, that the regularity of their structure can only be perceived from a distance; they conceal under a cold and tranquil coating of ice the fiery germs of terrific convulsions, sometimes bursting into dreadful activity, sometimes quiescent for ages. The most extensive of the two parallel ranges of Jockuls or Ice Mountains runs along the eastern side of the valley, and contains Oräfäjokel, 6405 feet high, the highest point in Iceland, seen like a white cloud from a great distance at sea: the western high land passes through the centre of the island.

Glaciers cover many thousand square miles in Iceland, descending from the mountains, and pushing far into the low lands. This tendency of the ice to encroach has very materially diminished the quantity of habitable ground, and the progress of the glaciers is facilitated by the influence of the ocean of subterranean fire, which heats the superincumbent ground, and loosens the ice.

The longitudinal space between the mountainous table-lands is a low valley 100 miles wide, extending from sea to sea, where a substratum of trachyte is covered with lava, sand, and ashes, studded with low volcanic cones. It is a tremendous desert, never approached without dread even by the natives—a scene of perpetual conflict between the antagonist powers of fire and frost, without a drop of water or a blade of grass; no living creature is to be seen —not a bird, nor even an insect. The surface is a confused mass of streams of lava rent by crevices; and rocks piled on rocks, and occasional glaciers, complete the scene of desolation. As herds of reindeer are seen browsing on the Iceland moss that grows plentifully at its edges, it is presumed that some unknown parts may be less barren. The extremities of the valley are more especially the seat of perpetual volcanic activity. At the southern end, which opens to the sea in a wide plain, there are many volcanoes, of which Hecla is most known, from its insulated position, its vicinity to the coast, and its tremendous eruptions. Between the years 1004 and 1766 twenty-three violent eruptions have taken place, one of which continued six years, spreading devastation over a country once the abode of a thriving colony, now covered with lava, scoria, and ashes: in the year 1846 it was in full activity. The eruption of the Skaptar Jockul, which broke out on the 8th of May, 1783, and continued till August, is one of the most dreadful recorded. The volcanic fire must have been in fearful commotion under Europe, for a tremendous earthquake ruined a wide extent of Calabria that year, and a submarine volcano had been burning fiercely for many weeks in the ocean, 30 miles from the south-west cape of Iceland. Its fires suddenly ceased, the island was shaken by earthquakes, when, at the distance of 150 miles, they burst forth with almost unexampled fury in Skaptar. The sun was hid many months by dense clouds of vapour, which extended to England and Holland, and clouds of ashes were carried many hundreds of miles to sea. The quantity of matter thrown out in this eruption was computed at fifty or sixty thousand millions of cubic yards. The lava flowed in a stream in some places from 20 to 30 miles broad, and of enormous thickness, which filled the beds of rivers, poured into the sea nearly 50 miles from the places of its eruption, and destroyed the fishing on the coast. Some rivers were heated to ebullition, others dried up; the condensed vapour fell in snow and torrents of rain; the country was laid waste; famine and disease ensued; and in the course of the two succeeding years 1300 people and 150,000 sheep and horses perished. The scene of horror was closed by a dreadful earthquake. Previous to the explosion an ominous mildness of temperature indicated the approach of the volcanic fire towards the surface of the earth; similar warnings had been observed before in the eruptions of Hecla.

A semicircle of volcanic mountains on the eastern side of the lake Myvatr is the focus of the igneous phenomena at the northern end of the great central valley. Leirhnukr and Krabla, on the N.E. of the lake, have been equally formidable. After years of quiescence they suddenly burst into violent eruption, and poured such a quantity of lava into the lake Myvatr, which is 20 miles in circumference, that the water boiled many days. There are many volcanoes in this district no less formidable. Various caldrons of boiling mineral pitch, the shattered craters of ancient volcanoes, occur at the base of this semicircle of mountains, and also on the flanks of Mount Krabla: these caldrons throw up jets of the dark matter, enveloped in clouds of steam, at regular intervals, with loud explosion. That which issues from the crater of Krabla must, by Mr. Henderson's description, be one of the most terrific objects, in nature.

The eruptive boiling springs of Iceland are perhaps the most extraordinary phenomena in this singular country. All the great aqueous eruptions occur in the trachytic formation; they are characterised by their high temperature, by holding siliceous matter in solution, which they deposit in the form of siliceous sinter, and by the discharge of sulphuretted hydrogen gas. Numerous instances of spouting springs occur at the extremities of the great central valley, especially at its southern end, where more than fifty have been counted in the space of a few acres—some constant, others periodical—some merely agitated, or stagnant. The Great Geyser and Strokr, 35 miles north-west from Hecla, are the most magnificent; at regular intervals they project large columns of boiling water 100 feet high, enveloped in clouds of steam, with tremendous noise. The tube of the Great Geyser whence the jet issues is about 10 feet in diameter and 75 feet deep; it opens into the centre of a basin 4 feet deep and between 46 and 50 feet in diameter: as soon as the basin is filled by the boiling water that rises through the tube, explosions are heard, the ground trembles, the water is thrown to the height of 100 or 150 feet, followed by large volumes of steam. No farther explosion takes place till the empty basin and tube are again replenished.

MM. Descloiseaux and Bunsen, who visited Iceland in 1846, found the temperature of the Great Geyser, at the depth of 72 feet, before a great eruption, to be 260½° of Fahrenheit, and after the eruption 251½°; an interval of 28 hours passed without any eruption. The Strokr (from stroka, to agitate), 140 yards from the Great Geyser, is a circular well, a little more than 44 feet deep, with an orifice of 8 feet, which diminishes to little more than 10 inches at a depth of 27 feet. The surface of the water is in constant ebullition, while at the bottom the temperature exceeds that of boiling water by about 24°. By the experiments of M. Donny of Ghent, water long boiled becomes more and more free from air, by which the cohesion of the particles is so much increased that when it is exposed to a heat sufficient to overcome the force or cohesion, the production of steam is so instantaneous and so considerable as to cause explosion. To this cause he ascribes the eruptions of the Geysers, which are in constant ebullition for many hours, and become so purified from air, that the strong heat at the bottom at last overcomes the cohesion of the particles, and an explosion takes place. The boiling spring of Tunquhaer, in the valley of Reikholt, is remarkable from having two jets, which play alternately for about four minutes each. Some springs emit gas only, or gas with a small quantity of water. Such fountains are not confined to the land or fields of ice; they occur also in the sea, and many issue from the crevices in the lava-bed of Lake Myvatr, and rise in jets above the surface of the water.

A region of the same character with the mountains of the Icelandic desert extends due west from it to the extremity of the long narrow promontory of the Snäfell Syssel, ending in the snow-clad cone of the Snäfell Jockel, 5970 feet high, one of the most conspicuous mountains in Iceland.

With the exception of the purely volcanic districts described, trap-rocks cover a great part of Iceland, which have been formed by streams of lava at very ancient epochs, occasionally 4000 feet deep.

The dismal coasts are torn in every direction by fiords, penetrating many miles into the interior, and splitting into endless branches. In these fissures the sea is still, dark, and deep, between walls of rock 1000 feet high. The fiords, however, do not here, as in Greenland, terminate in glaciers, but are prolonged in narrow valleys, through which streams and rivers run into the sea. In these valleys the inhabitants have their abode, or in meadows which have a transient verdure along some of the fiords, where the sea is so deep that ships find safe anchorage.

In the valleys on the northern coast, near as they approach to the Arctic Circle, the soil is wonderfully good, and there is more vegetation than in any other part of Iceland, with the exception of the eastern shore, which is the most favoured portion of this desolate land. Rivers abounding in fish are much more frequent there than elsewhere; willows and juniper adorn the valleys, and birch-trees, 20 feet high, grow in the vale of Lagerflest, the only place which produces them large enough for house-building, and the verdure is fine on the banks of those streams which are heated by volcanic fires.

The climate of Iceland is much less rigorous than that of Greenland, and it would be still milder were not the air chilled by the immense fields of ice from the Polar Sea which beset its shores.

The inhabitants are supplied with fuel by the Gulf Stream, which brings drift-wood in great quantities from Mexico, the Carolinas, Virginia, the river St. Lawrence, and some even from the Pacific Ocean is drifted by currents round by the northern shores of Siberia.[1] The mean temperature in the south of the island is about 39° of Fahrenheit, that of the central districts, 36°, and in the north it is rarely above the freezing point. The cold is most intense when the sky is clear, but that is a rare occurrence, as the wind from the sea covers mountain and valley with thick fog. Hurricanes are frequent and furious; and although thunder is seldom heard in high latitudes, Iceland is an exception, for tremendous thunder-storms are not uncommon there—a circumstance no doubt owing to the volcanic nature of that island, as lightning accompanies volcanic eruptions everywhere. At the northern end of the island the sun is always above the horizon in the middle of summer, and under it in midwinter, yet there is no absolute darkness.

The island of Jan Mayen lies midway between Iceland and Spitzbergen; it is the most northern volcanic country known. Its principal feature is the volcano of Beerenberg, 6870 feet high, whose lofty snow-capped cone, apparently inaccessible, has been seen to emit fire and smoke. It is flanked by enormous glaciers, like frozen cataracts, which occupy three hollows in an almost perpendicular cliff, which descends from the base of the mountain to the sea.

The group of New Siberian Islands, which lie north of the province of Yakutsk, and in about 78° of N. lat., have so rude a climate that they have no permanent inhabitants; they are remarkable for the vast quantity of fossil bones they contain: the elephants' tusks found there have for years been an article of commerce.

The south polar lands are equally volcanic, and as deeply icebound, as those to the north. Victoria Land, which from its extent seems to form part of a continent, was discovered by Sir James Ross, who commanded the expedition sent by the British government in 1839 to ascertain the position of the south magnetic pole. This extensive tract lies under the meridian of New Zealand; Cape North, its most northern point, is situate in 70° 31′ S. lat., and 165° 28′ E. long. To the west of that cape the northern coast of this new land terminates in perpendicular ice-cliffs, from 200 to 500 feet high, stretching as far as the eye can reach, with a chain of grounded icebergs extending for miles from the base of the cliffs, all of tabular form, and varying in size from one to nine or ten miles in circumference. A lofty range of peaked mountains rises in the interior at Cape North, covered with unbroken snow, only relieved from uniform whiteness by shadows produced by the undulations of the surface. The indentations of the coast are filled with ice many hundreds of feet thick, which makes it impossible to land. To the east of Cape North the coast trends first to S.E by E. and then in a southerly direction to $78\frac{1}{4}$° of S. lat., at which point it suddenly

[1] [See, a paper read before the National Institute, April 2nd, 1844, by M. F. Maury, Lieut. U.S.N., "On the Gulf Stream and Currents of the Sea."]

bends to the east, and extends in one continuous vertical ice-cliff to an unknown distance in that direction. The first view of Victoria Land is described as most magnificent. "On the 11th of January, 1841, about latitude 71° S. and longitude 171° E., the Antarctic continent was first seen, the general outline of which at once indicated its volcanic character, rising steeply from the ocean in a stupendous mountain-range, peak above peak enveloped in perpetual snow, and clustered together in countless groups, resembling a vast mass of crystallisation, which, as the sun's rays were reflected on it, exhibited a scene of such unequalled magnificence and splendour as would baffle all power of language to portray, or give the faintest conception of. One very remarkable peak, in shape like a huge crystal of quartz, rose to the height of 7867 feet, another to 9096, and a third to 8444 feet above the level of the sea. From these peaks ridges descended to the coast, terminating abruptly in bold capes and promontories, whose steep escarpments, affording shelter to neither ice nor snow, alone showed the jet black lava or basalt, which reposed beneath the mantle of eternal frost." . . . . "On the 28th, in lat. 77° 31′ and long. 167° 1′, the burning volcano, Mount Erebus, was discovered, covered with ice and snow from its base to its summit, from which a dense column of black smoke towered high above the other nnmerous lofty cones and crateriferous peaks with which this extraordinary land is studded from the 72nd to the 78th degree of latitude. Its height above the sea is 12,367 feet, and Mount Terror, an extinct crater near to it, which has doubtless once given vent to fires beneath, attains an altitude little inferior, being 10,884 feet in height, and ending in a cape, from which a vast barrier of ice extended in an easterly direction, checking all farther progress south. This continuous perpendicular wall of ice, varying in height from 200 to 100 feet, its summit presenting an almost unvarying level outline, we traced for 300 miles, when the pack-ice obstructed all farther progress."[1]

The vertical cliff in question forms a completely solid mass of ice about 1000 feet thick, the greater part of which is below the surface of the sea; there is not the smallest appearance of a fissure throughout its whole extent, and the intensely blue sky beyond, indicated plainly the great distance to which the ice-plains reach southward. Gigantic icicles hang from every projecting point of the icy cliffs, showing that it sometimes thaws in these latitudes, although in the month of February, which corresponds with August in England, Fahrenheit's thermometer did not rise above 14° at noon. In the North Polar Ocean, on the contrary, streams of water flow from every iceberg during the summer. The whole of this country is

[1] Remarks on the Antarctic Continent and Southern Islands, by Robert MacCormick, Esq., Surgeon of H.M.S. Erebus.

beyond the pale of vegetation; no moss, not even a lichen, covers the barren soil where everlasting winter reigns. Parry's Mountains, a lofty range, stretching south from Mount Terror to the 79th parallel, is the most southern land yet discovered. The South Magnetic Pole, one of the objects of the expedition, is situate in Victoria Land, in 75° 5′ S. lat., and 154° 8′ E. long., according to Sir James C. Ross's observations.

[British writers are prone to discredit American success. Charles Wilkes, Esq., commander of the U.S. Exploring Expedition, reports that he discovered this Antarctic Continent on the 16th of January, 1840, and again, on the 14th of February, 1840, long. 106° 18′ E., and lat. 65° 50′ S. This discovery by Lieut. Wilkes, U.S. Navy, was confirmed a year afterwards, in January, 1841, by the discovery of a part of the same continent, in long. 171° E., and lat. 71° S., by Captain Sir James Ross, of the British Navy. Commander Wilkes says:—"That land does exist within the Antarctic Circle is now confirmed by the united testimony of both French and English navigators. D'Urville, the celebrated French navigator, *within a few days after land was seen by the vessels of our squadron*, reports that his boats landed on a small point of rocks, at the place (as I suppose) which appeared accessible to us in Piner's Bay, whence the Vincennes was driven by a violent gale; this he called Claire Land, and testifies to his belief of the existence of a vast tract of land, where our view of it has left no doubt of its existence. Ross, on the other hand, penetrated to the latitude 79° S. in the succeeding year, coasted for some distance along a lofty country connected with our Antarctic Continent, and establishes beyond all cavil the correctness of our assertion that we have discovered, not a range of detached islands, but a vast Antarctic Continent. How far Captain Ross was guided in his search by our previous discoveries, will best appear by reference to the chart, with a full account of the proceedings of the squadron, which I sent to him, and which I have inserted in Appendix xxiv., and Atlas. Although I have never received any acknowledgment of their receipt from him personally, yet I have heard of their having reached his hands a few months prior to his Antarctic cruise."—*Wilkes' "Narrative of the U.S. Exploring Expedition,"* vol. ii. p. 281–2.]

[In relation to Arctic explorations and discoveries, we have a similar instance of injustice.

[Lieutenant De Haven, U.S. Navy, in command of an expedition fitted out at the expense of Henry Grinnell, of New York, to search for Sir John Franklin and his comrades in arctic regions, reached in the month of Sept. 1850, as far north as 75° 25′; and here discovered land to the east and west not previously known; and, he says, "to the channel which appeared to lead into the open sea, over which the cloud of 'frost smoke' hung as a sign, I gave the name of 'Maury,' after the distinguished gentleman at the head of our National Observatory, whose theory with regard to an open sea to the North, is likely to be realized through this channel.

"To the large mass of land visible between N.W. and N.N.E., I gave the name of 'Grinnell,' in honour of the head and heart of the man in whose philanthropic mind originated the idea of this Expedition, and to whose munificence it owes its existence.

"To a remarkable Peak bearing N.N.E. from us, distant about forty miles, was given the name of 'Mount Franklin.'"

In May, 1851, eight months after the discovery by Lieutenant De Haven, the same land was seen by Captain Penny and his parties, and new names have been given to De Haven's discoveries under the authority and sanction of the British Admiralty. Grinnell Land is called "Albert Land," and Mount Franklin is made the foundation for "Sir John Barrow's Monument." This effort of the English writers and English hydrographers to rob an American officer of the honours of a discovery, has been demonstrated in detail by Peter Force, Esq., of Washington City, D.C., before the National Institute, in May, 1852. The British Admiralty should not connive at an injudicious attempt to enhance the value of a Penny by bestowing upon it deceptive colours and deceptive devices.]

Various tracts of land have been discovered near the Antarctic Circle, and within it, though none in so high a latitude as Victoria Land. Whether they form part of one large continent remains to be ascertained. Discovery ships sent by the Russian, French, and American governments have increased our knowledge of these remote regions, and the spirited adventures of British merchants and captains of whalers have contributed quite as much.[1] The land within the Antarctic Circle is generally volcanic, at least the coast line, which is all that is yet known, and that, being covered with snow and ice, is destitute of vegetation.

[1] Captain Cook discovered Sandwich Land in 1772–5. — Captain Smith, of the brig William, discovered New South Shetland in 1819.—Captain Billingshausen discovered Peter's Island, and the coast of Alexander the First. —Captain Weddel discovered the Southern Orcades.—Captain Bisco discovered Enderby's Land and Graham's Land in 1832, Admiral Dumont d'Urville La Terra d'Adelie in 1841, and Sir James Ross, Victoria Land in the same year.

CHAPTER XV.

Nature and Character of Mineral Veins—Metalliferous Deposits—Mines—Their Drainage and Ventilation—Their Depth—Diffusion of the Metals—Gold—Silver—Lead—British Mines—Quicksilver—Copper—Tin—Cornish Mines—Coal—Iron—Most abundant in the Temperate Zones, especially in the Northern—European and British Iron and Coal—American Iron and Coal—Arsenic and other Metals—Salt—Sulphur—Diffusion of the gems.

THE tumultuous and sudden action of the volcano and the earthquake on the great masses of the earth is in strong contrast with the calm, silent operations on the minute atoms of matter by which Nature seems to have filled the fissures in the rocks with her precious gifts of metals and minerals, sought for by man from the earliest ages to the present day. Tubal-cain was "the instructor of every artificer in brass and iron." Gold was among the first luxuries, and even in our own country, from time immemorial, strangers came from afar to carry off the produce of the Cornish mines.[1]

The ancients scarcely were acquainted with a third of the thirty-five metals now known, and the metallic bases of the alkalis only date from the time of Sir Humphry Davy, having formed a remarkable part of his brilliant discoveries.[2]

Minerals are deposited in veins or fissures of rocks, in masses, in beds, and sometimes in gravel and sand, the detritus of water. Most

[1] The author owes her information on British mines to two publications on the Mining District of the North of England, by J. Sopwith, Esq., Civil Engineer, and Mr. Leithart, Mine Agent. On the Cornish Mines she has derived her information from the writings of John Taylor, Esq., and Sir Charles Lemon, Bart.; from a store of valuable materials contained in the 'Progress of the Nation,' by G. R. Porter, Esq.; from the Statistical Journal, and on the general distribution of minerals over the globe, from the 'Penny Cyclopædia,' and various other sources.

[2] The metals are gold, silver, platinum, copper, lead, tin, iron, zinc, arsenic, bismuth, antimony, nickel, quicksilver, manganese, cadmium, cerium, cobalt, iridium, uranium, chrome, lantanium, molybdenum, columbium, osmium, palladium, pelapium, tantalum, tellurium, rhodium, titanium, vanadium, tungsten, dydynium, ferbium, erbium. The three last are little known.

Sir Humphry Davy discovered that lime, magnesia, alumine, and other similar substances, are metals combined with oxygen. There are thirteen of these metalloids, namely—calcium, magnesium, aluminium, glucinium, thorium, yttrium, zirconium, strontium, barium, lithium, natrium, potassium, and silicium.

of the metals are found in veins; a few, as gold and tin, iron and copper, are disseminated through the rocks, though rarely. Veins are cracks or fissures in rocks, seldom in a straight line, yet they maintain a general direction, though in a zigzag form, striking downwards at a very high angle, seldom deviating from the perpendicular by so much as forty-five degrees, and extending to an unfathomable depth. When cutting through stratified rocks, they are for the most part accompanied by a subsidence of the beds on one side of their course, and by an elevation on the other; the throw, or perpendicular distance between the corresponding strata on the opposite sides of a vein, varies from a few inches to thirty, forty, even a hundred fathoms. The beginning or end of a vein is scarcely ever known; but, when explored, they are found to begin abruptly, and, after continuing entire to a greater or less distance, they branch into small veins or strings.

In the downward zigzag course, the bending of the strata upwards on one side and downwards on the other, and the chemical changes almost always observed on the adjacent rocks, veins bear a strong analogy to the course and effects of a very powerful electrical discharge.

Veins have been filled with substances foreign to them, which have probably been disseminated in atoms in the adjacent rocks, or by sublimation. Nothing can be more certain than that the minute particles of matter are constantly in motion from the action of heat, mutual attraction, and electricity. Prismatic crystals of salts of zinc are changed in a few seconds into crystals of a totally different form by the heat of the sun: casts of shells are found in rocks, from which the animal matter has been removed, and its place supplied by mineral matter; and the excavations made in rocks diminish sensibly in size in a short time if the rock be soft, and in a longer time when it is hard—circumstances which show an intestine motion of the particles, not only in their relative positions, but in space, which there is every reason to believe is owing to electricity—a power which, if not the sole agent, must at least have co-operated essentially in the formation and filling of mineral veins.[1]

The magnetism of the earth is presumed to be owing to electrical currents circulating through its mass in a direction at right angles to the magnetic meridians. Mr. Fox, so well known in the scien-

[1] This subject is ably discussed by Mr. Leithart in his work, already mentioned, on the formation and filling of metallic veins. Mr. Leithart is an instance of the intelligence that prevails among miners, notwithstanding the scanty opportunities of acquiring that knowledge which they are generally so eager to obtain. He was a working miner, whose only education was at a Sunday-school. There are eminent engineers in England, employed in the construction of railways, canals, bridges, and other important works, who began their career as working miners.

tific world, has long since shown, from observations in the Cornish mines, that such currents do flow through all metallic veins. Now, as the different substances of which the earth is composed are in different states of electro-magnetism, and are often interrupted by non-conducting rocks, the electric currents, being stopped in their course, act chemically on all the liquids and substances they meet with. Hence Mr. Fox has come to the conclusion that not only the nature of the deposits must have been determined by their relative electrical condition, but that the direction of the metallic veins themselves must have been influenced by the direction of the magnetic meridians; and, in fact, almost all the metallic deposits in the world are in parallel veins or fissures tending from east to west, or from north-east to south-west. Veins at right angles to these are generally non-metalliferous, and, if they do contain metallic ores, they are of a different kind. In some few cases both contain the same ore, but in very different quantities, as in the silver-mine at Pasco, in the Andes, and both veins are richer near the point of crossing than elsewhere.

Sir Henry de la Beche conceives that the continued expansion and elevation of an intensely heated mass from below would occasion numerous vertical fissures through the superincumbent strata, within which some mineral matters may have been drawn up by sublimation, and others deposited in them when held in solution by ascending and descending streams of water; but even on this hypothesis the direction of the rents and the deposition of the minerals would be influenced by the electrical currents. But if veins were filled from below, the richest veins would be lowest, which is not the case in Cornwall, Mexico, or Peru.[1] The primum mobile of the whole probably lies far beyond our globe: we must look to the sun's heat, if not as the sole cause of electrical currents, at least as combined with the earth's rotation in their evolution.[2]

When veins cross one another, the traversed veins are presumed to be of prior formation to those traversing, because the latter are dislocated and often heaved out of their course at the point of meeting; and such is the case with the metalliferous veins, which are therefore the most recent. Veins are rarely filled in every part with ore; they contain sparry and stony matter, called its matrix, with

[1] Mineral veins are generally richer near the surface than at great depths: this is particularly the case in the mines of the precious metals in America, where the greatest quantities of ore have been found near the surface—a fact that may be explained by supposing the mineral substances brought by sublimation from the interior of the earth, and deposited where the temperature was lowest at or near the surface in the rocks among which they are situated.

[2] Rotation alone produces electrical currents in the earth.—'Connexion of the Physical Sciences,' page 364, 7th edition.

here and there irregular masses of the metallic ores, often of great size and value. Solitary veins are generally unproductive, and veins are richer when near one another. The prevalence and richness of mineral veins are intimately connected with the proximity or junction of dissimilar rocks, where the electro-molecular and electro-chemical actions are most energetic. Granite, porphyry, and the plutonic rocks are often eminently metalliferous; but mineral deposits are also abundant in rocks of sedimentary origin, especially in and near situations where these two classes of rocks are in contact with one another, or where the metamorphic structure has been induced upon the sedimentary. This is remarkably the case in Cornwall, the north of England, in the Ural, and all the great mining districts.

Metalliferous deposits are peculiar to particular rocks: tin is most plentiful in granite and the rocks lying immediately above it; gold in the palæozoic rocks in the vicinity of porphyritic eruptions; copper is deposited in various slate formations, and in the trias; lead is particularly abundant in the mountain-limestone system, and is rare where iron and copper abound; iron abounds in the coal and oolitic strata, and in a state of oxidule and crystallized carbonate in the older rocks; and silver is found in almost all of these formations; its ores being frequently combined with those of other metals, especially of lead and copper. There is such a connexion between the contents of a vein and the nature of the rock in which the fissure is, that, when in the oldest rocks the same vein intersects clay-slate and granite, the contents of the parts enclosed in one rock differ very much from what is found in the other. It is believed that in the strata lying above the coal-measures none of the more precious metals have been found in England in such plenty as to defray the expense of raising them, although such a rule does not extend to the continent of Europe or to South America, where copper and silver ores abound in our new red sandstone series. In Great Britain no metal is raised in any stratum newer than the magnesian limestone. Metals exist chiefly in the primary and early secondary strata, especially near the junction of granite and porphyry with slates; and it is a fact that rich veins of lead, copper, tin, &c., abound only in and near the districts which have been greatly shaken by subterraneous movements. In other countries, as Auvergne and the Pyrenees, the presence of igneous rocks may have caused mineral veins to appear in more recent strata than those which contain them in Great Britain.

When a mine is opened, a shaft like a well is sunk perpendicularly from the surface of the ground, and from it horizontal galleries are dug at different levels according to the direction of the metallic veins, and gunpowder is used to blast the rocks when too hard for the pickaxe. When mines extend very far in a horizontal direction, it becomes necessary to sink more shafts, for ventilation as well as for facility in raising the ore. Such is the perfection of underground surveying in England, that the work can be carried on at the same time from above and below so exactly as to meet; and in order to accelerate the operation, the shaft is worked simultaneously from the different galleries or levels of the mine. In this manner a perpendicular shaft was sunk 204 fathoms deep, about nineteen years ago, in the Consolidated mines in Cornwall; it was finished in twelve months, having been worked in fifteen different points at once. In that mine there are ninety-five shafts, besides other perpendicular communications under-ground from level to level: the depth of the whole of these shafts added together amounts to about 25 miles; the galleries and levels extend horizontally about 43 miles, and 2500 people are employed in it: yet this is but one of many mines now in operation in the mining district of Cornwall alone.[1]

The infiltration of the rain and surface-water, together with subterranean springs and pools, would soon inundate a mine and put a stop to the work, were not adequate means employed to remove it. The steam-engine is often the only way of accomplishing what in many cases would otherwise be impossible, and the produce of mines has been in proportion to the successive improvements in that machine. In the Consolidated mines already mentioned there are nine steam-engines constantly pumping out the water; four of these, which are the largest ever made, together lift from thirty to fifty hogsheads of water per minute, from an average depth of 230 fathoms. The power of the steam-engines in draining the Cornish mines is equal to 44,000 horses—one-sixth of a bushel of coals performing the work of a horse. The largest engine is between 300 and 350 horse-power; but as horses must rest, and the engine works incessantly, it would require 1000 horses to do its work.[2]

Mines in high ground are sometimes drained to a certain depth by an adit or gallery dug from the bottom of a shaft in a sloping direction to a neighbouring valley. One of these adits extends through the large mining district of Gwennap, in Cornwall; it begins in a valley near the sea, and very little above its level, and goes through all the neighbouring mines, which it drains to that depth, and with all its ramifications is 30 miles long. Nent Force Level, in the north of England, forms a similar drain to the mines in Alston Moor: it is a stupendous aqueduct 9 feet broad, and in some places from 16 to 20 feet high; it passes for more than 3 miles under the course of the river Nent to Nentsbury engine-shaft,

---

[1] J. Taylor, Esq., on Cornish Mines.

[2] The total amount of steam-power in Great Britain in 1833 was equal to that of 2,000,000 of men.—J. Taylor, Esq., on Cornish Mines.—It is now nearly doubled.

and is navigated underground by long narrow boats. Daylight at its mouth is seen like a star at the distance of a mile in the interior. Most of the adits admit of the passage of men and horses, with rails at the sides for waggons.

The ventilation of mines is accomplished by burning fires in some of the shafts, which are in communication with the others, so that currents of air flow up one and down the others. In some cases fresh air is carried into the mines by streams that are made to flow down some of the shafts. Were this not done, the heat, which increases with the depth, would be insupportable; ventilation diminishes the danger from the fire-damp, for, even where Sir Humphry Davy's safety-lamp is used, accidents happen from the carelessness of the miners.[1]

The access to deep mines, as in Cornwall, is by a series of perpendicular or slightly-inclined ladders, sometimes uninterrupted, but generally broken at intervals by resting-places. It is computed that one-third of a miner's physical strength was exhausted in ascending and descending a deep mine: they are now drawn up by the steam-engine.

The greatest depth to which man has excavated is nothing when compared with the radius of the earth. The Eselschacht mine at Kuttenberg in Bohemia, now inaccessible, which is 3778 feet below the surface, is deeper than any other mine. Its depth is only 150 feet less than the height of Vesuvius, and it is eight times greater than the height of the pyramid of Cheops, or the cathedral of Strasburg. The Monkwearmouth coal-mine near Sunderland descends to 1500 feet below the level of the sea, so that the barometer stands there at 31·70, which is higher than anywhere on the earth's surface.[2] The salt-works of New Saltzwerk in Prussia are 2231 feet deep, and 1993 feet below the level of the sea. Mines on high ground may be very deep without extending to the sea-level: that of Valenciana, near Guanaxuato in Mexico, is 1686 feet deep, yet its bottom is 5960 feet above the surface of the sea; and the mines in the higher Andes must be much more. For the same reason the rich mine of Joachimsthal in Bohemia, 2120 feet deep, has not yet reached that level. The fire-springs at Tseu-lieu-tsing in China are

[1] The splendid discovery of Sir Humphry Davy, that flame does not pass through fine wire gauze, prevents the fatal explosion of inflammable air in the mines, by which thousands of lives have been lost. By means of a light enclosed in a wire-gauze lantern, a miner now works with safety surrounded by fire-damp. To the honour of the illustrious author of this discovery, be it observed that it was not, like that of gunpowder and others, the unforeseen result of chance by new combinations of matter, but the solution of a question based on scientific experiment and induction, which it required the genius of a philosophic mind like his to arrive at.

[2] Supposing the barometer to be 30 inches on the level of the sea.

3197 feet deep, but their relative depth is unknown.[1] How insignificant are all the works of man compared with nature!—A line of 27,600 feet long did not reach the bottom of the Atlantic Ocean.

The metals are very profusely diffused over the earth. Few countries of any extent do not contain some of them. A small number occur pure, but in general they are found in the form of ores, in which the metal is chemically combined with other substances, and the ore is often so mixed with earthy matter and rock that it is necessary to reduce it to a coarse powder in order to separate the ore, which is rarely more than a third or fourth part of the mass brought above ground.

Gold is found in almost every country, but in such minute quantities that it is often not worth the expense of working. It is almost always in a native state, and in the form of crystals, grains, or rolled masses. Sometimes it is combined with silver: [in Chile and in Peru, as well as in Georgia and other States, it occurs also in combination with iron pyrites—sulphuret of iron.] It is exhausted in several parts of Europe where it was formerly found. The united produce of the mines in Transylvania, Hungary, the north-western districts of Austria, and the bed of the Danube, is nearly 60,000 ounces annually. Gold is found in small quantities in Spain, in the Lead-hills in Scotland, and the Wicklow mountains in Ireland.

Gold abounds in Asia, especially in Siberia. The deposits at the foot of the Ural mountains are very rich. In 1826 a piece of pure gold weighing 23 pounds was found there, along with others weighing three or four pounds each, together with the bones of elephants. All the diluvium there is ferruginous; and more to the east, as already mentioned, a region as large as France has lately been discovered with a soil rich in gold-dust, resting on rocks which contain it. In 1834 the treasures in that part of the Altaï chain called the Gold Mountains were discovered, forming a mountain-knot nearly as large as England, from which a great quantity of gold has been extracted. Gold is found in Tibet, in the Chinese province of Yunnan, and abundantly in the mountains of the Indo-Chinese peninsula, in Japan, and Borneo. In the latter island it occurs near the surface in six different places.

Africa has long furnished a large supply to Europe. That part of the Kong mountains west of the meridian of Greenwich was one of the most auriferous regions in the world before the recent discoveries of California. The gold stratum lies from 20 to 25 feet below the surface, and increases in richness with the depth. It is found in particles and pieces in a reddish sand. Most of the streams from the table-land bring down gold, as well those that descend to

[1] Note to the English translation of Kosmos, by Colonel Sabine, on the depths below the surface of the earth attained by man.

the low ground to the north, as those that flow to the Atlantic. On the shores of the Red Sea it was found in sufficient quantity to induce the Portuguese to form a settlement there.

In South America the western Cordillera is poor in metals except in New Grenada, where the most westerly of the three chains of the Andes is rich in gold and platinum — a metal found only there, in Brazil, and on the European side of the Ural mountains — in alluvial deposits. The largest piece of platinum that has been found weighed 21 ounces. Gold is found in sand and gravel on the high plains of the Andes, on the low lands to the east of them, and in almost all the rivers that flow on that side. The whole country between Jaen de Bracamoros and the Guaviare is celebrated for its metallic riches. Almost all the Brazilian rivers bring down gold; and the mine of Gongo Soco, in the province of Minas Geraës, is said to yield several varieties of gold-ore. Central America, Mexico, and California are auriferous countries. The quantity of gold recently found near the surface of California is immense, greatly surpassing that of any other country.[1] A considerable quantity is found in Tennessee, [in Virginia,] the mountains of Georgia, and on 1000 square miles of North Carolina it occurs in grains and masses. [Gold has been found at intervals from Canada to Georgia, a distance of 1000 miles.

[WE find the most methodical account of the discovery of gold in Australia in a pamphlet by Captain John Elphinstone Erskine, R.N., whence the following details are condensed:

Among the convicts who were sent out to form the first settlement in New South Wales, several instances are known to have occurred of rewards being demanded for real or pretended discoveries of gold; but the applications were discouraged by the authorities.

In December, 1829, it is mentioned in a Sydney paper that a piece of gold in the quartz matrix was bought from a labouring man by Mr. Cohen, a silversmith.

For several years after, a shepherd named McGregor, perhaps the same individual, was in the habit of occasionally bringing pieces of gold to Sydney, by the sale of which he realized considerable property. He repeatedly offered to reveal the fortunate locality (supposed to be in the Wellington districts) for a large reward; but this person was in jail for debt at the time of the late discoveries.

The Rev. W. Clarke, well known in South Wales as an able geologist, brought specimens of the metal in 1841 from the basin of the very river (the Macquarie) now supplying it, and he also repeat-

[1] The reader is referred for further information on this subject to a very interesting article (Siberia and California), attributed to one of our most eminent British geologists, on the distribution of gold in different parts of the world, and particularly in the Ural Mountains and California; in the 174th Number of the Quarterly Review, September, 1850.

edly announced his conviction that gold existed in considerable abundance in the "schists and quartzites" of the mountain chain. In consequence of communications made by him to the Geological Society, Sir Roderick I. Murchison, in a letter addressed to Sir Charles Lemon, advised that a person well acquainted with the washing of mineral sands be sent to Australia, speculating on the probability of auriferous alluvia being abundant, and suggested "that such would be found at the base of the western flanks of the dividing ranges."

In September, 1850, it was remarked in the Quarterly Review: "The important point for Englishmen now to consider, is the extent to which our own great Australian colonies are likely to become gold-bearing regions. The works of Count Strezlecki and others have made known the facts, that the chief or eastern ridge of that continent consists of palæozoic rocks, cut through by syenites, granites, and porphyries; and that quartzose rocks occasionally prevail in this long meridian chain. Sir Roderick I. Murchison announced, first to the Geographical Society, (May, 1845,) and afterwards to the Geographical Society of Cornwall, his belief that wherever such contrasts occurred, gold might be expected to be found; and Colonel Henderson suggested the same idea at St. Petersburg. Very shortly afterwards, not only were several specimens of gold in fragments of quartz veins found in the Blue Mountains north of Sydney, but one of the British chaplains, himself a good geologist, in writing more recently, thus expressed himself: 'This colony is becoming a mining country, as well as South Australia. Copper, lead, and gold are in considerable abundance in the schists and quartzites of the Cordillera (Blue Mountains). Vast numbers of the population are going to California, but some day I think we shall have to recall them.'"

Mr. Montgomery Martin, in a pamphlet published in 1847, says: "Sir Thomas Mitchell, in his recent expeditions to the north-east, found a region like the Uralian Mountains, abounding in gold. The specimens I have seen of the gold are very rich. It is in large grains, or irregular veins, loosely embedded in white quartz."

About the beginning of 1849, a very fine specimen of gold in quartz was brought to Melbourne, Port Philip; it was said to have been found by a shepherd in the "Pyrenees," a day or two's journey from the town; but his specimen was at first suspected to be an artful fabrication.

In the same year, Thomas Icely, Esq., of Coombings, a member of the Legislative Council of New South Wales, exhibited specimens of quartz brought from his property on the Dulabula, in which gold was distinctly visible; and persons of good authority in England, to whom he also submitted them, expressed opinions favourable to their richness in the precious metal. Assertions were now confidently made, that by washing the alluvial deposits in the streams or gullies, flowing from the supposed auriferous ridges, gold in dust would certainly be procured. Strange to say, in spite of Californian experience, the above experiment was not made, and the subject was altogether disbelieved. A Mr. Trappit having found a lump, or, as it is now termed, a "pocket," of gold, at the root of an old tree, was derisively told by persons to whom he showed his treasure, that it was evidently the effect of a bush fire, fusing into an irregular mass some gold watches, which must have been stolen and planted (hidden) by a convict servant.

The Colonial Government, about this time, expressed a desire to secure the services of some eminent English geologist in exploring the mineral capabilities of New South Wales (with a view to the extension of copper-mining), and accordingly, in November or December, 1850, Mr. Stutchbury, who had been some time curator of Bristol Museum, arrived in Sydney; having been named geologist to the colony. Up to the beginning of May, 1851, however, no report holding out any hope of the existence of the precious metals had been received from this gentleman; although he was said to have visited some of the localities in which they were believed to be most abundant.

On May 2, 1851, a notice appeared in the Sydney Morning Herald (the leading paper of the colony), intimating it to be no longer a secret that gold had been found in the earth in several places in the western country; and that the fact was established on the 12th of February by Mr. E. Hargreaves, a resident of Brisbane Water, who had returned from California a few months previously. It was added, that while in California, Mr. Hargreaves felt persuaded that, from the similarity of the geological formation, there must be gold in several districts of New South Wales.

On May 8, Mr. Hargreaves delivered a lecture in Bathurst, stating, that after a careful examination of from two to three months, he had found that one large gold-field existed from the foot of the "Big Hill" to a considerable distance below Wellington; that the precious metal had been picked up in numberless places, and that indications of its existence were to be seen in every direction; adding that he had established a company of nine working miners, who were then digging at a point of the Summer-hill Creek (fresh-water stream), near its junction with the Macquarie, about 50 miles from Bathurst, and 30 from Guyongi, and that the name of "Ophir" had been given to the spot. Mr. Hargreaves exhibited to the people present samples of fine gold, weighing in all about four ounces; the produce, he stated, of three days' work. The amount thus earned by each man he represented to be £2 4*s*. 8*d*. per day; but from want of practical knowledge and proper implements, nearly one-half the gold actually dug had been lost. One of the samples was a solid piece, weighing about 2 ounces, which had been found attached to the root of a tree; another consisted of small pieces, weighing from several grains to a pennyweight, all elongated; and a third of small particles, principally oval.

Besides at Summer-hill and Lewis-pond Creeks, Mr. H. had also found gold at Dubbo, below Wellington, in powder fine as the finest flour; but he did not believe that it existed in sufficient quantity to pay for labour.

Mr. Stutchbury, the geological surveyor, was now directed to accompany Mr. Hargreaves to the Summer-hill Creek; and on arriving there, digging had already commenced. On May 10, two days after Mr. Hargreave's lecture, three persons left Bathurst, and on the 14th two of them returned, bringing one piece of gold, which weighed down 35 sovereigns; another about half an ounce in weight; and several small pieces, half an ounce altogether. The largest piece was described as of solid gold, about three inches long, and of varying thickness, with a small portion of quartz imbedded in its thickest part; the smallest was like spangles, but rough and uneven on the edges. On the following day, 2½ lbs. of gold in lumps, besides a quantity of dust, were brought into Bathurst. This good fortune naturally led to the formation of parties for mining, and the construction of machines, &c., for washing the soil.

On May 17, Mr. Stutchbury's Report reached the government; and this was so conclusive as to the existence of gold in large quantities, that a proclamation was issued, declaring the right of the Crown in all precious metals, and prohibiting all persons from searching for or carrying off the same, except under regulations, subsequently settled at 30*s*. for a charge or license fee to be paid by each individual to search for the precious metals, for every calendar month, or part of a month, to a Land Commissioner, who was also empowered to allot small portions of Crown-land to each worker, and to settle disputes, &c.

At this time, May 19, there were, even in this thinly populated country, from 500 to 600 persons at work on the Summer-hill and in Lewis-pond Creeks; but, from ignorance of mining or washing, and the want of implements, few earned more than they could at their respective trades, and many gave up the search in despair. Meanwhile, Mr. Hargreaves was rewarded by the government with £500 for his discovery; and he was appointed a land commissioner.

On May 24, news reached Sydney that the gold-diggers made from £3 to £4 per day; a party of four was said to have taken out thirty ounces in a day, and a piece of one pound weight had been found. One person was said to have accumulated, within three weeks, £1600 worth of gold! A large quantity of gold was lying in the bank at Bathurst, awaiting a safe conveyance to Sydney; and the whole of Mr. Wentworth's property near Bathurst (Fitzgerald's Valley) was found to be one large gold-field.

Before the end of May, the first shipment of gold had been made for London on board the Thomas Arbuthnot, the estimated value being about £800. Among the freight was one piece weighing about 40 ounces, which had previously been exhibited in Sydney. Pieces of the same description continued to be found at intervals: viz., one of 36 ounces, and another of 22, by a Mr. Lester, who sold the latter for £76. Two fine specimens (18 and 23 ounces) were bought by the Colonial Government, for presentation to Queen Victoria.

Among the places where gold was found were the Shoal-haven Gullies, and the Crookwell River, in the county of Argyle, south of Sydney; Fitzgerald's Valley and O'Connell's Plains, near Bathurst; Mudgee and Cassilis, in the county of Bligh, northward; and many tributary streams of the Macquarie; all which spots lie at but a short distance from the meridian which Mr. Clarke had pointed out as that near which auriferous deposits might confidently be looked for. On June 9, Mr. Stutchbury's report, that he had found gold by prospecting with a small pan, and without going any depth, at various points of the Turon, attracted the gold-seekers to that river, where the metal might be found with less trouble than at Ophir. Adventurers now flocked from Sydney, Maitland, and New England, as well as overland parties from Port Philip (Victoria); and a Bathurst paper of July 5, estimates the number of miners at 800 or 1000, stretching over 7 or 8 miles of the Turon River. On July 14, many parties had arrived in Bathurst, bringing with them large quantities of gold; one party of six had made £400 in ten days; another, of the same number, £500 in fourteen days, &c.

About the middle of July it was rumoured that a mass of quartz, weighing 3 cwt., and containing upwards of one hundred pounds of gold, had been found near Meroo or Merinda Creek. The Bathurst Mail, of July 15, confirmed this report, when there were found by Dr. Kerr, or rather his aboriginal shepherd, and brought to Sydney in a tin box, 106 pounds of gold, in pieces all disembowelled from the earth at one time. The largest of the blocks was about a foot in diameter, and weighed 75 pounds, out of which were taken 65 pounds of pure gold! The auriferoús mass, before it was broken, weighed from 2 to 3 cwt.; had it not been broken, it would have been invaluable as a specimen which the world had seen nothing like. The heaviest of the two large pieces resembled a honeycomb, or sponge, of crystalline particles, as did nearly the whole of the gold. The quartz block, when found, formed an isolated heap, about 100 yards from a quartz vein, stretching up the ridge from the Murroo Creek, about 53 miles from Bathurst, 18 from Mudgee, 30 from Wellington, and 18 to the nearest point of the Macquarie River.

This vast lump of gold was sold at Bathurst to the agent of Messrs. Thacker and Co., of Sydney, for £4160. On July 23, the Mary Bannatyne shipped for London nearly 280 pounds of gold, valued at £11,600, besides 800 ounces of California gold; and smaller amounts had been privately despatched in other vessels. On August 5, 400 persons left Sydney for the "diggings." On August 7, there was delivered at the treasury, in Sydney, 288 pounds of gold, valued at upwards of £11,500; and on the 12th, a shipment to the amount of £28,960 (including the shipment of Messrs. Thacker's gold) was made on board the barque Bondicar, for London. On August 15, there was delivered at Sydney about 240 pounds of gold, (government price, £3. 8*s*. 6*d*.,) valued at £9684.

The alluvial gold hitherto found in New South Wales is said by Mr. Clarke to be rather superior in fineness to that of California, and of Minok, in Russia, and yields somewhere about 90 per cent. of pure metal. One mugget, which weighed 51 oz. 14 dwts., with small bits of quartz in the indentations, was estimated by Mr. Hale to contain 51 ounces of clean gold, of 23 carats fine. The price first given at Bathurst by purchase, was £2 18*s*. an ounce: it gradually rose to £3 5*s*., and was, in August, at Sydney, £3 8*s*. 6*d*. Such is an outline of the Gold Discovery in New South Wales, from the time of the first available finding, to the advices from the colony, August 18. For further details, with Notes of an Excursion to the Gold-fields, the reader is referred to Captain Erskine's account.

The district of Bathurst lies at the foot of the Blue Mountains, about one hundred miles from Sydney: a range which comprehends among its rock formations a great variety of the crystalline or unstratified rocks; as granite, (both the porphyritic and common kinds,) sienite, quartz-rock, serpentine, and eurite. Mica slate and silicious slate form also a portion of the stratified rocks. This very extensive Alpine range stretches from the southern shore of Australia to the southern shore of Van Diemen's Land; and through its whole length the same geological conditions prevail. These bear a striking resemblance to those observed in the Uralian Mountains and in the ranges of California. Dr. Lhotsky, in describing his journey from Sydney to the Australian Alps, in February, 1834, says "In many places on Menero, my attention was fixed by the people upon the *gold*, which they said is to be found in the creek, &c. However, I knew it was nothing but the metallic scales of mica they were pointing out to me."

Gold has since been found in still greater abundance at Buninyong, about 80 miles from Melbourne, and in the Hunter's River district, on the Liverpool Plains, 200 miles from Maitland. At the Victoria (Port Philip) diggings, eight feet square of ground are stated to have produced 2360 ounces of gold. On September 25, were brought into Sydney 6456 ounces. A small portion of gold has also been discovered in Van Diemen's Land.]

[The total product of the Australian gold-fields, up to the end of August 1852, was 5,532, 422 ounces, or 105 tons, 10 cwts., and 2 ounces of gold. These astounding results have been obtained by unskilled labourers, working without either plan or concert.

The total product of gold in California up to June 30, 1852, was $174,780,877.]

A great deal of silver is raised in Europe. The mines of Hungary are the most productive, especially those in the mountains of Chemnitz. The metalliferous mountains of the Erzgebirge are also very rich, as also the mines near Christiania in Sweden. Silver is also found in Saxony, Transylvania, and Austria. In no part of the old continent is silver in greater abundance than in the Ural and Altaï mountains, especially in the district of Kolyvan. There are silver-mines in Armenia, Anatolia, Tibet, China, Cochin-China, and Japan.

The richness of the Andes in silver can hardly be conceived, but the mines are frequently on such high ground that the profits are diminished by the difficulty of carriage, the expense of living in a barren country, sometimes destitute of water, where the miners suffer from the cold and snow, and especially the want of fuel. This is particularly the case at the silver-mines of Copiapo in Chile, where the country is utterly barren, and not a drop of water is to be found in a circuit of nine miles. These mines were discovered by a poor man in 1832, who hit upon a mass of silver in rooting out a tree. They extend over 150 square leagues. Sixteen veins of silver were found in the first four days, and, before three weeks elapsed, forty more, not reckoning smaller ramifications. The rolled pieces which lay on the surface produced a large quantity of pure silver. A single mass weighed 5000 pounds.[1]

In Peru there are silver-mines along the whole range of the Andes, from Caxamarca to the confines of the desert of Atacama. The richest at present are those of Pasco, which were discovered by an Indian in 1630. They have been worked without interruption since the beginning of the seventeenth century, and seem to be still inexhaustible. The soil under the town of Pasco is metalliferous, the ores probably forming a series of beds contemporaneous with the strata. The richness of these beds is not everywhere the same, but the nests of ore are numerous. The mines of Potosí, 16,150 feet above the sea-level, are celebrated for riches, but the owners had to contend with all the difficulties which such a situation imposes. The small depth at which the silver lies on the high plains of the Andes, and the quantity of it on the surface, is probably owing, as has been already stated, to the greater deposition of the sublimed mineral from refrigeration near the surface. The ore in the mines at Chota is near the surface over an extent of half a square league, and the filaments of silver are sometimes even entwined with the roots of

[1] Dr. Pœppig's 'Travels in Chile and Peru.'

the grass. This mine is 13,300 feet above the level of the sea, and even in summer the thermometer is below the freezing-point in the night. In the district of Huantajaya, not far from the shores of the Pacific, there are mines where masses of pure silver are found, of which one weighed 800 pounds.[1]

[Silver has been observed at a mine about a mile south of Sing-Sing prison, in the State of New York, which was worked during the war of the Revolution; at the Bridgewater copper-mines in New Jersey; at Ring's mine in Davidson Co. N. C., and also at the copper-mines of Michigan. The United States have afforded but little native silver.]

According to Baron Humboldt, the quantity of the precious metals brought to Europe between the discovery of America and the year 1803 was worth 1257 millions sterling; and the silver alone taken from the mines during that period would form a ball 89 feet in diameter. The disturbed state of the South American republics and the high price of quicksilver have interfered with the working of the mines.

Lead-ore is very often combined with silver, and is then called Argentiferous Galena. It is one of the principal productions of the British mines, especially in the northern mining district, which occupies 400 square miles at the junction of Northumberland, Cumberland, Westmoreland, Durham, and Yorkshire. It comprises Alston Moor, the mountain-ridge of Crossfell, and the dales of Derwent, East and West Allendale, the Wear, and Tees. There are other extensive mining tracts separated from this by cultivated ground. The principal products of this rich district are lead and copper. The lead-mines lie chiefly in the upper dales of the Tyne, Wear, and Tees, and all of it contains more or less silver, though not always enough to indemnify the expense of refining or separating the silver. The deleterious vapours resulting from this process are conveyed in a tube along the surface of the ground for 14 miles: and instead of being, as formerly, a dead loss to the proprietor, they are condensed in their passage, and in one instance yield metal to the annual value of 10,000*l.*[2] The Hudgillburn lead-mine in that district has yielded treasures almost unexampled in the annals of mining. The veins, from ten to twelve, and in some places even twenty feet wide, were filled with ore, which is entirely obtained with the pickaxe, without blasting. In 1821 the galena of this mine yielded 32,000 ounces of silver.

Lead-mines are in operation in France, but not to any great amount: those of the south of Spain furnish large quantities of this metal; also in Saxony, Bohemia, and Carinthia, where they are

[1] Dr. Pœppig.
[2] Constructed under the direction of Thomas Sopwith, Esq.

very rich. Lead is not very frequently found in Siberia, though it does occur in the Nerchinsk mining district, in the basin of the river Amur. It is also a production of China, of the peninsula beyond the Ganges, and of America. It is also found in Lower Peru, Mexico, and in California.

[The northwest country, or Upper Mississippi Valley, is among the most remarkable in the world for the variety and abundance of its mineral deposits, and especially for those which are of most extensive use in the arts. The sulphuret of lead occupies about one degree of latitude, extending north from a point on the Mississippi, about eight miles below Galena, and lying on both sides, varying in width, till it covers as great an extent from east to west. On the east side of the river the lead-ore is found principally in a clay matrix, at a depth of sometimes only five or six feet from the surface; on the west side of the river it runs at the depth of one hundred feet or more, overlaid with magnesian limestone. To the south-west of the lead deposit is a very abundant bed of iron, about forty miles long by twenty-five broad. The copper region extends north from the lead deposits to Lake Superior; it embraces about three hundred square miles. To the south of the lead region is a vast bed of bituminous coal of good quality, at no great distance below the surface.

In the mineral district there are about four thousand persons employed in digging lead-ore. The value of the lead annually produced is estimated at $1,500,000. A considerable quantity was exported to China, before the emigration to California withdrew the miners, and thus diminished the product.

Lead-mines have been worked in the United States during more than half a century, the quantity produced constantly increasing. In the year 1839, according to the census returns, it was equal to nearly 14,000 tons; and in the year ending June 30th 1844, the quantity exported amounted to nearly 8200 tons, valued at $595,238.

The most extensive lead-mines known in the world are probably those found in the western section of the United States, in Washington, Jefferson, and Madison counties, Missouri; and at Galena, in the north-west part of Illinois; in Iowa; in Wisconsin, and in Michigan. Lead-ores also occur at various localities in the states of New York, Pennsylvania, Maine, New Hampshire, Virginia, &c.]

Quicksilver—a metal so important in separating silver from its ores, and in various arts and manufactures as well as in medicine—occurs either liquid in the native state, or combined with sulphur in that of cinnabar. It is found in the mines of Idria and some other places in the Austrian empire, in the Palatinate on the left bank of the Rhine, and in Spain. The richest quicksilver mines of Europe, at the present day, are those of Almaden, where the quicksilver is found in the state of sulphuret chiefly disseminated in the Silurian strata. These mines were worked 700 years before the Christian era, and as many as 1200 tons of the metal are extracted annually. It occurs in China, Japan, and Ceylon, at San Onofrio in Mexico, and in Peru, at Huancavelica, the mines of which, now almost abandoned, produced, up to the beginning of the present century, the enormous quantity of 54,000 tons of quicksilver.[1]

[Several very rich quicksilver mines have been discovered in California. There is one about twenty miles from San José, in the Santa Clara Valley, which probably has no rival at present in the world. At one time recently there were 2,400,000 pounds of ore lying at the mouth of the mine, from which it is brought in hide sacks on the shoulders of men. It is estimated, this ore will, at an average yield of fifty per cent., produce 1,200,000 pounds of pure quicksilver. There are three or four other mines of cinnabar—sulphuret of mercury—equally rich, in the same valley. They have been long known to the aborigines, who resorted to them to procure the ore as a pigment.]

Copper is of such common occurrence that it would be vain to enumerate the localities where it is found. It is produced in Africa and America, in Persia, India, China, and Japan. The Siberian mines are very productive both in ore and native copper. Malachite is the most beautiful of the ores, and the choicest specimens come from Siberia. Almost every country in Europe yields copper. The mines in Sweden, Norway, and Germany are very productive; and it forms a principal part of our own mineral wealth. It is raised in all the principal mining districts in England and Wales. It abounds in several localities in the United States. [The copper-mines of Lake Superior are unequalled in the world. Masses of 50 tons weight of the pure metal are found together. There is a shaft in the Cliff mine of upwards of 400 feet in depth, and a vein followed several hundred feet at that depth, and throughout the characteristics of the ore are the same.] In Cornwall it is very plentiful, and is often associated with tin. The period at which the Cornish mines were first worked goes far beyond history, or even tradition: certain, however, it is that the Phœnicians came to Britain for tin. Probably copper was also worked very early in small quantities, for its exportation was forbidden in the time of Henry VIII. It was only in the beginning of the eighteenth century that the Cornish copper-mines were worked with success, in consequence of the invention of an improved machine for draining them.[2]

[On the lands south of Lake Superior is a body of copper ore, supposed to be the richest in the world. It is almost pure in some

[1] Very rich mines of quicksilver are said to have been recently discovered in California.

[2] Sir Charles Lemon, Bart.

specimens: so that, as taken from the earth, it was wrought into church utensils by some of the French who first visited the place; and a portion of the large rock deposited on the grounds of the War Department, at Washington, has been polished so as to present the appearance of sheet-copper.

At a meeting of the "American Association for the Advancement of Science," held at Cambridge, Massachusetts, August 1849, Mr. J. S. Hodge, speaking of the mineral region of Lake Superior, said:—"The mines are wrought wholly for native copper. The veinstone with scattered particles, furnish what is called *stamp work;* which is crushed under heavy stamps and then washed; the lumps are called *barrel ore*, being packed in barrels for transportation; and the masses, after being cut up into pieces not exceeding two tons in weight, are shipped in bulk. The size of some of these masses is so enormous as almost to exceed belief. They have been broken up in the Cliff mine of 60 and even 80 tons in weight. Such pieces are reduced, in the mine, to fragments of seven tons weight and less, and after being hoisted to the surface are still further reduced.

"At the Minnesota mine, near the Ontonagon river, I had an opportunity of examining, in June, the most extraordinary mass yet met with. Two shafts had been sunk on the line of the vein 150 feet apart. At the depth of about 30 feet they struck massive copper, which lay in a huge sheet with the same underlay as that of the vein—about 55° towards the north. Leaving this sheet as a hanging wall, a level was run under it connecting the two shafts. For this whole distance of 150 feet the mass appears to be continuous, and how much further it goes on the line of the vein either way there is no evidence, nor beside to what depth it penetrates in the solid vein. I examined it with care, striking it repeatedly with my hammer in order to detect, if possible, by the sound, any break or interruption there might be in the mass—for a thin scale of stone encrusted it sometimes and concealed the face of the metal. Examinations had been made by drilling through this scale, where it attained the thickness of an inch or so; but in no place had any sign of a break been found. It forms the whole hanging wall of the level, showing a width of at least eight feet above the floor in which its lower edge was lost. It has been cut through in only one place, where a partial break afforded a convenient opportunity. Measuring the thickness here as well as the irregular shape of the gap admitted, it was found somewhat to exceed five feet. Assuming the thickness to *average* only one foot, there would be in this mass 1200 cubic feet, or about 250 tons—still it is not safe to assume even one foot, for the masses vary extremely in thickness.

"The mode adopted to remove these masses is to cut channels through them with cold chisels, after they are shattered by large sand blasts put in behind them. Grooves are cut with the chisels across their smallest places, one man holding, and another striking, as in drilling. A chip of copper three-quarters of an inch wide, and up to six inches in length, is taken out, and the process is repeated until the groove passes through the mass. The expense of this work is from eight to twelve dollars per superficial foot of the face exposed. Fragments of veinstone enclosed in the copper prevent the use of saws. A powerful machine, occupying little room, is much needed, which would perform more economically this work.

"The greatest thickness of any mass cut through at the Cliff mine has been about three feet. Their occurrence through the vein is not regular. Barren spots alternate with productive portions. The same is the case in all the mines. The total product of the Cliff mine for the year 1848 is estimated at 830 tons, averaging 60 per cent. During the present year more than half this amount has been already sent down, and there is enough more on the surface and in sight in the mine to warrant the belief that 1000 tons will be the product of the year's work, or 600 tons of copper. The whole amount of copper annually imported into the United States is about the value of two million dollars, or about 5400 tons. But little has been supplied from our own mines. Nine such mines, then, as the Cliff, would render us independent of foreign supplies. From present appearances, after careful examination of the region, and consideration of the progress made in mining since my last visit in 1846, I feel myself warranted in expressing a decided conviction that this amount of copper must be supplied in very few years, and this metal soon become, as lead already has, one of export instead of import. The recent failures of mining speculations, wildly undertaken, and ignorantly and extravagantly conducted, may for a time check the development of these mines; but their wonderfully rich character is now beginning to be properly appreciated, as well as the reliance which may be put in the surface-appearance of the veins. Some curious features in their character and distribution have been detected, which have heretofore escaped observation for want of sufficient data, and which will, I believe, be found of great consequence in the selection of the best localities. These, after farther examination, I may at another time make public. The history of these mines, so far, has remarkably proved the foresight and excellent judgment of the lamented Dr. Houghton, particularly so in his predictions of the disastrous effects that must result from such speculations as have caused the country to be overrun by hordes of adventurers.

"The silver found associated with the copper has not proved of much importance, perhaps for the reason that the greater part of it is purloined by the miners. The Cliff mine has probably yielded more than thirty thousand dollars worth, of which not more than a tenth part has been secured by the proprietors. I saw myself, the present season, no less than six pounds and eight ounces of lumps and bars of silver seized in the hands of an absconding workman."]

In Cornwall clay-slate rests upon granite, and is traversed by porphyritic dykes. The veins which contain copper or tin, or both, run east and west, and penetrate both the granite and the clay-slate. The non-metalliferous veins run north and south; and if veins in that direction do contain any metal, it never is tin or copper, but lead, silver, cobalt, or antimony, which with little exception are believed to be always in the clay-slate. No miner in Cornwall has ever seen the end or bottom of a vein; their width varies from the thickness of a sheet of paper to 30 feet; the average is from one to three feet. It rarely happens that either tin or copper is found nearer the surface than 80 or 100 feet. If tin be first discovered, it sometimes disappears after sinking the mine 100 feet deeper, when copper is found, and in some instances tin is found 1000 feet deep without a trace of copper; but if copper is first discovered, it is very rarely succeeded by tin. Tin is found in rolled pieces, in horizontal beds of sand and gravel, and is called stream-tin. The most valuable tin-mines on the continent of Europe are those in Saxony; it also occurs in France, Bohemia, and Spain. One of the richest deposits of tin known is in the province of Tenasserim, on the east side of the gulf of Martaban, in the Malayan peninsula. These deposits occur in several parts of that country; the richest is a layer eight or ten feet thick of sand and gravel, in which masses of oxide of tin are sometimes the size of a pigeon's egg. The best of all comes from the island of Banca, at the extremity of the peninsula of Malacca; a large portion of it is imported into Britain, and much goes to China. It is found in the alluvial tracts through every part of the island, rarely more than 25 feet below the surface. Great deposits occur also in the Siberian mining district of Nertshinsk, near the desert of the Great Gobi, and in Bolivia, near Oruro.

There are comparatively few coal-mines worked within the tropics; they are mostly in the temperate zones, especially between the Arctic Circle and the Tropic of Cancer; and as iron, the most useful of metals, is chiefly found in the carboniferous strata, it follows the same distribution. In fact, the most productive iron-mines yet known are in the temperate zones. In the eastern mining district of Siberia, in the valley of the river Vilui, the ores are very rich, and very abundant in many parts of the Altaï and Ural. In the latter the mountain of Blagod, at 1534 feet above the sea, is one mass of magnetic iron-ore.[1] Coal and iron are worked in so many parts of

[1] Mr. Erman's 'Travels in Siberia.'

Northern China, Japan, India, and Eastern Asia, that it would be tedious to enumerate them.

In Europe the richest mines of iron, like those of coal, lie chiefly north of the Alps. Sweden, Norway, Russia, Germany, Styria, Belgium, and France, all contain it plentifully. In Britain many of the coal-fields contain subordinate beds of a rich argillaceous iron-ore, interstratified with coal, worked at the same time and in the same manner; besides, there is a sub-stratum of limestone, which serves as a flux for melting the metal. The principal mines lie round Birmingham, in the Staffordshire coal-field, and the great coal-basin of South Wales, about Pontypool and Merthyr Tydvil. There are extensive iron-mines in Staffordshire, Shropshire, North and South Wales, Yorkshire, Derbyshire, and Scotland. Altogether there are about 220 mines, which yield iron sufficient for our own enormous consumption and for exportation. These productive mines would have been of no avail had it not been for the abundance of fuel with which the greater part of them in the north of England, Scotland, and Wales are associated—the great source of our national wealth, more precious than mines of gold. Most of the coal-mines would have been inaccessible but for the means which their produce affords of draining them at a small expense. A bushel of coals, which costs only a few pence, in the furnace of a steam-engine generates a power which in a few minutes will raise 20,000 gallons of water from a depth of 360 feet—an effect which could not be accomplished in a shorter time than a whole day by the continuous labour of twenty men working with the common pump. Yet this circumstance, so far from lessening the demand for human labour, has caused a greater number of men to be employed in the mines.[1]

The coal strata lie in basins, dipping from the sides towards the centre, which is often at a vast depth below the surface of the ground. The centre of the Liege coal-basin is 21,358 feet, or 3½ geographical miles deep, which is easily estimated from the dip, or inclination, of the strata at the edges, and the extent of the basin. The coal lies in strata of small thickness and great extent. It varies in thickness from 3 to 9 feet, though in some instances several layers come together, and then it is 20 and even 30 feet thick; but these layers are interrupted by frequent dislocations, which raise the coal-seam towards the surface. These fissures, which divide the coalfield into insulated masses, are filled with clay, so that an accumulation of water takes place, which must be pumped up.

There are three immense coalfields in England. The first lies north of the Trent, and occupies an area of 360 square miles; and although the quantity of coal annually raised in Northumberland

[1] In 1841 there were 196,921 persons employed in the mines of Great Britain and Ireland.

and Durham amounts to upwards of three millions of tons, there is enough to last 1000 years. London is chiefly supplied from it. The second or central coalfield, which includes, Leicester, Worcester, Stafford, and Shropshire, has an area of 1495 square miles, and supplies the manufactories round it, and the midland counties south and east of Derbyshire. The third or western coalfield includes South Wales, Gloucestershire, and Somersetshire. The coalfield of South Wales alone is 100 miles long, and 18 or 20 broad. The Workington and Whitehaven coal-mines extend a mile under the sea; several shafts in the latter are 100 fathoms deep: and it is one of the finest in England for extent and thickness of strata, some of the seams being nine feet thick.

The Scotch coalfield occupies the great central low land of Scotland, lying between the southern high lands and the Highland mountains; the whole of that wide tract is occupied by it, besides which there are others of less extent. Coal has been found in seventeen counties in Ireland, but the island contains only four principal coal districts — Leinster, Munster, Connaught, and Ulster. Thus there is coal enough in the British islands to last some thousands of years; and were it exhausted, our friends across the Atlantic have enough to supply the world for ages uncountable. Moreover, if science continues to advance at the rate it has lately done, a substitute for coal will probably be discovered before our own mines are worked out.[1]

[1] In the year 1829, the value of the mineral produce of Europe, including Asiatic Russia, amounted to—

Gold and Silver	£1,943,000
Other metals	28,519,000
Salts	7,640,000
Combustibles	18,050,000
Total	£56,148,000

England contributed more than half this amount, namely,—

Silver	£28,500
Copper	1,369,000
Iron	11,292,000
Lead	760,000
Tin	536,000
Salts	756,250
Vitriol	33,600
Alum	33,000
Coal	13,000,000
Total	£28,716,750

—nearly £29,000,000 sterling.—John Taylor, Esq., on the Cornish Mines

At present there are nearly 40,000,000 of tons of coals consumed in Great Britain annually, besides the quantity exported to our colonies and to foreign countries, amounting to nearly 1,410,000 of tons. 10,000,000 of tons are consumed in the working of iron alone. Between 500,000 and 600,000 tons are used in making gas.

The iron made in Britain in 1848 amounted to 2,093,736 tons. Iron is

The carboniferous strata are enormously developed in the states of North America. The Appalachian coalfield extends without in-

---

now applied to many uses instead of timber, especially in ship-building: between the years 1830 and 1847, 150 iron vessels were launched in Britain. 25 of the steam-ships of the East India Company are of iron.

The produce of our copper-mines has increased threefold within the last 60 years, and now reaches 15,000 tons of pure metal. The quantity of tin has also increased from our own mines to 4180 tons in 1848, and also from the extensive importation of that metal from Banca, where the country yielding stream-tin extends from 7° N. lat. to 3° S. lat. The produce from the latter country imported into Great Britain in 1849 amounted to 1781 tons of pure metal.—'Progress of the Nation, in its Social and Commercial Relations, since the Beginning of the Nineteenth Century,' by G. R. Porter, Esq., new edition, 1851.

In France there are 62 coal-mines, yielding 3,410,200 tons in 1841, and in 1838 the 12 iron districts in that country yielded to the value of 4,975,424*l.*

The British coal and metal imported into France amounted to 1,222,228*l.* —Progress of the Nation.

Belgium is next in importance to England as a coal-producing country. In Britain the coalfields occupy one 20th part of the area of the country—in Belgium one 22d part—in France one 210th part of its area.

The quantity of coal raised in one year is, according to 'The Statistics of Germany,' by R. Valpy, Esq.—

In Britain	34,700,000	In France	3,783,000
Belgium	4,000,000	Germany	3,000,000

[The following table exhibits the quantity and value of coal produced, in the six principal coal countries in the world, in the year 1845:—

Order in 1845.	COUNTRIES.	Square miles of Coal formations.	Tons of Fuel raised in the year 1845.	Relative parts of 1000.	Official estimated value at the places of production.	
					United States Dollars.	English Sterling.
1	Great Britain	11,859	31,500,000	642	$45,738,000	£9,450,000
2	Belgium	518	4,960,077	101	7,689,900	1,660,000
3	United States	133,132	4,400,000	89	6,650,000	1,373,963
4	France	1,719	4,141,617	84	7,663,000	1,603,106
5	Prussian States	Not defined.	3,500,000	70	4,122,945	856,370
6	Austrian States	Do.	659,340	14	800,000	165,290
	Total		49,164,034	1000	72,663,845	15,108,729

The coal trade appears to be increasing in all parts of the world.

There are no authentic data from which the increasing production of bituminous coal in the United States can be exactly deduced; but we have shown that it is very rapid. The production of *anthracite* may be said to be entirely confined to the State of Pennsylvania, which possesses a numerous and interesting group of coal basins, of various sizes and characters.

In the year 1820, the anthracite coal trade commenced with 365 tons, in 1827 it reached 48,047 tons; in 1837, 881,026 tons, and advanced to 3,000,000 tons in 1847: and 4,383,667 tons in the year 1851.

The following table exhibits the production of smelted or manufactured iron in different countries in the year 1845:—

1. Great Britain	2,200,000
2. United States	502,000
3. France	448,000

terruption 720 miles, with a maximum breadth of 280 miles, from the northern border of Pennsylvania to near Huntsville, in Alabama, occupying an area of 63,000 square miles. It is intersected by three great navigable rivers — the Monongahela, the Alleghany, and the Ohio — which expose to view the seams of coal on their banks. The Pittsburg seam, 10 feet thick, exposed on the banks of the Monongahela, extends, horizontally, 225 miles in length and 100 in breadth, and covers an area of 14,000 square miles, so that this seam of coal may be worked for ages almost on the surface, and in many places literally so. Indeed, the facility is so great, that it is more profitable to convey the coal by water to New Orleans, 1100 miles distant, than to cut down the trees with which the country is covered for fuel, and which may be had for the expense of felling. The coal is bituminous, similar to the greater part of the British coal; forty miles to the east, however, among the ridges of the Appalachian chain, there is an extensive outlying member of the great coalfield, which yields anthracite, a species of coal which has the advantage of burning without smoke.

In the western states, the Illinois coalfield, which occupies part of Illinois, Indiana, and Kentucky, is as large as England, and con-

4. Russia	400,000
5. Zollverein, or Prussian States	300,000
6. Austria	190,000
7. Belgium	150,000
8. Sweden	145,000
9. Spain (in 1841)	26,000
10. All other European countries	50,000
	4,411,000

The rapid increase in the number of railroads and locomotive engines, and the number of steam vessels employed in commerce, augments the demand, proportionally, for iron and fuel.[1]]

The number of miles of railway in operation upon the surface of the globe, up to January 1853, is 29,606; of which 15,436 miles are situated in the Eastern Hemisphere, and 14,170 in the Western. In the United States there are 13,586 miles; in the course of construction and in operation, there are 372 railroads in the Union, which have cost $400,713,907. The longest railway in the world is the New York and Erie, which is 467 miles in length.[2]

RAILWAYS OF THE WORLD, JANUARY 1853.

In the United States	13,586	In France	1,831
British Provinces	173	Belgium	532
Island of Cuba	359	Russia	422
Panama	22	Sweden	75
South America	30	Italy	170
Great Britain	6,976	Spain	60
Germany	5,340	India	30

[1] "Statistics of Coal." By Richard Cowling Taylor, Philadelphia, 1848.
[2] Hunt's Merchants' Magazine.

sists of horizontal strata, with numerous seams of rich bituminous coal. There is a vast coalfield also in Michigan. Large areas in New Brunswick and Nova Scotia abound in coal. Iron is worked in many parts of the United States, from Connecticut to South Carolina.[1]

The tropical regions of the globe have been so little explored that no idea can be formed of the quantity of coal or iron they contain; but as iron is so universal, it is probable that coal is not wanting. It is found in Formosa. Both abound in Borneo, and in various parts of tropical Africa and America. There is comparatively so little land in the southern temperate zone, that the mineral produce must be more limited than in the northern, yet New Holland, Van Diemen's Land, and New Zealand are rich in coal and iron.

Arsenic, used in the arts and manufactures, is generally found combined with other metals in many countries as well as our own. Manganese, zinc, bismuth, and antimony are raised to a considerable amount. As the qualities of the greater part of the more rare metals are little known, they have hitherto been interesting chiefly to the mineralogist.

The mines of rock-salt in Cheshire seem to be inexhaustible. Enormous deposits of salt extend 600 miles on each side of the Carpathian Mountains, and throughout wide districts in Austria, Gallicia, and Spain. It would not be easy to enumerate the places in Asia where rock-salt has been found. Armenia, Syria, and extensive tracts in the Punjab abound in it, also China and the Ural district; and the Andes contain vast deposits of rock-salt, some at great heights.

Volcanic countries in both continents yield sulphur. Sicily, where it is found in the tertiary marine strata, unconnected with the volcanic district, is the magazine which supplies the greater part of the manufactures of Europe. It is often found beautifully crystallized. Asphalt, nitre, alum, and naphtha are found in various parts of Europe and Asia, and natron is procured from small lakes in an oasis on the west of the Valley of the Nile.

The diffusion of precious stones is very limited. Diamonds are mostly found in a soil of sand and gravel, and in the beds of rivers. Brazil furnishes most of the diamonds in commerce; they are the produce of tracts on each side of the Sierra Espenhaço, and of a district watered by some of the affluents of the Rio San Francisco. During the century ending in 1822, diamonds were collected in Brazil to the value of three millions sterling, one of which weighed 138½ carats. The celebrated mines of Golconda have produced many splendid diamonds; they are also found in Borneo, which produced one weighing 367 carats, valued at £269,378. The eastern

[1] Sir Charles Lyell's 'Travels in the United States of North America.'

parts of the Thian-Tchan, on the great platform of Asia, and a wide district of the Ural Mountains, yield diamonds.

The ruby and sapphire, which have the same crystalline form, are found in Ceylon, in the gravel of streams. The rubies at Gharan, near to the river Oxus, are found in beds of limestone. The gravel of rivulets in the Birman empire contains the oriental, star, and opalescent rubies. The spinelle also occurs in that country in a district five days' journey from Ava. The Hungarian rubies are of inferior value. The blue, green, yellow, and white sapphires, are the produce of the Birman empire, and the spinelle is not uncommon in Brazil.

The finest emeralds come from veins in a blue slate, of the age of our lower chalk strata, in the valley of Muso, in New Grenada.[1] Beryls are found in Brazil, and in the old mines in Mount Zebarah, in Upper Egypt. Those of Hungary and of the Heubach Valley, near Saltzburg, are very inferior in colour and quality.

Mexico, Hungary, and Bohemia yield the finest opals; the most esteemed are opaque, of a pale brown, and shine with the most brilliant iridescence; some are white, transparent, or semi-transparent, and radiant in colours. The most beautiful garnets come from Bohemia and Hungary; they are found in the Hartz Mountains, Ceylon, and many other localities. The turquoise is a Persian gem, of which there are two varieties; one is supposed to be the enamel of the tooth of a fossilized mastodon, the other a mineral; it is also found in Tibet and in the Belor-Tagh in Badakshan, which is the country of the lapis lazuli, mined by heating the rock, and then throwing cold water upon it. This beautiful mineral is also found in several places of the Hindoo Coosh, in the hills of Istalif north of Cabool, in Tibet, and in the Baïkal Mountains in Siberia.

The cat's-eye is peculiar to Ceylon; the king of Kandy had one two inches broad. Topaz, beryl, and amethyst are of very common occurrence, especially in Brazil, Siberia, and other places. They are little valued, and scarcely accounted gems. Agates are so beautiful on the table-land of Tibet, and in some parts of the desert of the Great Gobi, that they form a considerable article of commerce in China; and some are brought to Rome, where they are cut into cameos and intaglios. But the greater part of the agates, cornelians, and chalcedonies used in Europe are found in the trap-rocks of Oberstein, in the Palatinate.

Thus, by her unseen ministers, electricity and reciprocal action, the great artificer, Nature, has adorned the depths of the earth and the heart of the mountains with her most admirable works, filling

[1] This curious geological fact has been recently established by the discoveries of Professor Lewy, who has sent to Paris specimens in which crystals of emerald and green-sand fossils are imbedded.—Dec. 1850.

the veins with metals, and building the atoms of matter, with the most elegant and delicate symmetry, into innumerable crystalline forms of inimitable grace and beauty. The calm and still exterior of the earth gives no indication of the activity that prevails in its bosom, where treasures are preparing to enrich future generations of man. Gold will still be sought for in the deep mine, and the diamond will be gathered among the débris of the mountains, while time endures.

CHAPTER XVI.

The Ocean—its Size, Colour, Pressure, and Saltness—Tides—Waves—their Height and Force—Currents—their Effect on Voyages—Temperature—the Stratum of Constant Temperature—Line of Maximum Temperature—North and South Polar Ice—Inland Seas.

THE ocean, which fills a deep cavity in the globe, and covers three-fourths of its surface, is so unequally distributed that there is three times more land in the northern than in the southern hemisphere. The torrid zone is chiefly occupied by sea, and only one twenty-seventh part of the land on one side of the earth has land opposite to it on the other. The form assumed by this immense mass of water is that of a spheroid, flattened at the poles; and as its mean level is nearly the same, for anything we know to the contrary, it serves as a base to which all heights of land are referred.

The bed of the ocean, like that of the land, of which it is the continuation, is diversified by plains and mountains, table-lands and valleys, sometimes barren, sometimes covered with marine vegetation, and teeming with life. Now it sinks into depths which the sounding-line has never fathomed, now it appears in chains of islands, or rises near to the surface in hidden reefs and shoals, perilous to the mariner. Springs of fresh water rise from the bottom, volcanoes eject their lavas and scoriæ, and earthquakes trouble the deep waters.

The ocean is continually receiving the spoils of the land, and from that cause would constantly be decreasing in depth, and, as the quantity of water is always the same, its superficial extent would increase. There are, however, counteracting causes to check this tendency: the secular elevation of the land over extensive tracts in many parts of the world is one of the most important. Volcanoes, coral islands, and barrier-reefs show that great changes of level are constantly taking place in the bed of the ocean itself—that symmetrical bands of subsidence and elevation extend alternately over an

17*

area equal to a hemisphere, from which it may be concluded that the balance is always maintained between the sea and land, although the distribution may vary in the lapse of time.

The Pacific, or Great Ocean, exceeds in superficies all the dry land on the globe. It has an area of 50 millions of square miles; including the Indian Ocean, its area is nearly 70 millions; and its breadth from Peru to the coast of Africa is 16,000 miles. Its length is less than the Atlantic, as it only communicates with the Arctic Ocean by Behring's Straits, whereas the Atlantic, as far as we know, stretches from pole to pole.

[The study of ocean currents, and of the habits and habitats of certain aquatic mammals—whales—has led Lieutenant M. F. Maury, U. S. Navy, to infer "that there is at times at least, an open-water communication through the polar regions between the Atlantic and Pacific oceans." In his opinion, the "right whale" of Behring's Straits and the right whale of Baffin's Bay are probably of one and the same species; and "this animal not being able to endure the warm waters of the equator, could not pass from one ocean to the other unless by way of the Arctic regions."]

The continent of Australia occupies a comparatively small portion of the Pacific, while innumerable islands stud its surface many degrees on either side of the equator, of which a great number are volcanic, showing that its bed has been, and indeed actually is, the theatre of violent igneous eruptions. So great is its depth, that a line five miles long has not reached the bottom in many places; yet as the whole mass of the ocean counts for little in the total amount of terrestrial gravitation, its mean depth is but a small fraction of the radius of the globe.

The bed of the Atlantic is a long deep valley, with few mountains, or at least but few that raise their summits as islands above its surface. Its greatest breadth, including the Gulf of Mexico, is 5000 miles, and its superficial extent is about 25 millions of square miles. This sea is exceedingly deep: in 27° 26′ S. latitude and 17° 29′ W. longitude Sir James Ross found the depth to be 14,550 feet; about 450 miles west from the Cape of Good Hope it was 16,062 feet, or 332 feet more the height of Mont Blanc; and 900 miles west from St. Helena, a line of 27,600 feet did not reach the bottom, a depth which is equal to the height of some of the most elevated peaks of the Himalaya; but there is reason to believe that many parts of the ocean are still deeper.[1] A great part of the Ger-

[1] The American papers recently (Nov. 8, 1850), have given an official report of a naval officer sent to discover the existence of the False Bermudas, in which it is stated that bottom was found, north of the real Bermudas, at the enormous depth of six statute miles, or 31,500 feet.

[See statement of deep sea-soundings, obtained by Lieutenant J. G. Walsh and Commander Barron, page 137.]

man Ocean is only 93 feet deep, though on the Norwegian side, where the coast is bold, the depth is 190 fathoms.

Immense sandbanks often project from the land, which rise from great depths to within a few fathoms of the surface. Of these, the Agulhas Banks, off the Cape of Good Hope, are amongst the most remarkable; those of Newfoundland are still greater in extent: they consist of a double sandbank, which is supposed to reach the north of Scotland. The Dogger Bank, in the North Sea, and many others, are well known. According to Mr. Stevenson, one-fifth of the German Ocean is occupied by sandbanks, whose average height is 78 feet, an area equal to about one-third of Great Britain. Currents are sometimes deflected from their course by sandbanks whose tops do not come within 50 or even 100 feet of the surface. Some on the coast of Norway are surrounded by such deep water that they must be submarine table-lands. All are the resort of fish.

The pressure at great depths is enormous. In the Arctic Ocean, where the specific gravity of the water is lessened, on account of the greater proportion of fresh water produced by the melting of the ice, the pressure at the depth of a mile and a quarter is 2809 pounds on a square inch of surface; this was confirmed by Captain Scoresby, who says, in his 'Arctic Voyages,' that the wood of a boat suddenly dragged to a great depth by a whale was found, when drawn up, so saturated with water forced into its pores, that it sank in water like a stone for a year afterwards. Even sea-water is reduced in bulk from 20 to 19 solid inches at the depth of 20 miles. The compression that a whale can endure is wonderful. Many species of fish are capable of sustaining great pressure, as well as sudden changes of pressure. Divers in the pearl-fisheries exert great muscular strength, but man cannot bear the increased pressure at great depths, because his lungs are full of air, nor can he endure the diminution of it at great altitudes above the earth.

The depth to which the sun's light penetrates the ocean depends upon the transparency of the water, and cannot be less than twice the depth to which a person can see from the surface. In parts of the Arctic Ocean shells are distinctly seen at the depth of 80 fathoms; and among the West India islands, in 80 fathoms water, the bed of the sea is as clear as if seen in air; shells, corals, and sea-weeds of every hue display the tints of the rainbow.

The purest spring is not more limpid than the water of the ocean; it absorbs all the prismatic colours, except that of ultramarine, which being reflected in every direction, imparts a hue approaching the azure of the sky. The colour of the sea varies with every gleam of sunshine or passing cloud, although its true tint is always the same when seen sheltered from atmospheric influence. The reflection of a boat on the shady side is often of the clearest blue, while the surface of the water exposed to the sun is bright as burnished gold. The waters of the ocean also derive their colour from animalcules of the infusorial kind, vegetable substances, and minute particles of matter. It is white in the Gulf of Guinea, black round the Maldives; off California the Vermilion Sea is so called on account of the red colour of the infusoria it contains; the same red colour was observed by Magellan near the mouth of the river la Plata. The Persian Gulf is called the Green Sea by eastern geographers, and there is a strip of green water off the Arabian coast so distinct that a ship has been seen in green and blue water at the same time. Rapid transitions take place in the Arctic Sea, from ultramarine to olive-green, from purity to opacity. These appearances are not delusive, but constant as to place and colour; the green is produced by myriads of minute insects, which devour one another and are a prey of larger animals. The colour of clear shallow water depends upon that of its bed; over chalk or white sand it is apple-green, over yellow sand dark-green, brown or black over dark ground, and grey over mud.

The sea is supposed to have acquired its saline principle when the globe was in the act of subsiding from a gaseous state. The density of sea-water depends upon the quantity of saline matter it contains: the proportion is generally a little above 3 per cent., though it varies in different places; the ocean contains more salt in the southern than in the northern hemisphere, the Atlantic more than the Pacific. The greatest proportion of salt in the Pacific is in the parallels of 22° N. lat. and 17° S. lat.; near the equator it is less, and in the Polar Seas it is least, from the melting of the ice. The saltness varies with the seasons in these regions, and the fresh water, being lightest, is upppermost. Rain makes the surface of the sea fresher than the interior parts, and the influx of rivers renders the ocean less salt at their estuaries; the Atlantic is brackish 300 miles from the mouth of the Amazon. Deep seas are more saline than those that are shallow, and inland seas communicating with the ocean are less salt, from the rivers that flow into them; to this, however, the Mediterranean is an exception, occasioned by the great evaporation, and the influx of salt currents from the Atlantic. The water in the Straits of Gibraltar at the depth of 670 fathoms is four times as salt as that at the surface.[1]

Fresh water freezes at the temperature of 32° of Fahrenheit; the point of congelation of salt water is much lower [28° F.]. As the specific gravity of the water of the Greenland Sea is about 1·02664, it does not freeze till its temperature is reduced to 28½° of Fahrenheit, so that the saline principle preserves the sea in a liquid

[1] This anomalous result, given on the authority of Dr. Wollaston, has not been confirmed by the recent analysis of water taken near the Straits by M. Coupvent de Bois, and examined by the eminent chemist, M. Barral.

state to a much higher latitude than if it had been fresh, while it is better suited for navigation by its greater buoyancy. The healthfulness of the sea is ascribed to the mixing of the water by tides and currents which prevents the accumulation of putrescent matter.

Besides its saline ingredients, the sea contains bromine and iodine in very minute quantities, and, no doubt, portions of other substances too small to be detected by chemical analysis,[1] since it has constantly received the débris of the land and all its organised matter.

Raised by the moon and modified by the sun, the area of the ocean is elevated into great tidal waves which keep time with the attractions of these luminaries at each return to the upper and lower meridian. The water under the moon is drawn from the earth by her attraction, at the same time that she draws the earth from the water diametrically opposite to her, in both cases producing a tide of nearly equal height. The height to which the tides rise depends upon the relative positions of the sun and moon, upon their declination and distance from the earth, but much more upon local circumstances. The spring tides happen at new and full moon, consequently, twice in each lunar month, because in both cases the sun and moon are in the same meridian; for when the moon is new they are in conjunction, and when she is full they are in opposition, and in each of these positions their attraction is combined to raise the water to its greatest height; while, on the contrary, the neap or lowest tides happen when the moon is in quadrature, or 90° distant from the sun, for then they counteract each other's attraction to a certain degree.

The tides ordinarily happen twice in 24 hours, because the rotation of the globe brings the same point of the ocean twice under the meridian of the moon; but peculiar local circumstances sometimes affect the tides, so as to produce only one tide in 24 hours, while on the other hand there have been known three and even four tides in the same space of time.

As the earth revolves, a succession of tides follow one another, and are diffused over the Pacific, Indian, and Atlantic Oceans, giving birth to the tides which wash the shores of the vast continents and islands which rise above their surfaces; but in what manner these marginal tides branch off from the parent wave, science has not yet determined: we know only their course along each shore, but are unable to connect these curves with the great ridge of the tidal wave.

In the Atlantic the marginal wave travels towards the north, and impinges upon the coasts of North America and of Europe. In

---

[1] It has been recently stated by a very learned chemist, M. Malaguti, that sea-water contained silver in very minute portions.—'Comptes Rendus,' 1849–50.

the Indian Ocean it also pursues a northerly course, and finally washes the shores of Hindostan, the Bay of Bengal, and the Arabian Gulf; while in the Pacific, on the contrary, the waves diverge from the equator towards the poles — but in all they partake also of the westerly course of the moon.

Although such are the directions in which the tides unquestionably proceed along *the shores* of those seas, yet observations at islands in the open sea and towards the centres of the oceans contradict the idea of corresponding progressive waves throughout the entire area of those seas.

Upon the coasts of Britain and New Brunswick the tides are high, from the local circumstances of the coast and bottom of the sea; while in the centre of the ocean, where they are due to the action of the sun and moon only, they are remarkably small. The spring-tides rise more than 40 feet at Bristol, and in the Bay of Fundy; in Nova Scotia, they rise upwards of 50 feet; the general height in the North Atlantic is 10 or 12 feet, but in the open and deep sea they are less; and at St. Helena they are not more than 3 feet, whilst among the islands in the Pacific they are scarcely perceptible.

The mean height of the tides will be increased by a very small quantity for ages to come, in consequence of the decrease in the mean distance of the moon from the earth; the contrary effect will take place after that period has elapsed, and the moon's mean distance begins to increase again, which it will continue to do for many ages. Thus, the mean distance of the moon, and the consequent minute increase in the height of the tides, will oscillate between fixed limits for ever.[1]

The tidal wave extends to the bottom of the ocean, and moves uniformly and with great speed in very deep water, variably and slow in shallow water; the time of propagation depends on the depth of the water as well as on the nature and form of the shores. Its velocity varies inversely as the square of the depth — a law which theoretically affords the means of ascertaining the proportionate depth of the sea in different parts; it is one of the great constants of nature, and is to fluids what the pendulum is to solids—a connecting link between time and force.

The great oceanic wave that twice a-day brings the tides to our shores, has occupied a day and a half in travelling from the place where it was generated. The wave first impinges on the west coast of Ireland and England, and then passes round the north of Scotland, up the North Sea, and enters the Thames, having made the tour of Great Britain in about 18 hours.[1]

---

[1] For the reason of this secular variation in the moon's distance, see page 42 of 'The Connexion of the Physical Sciences.'

[1] For illustration of the course of tidal waves, see plates, in Johnston's Physical Atlas,' in folio.

At the equator the tide-wave follows the moon at the rate of 1000 miles an hour [through space, as the earth revolves]; it moves very slowly in the northern seas on account of the shallowness of the water; but the tides are so retarded by the form of the coasts and irregularities of the bottom of the sea, that a tide is sometimes impeded by an obstacle till a second tide reaches the same point by a different course, and the water rises to double the height it would otherwise have attained. A complete extinction of the tide takes place when a high water interferes in the same manner with a low water, as in the centre of the German Ocean—a circumstance predicted by theory, and confirmed by Captain Hewett, who was not aware that such interference existed. When two unequal tides of contrary phases meet, the greater overpowers the lesser, and the resulting height is equal to their difference; such varieties occur chiefly in channels among islands and at the estuaries of rivers. When the tide flows suddenly up a river encumbered with shoals, it checks the descent of the stream: the water spreads over the sands, and a high crested wave, called a bore, is driven with force up the channel. This occurs in the Ganges; in the Amazon, at the equinoxes, where, during three successive days, five of these destructive waves, from 12 to 15 feet high, follow one another up that river daily; and in a lesser degree in some of our own rivers.

There may be some small flow of *stream* with the oceanic tide; but that does not necessarily follow, since the tide in the open ocean is merely an alternate rise and fall of the surface; so that the wave, not the stream, follows the moon. A bird resting on the sea is not carried forward as the waves rise and fall; indeed, if so heavy a body as water were to move at the rate of 1000 miles in an hour, it would cause universal destruction, since in the most violent hurricanes the velocity of the wind hardly exceeds 100 miles an hour.

During the passage of the great tidal wave in deep water, the particles of the fluid glide for the moment over each other into a new arrangement, and then retire to their places; but this motion is extremely limited and momentary. Over shallows, however, and near the land, both the water and the waves advance during the flow of the tide, and roll on the beach.[1]

The friction of the wind combines with the tides in agitating the surface of the ocean, and, according to the theory of undulations,

---

[1] Every undulating motion consists of two distinct things—an advancing form and a molecular movement. The motion of each particle is in an ellipse lying wholly in the vertical plane, so that, after the momentary disturbance during the passage of the wave, they return to their places again.—'Theory of Waves,' by J. Scott Russell, Esq.

[For a very lucid and concise explanation of undulation or wave-motion in liquids, the reader is recommended to consult Dr. Lardner's 'Hand Books of Natural Philosophy.']

each produces its effect independently of the other; wind, however, not only raises waves, but causes a transfer of superficial water also. Attraction between the particles of air and water, as well as the pressure of the atmosphere, brings its lower stratum into adhesive contact with the surface of the sea. If the motion of the wind be parallel to the surface, there will still be friction, but the water will be smooth as a mirror; but if it be inclined, in however small a degree, a ripple will appear. The friction raises a minute wave, whose elevation protects the water beyond it from the wind, which consequently impinges on the surface at a small distance beyond; thus, each impulse, combining with the other, produces an undulation which continually advances.

Those beautiful silvery streaks on the surface of a tranquil sea, called cats'-paws by sailors, are owing to a partial deviation of the wind from a horizontal direction. The resistance of the water increases with the strength and inclination of the wind. The agitation at first extends little below the surface, but in long-continued gales even the deep water is troubled: the billows rise higher and higher, and, as the surface of the sea is driven before the wind, their "monstrous heads," impelled beyond the perpendicular, fall in wreaths of foam. Sometimes several waves overtake one another, and form a sublime and awful sea. The highest waves known are those which occur during the north-west gale off the Cape of Good Hope, aptly called by the ancient Portuguese navigators the Cape of Storms: Cape Horn also seems to be the abode of the tempest. The sublimity of the scene, united to the threatened danger, naturally leads to an over-estimate of the magnitude of the waves, which appear to rise mountains high, as they are proverbially said to do: there is, however, reason to doubt if the highest waves off the Cape of Good Hope exceed 40 feet from the hollow trough to the summit.[1] The waves are short and abrupt in small shallow seas, and on that account are more dangerous than the long rolling billows of the wide ocean.

"The sea-shore after a storm presents a scene of infinite grandeur. It exhibits the expenditure of gigantic force, which impresses the mind with the presence of elemental power as sublime as the water-fall or the thunder. Long before the waves reach the shore they may be said to feel the bottom as the water becomes shallower, for they increase in height, but diminish in length. Finally the wave becomes higher, more pointed, assumes a form of unstable equilibrium, totters, becomes crested with foam, breaks with great

[1] Dr. Scoresby's late Observations in the Atlantic, made with greater care than had been hitherto employed, appear to confirm this result.—Proceedings of British Association, 1850; in 'Athenæum,' August, 1850.

violence, and, continuing to break, is gradually lessened in bulk till it ends in a fringed margin."[1]

The waves raised by the wind are altogether independent of the tidal waves; each maintains its undisturbed course; and as the inequalities of the coasts reflect them in all directions, they modify those they encounter and offer new resistance to the wind, so that there may be three or four systems or series of coexisting waves, all going in different directions, while the individual waves of each maintain their parallelism.

The undulation called a ground-swell, occasioned by the continuance of a heavy gale, is totally different from the tossing of the billows, which is confined to the area vexed by the wind; whereas the ground-swell is rapidly transmitted through the ocean to regions far beyond the direct influence of the gale that raised it, and it continues to heave the smooth and glassy surface of the deep long after the wind and the waves are at rest. In the South Pacific, billows which must have travelled 1000 miles against the trade-wind from the seat of the storm, expend their fury on the lee side of the many coral islands which bedeck that sunny sea.[2] Thus a swell sometimes comes from a quarter in direct opposition to the wind, and occasionally from various points of the compass at the same time, producing a vast commotion even in a dead calm, without ruffling the surface. They are the heralds that point out to the mariner the distant region where the tempest has howled, and not unfrequently they are the harbingers of its approach. At the margin of the polar ice, in addition to other dangers, there is generally a swell which would be very formidable to the mariner in thick weather, did not the loud grinding noise of the ice warn him of his approach.

Heavy swells are propagated through the ocean till they gradually subside from the friction of the water, or till the undulation is checked by the resistance of land, when they roll in surf to the shore, or dash in spray and foam over the rocks. The rollers at the Cape de Verde Islands are seen at a great distance approaching like mountains. When a gale is added to a ground-swell the commotion is great and the force of the surge tremendous, tossing huge masses of rock and shaking the cliffs to their foundations. During heavy gales on the coast of Madras the surf breaks in nine fathoms water at the distance of four and even four and a half miles from the shore. The violence of the tempest is sometimes so intense as to quell the billows and scatter its surface in a heavy shower called by sailors spoon-drift. On such occasions saline particles have impregnated the air to the distance of fifty miles inland.

The force of the waves in gales of wind is tremendous; from ex-

[1] J. Scott Russell, Esq., on Waves.
[2] Beechey's Voyage to the Pacific.

periments made by Mr. Stevenson, civil engineer, on the west coast of Scotland, exposed to the whole fury of the Atlantic, it appears that the average pressure of the waves during the summer months was equal to 611 pounds weight on a square foot of surface, while in winter it was 2086 pounds, or three times as great. During the storm that took place on the 9th of March, 1845, it amounted to 6083 pounds. Now as the pressure of a wave 20 feet high not in motion is only about half a ton on a square foot, it shows how much of their force waves owe to their velocity. The rolling breakers on the cliffs on the west coast of Ireland are magnificent: Lord Adare measured some the spray of which rose as high as 150 feet.

In the Isle of Man a block which weighed about 10 stone was lifted from its place and carried inland during a north-westerly gale; and in the Hebrides a block of 42 tons weight was moved several feet by the force of the waves. The Bell Rock lighthouse in the German Ocean, though 112 feet high, is literally buried in foam and spray to the very top during ground-swells when there is no wind. On the 20th of November, 1827, the spray rose 117 feet, so that the pressure was computed by Mr. Stevenson to be nearly three tons on a square foot.

The effect of a gale descends to a comparatively small distance below the surface; the sea is probably tranquil at the depth of 200 or 300 feet; were it not so, the water would be turbid and shellfish would be destroyed. Anything that diminishes the friction of the wind smooths the surface of the sea — for example, oil or a small stream of packed ice, which suppresses even a swell. When the air is moist, its attraction for water is diminished, and consequently so is the friction; hence the sea is not so rough in rainy as in dry weather.

Currents of various extent, magnitude, and velocity disturb the tranquillity of the ocean; some of them depend upon circumstances permanent as the globe itself, others on ever-varying causes. Constant currents are produced by the combined action of the rotation of the earth, the heat of the sun, and the trade-winds; periodical currents are occasioned by tides, monsoons, and other long-continued winds; temporary currents arise from the tides, melting ice, and from every gale of some duration. A perpetual circulation is kept up in the waters of the main by these vast marine streams; they are sometimes superficial and sometimes submarine, according as their density is greater or less than that of the surrounding sea.

[Ocean currents are due to the influence of several principles or causes which are in ceaseless operation. There are other forces than those of temperature, wind, planetary attraction and rotation of the earth which contribute to keep the waters in constant circulation.

Speaking in general terms, the ocean contains 3½ per cent. of matter dissolved in it. This matter consists of all substances existing on land, capable of solution in water under the ordinary temperatures and pressures of the atmosphere. Rains fall upon the surface of the land; the water while flowing towards river beds and percolating through the earth, dissolves some of the constituents, (though in very small quantity) and carries them finally to mingle with the ocean. Sea water contains chlorides of sodium, of potassium, and of magnesium; sulphates of magnesia and of lime; carbonate of lime; iodine, and bromide of magnesium, and, perhaps, other matters in solution.

As the rivers are constantly conveying additions of saline matter into the ocean, and as none of this matter is removed by the process of evaporation which never ceases, it might be supposed that the per-centage of matter dissolved in the ocean would increase regularly, and in time, its water would become saturated, incapable of holding any more matter in solution, and then deposition would commence and at last fill up the bed of the ocean with solid matter. But the ocean and seas are inhabited by organic forms, to the nutrition and existence of which, the matters dissolved in their waters are essential. The shell-bearing mollusk, and the minute coral animal; the sponges and crustaceans, all require lime and magnesia and soda to preserve their structure and existence. Each animal has an instinct and a wisely contrived apparatus through the means of which it derives from the waters all the materials necessary for its nourishment and growth. Then, there are still higher organic structures whose nutrition is in part supplied from the ocean; to say nothing of the aquatic mammals and fishes which consume some of the substances dissolved in the sea, there are the various algæ and fuci which live upon what they derive from the water.

If we imagine for a moment, an ocean containing 3½ per cent. of matter of various kinds and in various proportions, in a state of repose or perfect equilibrium, it will be perceived that such equilibrium would be disturbed and movement set up, from the instant that the life-organs of the animals and vegetables inhabiting it, began to draw from the waters the materials, (the lime and magnesia, the iodine and bromine,) they require respectively for support. It is not to be supposed that even a single atom or molecule could be appropriated through the function of an organ of an animal or of a plant, without imposing a necessity of movement in the surrounding atoms of the mass. This movement would be imperceptible; yet when we admit that there are countless millions of organised beings in the depths of the ocean, deriving from its waters the materials of their nutrition, we cannot fail to perceive in the aggregate of their actions a force adequate to destroy equilibrium and sustain a ceaseless movement in the vast expanse of the waters of the earth. Even if the temperature of the ocean was everywhere uniform, the functions of organic life, to say nothing of the locomotion of animals in it, would prevent stagnation.

The presence of various salts does not seem to be essential to prevent the decomposition of water in the ocean. The vast bodies of fresh water which constitute the North American lakes, show that salt is not necessary to its preservation. Indeed, pure water is as unchangeable and as indestructible as the granite of the everlasting hills.

The substances constantly found dissolved in the ocean were not placed there to prevent its putrefaction. They are there for a wise purpose; and they doubtlessly play an essential part in the economy of nature; they may exert an important influence in the production of ocean currents, in connection with alternation of temperature and evaporation from the surface.

"In every department of Nature," says Lieut. Maury, "there is to be found this self-adjusting principle—this beautiful and exquisite system of compensation by which the operations of the grand machinery of the universe are maintained in the most perfect order.

"Thus we behold the sea-shells [mollusks] and animalcula in a new light. We may not now cease to regard them as beings which have little or nothing to do in maintaining the harmonies of the creation? On the contrary, do we not see in them the principles of the most admirable compensation in the system of oceanic circulation? We may even regard them as regulators, to some extent, of climates in parts of the earth far removed from their presence. There is something suggestive, both of the grand and beautiful, in the idea that while the insects of the sea are building up their coral islands in the perpetual summer of the tropics, they are also engaged in dispensing warmth to distant parts of the earth, and in mitigating the severe cold of the polar winter."]

The exchange of water between the poles and the equator affects the great currents of the ocean. Although these depend upon the same causes as the trade-winds, they differ essentially in this respect—that whereas the atmosphere is heated from below by its contact with the earth, and transmits the heat to the strata above, the sea is heated at its surface by the direct rays of the sun, which diminish the specific gravity of the upper strata, especially between the tropics, and also occasion strong and rapid evaporation, both of which causes disturb the equilibrium of the ocean. The rotation of the earth also gives the water a tendency to take an oblique direction in its flow towards the equatorial regions, as, in order to restore the equilibrium, deranged by so many circumstances, great streams perpetually descend from either pole. When these currents leave the poles, they flow towards the equator; but, before proceeding far, their motion is deflected by the diurnal rotation of the earth. At the poles they have no rotary motion; and although they gain it more and more in their progress to the equator, which revolves at the rate of 1000 miles an hour, they arrive at the tropics before they have acquired the same velocity of rotation with the inter-tropical ocean. On that account they are left behind, and consequently seem to flow in a direction contrary to the diurnal rotation of the earth. For that reason the whole surface of the ocean, for 30 degrees on each side of the equator, has an apparent tendency from east to west, which produces all the effects of a great current or stream flowing in that direction. The trade-winds, which blow constantly in one direction, combine to give this current a mean velocity of 10 or 11 miles in 24 hours.[1]

It has been supposed that the primary currents, as well as those derived from them, are subject to periodical variations of intensity occasioned by the melting of the ice at each pole alternately.

In consequence of the uninterrupted expanse of ocean in the southern hemisphere, the prevalence of westerly winds, and the tendency of the polar water towards the equator, a great oceanic current is originated in the Antarctic Sea. Driven by the prevailing winds, the waters take an easterly direction inclining to the northward, and one part sets upon the American coast, where it is divided. A small part doubles Cape Horn, while the main cold stream flows down the American shore; then turning suddenly to the west, it loses itself in the great equatorial current of the Pacific, which crosses that ocean between the parallels of 26° S. and 24° N. in a vast stream nearly 3500 miles broad. In the north this stream is interrupted by the coast of China, the Eastern Peninsula, and the islands of the Indian Archipelago; but a part forces its way between the islands, and joins the great equatorial current of the Indian Ocean, which, impelled by the S.E. trade-wind maintains a westerly course between the 10th and 20th parallels of south latitude; as it approaches the Island of Madagascar, the stream is divided; one part runs to the north-west, bends round the northern end of Madagascar, flows through the Mosambique Channel, and, being joined by the other branch, it doubles the Cape of Good Hope outside of the Agulhas Bank, and, under the name of the South Atlantic Current, it runs along the west coast of Africa to the parallel of St. Helena. There it is deflected by the coast of Guinea, and forms the Great Atlantic Equatorial Current, which flows westward and divides upon Cape St. Roque in Brazil. One branch of the stream setting southward along the continent of South America, becomes insensible before it reaches the Straits of Magellan; but an offset from it stretches directly across the Atlantic to the Cape of Good

[1] Winds are named from the points whence they blow, currents exactly the reverse. An easterly wind comes from the east; whereas an easterly current comes from the west, and flows towards the east.

Hope, having made the circuit of the South Atlantic Ocean, and keeping 150 miles outside of the Cape or Agulhas current, which runs in the opposite direction, it pursues its course into the Indian Ocean, where traces of it are met with 2000 miles from the Cape.

The principal branch of the great equatorial current takes a northerly course from off Cape St. Roque, and rushes along the coast of Brazil with such force and depth that it suffers only a temporary deflection by the powerful streams of the river Amazon and of the Orinoco. Though much weakened in passing among the West Indian islands, it acquires new strength in the Caribbean Sea. From thence, after sweeping round the Gulf of Mexico with the high temperature of 88° 52′ of Fahrenheit, it flows through the Straits of Florida, and along the North American coast to Newfoundland under the name of the Gulf-stream: it is there deflected eastward by the form of the land and the prevalent wind, and after passing Newfoundland by a current from Baffin's Bay. From the Azores it bends southward, and aided by the north-east trade rejoins the equatorial current, having made a circuit of 3800 miles with various velocity, leaving a vast loop or space of water nearly stagnant in its centre, which is thickly covered with sea-weed. The bodies of men, animals, and plants of unknown appearance, brought to the Azores by this stream, suggested to Columbus the idea of land beyond the Western Ocean, and thus led to the discovery of America. The Gulf-stream is more salt, warmer, and of a deeper blue than the rest of the ocean, till it reaches Newfoundland, where it becomes turbid from the shallowness of that part of the sea. Its greatest velocity is 78 miles a-day soon after leaving the Florida Strait; and its breadth increases with its distance from the strait until the warm water spreads over a large surface of the ocean. An important branch leaves the current near Newfoundland, setting towards Britain and Norway; which is again subdivided into many branches, whose origin is recognised by their greater warmth, even at the edge of perpetual ice in the Polar Ocean; and in consequence of some of these branches the Spitzbergen Sea is 6 or 7 degrees warmer at the depth of 200 fathoms than at its surface. Though the warmth of the Gulf-stream diminishes as it goes north, Lieut. Maury says "that the quantity of heat which it spreads over the Atlantic in a winter's day would be sufficient to raise the whole atmosphere that covers France and Great Britain from the freezing point to summer heat; and it really is the cause of the mildness and of the damp of Ireland and the south of England."

These oceanic streams exceed all the rivers in the world in breadth and depth as well as length. The equatorial current in the Atlantic is 160 miles broad off the coast of Africa, and towards its mid-course across the Atlantic its width becomes nearly equal to the length of Great Britain: but as it then sends off a branch to the

N.W., it is diminished to 200 miles before reaching the coast of Brazil. The depth of this great stream is unknown; but the Brazilian branch must be very profound, since it is not deflected by the river La Plata, which crosses it with so strong a current that its fresh muddy waters are perceptible 500 miles from its mouth. When currents pass over banks and shoals, the colder water rises to the surface and gives warning of the danger.

In summer, the great north polar current coming along the coasts of Greenland and Labrador, together with the current from Davis's Straits, brings icebergs to the margin of the Gulf-stream. The difference between the temperatures of these two oceanic streams brought into contact is the cause of the dense fogs that brood over the banks of Newfoundland. The north polar current runs inside of the Gulf-stream, along the coast of North America to Florida, and beyond it—since it sends an under-current into the Caribbean Sea. Counter-currents on the surface are of such frequent occurrence that there is scarcely a strait joining two seas that does not furnish an example—a current running in along one shore, and a counter-current running out along the other. One of the most remarkable occurs in the Atlantic: it begins off the coast of France, and, after sending a mass of water into the Mediterranean, it holds a southerly direction at some distance from the continent of Africa; till, after passing Cape Mesurada, it flows rapidly for 1000 miles east to the Bight of Biafra in immediate contact with the equatorial current, running with great velocity in the opposite direction, and seems to merge in it at last.

Periodical currents are frequent in the eastern seas: one flows into the Red Sea from October to May, and out of it from May to October. In the Persian Gulf this order is reversed; in the Indian Ocean and China Sea the waters are driven alternately backwards and forwards by the monsoons. It is the south-westerly monsoon that causes inundations in the Ganges, and a tremendous surf on the coast of Coromandel. The tides also produce periodical currents on the coasts and in straits, the water running in one direction during the flood, and the contrary way in the ebb. The Roost of Sumburgh, at the southern promontory of Shetland, runs at the rate of 15 miles an hour; indeed, the strongest tidal currents known are among the Orkney and Shetland islands; their great velocity arises from local circumstances. Currents in the wide ocean move at the rate of from one to three miles an hour, but the velocity is less at the margin and bottom of the stream from friction.

Whirlpools are produced by opposing winds and tides; the whirlpool of Maelström, on the coast of Norway, is occasioned by the meeting of tidal currents round the islands of Lofoden and Moskoe; it is a mile and a half in diameter, and so violent that its roar is heard at the distance of several leagues.

Although with winds, tides, and currents, it might seem that the ocean is ever in motion, yet in the equatorial regions, far from land, dead calms prevail; the sea is of the most perfect stillness day after day; partaking of the universal quiet, and heaving its low flat waves in noiseless and regular periods as if nature were asleep.

The safety and length of a voyage depend upon the skill with which a seaman avails himself of the set of the different currents, and the direction of the permanent and periodical winds; it is frequently shortened by following a very circuitous track to take advantage of them if favourable, or to avoid them if unfavourable. From Acapulco, in Mexico, across the Pacific to Manilla or Canton, the trade-wind and the equatorial current are so favourable that the voyage is accomplished in 50 or 60 days; whereas, in returning, 90 or 100 are required. Within the Antilles navigation is so difficult from winds and currents, that a vessel, going from Jamaica to the lesser Antilles, cannot sail directly across the Caribbean Sea, but must go round about through the windward passage between Cuba and Haiti to the ocean; nearly as many weeks are requisite to accomplish this voyage as it takes days to return. On account of the prevalence of westerly winds in the North Atlantic, the voyage from Europe to the United States is longer than that from the latter to Europe; but the Gulf-stream is avoided in the outward voyage, [i e. from Europe to the United States,] because it would lengthen the time by a fortnight. Ships going to the West Indies, Central or South America, from Europe, generally make the Canary Islands in order to fall in with the N.E. trade-winds.

The passage to the Cape of Good Hope from the British Channel may be undertaken at any season, and is accomplished in 50 or 60 days; but it is necessary to regulate the voyage from the Cape to India and China according to the seasons of the monsoons. There are various courses adopted for that purpose, but all of them pass through the very focus of the hurricane district, which includes the islands of Rodriguez, the Mauritius and Bourbon, and extends from Madagascar to the island of Timor.

The extensive deposits of coal discovered in Australia, New Zealand, in the British settlement at Labuan, and on the neighbouring shores of Borneo, and in Vancouver's Island, will be the means of increasing the steam navigation of the Pacific, and shortening the voyages upon that ocean.

Sea-water is a bad conductor of heat, therefore the temperature of the ocean is less liable to sudden changes than the atmosphere; the influence of the seasons is imperceptible at the depth of 300 feet; and as light probably does not penetrate lower than 700 feet, the heat of the sun cannot affect the bottom of a deep sea. It has been established beyond a doubt that in all parts of the ocean the water has a constant temperature of about 39°·5 of Fahrenheit, at a certain depth, depending on the latitude. At the equator the stratum of water at that temperature is at the depth of 7200 feet; from thence it gradually rises till it comes to the surface in S. lat. 56° 26′, where the water has the temperature of 39°·5 at all depths; it then gradually descends till S. lat. 70°, where it is 4500 feet below the surface. In going north from the equator the same law is observed. Hence, with regard to temperature, there are three regions in the ocean; one equatorial and two polar. In the equatorial region the temperature of the water at the surface of the ocean is 80° of Fahrenheit, therefore higher than that of the stratum of 39°·5; while in the polar regions it is lower. Thus the surface of the stratum of constant temperature is a curve which begins at the depth of 4500 feet in the southern basin, from whence it gradually rises to the surface in S. lat. 56° 26′; it then sweeps down to 7200 feet at the equator, and rises up again to the surface in the corresponding northern latitude, from whence it descends again to the depth of 4500 feet in the northern basin.

The temperature of the surface of the ocean decreases from the equator to the poles. For 10 degrees on each side of the line the maximum is 80° of Fahrenheit, and remarkably stable; from thence to each tropic the decrease does not exceed 3°·7. The tropical temperature would be greater were it not for the currents, because the surface reflects much fewer of the sun's rays which fall on it directly, than in higher latitudes where they fall obliquely. In the torrid zone the surface of the sea is about 3°·5 of Fahrenheit warmer than the air above it; because the polar winds, and the great evaporation which absorbs the heat, prevent equilibrium; and as a great mass of water is slow in following the changes in the atmosphere, the vicissitude of day and night has little influence, whereas in the temperate zones it is perceptible.

The line of maximum temperature, or that which passes through all the points of greatest heat in the ocean, is very irregular, and does not coincide with the terrestrial equator; six-tenths of its extent lies on an average 5° to the north of it, and the remainder runs at a mean distance of 3° on its southern side. It cuts the terrestrial equator in the middle of the Pacific Ocean in 21° E. longitude in passing from the northern to the southern hemisphere, and again between Sumatra and the peninsula of Malacca in returning from the southern to the northern. Its maximum temperature in the Pacific is 88°·5 of Fahrenheit on the northern shores of New Guinea, where it touches the terrestrial equator, and its highest temperature in the Atlantic, which is exactly the same, lies in the Gulf of Mexico, which furnishes the warm water of the Gulf-stream.

The superficial water of the Pacific is much cooled on the east by the Antarctic current; it sends a cold stream along the coasts of Chile and Peru, which has great influence on the climate of both countries; it was first observed by Baron Humboldt, and is known as Humboldt's current.

It is more than 14° colder than the adjacent ocean, and renders the air 11° cooler than the surrounding atmosphere.

In the Indian Ocean the highest temperature of the surface-water (87°·4) is in the Arabian Sea, between the Strait of Bab-el-Mandeb and the coast of Hindostan; it decreases regularly from south to north in the Red Sea.

The superficial temperature diminishes from the tropics with the increase of the latitude more rapidly in the southern than in the northern hemisphere, till towards the poles the sea is never free from ice. In the Arctic Ocean the surface is at the freezing point even in summer; and during the eight winter months a continuous body of ice extends in every direction from the pole, filling the area of a circle between 3000 and 4000 miles in diameter. The outline of this circle, though subject to partial variations, is found to be nearly similar at the same season of each succeeding year, yet there are periodical changes in the polar ice which are renewed after a series of years. The freezing process itself is a bar to the unlimited increase of the oceanic ice. Fresh water congeals at the temperature of 32° of Fahrenheit, but sea-water must be reduced to 28°·5 before it deposits its salt, and begins to freeze: the salt thus set free, and the heat given out, retard the process of congelation more and more below.

The ice from the North Pole comes so far south in winter as to render the coast of Newfoundland inaccessible: it envelopes Greenland, sometimes even Iceland, and always invests Spitzbergen and Nova Zembla. As the sun comes north the ice breaks up into enormous masses of what is called packed ice. In the year 1806 Captain Scoresby forced his ship through 250 miles of packed ice, in imminent danger, until he reached the parallel of 81° 50′, his nearest approach to the pole: the Frozen Ocean is rarely navigable so far.

In the year 1827, Sir Edward Parry arrived at the latitude of 82° 45′, which he accomplished by dragging a boat over fields of ice, but he was obliged to abandon the bold and hazardous attempt to reach the pole, because the current drifted the ice southward more rapidly than he could travel over it to the north.

The following considerations have induced some persons to believe that there is sea instead of land at the north pole. The average latitude of the northern shores of the continent is 70°, so that the Arctic Ocean is a circle whose diameter is 2400 geographical miles, and its circumference 7200. On the Asiatic side of this sea are Nova Zembla and the New Siberian islands, each extending to about 76° N. latitude. On the European and American sides are Spitzbergen, extending to 80°, and a part of Old Greenland, whose northern termination is unknown. Facing America is a large island — Melville Island — with some others not extending so far north as those mentioned; consequently all of them may be considered continental islands. As there are no large islands very far from land in the other great oceans, there is reason to presume that the same structure may prevail here also, and consequently it may be open sea at the north pole. Possibly also it may be free from ice, for Admiral Wrangel found a wide and open sea, free from ice and navigable, beginning 16 miles north of the island of Kotelnoi, and extending to the meridian of Cape Jackan. In fine summers the ice suddenly clears away and leaves an open channel of sea along the western coast of Spitzbergen from 60 to 150 miles wide, reaching to 80° or even to 80½° N. latitude, probably owing to warm currents from low latitudes. It was through this channel that captain Scoresby made his nearest approach to the pole. A direct course from the Thames, across the pole to Behring's Straits, is 3570 geographical miles, while by Lancaster Sound it is 4660 miles. The Russians would be saved a voyage of 18,800 geographical miles could they go across the pole and through Behring's Straits to their North American settlements, instead of going by Cape Horn.

Floating fields of ice, 20 or 30 miles in diameter, are frequent in the Arctic Ocean: sometimes they extend 100 miles, so closely packed together that no opening is left between them; their thickness, which varies from 10 to 40 feet, is not seen, as there is at least two-thirds of the mass below water. Sometimes these fields, many thousand millions of tons in weight, acquire a rotatory motion of great velocity, dashing against one another with a tremendous collision. Packed ice always has a tendency to drift southwards, even in the calmest weather; and in their progress the ice-fields are rent in pieces by the swell of the sea. It is computed that 20,000 square miles of drift-ice are annually brought by the current along the coast of Greenland to Cape Farewell. In stormy weather the fields and streams of ice are covered with haze and spray from constant tremendous concussions; yet our seamen, undismayed by the appalling danger, boldly steer their ships amidst this hideous and discordant tumult.

Huge icebergs, and masses detached from the glaciers, which extend from the Arctic lands into the sea, especially in Baffin's Bay, are drifted southwards 2000 miles from their origin to melt in the Atlantic, where they cool the water sensibly for 30 or 40 miles around, and the air to a much greater distance. They vary from a few yards to miles in circumference, and rise hundreds of feet above the surface. Seven hundred such masses have been seen at once in the polar basin. When there is a swell, the loose ice dashing against them raises the spray to their very summits; and as they waste away they occasionally lose equilibrium and roll over, causing a swell which breaks up the neighbouring field-ice; the commotion spreads far and wide, and the uproar resounds like thunder.

Icebergs have the appearance of chalk-cliffs with a glittering surface and emerald green fractures: pools of water of azure-blue lie on their surface or fall in cascades into the sea. The field-ice also, and the masses that are heaped up on its surface, are extremely beautiful from the vividness and contrast of their colouring. A peculiar blackness in the atmosphere around a bright haze at the horizon indicates their position in a fog, and their place and character are shown at night by the reflection of the snow-light on the horizon. An experienced seaman can readily distinguish by the blink, as it is termed, whether the ice is newly formed, heavy, compact, or open. The blink or snow-light of field-ice is the most lucid, and is tinged yellow; of packed ice it is pure white: ice newly formed has a greyish blink, and a deep yellow tint indicates snow on land.

Icebergs come to a lower latitude by 10° from the south pole than from the north, and appear to be larger; they have been seen near the Cape of Good Hope, and are often of great size; one observed by Captain Dumont d'Urville was 13 miles long, with perpendicular sides 100 feet high: they are less varied than those on the northern seas; a tabular form is the most prevalent. The discovery ships under the command of Sir James Ross met with multitudes with flat surfaces, bounded by perpendicular cliffs on every side, from 100 to 180 feet high, sometimes several miles in circumference. Their size must have been enormous, since more than two-thirds of their mass was below water. From the condensation of moisture in the surrounding air by their cold, they are often enveloped in mist, which makes them still more formidable to navigators. On one occasion they fell in with a chain of stupendous bergs close to one another, extending farther than the eye could reach even from the mast-head. Packed ice too is often in immense quantities: these ships forced their way through a pack 1000 miles broad, often under the most appalling circumstances. It generally consists of smaller pieces than the packs in the comparatively tranquil North Polar seas, where they are often several miles in diameter, and where fields of ice extend beyond the reach of vision. The Antarctic Ocean, on the contrary, is almost always agitated; there is a perpetual swell, and terrific storms are common, which break up the ice and render navigation perilous. The floe pieces are rarely a quarter of a mile in circumference, and generally much smaller.

A more dreadful situation can hardly be imagined than that of ships beset during a tempest in a dense pack of ice in a dark night, thick fog, and drifting snow, with the spray beating perpetually over the decks, and freezing instantaneously. Sir James Ross's own words can alone give an idea of the terrors of one of the many gales which the two ships under his command encountered :—" Soon after midnight our ships were involved in an ocean of rolling fragments of ice, hard as floating rocks of granite, which were dashed against them by the waves with so much violence, that their masts quivered as if they would fall at every successive blow; and the destruction of the ships seemed inevitable from the tremendous shocks they received. In the early part of the storm the rudder of the Erebus was so damaged as to be no longer of any use; and about the same time I was informed by signal that the Terror's was completely destroyed and nearly torn away from the stern-post. Hour passed away after hour without the least mitigation of the awful circumstances in which we were placed. The loud crashing noise of the straining and working of the timbers and decks, as they were driven against some of the heavier pieces of ice, which all the exertions of our people could not prevent, was sufficient to fill the stoutest heart, that was not supported by trust in Him who controls all events, with dismay; and I should commit an act of injustice to my companions if I did not express my admiration of their conduct on this trying occasion. Throughout a period of 28 hours, during any one of which there appeared to be little hope that we should live to see another, the coolness, steady obedience, and untiring exertions of each individual were every way worthy of British seamen.

"The storm gained its height at 2 P.M., when the barometer stood at 28·40 inches, and after that time began to rise. Although we had been forced many miles deeper into the pack, we could not perceive that the swell had at all subsided, our ships still rolling and groaning amidst the heavy fragments of crushing bergs, over which the ocean rolled its mountainous waves, throwing huge masses upon one another, and then again burying them deep beneath its foaming waters, dashing and grinding them together with fearful violence." For three successive years were these dangers encountered during this bold and hazardous enterprise. It was impossible to pass the winter in these southern seas, but in the various expeditions to the North Polar Ocean the ships were frozen fast in boundless fields of ice for many months, ready to continue their perilous voyage as soon as the late and short summer should break up the ice. The stillness and dead silence of these sunless islands and frozen seas was strongly contrasted with the wild tumult of the floating ice, through which they had to tread.

The ocean is one mass of water, which, entering into the interior of the continents, has formed seas and gulfs of great magnitude, which afford easy and rapid means of communication, while they temper the climates of the widely expanding continents.

The inland seas communicating with the Atlantic are larger, and penetrate more deeply into the continents, than those connected with the great ocean; a circumstance which gives a coast of 48,000 miles to the former, while that of the great ocean is only 44,000. Most of these internal seas have extensive river domains, so that by inland navigation the Atlantic virtually enters into the deepest recesses of the land, brings remote regions into contact, and improves the condition of the less cultivated races of mankind by commercial intercourse with those that are more civilized.

The Baltic, which occupies 125,000 square miles in the centre of northern Europe, is one of the most important of the inland seas connected with the Atlantic, and, although inferior to the others in size, the drainage of more than a fifth of Europe flows into it. Only about a fourth part of the boundary of its enormous basin of 900,000 square miles is mountainous; and so many navigable rivers flow into it from the watershed of the great European plain, that its waters are one-fifth less salt than those of the Atlantic: it receives at least 250 streams. Its depth nowhere exceeds 167 fathoms,[1] and generally it is not more than 40 or 50. From that cause, together with its freshness and northern latitude, the Baltic is frozen five months in the year. From the flatness of the greater part of the adjacent country, the climate of the Baltic is subject to influences coming from regions far beyond the limits of its river-basin. The winds from the Atlantic bring warmth and moisture, which, condensed by the cold blasts from the Arctic plains, falls in rain in summer, and deep snow in winter, which also makes the sea less salt. The tides are imperceptible; but the waters of the Baltic occasionally rise more than three feet above their usual level from some unknown cause—possibly from oscillations in its bed, or from changes of atmospheric pressure.

The Black Sea, which penetrates most deeply into the continent of all the seas in question, has, together with the Sea of Azov, an area of 190,000 square miles: it was at a remote period probably united with the Caspian Lake, their united waters covering all the steppe of Astracan. It receives some of the largest European rivers, and drains about 950,000 square miles, consequently its waters are brackish and freeze on its northern shores in winter. It is very deep, no bottom having been reached with a line of 140 fathoms: on the melting of the snow, such a body of water is poured into it by the great European rivers that a rapid current is produced, which sets along the western shore from the mouth of the Dnieper to the Bosphorus.

Of all the branches of the Atlantic that enter deeply into the bosom of the land, the Mediterranean is the largest and most beautiful, covering with its dark-blue waters more than 760,000 square miles. Situate in a comparatively low latitude, exposed to the heat

[1] By Captain Albrecht's soundings.

of the African deserts on the south, and sheltered on the north by the Alps, the evaporation is great; on that account the water of the Mediterranean is salter than that of the ocean, and for the same reason the temperature at its surface is $3\frac{1}{2}°$ degrees of Fahrenheit higher than that of the Atlantic; it does not decrease so rapidly downwards as in tropical seas, and it becomes constant at depths of from 340 to 1000 fathoms, according to the situations.[1] Although its own river domain is only 250,000 square miles, the constant current that sets in through the Dardanelles brings a great part of the drainage of the Black Sea, so that it is really fed by the melted snow and rivers from the Caucasus, Asia Minor, Abyssinia, the Atlas, and the Alps. Yet the quantity of water that flows into the Mediterranean from the Atlantic, by the superficial current in the Straits of Gibraltar, exceeds that which goes out by the inferior currents.

Near Alexandria the surface of this sea is 26 feet 6 inches lower than the level of the Red Sea at Suez at low water, and about 30 feet lower at high water.[2]

On the shore of Cephalonia there is a cavity in the rocks, into which the sea has been flowing for ages.[3]

The Mediterranean is divided into two basins by a shallow that runs from Cape Bon on the African coast to the Strait of Messina, on each side of which the water is exceedingly deep, and said to be unfathomable in some parts. M. Bérard has sounded to the depth of more than 1000 fathoms in several places without reaching the bottom. At Nice, within a few yards of the shore, it is nearly 700 fathoms deep; and Captain Smyth, R.N., ascertained the depth to be 960 fathoms between Gibraltar and Ceuta. This sea is not absolutely without tides; in the Adriatic they rise five feet in the port of Venice, and at the Great Syrtis to five feet at new and full moon, at Naples about 12 inches, but in most other places they are scarcely

[1] The anomaly of the waters of the Mediterranean being at a higher temperature at great depths than in the ocean, is explained by the existence of a constant current of heated water setting towards the Atlantic, preventing the entrance of the cold polar current to replace the upper one which enters the Mediterranean from the ocean, through the Straits of Gibraltar.

It may be regarded as a general rule, that the temperature of all inland seas, at great depths, represents the mean temperature of the earth in the latitudes where they are situated; whilst in the ocean, the low temperature at the bottom, in every latitude, is produced by the cold currents setting eternally from the polar regions, and which maintain the water at an almost constant temperature, that of its maximum density, 39° Fahrenheit.

[2] By the measurement of M. Lepère during the French expedition to Egypt. It would appear, however, from surveys recently executed, that the difference of level between the two seas, if any, is very trifling.

[3] Proceedings of the Royal Geological Society, vol. ii. p. 210

perceptible. The surface is traversed by various currents, two of which, opposing one another, occasion the celebrated whirlpool of Charybdis, whose terrors were much diminished by the earthquake of 1783. Its bed is subject to violent volcanic paroxysms, and its surface is studded with islands of all sizes, from the magnificent kingdom of Sicily to mere barren rocks — some actively volcanic, others of volcanic formation, and many of the secondary geological period.

Various parts of its coasts are in a state of great instability; in some places they have sunk down and risen again more than once within the historical period.

Far to the north the Atlantic penetrates the American continent by Davis's Straits, and spreads out into Baffin's Bay, twice the size of the Baltic, very deep, and subject to all the rigours of an arctic winter — the very storehouse of icebergs — the abode of the walrus and the whale. Hudson's Bay, though without the Arctic Circle, is but little less dreary.

Very different is the character of those vast seas where the Atlantic comes "cranking in" between the northern and southern continents of America. The surface of the sea in Baffin's Bay is seldom above the freezing point; here, on the contrary, it is always 88°·5 of Fahrenheit, while the Atlantic Ocean in the same latitude is not above 77° or 78°. Of that huge mass of water, partially separated from the Atlantic by a long line of islands and banks, the Caribbean Sea is the largest; it is as long from east to west as the distance between Great Britain and Newfoundland, and occupies a million of square miles. Its depth in many places is very great, and its water is limpid. The Gulf of Mexico, fed by the Mississippi, one of the greatest of rivers, is more than half its size, or about 800,000 square miles, so that the whole forms a sea of great magnitude. Its shores, and the shores of the numerous islands, are dangerous from shoals and coral-reefs, but the interior of these seas is not. The trade-winds prevail there; they are subject to severe northern gales, and some parts are occasionally visited by tremendous hurricanes.

By the levelling across the Isthmus of Panamá by Mr. Lloyd, in 1828, the mean height of the Pacific above that of the Atlantic was found to be about three feet. The rise of the tide on the Atlantic side does not exceed two feet, while at Panamá it is more than eighteen; and it is high water at the same time on both sides of the Isthmus.

The Pacific does not penetrate the land in the same manner that the Atlantic does the continent of Europe. The Red Sea and Persian Gulf are joined to it by very narrow straits; but almost all the internal seas on the eastern coast of Asia, except the Yellow Sea, are great gulfs shut in by islands, like the Caribbean Sea and the

Gulf of Mexico, to which the China Sea, the Sea of Japan, and that of Okhotsk are perfectly analogous.

The set of the great oceanic currents has scooped out and indented the southern and eastern coasts of the Asiatic continent into enormous bays and gulfs, and has separated large portions of the land, which now remain as islands—a process which probably has been ncreased by the submarine fires extending along the eastern coast rom the equator nearly to the Arctic Circle.

The perpetual agitation of the ocean by winds, tides, and currents is continually, but slowly, changing the form and position of the land—steadily producing those vicissitudes on the surface of the earth to which it has been subject for ages, and to which it will assuredly be liable in all time to come.

CHAPTER XVII.

Springs—Basins of the Ocean—Origin, Course, and Heads of Rivers—Hydraulic Systems of Europe—African Rivers—the Nile, Niger, &c.

THE vapour which rises invisibly from the land and water ascends in the atmosphere till it is condensed by the cold into clouds, which restore it again to the earth in the form of rain, hail, and snow; hence there is probably not a drop of water on the globe that has not been borne on the wings of the wind. Part of this moisture restored to the earth is re-absorbed by the air, part supplies the wants of animal and vegetable life, a portion is carried off by the streams, and the remaining part penetrates through porous soils till it arrives at a stratum impervious to water, where it accumulates in subterranean lakes often of great extent. The mountains receive the greatest portion of the aërial moisture, and, from the many alternations of permeable and impermeable strata they contain, a complete system of reservoirs is formed in them, which, continually overflowing, form perennial springs at different elevations, which unite and run down their sides in incipient rivers. A great portion of the water at these high levels penetrates the earth till it comes to an impermeable stratum below the plains, where it collects in a sheet, and is forced by hydraulic pressure to rise in springs, through cracks in the ground, to the surface. In this manner the water which falls on hills and mountains is carried through highly-inclined strata to great depths, and even below the bed of the ocean, in many parts of which there are springs of fresh water. In boring Artesian wells the water often rushes up with such impetuosity by the hydrostatic pressure as to form jets 40 or 50 feet high. In this operation several successive reservoirs have been met with; at St. Ouen, near Paris, five sheets of water were found; the water in the first four not being good, the operation was continued to a greater depth; it consists merely in boring a hole of small diameter, and lining it with a metallic tube. It rarely happens that water may not be procured in this way; and as the substratum in many parts of deserts is an argillaceous marl, it is probable that Artesian wells might be bored with success in the most arid regions.

A spring will be intermittent when it issues from an opening in the side of a reservoir fed from above, if the supply be not equal to the waste, for the water will sink below the opening, and the spring will stop till the reservoir is replenished. Few springs give the same quantity of water at all times; they also vary much in the quantity of foreign matter they contain. Mountain-springs are generally very pure: the carbonic acid gas almost always found in them escapes into the atmosphere, and their earthy matter is deposited as they run along, so that river-water from such sources is soft, while wells and springs in the plains are hard, and more or less mineral.

The water of springs takes its temperature from that of the strata through which it passes; mountain-springs are cold, but if the water has penetrated deep into the earth, it requires a temperature depending on that circumstance.

The temperature of the surface of the earth varies with the seasons to a certain depth, where it becomes permanent and equal to the mean annual temperature of the air above. It is evident that the depth at which this stratum of invariable temperature lies must vary with the latitude. At the equator the effect of the seasons is imperceptible at the depth of a foot below the surface: between the parallels of 40° and 52° the temperature of the ground in Europe is constant at the depth of from 55 to 60 feet: and in the high Arctic regions the soil is perpetually frozen a foot below the surface. Now, in every part of the world where experiments have been made, the temperature of the earth increases with the depth below the constant stratum at the rate of 1° of Fahrenheit for every 50 or 60 feet of perpendicular depth; hence, should the increase continue to follow the same ratio, even granite must be in fusion at little more than five miles below the surface. In Siberia the stratum of frozen earth is some hundred feet thick, but below that the increase of heat with the depth is three times as rapid as in Europe. The temperature of springs must therefore depend on the depth to which the water has penetrated before it has been forced to the surface, either by the hydraulic pressure of water at higher levels or by steam. If it never goes below the stratum of invariable temperature, the heat of the spring will vary with the seasons, more or less, according to the depth below the surface: should the water come from the constant stratum itself, its temperature will be invariable; and if from below it, the heat will be in proportion to the depth to which it has penetrated. Thus, there may be hot and even boiling springs hundreds of miles distant from volcanic action and volcanic strata, of which there are many examples, though they are more frequent in volcanic countries and those subject to earthquakes. The temperature of hot springs is very constant, and that of boiling springs has remained unchanged for ages: shocks of earthquakes sometimes affect the temperature, and have even stopped them altogether. Jets of steam of high tension are frequent in volcanic countries, as in Iceland.

Both hot and cold water dissolves and combines with many of the mineral substances it meets with in the earth, and comes to the surface from great depths as medicinal springs, containing various ingredients. So numerous are they that in the Austrian dominions alone there are 1500; and few countries of any extent are destitute of them. They contain hydro-sulphuric and carbonic acids, sulphur, iron, magnesia, and other substances. Boiling springs deposit silex, as in Iceland and in the Azores; and others of lower temperature deposit carbonate of lime in great quantities all over the world. Springs of pure brine are rare; those in Cheshire are rich in salt, and have flowed unchanged 1000 years, a proof of the tranquil state of that part of the globe. Many substances that lie beyond our reach are brought to the surface by springs, as naphtha, petroleum, and boracic acid; petroleum is particularly abundant in Persia, and numberless springs and lakes of it surround some parts of the Caspian Sea. It is found in immense quantities in various parts of the world.

RIVERS.

Rivers have had a greater influence on the location and fortunes of the human race than almost any other physical cause, and, since their velocity has been overcome by steam-navigation, they have become the highway of the nations.

They frequently rise in lakes, which they unite with the sea; in other instances they spring from small elevations in the plains, from perennial sources in the mountains, alpine lakes, melted snow and glaciers; but the everlasting storehouses of the mightiest floods are the ice-clad mountains of table-lands.

Rivers are constantly increased, in descending the mountains and traversing the plains, by tributaries, till at last they flow into the ocean, their ultimate destination and remote origin. "All rivers run into the sea, yet the sea is not full," because it gives in evaporation an equivalent for what it receives.

The Atlantic, the Arctic, and the Pacific Oceans are directly or indirectly the recipients of all the rivers, therefore their basins are bounded by the principal watersheds of the continents; for the basin of a sea or ocean does not mean only the bed actually occupied by the water, but comprehends also all the land drained by the rivers which fall into it, and is bounded by an imaginary line passing through all their sources. These lines generally run through the elevated parts of a country that divide the streams which flow in one direction from those that flow in another. But the watershed does not coincide in all cases with mountain-crests of great elevation, as the mere convexity of a plain is often sufficient to throw the streams into different directions.

From the peculiar structure of the high land and mountain-chains, by far the greater number of important rivers on the globe flow into the ocean in an easterly direction, those which flow to the south and north being the next in size, while those that flow in a westerly direction are small and unimportant.

The course of all rivers is changed when they pass from one geological formation to another, or by dislocations of the strata: the sudden deviations in their directions are generally owing to these circumstances.

None of the European rivers flowing directly into the Atlantic exceed the fourth or fifth magnitude, except the Rhine; the rest of the principal streams come to it indirectly through the Baltic, the Black Sea, and the Mediterranean. It nevertheless receives nearly half the waters of the old continent, and almost all of the new, because the Andes and Rocky Mountains, which form the watershed of the American continent, lie along its western side, and the rivers which rise on their western slopes flow to the east, whilst the Alleghanies are tributaries to the Mississippi, which comes indirectly into the Atlantic by the Gulf of Mexico.

The Arctic Ocean drains the high northern latitudes of America, and receives those magnificent Siberian rivers that originate in the Altaï range from the Steppe of the Kerghis to the extremity of Kamtchatka, as well as the very inferior streams of North European Russia. The running waters of the rest of the world flow into the Pacific. The Caspian and Lake Aral are mere salt-water lakes, which receive rivers, but emit none. However, nearly one-half of all the running water in Europe falls into the Black Sea and the Caspian.

Mountain torrents gradually lose velocity in their descent to the low lands by friction, and when they enter the plains their course becomes still more gentle, and their depth greater. A slope of one foot in 200 prevents a river from being navigable, and a greater inclination forms a rapid or cataract. The speed, however, does not depend entirely upon the slope, but also upon the height of the source of the river, and the pressure of the body of water in the upper part of its course; consequently, under the same circumstances, large rivers run faster than small, but in each individual stream the velocity is perpetually varying with the form of the banks, the winding of the course, and the changes in the width of the channel. The Rhone, one of the most rapid European rivers, has a declivity of one foot in 2620, and flows at the rate of 120 feet in a minute; the sluggish rivers in Flanders have only one-half that velocity. The Danube, the Tigris, and the Indus are among the most rapid of the large rivers. In flat countries rivers are generally more meandering, and thus they afford a greater amount of irrigation; the windings of the Vistula are nearly equal to nine-tenths of its direct course from its source to its mouth.

When one river falls into another, the depth and velocity are increased, but not always proportionally to the width of the channel, which sometimes even becomes less, as at the junction of the Ohio with the Mississippi. When the angle of junction is very obtuse, and the velocity of the tributary stream great, it sometimes forces the waters of its primary to recede a short distance. The Arve, swollen by a freshet, occasionally drives the water of the Rhone back into the Lake of Geneva; and it once happened that the force was so great as to make the mill-wheels revolve in a contrary direction.

Streams sometimes suddenly vanish, and after flowing underground to some distance reappear at the surface, as in Derbyshire. Instances have occurred of rivers suddenly stopping in their course for some hours, and leaving their channels dry. On the 56th of November, 1838, the water failed so completely in the Clyde, Nith, and Teviot, that the mills were stopped eight hours in the lower part of their streams. The cause was the coincidence of a gale of wind and a strong frost, which congealed the water near their sources. Exactly the contrary happens in the Siberian rivers, which flow from south to north over so many hundreds of miles; the upper parts are thawed, while the lower are still frozen, and the water, not finding an outlet, inundates the country.

The alluvial soil carried down by streams is gradually deposited as their velocity diminishes; and if they are subject to inundations, and the coast flat, it forms deltas at their mouths; there they generally divide into branches, which often join again, or are united by transverse channels, so that a labyrinth of streams and islands is formed. Deltas are sometimes found in the interior of the continents at the junction of rivers, exactly similar to those on the ocean, though less extensive: deltas are said to be maritime, lacustrine, or fluviatile, according as the stream that forms them falls into the sea, a lake, or another river.

Tides flow up rivers to a great distance, and to a height far above the level of the sea: the tide is perceptible in the river of the Amazon 576 miles from its mouth, and it ascends 255 miles in the Orinoco.

In the temperate zones rivers are subject to floods from autumnal rains, and the melting of the snow, especially on mountain-ranges. The Po, for example, spreads desolation far and wide over the plains of Lombardy; but these torrents are as variable in their recurrence and extent as the climate which produces them. The inundations of the rivers in the torrid zone, on the contrary, occur with a regularity peculiar to a region in which meteoric phenomena are uniform in all their changes. These floods are due to the periodical rains, which, in tropical countries, follow the cessation of the trade-winds after the vernal equinox and at the turn of the monsoons, and are thus dependent on the declination of the sun, the immediate cause of all these variations. The melting of the snow no doubt adds greatly to the floods of the tropical rivers which rise in high mountain-chains, but it is only an accessary circumstance; for although the snow-water from the Himalaya swells the streams considerably before the rains begin, yet the principal effect is owing to the latter, as the southern face of the Himalaya is not beyond the influence of the monsoon, and the consequent periodical rains, which besides prevail all over the plains of India traversed by the great rivers and their tributaries.

Under like circumstances, the floods of rivers, whose sources have the same latitude, take place at the same season; but the periods of the inundations of rivers on one side of the equator are exactly the contrary of what they are in rivers on the other side of it, on account of the declination of the sun. The flood in the Orinoco is at its greatest height in the month of August, while that of the river Amazon, south of the equinoctial line, is at its greatest elevation in March.[1] The commencement and end of the annual inundations in each river depend upon the average time of the beginning, and on the duration of the rains in the latitudes traversed by its affluents. The periods of the floods in such rivers as run towards the equator are different from those flowing in an opposite direction; and as the rise requires time to travel, it happens at regular but different periods in various parts of the same river, if very long. The height to which the water rises in the annual floods depends upon the nature of the country, but it is wonderfully constant in each individual river where the course is long; for the inequality in the quantity of rain in a district drained by any of its affluents is imperceptible in the general flood, and thus the quantity of water carried down is a measure of the mean humidity of the whole country comprised in its basin from year to year. By the admirable arrangement of these

[1] Baron Humboldt's Personal Narrative.

periodical inundations the fresh soil of the mountains, borne down by the water, enriches countries far remote from their source. The waters from the high lands designated as the Mountains of the Moon, and of Abyssinia, have fertilized the banks of the Nile through a distance of 2500 miles for thousands of years.

When rivers rise in mountains, water communication between them in the upper parts of their course is impossible; but when they descend to the plains, or rise in the low lands, the boundaries between the countries drained by them become low, and the different systems may be united by canals. It sometimes happens in extensive and very level plains, that the tributaries of the principal streams either unite or are connected by a natural canal by which a communication is formed between the two basins — a circumstance advantageous to the navigation and commerce of both, especially where the junction takes place far inland, as between the Orinoco and Amazon in the interior of South Amersca. The Rio Negro, one of the largest affluents of the latter, is united to the Upper Orinoco in the plains of Esmeralda by the Cassiquiare — a stream as large as the Rhine, with a velocity of 12 feet in a second. Baron Humboldt observes that the Orinoco, sending a branch to the Amazon, is, with regard to distance, as if the Rhine should send one to the Seine or Loire. At some future period this junction will be of great importance. These bifurcations are frequent in the deltas of rivers, but very rare in the interior of continents. The Chiana, which connects the upper branches of the Tiber and the Arno, is the most remarkable instance of this kind of junction in Europe. The Mahanuddy and Godavery, in Hindostan, offer something of the kind; and there are several instances in the great rivers of the Indo-Chinese peninsula.

The hydraulic system of Europe is eminently favourable to inland navigation, small as the rivers are in comparison with those in other parts of the world; but the flatness of the great plain, and the lowness of its watershed, are very favourable to the construction of canals. In the west, however, the Alps and German mountains divide the waters that flow to the Atlantic on one side, and to the Mediterranean and Black Sea on the other; but in the eastern parts of Europe the division of the waters is merely a more elevated ridge of the plain itself, for in all plains such undulations exist, though often imperceptible to the eye. This watershed begins on the northern declivity of the Carpathian Mountains about the 23rd meridian, in a low range of hills running between the sources of the Dnieper and the tributaries of the Vistula, from whence it winds in a tortuous course along the plain to the Valdai table-land, which is its highest point, 1200 feet above the sea; it then declines northward towards Onega, about the 60th parallel, and lastly turns in a very serpentine line to the sources of the Kama in the Ural moun tains near the 62nd degree of north latitude. The waters north of this line run into the Baltic and White Sea, and, on the south of it, into the Black Sea and the Caspian.

Thus Europe is divided into two principal hydraulic systems; but since the basin of a river comprehends all the plains and valleys drained by it and its tributaries from its source to the sea, each country is subdivided into as many natural divisions or basins as it has primary rivers, and these generally comprise all the rich and habitable parts of the earth, and are the principal centres of civilization, or are capable of becoming so.

The streams to the north of the general watershed are very numerous; those to the south are of greater magnitude. The systems of the Volga and Danube are the most extensive in Europe; the former has a basin comprising 640,000 square miles, and is navigable throughout the greater part of its course of 1900 miles. It rises in a small lake on the slopes of the Valdai table-land, 550 feet above the level of the ocean, and falls into the Caspian, which is 83 feet 7 inches below the level of the Black Sea, so that it has a fall of 633 feet in a course of more than 2400 miles. It carries to the Caspian one-seventh of all the river-water of Europe.

The Danube drains 300,000 square miles, and receives 60 navigable tributaries. Its quantity of water is nearly as much as that of all the rivers that empty themselves into the Black Sea taken together. Its direct course is 900 miles, its meandering line is 2400. It rises in the Black Forest at an elevation of 2850 feet above the level of the sea, so that it has considerable velocity, which, as well as rocks and rapids, impedes its navigation in many places, but it is navigable downwards, through Austria, for 600 miles, to New Orsova, from whence it flows in a gentle current to the Black Sea. The commercial importance of these two rivers is much increased by their flowing into inland seas. By canals between the Volga and the rivers north of the watershed, the Baltic and White Seas are connected with the Black Sea and the Caspian; and the Baltic and Black Sea are also connected by a canal between the Don and the Dnieper. Altogether the water system of Russia is the most extensive in Europe.

The whole of Holland is a collection of deltoid islands, formed by the Rhine, the Meuse, and the Scheldt — a structure very favourable to commerce, and which has facilitated an extensive internal navigation. The Mediterranean is already connected with the North Sea by the canal which runs from the Rhone to the Rhine; and this noble system, extended over the whole of France by 7591 miles of inland navigation, has conduced mainly to the improved state of that great country.

Many navigable streams rise in the Spanish mountains; of these the Tagus has depth enough for the largest ships as high as Lisbon.

Its actual course is 480 miles, but its direct line much less. In point of magnitude, however, the Spanish rivers are of inferior order, but canals have rendered them beneficial to the country. Italy is less favoured in her rivers, which only admit vessels of small burthen; those on the north are by much the most important, especially the Po and its tributaries, which by canals connect Venice and Milan with various fertile provinces of Northern Italy; but whatever advantages nature has afforded to the Italian states have been improved by able engineers, both in ancient and modern times.

The application of the science of hydraulics to rivers took its rise in Northern Italy, which has been carried to such perfection in some points, that China is the only country which can vie with it in the practice of irrigation. The lock on canals was in use in Lombardy as early as the 13th century, and in the end of the 15th it was applied to two canals which unite the Ticino and the Adda, by that great artist and philosopher Leonardo da Vinci; about the same time he introduced the use of the lock into France.[1]

Various circumstances combine to make the British rivers more useful than many others of greater magnitude. The larger streams are not encumbered with rocks or rapids; they all run into branches of the Atlantic; the tides flow up their channels to a considerable distance; and above all, though short in their course, they end in wide estuaries and sounds, capable of containing whole navies — a circumstance that gives an importance to streams otherwise insignificant, when compared with the great rivers of either the old or new continent.

The Thames, whose basin is only 5027 square miles, and whose whole length is but 240 miles, of which, however, 204 are navigable, spreads its influence over the remotest parts of the earth; its depth is sufficient to admit large vessels even up to London, and throughout its navigable course a continued forest of masts display the flags of every nation: its banks, which are in a state of perfect cultivation, are the seat of the highest civilization, moral and political. Local circumstances have undoubtedly been favourable to this superior development, but the earnest and energetic temperament of the Saxon races has rendered the advantages of their position available. The same may be said of other rivers in the British islands, where commercial enterprise and activity vie with that on the Thames. There are 2790 miles of canal in Britain, and, including rivers,

[1] Leonardo da Vinci was appointed Director of Hydraulic Operations in Lombardy by the Duke of Milan, and during the time he was painting the "Last Supper" he completed the Canal of Martesana, extending from the Adda to Milan, and improved the course of the latter river from where it emerges from the Lake of Como to the Po. By means of the Naviglio Grande, the Martesana Canal establishes a water communication between the Adda and the Ticino, the Lakes of Como and Maggiore.

5430 miles of inland navigation, which, in comparison with the size of the country, is very great; it is even said that no part of England is more than 15 miles distant from water communication.

On the whole, Europe is fortunate with regard to its water systems, and its inhabitants are for the most part alive to the bounties which Providence has bestowed.

AFRICAN RIVERS.

In Africa the tropical climate and the extremes of aridity and moisture give a totally different character to its rivers. The most southerly part is comparatively destitute of them, and those that do exist are of inferior size, except the Gariep, or Orange River, which has a long course on the table-land, but is nowhere navigable. From the eastern edge of the table-land of South Africa, which is very abrupt, rise all those rivers which flow across the plains of Mozambique and Zanguebar to the Indian Ocean. Of these the Zambesi, or Quillimane, is probably the largest: it is said to have a course of 900 miles, and to be navigable during the rains for 200 or 300 miles from its mouth. The Ozay, not far south of the equator, is also believed to be of great extent, and the Juba, more to the north; all these streams have little water at their mouths during the dry season, but in the rainy season they are navigable. Some of those still farther north do not reach the sea at all times of the year, but end in lakes and marshes, as the Haines, or Webbi, and Hawash. The first, after coming to within a small distance of the Indian Ocean, runs southward parallel to the coast, and falls into a very large and deep lake about a degree north of the equator. Between the Hawash and the Straits of Bab-el-Mandeb there is no river of any note. In many parts of the coast, near the rivers, grain ripens all the year, and every eastern vegetable production might be raised. The Hawash runs through a low desert country inhabited by the Dankali Beduins: that river is the recipient of the waters which come from the eastern declivity of the table-land of Abyssinia, while the Nile receives those of the opposite slope.

The part of the table-land between the 18th parallel of south latitude and the equator is the origin from whence the waters flow to the Atlantic on one hand, and to the Mediterranean on the other. Those which go to the Atlantic rise south of Lake N'yassi, chiefly in a ridge of no great elevation which runs from S.W. to N.E. to the west of the dominions of the Cambeze, and, after falling in cascades and rapids through the chains that border the table-land on the west, fertilize the luxuriant maritime plains of Benguela, Congo, Angola, and Loando. The Zaire, or Congo, by much the largest of these, is navigable for 140 miles, where the ascent of the tide is stopped by cataracts. The lower course of this river is 5 or 6 miles broad, full of islands, and 160 fathoms deep at its mouth. Its upper course, like that of most of these rivers, is unknown; the greater number are fordable on the table-land, but, from the abrupt descent of the high country to the maritime plains, none of them afford access to the interior of South Africa.

The mountainous edge of the table-land, with its terminal projections, Senegambia and Abyssinia, which separate the northern from the southern deserts, are the principal source of running water in Africa. Various rivers have their origin in these mountainous regions, of which the Nile and the Niger yield in size only to some of the great Asiatic and American rivers. In importance and historical interest the Nile is inferior to none.

Two large rivers unite their streams to form the Nile—the Bahr-el-Abiad, or White River, and the Bahr-el-Azrek, or Blue River; but the latter is so far inferior to the Bahr-el-Abiad that it may almost be regarded as a tributary. The main stream has never been ascended by any traveller above 4° 9′ north latitude, the point reached recently by the missionary Knoblecher, and who could see the river for 30 miles farther coming from the south-west. Bahr-el-Abiad, or the true Nile, was supposed, from the report of the natives, to rise, under the name of the Tubiri, at a comparatively small distance from the sea, in the country of Mono Moézi, which is a continuation of the high plateau of Abyssinia, situate to the north of the great Lake Zambéze, or N'yassi. The natives say that it flows from the lake itself; at all events it seems to be pretty certain that its origin is in the mountainous or hilly country of Mono Moézi, a word which in all the languages of that part of Africa signifies the Moon: hence the Nile has been said, since the days of Ptolemy, to rise in the Mountains of the Moon. Dr. Beke even supposes that it may have its upper sources in the snowy range of Kilimanjaro, situated south of the equator. Amidst many windings it takes a general direction towards the N.E. to the 14th northern parallel, whence it follows the same course till its junction near Khartum with the Blue Nile in the plains of Sennaar.

One of the largest affluents of the White River, if it be not its highest branch, rises by numerous heads in the mountainous countries of Enarya and Kaffa, between 7° and 9° North. The Gojab and Borora are its chief tributaries; the latter, which encircles the country of Enarea, is, according to M. d'Abbadie, the principal source of the White River, and rises in the forest of Babya, in latitude 8° N., at an elevation of nearly 6000 feet above the level of the sea. These united streams form the river Uma, and perhaps the Shoaberri; but scarcely anything is known of the latter between the high lands of Ethiopia and where it is said to empty itself into the Bahr-el-Abiad.

The Abyssinian branch of the Nile, known as the Bahr-el-Azrek, or Blue River, rises under the name of the Didhesa in the Galla country, south of Abyssinia, about 73 miles west of Saka, the capital of Enarea. It springs from a swampy meadow in the same elevated plains where the Godjeb and other affluents of the White Nile originate, in which it separates the kingdoms of Guma and Enarea, and maintains a general north-westerly direction till it joins the White Nile at Khartum. Of the many tributaries to the Blue River, the Abái, the Nile of Bruce, is the greatest and most celebrated. Its sources are in a swampy meadow near Mount Giesk, in the district of Sákkata, from whence it takes a circular direction round the peninsula of Gojam, passing through Lake Dembea, and receiving many affluents from the mountain-chain that forms the centre of the peninsula, and at last falls into the Didhesa or Bahr-el-Azrek, in about 11° N. latitude. From that point the only streams of any consequence that join either the Blue River or the united streams of the Blue and White Rivers, are the Rahad and Dender, which rise in Abyssinia, 160 miles below their confluence, where the Atbarah, formed by the junction of the Gwang and Takkazie, falls into it. This river, which is the principal tributary of the Nile, is formed by two branches. The Takkazie rises in the mountains of Lasta, a day's journey from Lalíbala, one of the most celebrated places in Abyssinia, remarkable for its churches hewn out of the solid rock, and the Tselari, which springs from Mount Biála, the northern extremity of the high land of Lasta, which divides the head waters of the two branches. The united stream, after winding like the other rivers of this country, joins the Nile in 18° N. latitude, the northern limit of the tropical rains.

The Abyssinian rivers in the upper part of their course are little more than muddy brooks in the dry season, but during the rains they inundate the plains. They break from the table-lands through fissures in the rocky surface, which are at first only a few yards wide, but gradually increase to several miles; the streams form cataracts from 80 to more than 100 feet high, and then continue to descend by a succession of falls and rapids, which decrease in height as they go northwards to join the main streams. The Takkazie takes its name of "The Terrible" from the impetuosity with which it rushes through the chasms and over the precipices of the mountains.[1]

A peculiarity of most of the principal affluents of the Nile is their spiral course, so that, after forming a curve of greater or less extent, generally round insulated mountain masses, they return upon themselves at a short distance from their sources. It is by no means improbable that the head stream of the Nile itself takes a spiral course

---

[1] According to M. d'Abbadie, Takkazie is the ancient Abyssinian name for river. See Exod. vii. 15.

round a lofty mountain mass, similar to the snow-clad mountains of Sámien and Kaffa.[1]

From the Takkazie down to the Mediterranean, a distance of 1200 miles, the Nile does not receive a single brook. The first part of that course is interrupted by cataracts, from the geological structure of the Nubian desert, which consists of a succession of broad sterile terraces, separated by ranges of rocks running east and west. Over these the Nile falls in nine or ten cataracts, the last of which is at Es-Souan (Syene), where it enters Egypt. Most of them are only rapids, where each successive fall of water is not a foot high. That they were higher at a former period has recently been ascertained by Dr. Lepsius, the very intelligent traveller sent by the King of Prussia at the head of a mission to explore that country. He found a series of inscriptions on the rocks at Sennaar, marking the height of the Nile at different periods: and it appears from these, that in that country the bed of the river had been 30 feet higher than it is now.

Fifteen miles below Cairo, and at 90 miles from the sea, the Nile is divided into two branches, of which one, running in a northerly direction, enters the Mediterranean below Rosetta; the other, cutting Lower Egypt into two nearly equal parts, enters the sea above Damietta, so that the delta between these two places has a sea-coast of 187 miles. The fall from the great cataract to the sea is two inches in a mile.

The basin of the Nile, occupying an area of 500,000 square miles, has an uncommon form; it is wide in Ethiopia and Nubia, but for the greater part of a winding course of 2750 miles[2] it is merely a verdant line of the softest beauty, suddenly and strongly contrasted with the dreary waste of the Red Desert. Extending from the equatorial far into the temperate zone, its aspect is less varied than might have been expected on account of the parched and showerless country it passes through. Nevertheless, from the great elevation of the origin of the river, the upper part has a perpetual spring, though within a few degrees of the equator. At the foot of the table-land of Abyssinia the country is covered with dense tropical jungles, while the rest of the valley is rich soil, the detritus of the mountains for thousands of years.

---

[1] Dr. Beke on the Nile and its affluents. See also Researches of M. d'Abbadie on the higher branches of the Nile, in the Journal de la Société de Géographie, 1849; and in the Athenæum.

[2] If we consider the Uma as the highest branch of the Nile, and adopt M. Arnaud's estimation of the windings of the Bahr el Abiad from Khartum upwards, it is probable that the winding course of the river will be found much greater than that given in the text: indeed M. d'Abbadie has calculated from these data that the course of the Nile, if developed on a meridian line, would reach from the Equator to Tornea, in Lapland, 3950 geographical miles.

As the mean velocity of the Nile, when not in flood, is about two miles and a half an hour, a particle of water would take twenty-two days and a half to descend from the junction of the Takkazie to the sea; hence the retardation of the annual inundations of the Nile in its course is a peculiarity of this river, owing to some unknown cause towards its origin which affects the whole stream. In Abyssinia and Sennaar the river begins to swell in April,[1] yet the flood is not sensible at Cairo till towards the summer solstice; it then continues to rise about a hundred days, and remains at its greatest height till tne middle of October, when it begins to subside, and arrives at its lowest point in April and May. The height of the flood in Upper Egypt varies from 30 to 35 feet; at Cairo it is 23, and in the northern part of the delta only 4 feet.

Anubis, or Sirius, the Dog-star, was worshipped by the Egyptians, from its supposed influence on the rising of the Nile. According to Champollion, their calendar commenced when the heliacal rising of that star coincided with the summer solstice—the time at which the Nile began to swell at Cairo. Now this coincidence made the nearest approach to accuracy 3291 years before the Christian era; and as the rising of the river still takes place precisely at the same time and in the same manner, it follows that the heat and periodical rains in Upper Ethiopia have not varied for 5000 years. In the time of Hipparchus the summer solstice was in the sign of Leo, and probably about that period the flowing of the fountains from the mouths of lions of basalt and granite was adopted as emblematical of the pouring forth of the floods of the Nile. The emblem is still common in Rome, though its origin is probably forgotten, and the signs of the Zodiac have moved backwards more than 30°.

The two greatest African rivers, the Nile and the Niger, are dissimilar in almost every circumstance; the Nile discharging for ages into the sea, the centre of commerce and civilization, has been renowned by the earliest historians, sacred and profane, for the exuberant fertility of its banks, and for the learning and wisdom of their inhabitants, who have left magnificent and imperishable monuments of their genius and power. Egypt was for ages the seat of science, and by the Red Sea it had intercourse with the most highly cultivated nations of the East from time immemorial. The Niger, on the contrary, though its rival in magnitude, and running through

[1] The April rains in Abyssinia are slight, and coincide with the passage of the sun in the prime vertical, and a partial rise of the Nile corresponding to them has been observed at Cairo, but the principal rains, the probable cause of the great rise in the waters of the Nile, take place at a later period in Enarea, and probably throughout all Ethiopia, between 7° and 9° N. It rains there every day in September, and as the maximum rise of the Nile at Cairo is in October, these two phenomena are evidently connected.—*D'Abbadie.*

a country glowing with all the brilliancy of tropical vegetation, has ever been inhabited by barbarous or semi-barbarous nations; and its course till lately was little known, as its source still is. In early ages, before the Pillars of Hercules had been passed, and indeed long afterwards, the Atlantic coast of Africa was an unknown region, and thus the flowing of the Niger into that lonely ocean kept the natives in their original rude state. Such are the effects of local circumstances on the intellectual advancement of man.

The sources of the Niger, Joliba, or Quorra, are supposed to be on the northern side of the Kong Mountains, in the country of Bambarra, more than 1600 feet above the level of the sea. From thence it runs north, and, after passing Lake Debo, makes a wide circuit in the plains of Soudan to Timbuctoo through eight or nine degrees of latitude; then bending round, it again approaches the Kong Mountains, at the distance of 1000 miles in a straight line from its source; and having threaded them, it flows across the low lands into the Gulf of Guinea, a course of 2300 miles. In the plains of Soudan it receives many very large affluents from the high land of Senegambia on the west, and the Tchadda on the east—a navigable river larger than itself, probably the outlet of the great lake Tchad, and falls into it a little below Fundah, after a course of some hundred miles: thus the Niger probably affords an uninterrupted water-communication from the Atlantic to the heart of Africa.[1] Long before leaving the plains of Soudan it becomes a noble river with a smooth stream, running at the rate of from 5 to 8 miles an hour, varying in breadth from 1 to 8 miles. Its banks are studded with densely populous towns and villages, groves of palm-trees, and cultivated fields.

This great river divides into three branches near the head of a delta which is equal in area to Ireland, intersected by navigable branches of the principal stream in every direction. The soil is rich in mould, and the vegetation so rank that the trees seem to grow out of the water. The Nun, which is the principal or central branch, flows into the sea near Cape Formosa, and is that which the brothers Lander descended. There are, however, six rivers which run into the Bight of Benin, all communicating with the Niger, and with one another. The old Calabar is the most eastern; it rises in the high land of Calbongos, and is united to the Niger by a natural canal. The Niger, throughout its long winding course, lies entirely within the tropic of Cancer, and is consequently subject to periodical inundations, which reach their greatest height in August, about 40 or 50 days after the summer solstice. The plains of Soudan are then covered with water and crowded by boats. These fertile regions are inaccessible to Europeans from the pernicious cli-

[1] Captain W. Allen, R.N.

mate, and dangerous from the savage condition of many of the tribes.

The coast of Guinea, west from the Niger, is watered by many streams, of no great magnitude, from the Kong Mountains. The table-land of Senegambia is the origin of the Rio Grande, the Gambia, the Senegal, and others of great size; and also of many of an inferior order that fertilize the luxuriant maritime plains on the Atlantic. Their navigable course is cut short by a semi-circular chain of mountains which forms the boundary of the high land, through which they thread their way in rapids and cataracts. The Gambia rises in Foula Toro, and after a course of about 600 miles enters the Atlantic by many branches connected by natural channels, supposed at one time to be separate rivers. The Senegal, the largest river in this part of Africa, is 850 miles long. It receives many tributaries in the upper part of its course, and the lower is full of islands. It drains two lakes, and is united to the basin of the Gambia by the river Neriko.

CHAPTER XVIII.

Asiatic Rivers—Euphrates and Tigris—River Systems South of the Himalaya—Chinese Rivers—Siberian Rivers.

THE only river system of importance in Western Asia is that of the Euphrates and Tigris, in the basin of which, containing an area 230,000 square miles, immense mounds of earth, in a desolate plain, point out the sites of some of the most celebrated cities of antiquity —of Nineveh and Babylon. Innumerable remains and inscriptions, the records of times very remote, have been discovered by adventurous travellers, and bear testimony to the truth of some of the most interesting pages of history. The Euphrates, and its affluent the Merad-Chaï (supposed to be the stream forded, as the Euphrates, by the Ten Thousand in their retreat), rise in the heart of Armenia, and, after running 1800 miles on the table-land to 38° 41′ of north latitude, they join the northern branch of the Euphrates, which rises in the Gheul Mountains, near Erzeroum. The whole river then descends in rapids through the Taurus chain, north of Romkala, to the plains of Mesopotamia.

The Tigris rises in the mountains to the N. and W. of Dyar-bekir, and after receiving several tributaries from the high lands of Kurdistan, it pierces the Taurus range about 100 miles above Mosul, from whence it descends in a tortuous course through the plain of ancient Assyria, receiving many streams from the Tyari Mountains, inhabited by the Nestorian Christians, and, farther south, from those of Luristan. The country through which it flows is rich in corn-fields, date-groves, and forest-trees.[1] Near to the city of Bagdad the Tigris and Euphrates approach to within 12 miles of each other, where they were once connected by two great canals. From this point they run nearly parallel for more than 100 miles, encircling the plain of Babylon or Southern Mesopotamia — the modern Irak-Arabi. The two rivers unite at Korna, and form one stream, which, under the name of Shat el Arab, runs for 150 miles before it falls into the Persian Gulf. The banks of the Tigris and Euphrates, once the seat of an extensive population, and of art, civilization, and industry, are now nearly deserted, covered with brushwood and grass, dependent on the rains alone for that luxuriant vegetation which, under an admirable system of irrigation, formerly covered them. Excepting the large centres of population, Bagdad and Mosul, the inhabitants consist of nomade Kurdish tribes. What remains of civilization has taken refuge in the mountains, where the few traces of primitive and most ancient Christianity, under the misapplied denomination of Nestorian Christians, are to be found in the Tyari range. The floods of the rivers are very regular in their period; beginning in March, they attain their greatest height in June.

The Persian Gulf may be navigated by steam all the year, the Euphrates only eight months; it might, however, afford easy inter-course with eastern Asia, as it did in former times. The distance from Aleppo to Bombay by the Euphrates is 2870 miles, of which 2700, from Bir to Bombay, are by water; in the time of Queen

---

[1] It is in the space comprised between two of the eastern tributaries of the Tigris, the Khaus and the Great Zab, or Abou Selman of the Arabs, that the extensive ruins of Koyunjik, Khorsabad, and especially of Nim-roud, are situated, the last of which have been so satisfactorily identified with the capital of Assyria—the ancient Nineveh—by our enterprising and talented countryman Mr. Layard, to whose exertions, under circumstances of peculiar difficulty, surrounded by every privation, our national Museum is indebted for that magnificent collection of Assyrian monuments which at this moment forms the admiration of the British public. In the former edition of this book we expressed a hope that our Government would follow up the researches commenced by Mr. Layard, and that several of the gi-gantic sculptures removed by him with such perseverance and labour, to Bussorah, would ere long be added to the riches of the British Museum. These hopes have been partially responded to; Mr. Layard being again enabled to return to the scene of his former labours by the liberality of Her Majesty's Government. But much yet remains to be effected; the field of research is so vast, and pecuniary assistance only wanting to reap in it.

See Mr. Layard's work on 'Nineveh and its Remains,' 2 vols. 8vo., and his illustrated work in folio — the former one of the most interesting nar-ratives ever published on the antiquities of Central Asia.

Elizabeth this was the common route to India, and a fleet was then kept at Bir, expressly for that navigation.

Six rivers of the first magnitude descend from the southern side of the table-land of eastern Asia and its mountain barriers, all different in origin, direction, and character, while they convey to the ocean a greater volume of water than all the rivers of the rest of the continent conjointly. Of these, the Indus, the double system of the Ganges and Brahmapootra, and the three parallel rivers in the Indo-Chinese peninsula, water the plains of southern Asia; the great system of rivers that descend from the eastern terraces of the table-land irrigates the fertile land of China; and lastly the Siberian rivers, not inferior to any in magnitude, carry the waters of the Altaï and northern slope of the table-land to the Arctic Ocean.

The hard-fought battles and splendid victories recently achieved by British valour over a bold and well-disciplined foe have added to the historical interest of the Indus and its tributary streams, now the western boundaries of our Asiatic empire.

The sources of the Indus were only ascertained in 1812; the Ladak, the largest branch of the Indus, has its origin in the snowy mountains of Karakorum; and the Shyook, which is the smaller stream, rises in the Kentese or Gangri range, a ridge parallel to the Himalaya, which extends along the table-lands of Tibet, north and west of the sacred lakes of Mànasarowar. These two streams join north-west of Ladak, and form the Indus; the Sutlej, its principal tributary, springs from the lake of Rakas Tal, which communicates with that of Mànasarowar, both situated in a valley between the Himalaya and Gangri chain at the great elevation of 15,200 feet. These rivers, fed by streams of melted snow from the northern side of the Himalaya, both flow westward along the extensive longitudinal valley of western Tibet. The Sutlej breaks through the Himalaya about the 75th meridian, and traverses the whole breadth of the chain, in frightful chasms and clefts in the rocks, to the plains of the Punjab; the Indus, after continuing its course on the table-land through several degrees of longitude farther, descends near the junction of the Himalaya and the Hindoo Coosh, west of the valley of Cashmere, to the same plain. Three tributaries—the Jelum or Hydaspes, the Chenab or Acescines, and the Ravee or Hydraötes, all superior to the Rhone in size—flow from the southern face of the Himalaya, and with the Sutlej (the ancient Hyphasis) join the Indus before it reaches Mittun; hence the name Punjab, "the plain of the five rivers," now one of the most valuable countries in the East. From Mittun to the ocean, the Indus, like the Nile, does not receive a single accessary, from the same cause—the sterility of the country through which it passes. The Cabul river, which rises near Guzni, and is joined by a larger affluent from the southern declivities of the Hindoo Coosh, flows through picturesque and dangerous defiles, and joins the Indus at the town of Attock, and is the only tributary of any magnitude that comes from the west.

The Indus is not favourable to navigation: for 70 miles after it leaves the mountains the descent in a boat is dangerous, and it is only navigable for steam-vessels of small draught of water; yet, from the fertility of the Punjab, and the near approach of its basin to that of the Ganges at the foot of the mountains, it must ultimately be a valuable acquisition, and the more especially because it commands the principal roads between Persia and India, one through Cabul and Peshawer, and the other from Herat through Candahar. The delta of the Indus, formerly celebrated for its civilization, has long been a desert; but from the luxuriance of the soil, and the change of political circumstances, it may again resume its pristine aspect. It is 60 miles long, and presents a face of 120 miles to the sea in the Gulf of Oman, where the river empties itself by many mouths, of which only three or four are navigable; one only can be entered by vessels of 50 tons, and all are liable to change. The tide ascends them with extraordinary rapidity for 75 miles, and so great is the quantity of mud carried by it, and the absorbing violence of the eddies, that a vessel wrecked on the coast was buried in sand and mud in two tides. The annual floods begin with the melting of the snow in the Himalaya in the end of April, come to their height in July, and end in September. The length of this river is 1500 miles, and it drains an area of 400,000 square miles.

The second group of South Indian rivers, and one of the greatest, is the double system of the Ganges and the Brahmapootra. These two rivers, though wide apart at their courses, have their sources little removed from each other, on opposite sides of the central ridge of the Himalaya, and which, converging to a common delta, constitute one of the most important river systems on the globe.

Mr. Alexander Elliot, of the Body Guard in Bengal, son of Admiral Elliot, with his friends, are the first who have accomplished the arduous expedition to the sources of the Ganges. The river flows at once in a very rapid stream not less than 40 yards across, from a huge cave in a perpendicular wall of ice at the distance of about three marches from the Temple of Gungootree, to which the pilgrims resort. Mr. Elliot says, "The view from the glacier was perfectly amazing; beautiful or magnificent is no word for it, — it was really quite astonishing. If you could fancy a bird's-eye of all the mountains of the world in one cluster, and every one of them covered with snow, it would hardly give you an idea of the sight which presented itself."

Many streams from the southern face of the Himalaya unite at Hurdwar to form the great body of the river. It flows from thence in a south-easterly direction through the plains of Bengal, receiving in its course the tribute of 19 or 20 rivers, of which 12 are larger than the Rhine. About 220 miles in a direct line from the Bay of Bengal, into which the Ganges flows, the innumerable channels and branches into which it splits form an intricate maze over a delta twice as large as that of the Nile.

The Brahmapootra, a river equal in the volume of its waters to the Ganges, may be considered as the continuation of the Dzangho Tchou or river of Lassa, which rises near the sources of the Sutlej and the Indus, in long. 82° E. After watering the great longitudinal valley of eastern Tibet, it makes a sudden bend to the south in long. 90° E., cutting through the Himalaya chain, as the Indus does at its opposite extremity between Iskardo and Attock; after which it receives several tributaries from the northern mountains of the Birman empire; but very little is known of this part of its basin. The upper part of the Brahmapootra is parallel to the Himalaya chain, until it enters Upper Assam, where, passing through the sacred pool of Brahma-Koond, it receives the name which it bears in the lower part of its course — Brahmapootra, the "offspring of Brahma:" the natives call it the Lahit, Sanscrit for the "Red River." In Upper Assam, through which it winds 500 miles and forms some extensive channel islands, it receives six very considerable accessories, of which the origin is unknown, though some are supposed to come from the table-land of Tibet. They are only navigable in the plains, but vessels of considerable burthen ascend the parent stream as far as Sundiva. Before it enters the plains of Bengal, below Goyalpara, the Brahmapootra runs with rapidity and in great volume, and, after receiving the rivers of Bhotan and other streams, branches of it unite with those of the Ganges about 40 miles from the coast, but the two rivers enter the sea by different mouths, though they sometimes approach within two miles. The length of the Brahmapootra is estimated at 1500 miles, or nearly the same as that of the Ganges: the volume of water discharged by it during the dry season is about 146,188 cubic feet in a second; the quantity discharged by the Ganges in the same time and under the same circumstances is only 80,000 cubic feet. In the perennial floods the quantity of water poured through the tributaries of the Brahmapootra through their snowy sources is incredible; the plains of Upper Assam are an entire sheet of water from the 15th of June to the 15th of September, and there is no communication but by elevated causeways eight or ten feet high: the two rivers, with their branches, lay the plain of Bengal under water for hundreds of miles annually. They begin first to swell from the melting of the snow on the mountains; but, before their inferior streams overflow from that cause, all the lower parts of Bengal adjacent to the Ganges and Brahmapootra are under water from the swelling of these rivers by the rains. The increase is arrested before the middle of August, by the cessation of the rains in the mountains, though they continue to fall longer on the plains. The delta is traversed in every direction by arms of the rivers. The Hoogly branch, at all times navigable, passes Calcutta and Chandernagor; and the Hauringotta arm is also navigable, as well as the Ganges properly so called. The channels, however, are perpetually changing, from the strength of the current, and the prodigious quantity of matter washed from the high lands; the Ganges alone carries to the sea 6,000,000,000 cubic feet of mud annually, the effects of which are perceptible 60 miles from the coast. The elevation of the mountains, and indeed of the land generally, must have been enormous, since it remains still so stupendous after ages of such degradation. The Sunderbunds, a congeries of innumerable river islands formed by the endless streams and narrow channels of the rivers, as well as by the indentations of arms of the sea, line the coast of Bengal for 180 miles, a wilderness of jungle and heavy timber. The united streams of the Ganges and Brahmapootra drain an area of 650,000 square miles, and there is scarcely a spot in Bengal more than 20 miles distant from a river navigable even in the dry season.

These three great rivers of Southern India do not differ more widely in their physical circumstances than in the races of men who inhabit their banks, yet from their position they seem formed to unite nations the most varied in their aspect and speech. The tributaries of the Ganges and Indus come so near to each other at the foot of the mountains, that a canal only two miles long would unite them, and thus an inland navigation from the Bay of Bengal to the Gulf of Oman might be established.

An immense volume of water is poured in a series of nearly parallel rivers of great magnitude, and running in the direction of the meridian through the Indo-Chinese peninsula, to empty themselves into the ocean on either side of the peninsula of Malacca. They rise in those elevated regions at the south-eastern angle of the table-land of Tibet, the lofty but unknown province of the Chinese empire, and water the great valleys that extend from north to south with perfect uniformity, between chains of mountains no less uniform, which spread out like a fan as they approach the sea. Scarcely anything is known of the origin or upper parts of these rivers, and with a few exceptions almost as little of the lower.

Their number amounts to six or seven, all large, though three surpass the rest—the Irrawady, which waters the Birman empire, and falls into the Bay of Bengal at the Gulf of Martaban; the Menam, or river of Siam; and the river Cambodja; which flows through the empire of Annam: the last two fall into the Gulf of Siam and the China Sea.

The sources of the Irrawady are in the same chain of mountains with the eastern affluents of the Brahmapootra more to the south. Its course is through countries hardly known to Europeans, but it seems to be navigable by boats before coming to the city of Amarapoora, south of which it enters the finest and richest plain of the empire, containing its four capital cities. There it receives two large affluents, one from the Chinese province of Yunnan, which flows into the Irrawady at the city of Ava, 446 miles from the sea, the highest point attained by the British forces during the Burmese war.

From Ava to its delta the Irrawady is a magnificent river, more than four miles broad in some places, but encumbered with channel islands. In this part of its course it receives its largest tributary, and forms in its delta one of the most extensive systems of internal navigation. The Rangoon is the only one of its 14 mouths that is always navigable, and in it the commerce of the empire is concentrated. The internal communication is extended by the junction of the two most navigable deltoid branches with the rivers Salüaen and Pegu by natural canals: that joining the former is 200 miles long; the canal uniting the latter is only navigable at high water.

The Menam, one of the largest Asiatic rivers, is less known than the Irrawady; it comes from the Chinese province of Yunnan, and runs through the kingdom of Siam, which it cuts into several islands by many diverging branches, and enters the Gulf of Siam by three principal arms, the most easterly of which forms the harbour of Bangkok.[1] It is joined to the Menam Kong, or Cambodja, by the small river Anan-Myit.

The river of Cambodja has the longest course of any in the peninsula: it is supposed to be the Lantsan-Kiang, which rises in the high land of K'ham, in eastern Asia, not far from the sources of the great Chinese river, the Yang-tse-Kiang. After traversing the elevated plain of Yunnan, where it is navigable, it rushes through the mountain barriers, and, on reaching a wider valley, about 300 miles from its mouth, it is joined to the Menam by the natural canal of the Anan-Myit. More to the south it is said to split into branches which unite again.

The ancient capital of Annam is situate on the Cambodja, about 150 miles from the sea; a little to the south its extensive delta begins, projects far into the ocean, and is cut in all directions by arms of the river, navigable during the floods; three of its mouths are permanently so for large vessels up to the capital. The Saüng, more to the east, is much shorter than the Cambodja, though said to be 1000 miles long, but Europeans have not ascended higher than the town of Sai-Gon. Near its mouth it sends off several branches to the eastern arm of the Cambodja. All rivers of this part of Asia

[[1] The city of Bangkok is upon the river Menam, about 20 miles from its mouth, and is accessible to vessels of 300 tons burthen; technically, there is no harbour of Bangkok.]

are subject to periodical inundations, which fertilize the plains at the expense of the mountains.

The parallelism of the mountain chains constitutes formidable barriers between the upper basins of the Indo-Chinese rivers, and decided lines of separation between the inhabitants of the intervening valleys; but this inconvenience is in some degree compensated by the natural canals of junction and the extensive water communication towards the mouths of the rivers.

Four great systems of rivers take their origin on the eastern declivity of the great table-land of central Asia, and running from west to east, traverse the Chinese empire:—the Hong-Kiang, which, rising in the province of Yunnan, empties itself into the bay of Canton; the Yang-tse-Kiang, or Son of the Ocean; the Hoang-Ho; and the great river of Amur.

The length of the Hoang-Ho or Yellow River is 2000 miles, that of the Yang-tse-Kiang 2900. Though near their sources they are widely separated by the mountain-chains that border the table-land, they approach as they proceed on their eastern course, and are not more than 100 miles apart when they enter the Yellow Sea. From a map constructed by the Jesuit missionaries in the 18th century it appears that the mouth of the Hoang-Ho has shifted to the enormous distance of 126 leagues from its former position. The Yang-tse-Kiang and the Yellow River in the lower part of their course are united by innumerable canals, forming the grandest system of irrigation and of internal navigation in existence.

The Hoang-Ho has derived its name of "Yellow" River from the large quantity of earthy matter it brings down with it to the sea, like the Tiber of old.

Strong tides ascend these rivers to the distance of 400 miles, and for the time prevent the descent of the fresh water, which forms large interior seas, frequented by thousands of trading-vessels, and they irrigate the productive lands of central China, from time immemorial the most highly cultivated and the most densely peopled region of the globe.

Almost all the Chinese rivers of less note—and they are numerous—feed these giant streams, with the exception of the Ta-si or Hong-Kiang and the Pee-ho or White River, which have their own basins. The former, rising to the east of the town of Yunnan, flows through the plains of Canton eastward to the Gulf of Canton, into which it discharges itself, increased in its course by the Sekiang.

The White River, rising in the mountains near the Great Wall, becomes navigable a few miles east of Pekin, unites with the Eu-ho, joins the Great Canal, and, as the tide ascends it for 80 miles, it is crowded with shipping.

The Amur, the sources of which are partly in the Russian dominions, though its course is chiefly in the Mantchourian territory of

China, is 2000 miles long, including its windings, and has a basin of 853,000 square miles. Almost all its tributaries come from that part of the Baikalian group called the Yablonnoi Khrebit by the Russians, and Khing-Khan-Oola by the Chinese. The river Onon, which is the parent stream, has its origin in the Khentai Khan, a branch of the latter; and though its course is through an uninhabited country, it is celebrated as the birthplace and the scene of the exploits of Tshingis Khan. After passing through the lake of Dalai-nor, which is 210 miles in circumference, it takes the name of Argun, and forms the boundary between the Chinese and Russians for 400 miles; it is then joined by the Shilka, where it assumes the Tunguse name of the Amur or Great River: the Mantchoos call it the Saghalin or Black Water. It receives most of the unknown rivers which come from the mountain-slopes of the Great Gobï, and falls into the Pacific opposite to the island of Saghalin, after having traversed three degrees of latitude and thirty-three of longitude.

Three great rivers, the Lena, the Yenessei, and the double system of the Irtish and Oby, not inferior in size to any of the rivers of Asia, carry off the waters of the Altaï chain, and of the mountains which bound the northern border of the great Asiatic table-land. The Lena, whose basin occupies 800,000 square miles, springs from mountains north of the Lake of Baikal, and runs north-east through more than half its course to the Siberian town of Yakutzk, the coldest town on the face of the earth, receiving in its course the Vitim and the Olekma, its two principal affluents, the former from the Baikal mountains, the latter from Stannovoi Khrebit, the most southerly part of the Aldan range. North of Yakutzk, about the 63rd parallel of latitude, the Lena receives the Aldan, its greatest tributary, which also comes from the Stannovoi Khrebit; it then goes to the Arctic Ocean, between banks of frozen mud, prodigious masses of which are hurled down by the summer floods, and bring to view the bones of those huge extinct species of elephant and rhinoceros, which at some remote period had found their nourishment in these desert plains.[1] The length of the Lena, including its windings, is 1900 miles.

The Yenessei, a much larger river than the Lena, drains about 1,000,000 square miles, and is formed by the union of the Great

[1] The elephant and rhinoceros of Siberia belong to species that are widely scattered over the whole of Europe, and one of which is even found in America. The Siberian individuals were covered with a thick coating of hair and fur, so different from any of their living congeners, which suggested to Cuvier the explanation of their being able to exist in so cold a climate, where, from their extraordinary state of preservation, they must evidently have lived, by their hairy coats enabling them to brave an excessive climate, whilst they found nourishment in the birch and pine forests of these high latitudes. See Cuvier, Ossemens Fossiles, article Eléphants Fossiles.

and Little Kem. The former rises at the junction of the Sayansk range with the Baikalian mountains to the north-west of Lake Kassagol; the latter comes from the Egtag or Little Altaï, in quite an opposite direction, so that these two meet nearly at right angles, and take the name of Yenessei; it then crosses the Sagaetses range in cataracts and rapids, entering the plains of Siberia below the town of Krasnojarsk. Below this many rivers join it, chiefly the Angara from the Lake Baikal; but its greatest tributaries, the Upper and Lower Tunguska, both large rivers from the Baikalian mountains, join it lower down, the first to the south, the latter to the north of the town of Yeniseisk, whence it runs north to the Icy Ocean, there forming a large gulf, its length, measured along its bed, being 2500 miles.

The Oby rises in the Lake of Toleskoi, "the Lake of Gold," in Great Tartary; all the streams of the Lesser Altaï unite to swell it and its great tributary the Irtish. The rivers which come from the northern declivity of the mountains go to the Oby, those from the western side to the Irtish, which springs from numerous streams on the south-western declivity of the Little Altaï, and run westward into Lake Zaidzan, 200 miles in circumference. Issuing from thence, it takes a westerly course to the plain on the north of Semipolatinsk. In the plain it is joined by the Tobol, which crosses the steppe of the Kirghiz Cossacks from the Ural Mountains, and soon unites with the Oby; the joint stream then proceeds to the Arctic Ocean in 67° N. lat. The Oby is 2000 miles long, and the basin of these two rivers occupies a third part of Siberia.

Before the Oby leaves the mountains, at a distance of 1200 miles from the Arctic Ocean, its surface has an absolute elevation of not more than 400 feet, and the Irtish, at the same distance, is only 72 feet higher; both are consequently sluggish. When the snow melts, they cover the country like seas; and as the inclination of the plains in the middle and lower parts of their course is not sufficient to carry off the water, those immense lakes and marshes are formed which characterize this portion of Siberia.

The bed of the Oby is very deep, and there are no soundings at its mouth; hence the largest vessels might ascend at least to its junction with the Irtish. Its many affluents also might admit ships, did not the climate form an insurmountable obstacle the greater part of the year. Indeed all Siberian rivers are frozen annually for many months, and even the ocean along the Arctic coasts is rarely disencumbered from ice; therefore these vast rivers never can be important as navigable streams. They abound in fish and water-fowl, for which the Siberian peasant braves the extremest severity of the climate.

Local circumstances have nowhere produced a greater difference in the human race than in the basins of the great rivers north and

21 *

south of the table-land of eastern Asia. The Indian, favoured by the finest climate, and a soil which produces the luxuries of life, intersected with rivers navigable at all seasons, and affording easy communication with the surrounding nations, attained early a high degree of civilization; while the Siberian and Samoide, doomed to contend with the rigours of the polar blasts in order to maintain mere existence, have never risen beyond the lowest grade of humanity; but custom softens the rigour of this stern life, so that even here a share of happiness is enjoyed.

CHAPTER XIX.

River Systems of North America—Rivers of Central America—Rivers of South America and of Australia.

North America is divided into four distinct water systems by the Rocky Mountains, the Alleghanies, and a table-land which contains the great lakes, and separates the rivers that flow into the Arctic Ocean from those which go to the Gulf of Mexico. This table-land, which is a level, nowhere more than 1200 or 1500 feet above the surface of the sea, is the watershed of the Mackenzie, the Mississippi, the St. Lawrence, and of the rivers that flow into Hudson's Bay. The St. Lawrence rises under the name of the St. Louis in 47° 43′ N. lat. and 93° W. long.; after joining the Lakes Superior, Huron, Erie, and Ontario, it issues from the last by the name of the Iroquois, and, expanding in its north-easterly course into the Lakes of St. Francis, St. Louis, and St. Peter, it is first known as the St. Lawrence at Montreal, from whence it runs north-east into the Atlantic and ends in an estuary 100 miles wide. It has a basin of 297,600 square miles, of which 94,000 are covered with water, exclusive of the many lesser lakes with which it is in communication.

North of the watershed there is an endless and intricate labyrinth of lakes and rivers, almost all connected with one another. But the principal streams of these Arctic lands are—the Great Fish River, which flows north-east in a continued series of dangerous and all but impassable rapids to the Arctic Ocean at Melville Strait; the Copper-mine River, of much the same character, which, after traversing many lakes, enters the Icy Sea at George the Fourth's Gulf; the Mackenzie River, a stream of greater magnitude, formed by the confluence of the Peace River and the Athabasca from the Rocky Mountains, which, after flowing north over 16 degrees of latitude, enters the Frozen Ocean in the Esquimaux country beyond the Arctic Circle; and the Colville, a very large river, the upper course of which in the Russian possessions is very little known, enters the sea near Point Barrow, in 152° W. longitude. All these rivers are frozen more than half the year, and the Mackenzie, in consequence of its length and direction from south to north, is subject to floods like the Siberian rivers, because its lower course remains frozen for several hundred miles long after the upper part is thawed, and the water, finding no outlet, flows over the ice and inundates the plains.

South of the table-land the valley of the Mississippi extends for 1000 miles, and this greatest of North American rivers has its origin in the junction of the streams from the small lakes Itasca and Ussawa, on the table-land at no greater height than 1500 feet above the sea. Before their junction these streams frequently spread out into sheets of water, and the Mississippi does the same in the upper part of its course. This river flows from north to south through more degrees of latitude than any other, and receives so many tributaries of the higher order, that it would be difficult even to name them. Among those that swell its volume from the Rocky Mountains, the Missouri, the Arkansas, and the Red River are the largest, each being in itself a mighty stream, receiving tributaries without number. Before their junction the Missouri is a stream much superior to the Mississippi both in length and volume, and has many affluents larger than the Rhine. It rises in about 44° N. lat., and runs partly in a longitudinal valley of the Rocky Mountains, and partly at their foot, and drains the whole of the country on the right bank of the Mississippi between the 49th and 40th parallels of north latitude. It descends in cataracts through the mountain regions, and in the plains it sometimes passes through large prairies and sometimes through dense forests, in all accomplishing 3000 miles in a very tortuous and generally south-eastern direction till it joins the Mississippi near the town of St. Louis. Lower down, the Mississippi is joined by the Arkansas, 2000 miles long, with many tributaries, and then by the Red River, the former from the Rocky Mountains; the latter, which rises in the table-land of New Mexico, is fed by rivers from the Sierra del Sacramento, and enters the main stream not far from the beginning of the delta, at the head of which the Mississippi sends off a large branch called the Atchafalaya to the south, and then turning to the east it discharges itself by five mouths at the extremity of a long tongue of land which stretches 50 miles into the Gulf of Mexico, having formed a delta considerably larger than that of the Nile. The shore is lined with shallow salt lagoons; the greater part of the delta is covered with water and unhealthy marshes, the abode of the crocodile, and during the floods it is a muddy sea. This river is navigable for 2240 miles. Its valley is of variable width, but at its greatest width, at the junction of the White River, it is 80 miles.

The tributaries from the Rocky Mountains, though much longer, run through countries of less promise than those which are traversed by the Ohio and the other rivers that flow into the Mississippi on the east, which offer advantages unrivalled even in this wonderful country, only beginning to be developed.

The Ohio is formed by the union of the rivers Alleghany and Monongahela, the latter from the Laurel ridge of the Alleghany chain in Virginia; the former comes from sources near Lake Erie, and the two unite at Pittsburg, from whence the river winds 948 miles through some of the finest states of the Union, till its junction with the Mississippi, having received many accessories, six of which are navigable streams. There are some obstacles to navigation in the Ohio, but they have been avoided by canals. Other canals join both the Mississippi and its branches with Lake Erie, so that there is an internal water communication between the St. Lawrence and the Gulf of Mexico. The whole length of the Mississippi is 3160 miles, but, if the Missouri be considered the main stem, it is 4265, and the joint stream drains an area of about a million and a quarter of square miles. The breadth of the river nowhere corresponds with its length. At the confluence of the Missouri each river is half a mile wide, and after the junction of the Ohio it is not more. A steamer may ascend the Mississippi for 2000 miles from Balize without any perceptible difference in its breadth. The depth is 168 feet where it enters the Gulf of Mexico at New Orleans: the fall of the river at Cape Girardeau is four inches in a mile. The river is a rapid desolating torrent loaded with mud; its violent floods, from the melting of the snow in the high latitudes, sweep away whole forests, by which the navigation is rendered very dangerous, and the trees, being matted together in masses many yards thick, are carried down by the spring floods, and deposited over the delta and Gulf of Mexico for hundreds of square miles.

North America can boast of two other great water systems, one from the eastern versant of the Alleghanies, which flows into the Atlantic, and another from the western declivity of the Rocky Mountains, which runs into the Pacific.

All the streams that flow eastward through the United States to the Atlantic are short, and comparatively small, but of the highest utility, because many of them, especially those to the north, end in gulfs of vast magnitude, and the whole are so united by canals that few places are not accessible by water — one of the greatest advantages a country can possess. There are at least 24 canals in the United States, the length of which is 3101 miles. [At the close of the year 1845, the aggregate distance navigable in the United

States by canal was 3,450 miles: 3000 miles of canal in the free states cost about $89,000,000, and the 450 miles in the slave states cost about $22,000,000 in their construction.]

Many of the streams which ultimately come to the Atlantic rise in the western ridges of the Alleghany chain, and traverse its longitudinal valleys before leaving the mountains to cross the Atlantic slope, which terminates in a precipitous ledge for 300 miles parallel to the range. By falling over this rocky barrier in long rapids and picturesque cascades they afford an enormous and extensive water-power; and as the rivers are navigable from the Atlantic quite across the maritime plains, these two circumstances have determined the location of most of the principal cities of the United States at the foot of this rocky ledge, which, though not more than 300 feet high, has had a greater influence on the political and commercial interests of the Union than the highest chains of mountains have had in other countries. The Hudson in the north is navigable to Albany; the Delaware and Susquehanna, ending in bays, are important rivers: and the Potomac, which falls into Chesapeake Bay, passes Washington, the capital of the United States, to which the largest ships can ascend.

The watershed of the Rocky Mountains lies at a greater distance from the Pacific than that of the Alleghanies from the Atlantic; consequently the rivers are longer, but they are few, and little known; the largest are, the Oregon or Columbia, and the Rio Colorado. The former has its sources not far from those of the Missouri and of the Rio del Norte; and after an exceedingly tortuous course, in which it receives many tributaries, it falls into the Pacific below Astoria. The Colorado is a considerable stream, which comes from the Serra Verde and falls into the Gulf of California. The Sacramento with its tributaries, a Californian stream, lying between the two, and much inferior to either, has been brought into notice of late from the extensive and rich auriferous country through which it flows in its course to the Bay of San Francisco on the Pacific.

On the table-land of Mexico there is a basin of continental streams, which, rising from springs on the eastern side of the Serra Madre, and fed by the periodical rains, flow northward and terminate in lakes, which part with their superfluous water by evaporation. Of these the Rio Grande, which, after a course of 300 miles, falls into the Parras, is the greatest.

The largest river in the isthmus of Mexico is the Rio de Lerma or Rio Grande Santiago, which rises on the table-land of Toluca, passes through Lake Chapāla, forms numerous cascades, and falls into the Pacific after a course of 400 miles. The River Guasacualco, which traverses the Isthmus nearly from sea to sea, emptying itself into the Gulf of Mexico, has by some been considered as the best point for a sea canal between the two oceans. There are many streams in Central America, and above 10 rivers that are navigable for some miles; six of these fall into the Gulf of Mexico and Caribbean Sea, and four into the Pacific. Of these the Rio Montagua, which rises in the mountains near Guatemala, flows into the Bay of Honduras, and the Blewfields river, the greater part of whose course is in the Mosquito territory.

In the southern part of the State of Guatemala is situated the River of San Juan, which drains the Lakes of Nicaragua and Leon, and by which it is supposed a water communication could be easily effected between the Atlantic and the Pacific.

The Andes, the extensive watershed of South America, are so close to the sea, that there are no rivers of considerable size which empty themselves into the Pacific; even some of the streams that rise in the western Cordilleras find their way to the eastern plains.

The Magdalena, at the northern end of the Andes, though a secondary river in America, is 620 miles long. It rises in the central chain, at the divergence of the Cordilleras of Suma Paz and Quindiu, and enters the Caribbean Sea by various channels: it is navigable as far as Honda. The Cauca, its only feeder on the west, comes from Popayan, and is nearly as large as its primary, to which it runs parallel the greater part of its course. Many streams join the Magdalena on the right, as the stream which waters the elevated plain of Bogotá, and forms the cataract of Tequendama, one of the most beautiful and wildest scenes in the Andes. The river rushes through a chasm 30 feet wide, which appears to have been formed by an earthquake, and at a double bound descends 530 feet into a dark gloomy pool, illuminated only at noon by a few feeble rays. A dense cloud of vapour rising from it is visible at the distance of 15 miles. At the top the vegetation is that of a temperate climate, while palms grow at the bottom.

The river Atrato, parallel to the Cauca and Magdalena, but less considerable, empties itself into the Gulf of Darien. The rivers of Patia, of San Juan, of Las Esmeraldas, and of Guayaquil, all rise on the western declivity of the Andes to flow into the Pacific. With these exceptions all the water from the inexhaustible sources of the Andes north of Chile is poured into the Orinoco, the River Amazon, and the Rio de la Plata, which convey it eastward across the continent to the Atlantic. In the far south, indeed, there are the Colorado and Rio Negro, but they are insignificant when compared with these giant floods.

The basins of these three rivers are separated in their lower parts by the mountains and high lands of the Parima and Brazil; but the central parts of the basins of all three, toward the foot of the Andes, form an extensive level, and are only divided from one another by imperceptible elevations in the plains, barely sufficient to form the watersheds between the tributaries of these majestic rivers. This peculiar structure is the cause of the natural canal of the Cassiquiare, which joins the Upper Orinoco with the Rio Negro, a principal affluent of the Amazon. Ages hence, when the wilds are inhabited by civilized man, the tributaries of these three great rivers, many of which are navigable to the foot of the Andes, will, by means of canals, form a water system infinitely superior to any that now exists.

The Orinoco, altogether a Colombian river, rises in the Serra del Parima, 200 miles east of the elevated Peak of Duida, and maintains a westerly course to San Fernando de Atabapo, where it receives the Atabapo, and Guaviare, which is larger than the Danube; here ends the Upper Orinoco. The river then forces a passage through the Serra del Parima, and runs due north for three degrees of latitude, between banks almost inaccessible; its bed is traversed by dykes, and filled with boulders of granite and islands clothed with a variety of magnificent palm-trees. Large portions of the river are here engulfed in crevices, forming subterranean cascades; and in this part are the celebrated falls of the Atures and Apures, 36 miles apart, which are heard at the distance of many miles. At the end of this tumultuous part of its course it is joined by the Meta, and farther north by the Apure, two very large rivers, which drain the whole eastern side of the Andes in an extent of 10 degrees of latitude, and then runs eastward to its mouth, where it forms an extensive delta and enters the Atlantic by many channels. As the Upper Orinoco runs west, and the Lower Orinoco east, it makes a complete circuit round the Parima mountains, so that its mouth is only two degrees distant from the meridian of its sources.

The Cassiquiare leaves the Orinoco near the south base of the Peak of Duida, and joins the Rio Negro, a chief tributary of the Amazon, at the distance of 180 miles.

The Orinoco is navigable for 1000 miles at all seasons; a fleet might ascend it from the Dragon's Mouth to within 45 miles of Santa Fé de Bogotá. It receives many navigable rivers, of which the Guaviare, the Atures, and the Meta are each larger than the Danube. The Meta may be ascended to the foot of the Andes; its mean depth is 36 feet, and in many places 80 or 90. It rises so high in the Andes that Baron Humboldt says the vegetable productions at its source differ as much from those at its confluence with the Orinoco, though in the same latitude, as the vegetation of France does from that of Senegal. The larger feeders of the Orinoco come from the Andes, though many descend to it from both sides of the Parima, in consequence of its long circuit among these mountains.

The basin of the Orinoco has an area of 300,000 square miles, of which the upper part is impenetrable forest, the lower is Llanos.

The floods of the Orinoco, like those of all rivers entirely within the torrid zone, are very regular, and attain their height nearly at the same time with those of the Ganges, the Niger, and the Gambia. They begin to swell about the 25th of March, and arrive at their full and begin to decrease on the 25th of August. The inundations are very great, owing to the quantity of rain that falls in the wooded regions, which exceeds 100 inches in a year.

Below the confluence of the Apure the river is three miles and a quarter broad, but during the floods it is three times as much. By the confluence of four of its greatest tributaries at the point at which it bends to the east, a low inland delta is formed, in consequence of which 3600 square miles of the plain are under water during the inundation. The Orinoco in many places smells of musk from the number of dead crocodiles.

Upper Peru is the cradle of the Amazon, the greatest of rivers, which drains the chain of the Andes from the equator to the 20th parallel of southern latitude. Its highest branch, which bears the name of Marañon, issues in two streams from the Lake of Lauricocha in the plain of Bonbon, [latitude 10° 14′ S., distant 32 leagues in a direction north-west from the city of Lima,] at a great elevation in the Andes: it runs in a deep longitudinal valley from south to north, till it bursts through the eastern ridge at the Pongo de Manseriche, near the town of San Borja, [latitude 5° 30′ S.,] from whence it follows an uniform eastern course of nearly 4000 miles including its windings, till it reaches the Atlantic. West of San Borja and on its southern bank it receives the Huallaga and Ucayali, the latter a river of great size which rises in the Andes of Vilcañota, S. of Cusco, where its source was visited and its position determined by Mr. Pentland. The Amazon is supposed to drain an area of two millions and a half of square miles, which is ten times the size of France. In some places it has a great depth; it is navigable 2200 miles from its mouth, where it is 96 miles wide.

The name of the river is three times changed in its course: [at its head waters it is named the Taguaragua;] it is known as the Marañon from its source to the confluence of the Ucayali; from that point to its junction with the Rio Negro it is called the Solimöes; and from the Rio Negro till it enters the ocean it is the River Amazon.

[The Amazon was navigated in its whole length by the Fray Manuel Sobreviela, in the year 1790. It was then supposed that through this channel the Viceroy at Lima might communicate with the court at Madrid in three months.]

The number, length, and volume of its tributaries are in proportion to its magnitude; even the affluents of its affluents are noble streams. More than 20 superb rivers, navigable almost to their sources, pour their waters into it, and streams of less importance are numberless. Two of the largest are the Huallaga and the Ucayali: like their primary, the former has its origin near the mining district of the Cerro Pasco, and after a long northern course between the Cordilleras it breaks through a gorge similar to that of Manseriche and joins the Marañon in the plains; it is almost a mile broad above its junction. The Spanish governor of Peru sent Pedro de Orsoa down this river in the year 1560 to search for the Lake of Parima and the city of El Dorado. The Ucayali, not inferior to the Marañon itself, is believed by some eminent geographers to be the true Marañon. [The Ucayali has its principal source in the Apurimac, which arises in the Peruvian province of Tinta, latitude 16° S. and pursues an easterly direction. In latitude 12° 6′ S. it receives the river Jauja, which arises from the Lake Chincha-y-cocha, on the plains of Bonbon. About latitude 10° 45′ S. it is joined by the Beni, which has its origin in the Cordilleras in the neighbourhood of Cusco. More than 40 streams empty into the Apurimac before the Beni reaches it; at its junction with this river it takes the name of Ucayali, which pursues a north-easterly direction. It was explored in the year 1790, by the Fray Narciso Girbal y Barcelo; an interesting account of his perigrinations was published the following year, in the "Mercuris Peruano."] In a course of 1080 miles it is fed by accessaries from a wide extent of country, and at its junction with the main stream, near the mission of San Joaquin de Omaguas, [latitude 4° 45′ S.,] a line of 50 fathoms does not reach the bottom, and in breadth it is more like a sea than a river. By these streams there is access to Peru, and there is communication between the Amazon and the most distant regions around by other navigable feeders. Little is known of the rivers that empty themselves into the Amazon on its southern bank, between the Ucayali and the Madeira; the latter, which is its greatest affluent, rises near the sources of the Paraguay, the principal tributary of the Rio de la Plata. The River Amazon is not less extensively connected on the north. The high lands of Colombia are accessible by the Putumayo, the Japura, and other great navigable rivers; the Rio Negro, nearly nine miles broad, a little way above its junction with the Amazon, unites it with the Orinoco by the Cassiquiare; and lastly, the sources of the Rio Branco come very near to those of the Essequibo, an independent river of Demerara.

The main stream, from its mouth nearly throughout its length, is full of river islands, and most of its tributaries have deltoid branches at their junction with it. The annual floods of the Amazon are less regular than those of the Orinoco, and, as the two rivers are in different hemispheres, [the northern and southern,] they occur at opposite seasons. The Amazon begins to rise in December, is at its greatest height in March, and its least in July and August. The quantity of rain that falls in the deep forests traversed by this river is so great that, were it not for the enormous evaporation, and the streams that carry it off, the country would be flooded annually to the depth of eight feet. The Amazon is divided into two branches at its mouth, of which one joins the Parà south of the island of Das Johannes or Marajo, the other enters the ocean to the north of it.

The water of some of the rivers in equatorial America is white; in others it is of a deep coffee-colour, or dark green when seen in the shade, but perfectly transparent, and, when ruffled by a breeze, of a vivid green, like some of the Swiss lakes. In Scotland the brown waters come from peat-mosses; but it is not so in America, since they occur as often in forests as in savannahs. Sir Robert Schomburgk thinks they are stained by the iron in the granite; however, the colouring matter has not been chemically ascertained. The Orinoco and the Cassiquiare are white; Rio Negro, as its name implies, is black, yet the water does not stain the rocks, which are of a dazzling white. Black waters are sometimes, though rarely, found on the table-lands of the Andes.

The Rio de la Plata forms the third great water system of South America. The Rio Grande, its principal stream, rises in the mountains of Minas Geraes, in Brazil, and runs 500 miles on the table-land from north to south before it takes the name of Paranà. For more than 100 miles it is a continued series of cataracts and rapids, the greatest of which, El Salto Grande, is in about 24° 5′ lat. Above the fall the river is three miles broad, when all at once it is confined in a rocky pass only 60 yards wide, through which it rushes over a ledge with thundering noise, heard at the distance of many miles. The Paranà receives three large rivers on the right—the Paraguay, the Pilcomayo, and the Vermejo: all generally tend to the south, and unite at different distances before entering their primary at Corrientes. The Paraguay, 1200 miles long, is the finest of the three . in its upper part it is singularly picturesque, adorned with palms and other tropical vegetation, and its channel islands are covered with orange-groves. It springs from a chain of seven lakes, on the southern slopes of the Campos Pareicis, in Brazil, and may be ascended by vessels of considerable burthen through nineteen degrees of latitude. The Pilcomayo and Vermejo both come from Bolivia; the former traverses the desert of El Gran Chaco, the latter the district of Tarija. At Santa Fé the La Plata turns eastward, and before entering the Atlantic is augmented by the Uraguay from the north, which takes its name from the turbulence of its streams.

The Rio de la Plata is 2700 miles long, and for 200 miles from its mouth, up to Buenos Ayres, it is never less than 170 miles broad. Were it not for the freshness of its water, it might be mistaken for the ocean: it is, however, shallow, and loaded with mud, which discolours the Atlantic for 200 miles from its mouth.

The Paraguay is subject to dreadful floods. In 1812 the atmosphere was poisoned by the putrid carcases of drowned animals. The ordinary annual inundations of the Paranà, the principal or upper branch of the La Plata, cover 36,000 square miles.

In consequence of the vast extent of the very level plains along the base of the Andes, the basins of the three great rivers are apparently united. So small are the elevations that determine their direction, that with the exception of a portage of three miles, a water conveyance might be established from Buenos Ayres in 35° S. lat. to the mouth of the Orinoco in 9° N. lat. by inland navigation.

The Colorado, which runs in a long shallow stream through the Pampas of Buenos Ayres to the Atlantic, is formed of two principal branches, one from the west, and the other from the north, which unite at a great distance from the Atlantic, into which the river flows.

The Rio Negro, or Cusu-Lebu, rises at a great elevation, and separates the Pampas from Patagonia. In its long course through arid deserts to the Atlantic it does not receive a single adjunct, but it forms a communication between that ocean and Chile, as it reaches a pass in the Andes that is free from snow. There is some vegetation in its immediate neighbourhood; it has a bar at its mouth, and is navigable only for four miles above Carmen; it has floods twice in the year, one from the rains, the other from the melting of the snow in the Andes.

Some other streams from the Chilian Andes run through, but do not fertilize, the desolate plains of Patagonia.

There are various rivers in South America, unconnected with those described, which in any other country would be esteemed of a high order. Of many which descend from the mountains of Guiana, the Essequibo is the largest; its general width is a mile and a quarter; its water, though black, is transparent; and on its banks, and those of all its adjuncts, the forest reigns in impenetrable thickness. It rises in the Serra Acaray, which separates its basin from that of the Amazon, and, after a northerly course, falls into the Atlantic near 7° N. lat. by an outlet 14 miles broad, separated by three low islands into four branches. Sir Robert Schomburgk, whose scientific journeys have made us acquainted with a country of which so little was known, has shown that, by cutting a canal three miles long between the Madeira and the Guapore, an affluent of the Mamore, an inland navigation might be opened from Demerara to Buenos Ayres, over an extent of 42 degrees of latitude, with the exception of a portage of only 800 yards in the rainy season between Lake Amucu and the Quatata, a branch of the Rupununi, which flows into the Essequibo. But that is not the only water communication between Guiana and remote countries, great though the distance be, for the Napo, a tributary of the Solimöes, offers communication with Quito, the Huallaga with Peru and countries not far distant from the Pacific Ocean. By the Rio Negro, the Orinoco, the Cassiquiare, and its tributary the Meta, there is uninterrupted navigation to New Grenada and to within eight miles of Santa Fé de Bogotá. "If," says the distinguished traveller already mentioned, "British Guiana did not possess the fertility which is such a distinguishing feature, this water communication alone would render it of vast importance; but, blessed as it is with abundant fruitfulness, this extensive inland navigation heightens its value as a British colony; and, if emigration sufficient to make its resources available were properly directed thither, the port of Demerara would rival any in the vast continent of South America." It is certainly very remarkable that the tide of emigration has never set towards a country of such promise, abounding in valuable natural productions, and so much nearer to Great Britain than her colonies in the Pacific.

The Parà and San Francisco are the chief Brazilian rivers: both rise on the table-land; the former results from the union of the Tocantins and Araguay; it descends from the high lands in rapids in its northerly course, and, after running 1500 miles, joins the southern branch of the Amazon before entering the Atlantic south of the island of Marajo. The San Francisco is only 1275 miles long: it rises in the Serra Canastra in the province of Minas Geraes, and, after travelling northward between mountain ranges parallel to the coast, it breaks through them and reaches the ocean about the 11th degree of S. lat. As in the Appalachian chain, so here, many rivers come down the edge of the table-land to the level maritime plains of the Atlantic.

The historical renown and high civilization of Asia and Europe, their great wealth and population, may be attributed in a very great degree to the facility of transport afforded by their admirable river systems, and still more to the genius of the people who knew how to avail themselves of them; the same may be said of the inhabitants of the United States of America, while the Indians who have possessed these countries for ages never took advantage of the noble streams with which Providence had enriched and embellished them.

RIVERS OF AUSTRALIA.

After America, the land of the river and the flood, Australia appears in more than its usual aridity. The absence of large rivers is one of the greatest impediments to the improvement of this continent. What it may possess in the interior is not known, but it is certain that no large river discharges its water into the ocean, and most of the small ones are absorbed before they reach it.

The streams from the mountains on the eastern side of the continent are mere torrents, and would have short courses did they not run into longitudinal valleys, as for example the Hawkesbury. The Murrumbigee, the Lachlan, and the Macquarrie, formed by the accumulation of mountain torrents, are the largest.

The Murrumbigee rises in the ranges west of St. George's Lake, and, running south-west, meets the Lachlan, of unknown origin, coming from the east. After their junction they run into the Murray, a much larger stream, though only 350 feet broad, and not more than 20 feet deep: before entering the ocean in Encounter Bay, it passes through the Alexandrine Marsh: it is too shallow even for boats. The Darling is supposed to be merely the upper part of the Murray, probably rising towards the head of St. Vincent's Gulf. The origin of the Macquarrie is unknown; it is called the Fish River between Bathurst and Sydney; after running 300 miles north-west, it is lost in the marshes.

Swan River, on the western side of the continent, has much the same character; and from that river to the Gulf of Carpentaria, along the whole of the western and northern shores of the continent, there are none. The want of water makes it hardly possible to explore the interior of this continent. No country stands more in need of a complete system of irrigation, which could easily be accomplished from the nature of the rivers, which lie in deep channels, and might be converted into canals by dams, whence the water might be conveyed by channels over the surrounding country, as in Lombardy.

CHAPTER XX.

Lakes in general—European Lakes—Northern Europe—Of the Pyrenees, Alps, and Italy—Lake of Tiberias and Dead Sea—Asiatic Lakes—Caspian—Lakes of Aral, Baikal, and of the Himalaya—Sacred Lakes of Manasarowar—African Lakes—Bahr Assal—Zambeze—Ngami—American Lakes in Canada—Nicaragua—Titicaca.

THE hollows formed on the surface of the earth by the ground sinking or rising, earthquakes, streams of lava, craters of extinct volcanoes, the intersection of strata, and those that occur along the edges of the different formations, are generally filled with water, and constitute systems of lakes, some salt and some fresh. Many of the former may be remnants of an ancient ocean left in the depressions of its bed as the waters retired when the continents were raised above its surface.

Almost all lakes are fed by springs rising at the bottom, and they are occasionally the sources of the largest rivers. Some have neither tributaries nor outlets; the greater number have both. The quantity of water in lakes varies with the seasons everywhere, especially from the melting snow on mountain-chains and in high latitudes, and from periodical rains between the tropics. Small lakes occur in mountain-passes, formed by water which runs into them from the surrounding peaks; they are frequently, as in the Alps, very transparent, of a bright green or azure hue. Large lakes are common on table-lands, and in the valleys of mountainous countries, but the largest are on extensive plains. The basin of a lake comprehends all the land drained by it; consequently it is bounded by an imaginary line passing through the sources of all the waters that fall into it.

There are more lakes in high than in low latitudes, because evaporation is much greater in low latitudes than in high, and in this respect there is a great analogy between the northern plains of the two principal continents. Sheets of water of great beauty occur in the mountain valleys of the British islands, of Norway, and Sweden, countries similar in geological structure; and besides these there are two regions in the old world in which lakes particularly abound. One begins on the low coast of Holland, goes round the southern and eastern sides of the Baltic, often passing close to its shores, along the Gulf of Bothnia, and through the Siberian plains to Behring's Straits. The lakes which cover so much of Finland and the great lakes of Ladoga and Onega lie in a parallel direction; they occupy transverse rents which had taken place across the palæozoic strata, while rising in a direction from S.W. to N.E., between the Gulf of Finland and the White Sea; that elevation was, perhaps, the cause of the cavities now occupied by these two seas. Ladoga is the largest lake in this zone, having a surface of nearly 1000 square miles. It receives tributary streams, and sends off its superfluous water by rivers, and Onega does the same; but the multitude of small steppe lakes among the Ural Mountains and in the basin of the river Obi neither receive nor emit rivers, being for the most part mere ponds, though of great size, some of fresh and some of salt water, lying close together—a circumstance which has not been accounted for: those on the low Siberian plains have the same character.[1]

The second system of lakes in the old continent follows the zone of the mountain mass, and comprehends those of the Pyrenees, Alps, Apennines, Asia Minor, the Caspian, the Lake Aral, together with those on the table-land and in the mountains of central Asia.

In the Pyrenees lakes are most frequent on the French side; many are at such altitudes as to be perpetually frozen; one on Mont Perdu, 8393 feet above the sea, has the appearance of an ancient

[1] The salt-water lakes may possibly be the remains of the Ancient Ocean, while the hollows containing fresh water may be of subsequent formation.

volcanic crater. There is scarcely a valley in the Alpine range and its offsets that has not a sheet of water, no doubt owing to the cavities formed during the elevation of the ridges, and in some instances to subsidence of the soil: Lake Trüb, 7200 feet above the level of the sea, is the most elevated. There are more lakes on the north than on the south side of the Alps—the German valleys are full of them. In Bohemia, Gallicia, and Moravia there are no less than 30,000 sheets of water, besides great numbers throughout the Austrian empire.

Of the principal lakes on the northern side of the Alps, the Lake of Geneva, or Lake Leman, is the largest and most beautiful, from its situation, the pure azure of the waters, and the sublime mountains that surround it. Its surface, of about 240 square miles, is 1230 feet above the sea, and near Meillerie it is 1012 deep. The Lake of Lucerne is 1407 feet above the sea, and the lakes of Brienz 1900 feet. The Italian Lakes are at a lower level; the Lago Maggiore has only 678 feet of absolute altitude; they are larger than most of those on the north of the Alps, and, with the advantage of an Italian climate, sky, and vegetation, they surpass the others in beauty, though the mountains that surround them are less lofty.

These great lakes are fed by rivers rising in the glaciers of the higher Alps, and many large rivers issue from them. In this respect they differ from most of the lakes in Lower Italy, some of which are craters of ancient volcanoes, or perhaps ancient craters of elevation, where the earth had been swelled up by subterranean vapour without bursting, and had sunk down again into a hollow when the internal pressure was removed.[1]

In Syria, the Lake of Tiberias and the Dead Sea, sacred memorials to the Christian world, are situate in the deepest cavity on the earth. The surface of the Lake Tiberias is 329 feet below the level of the Mediterranean, surrounded by verdant plains bearing aromatic shrubs; while the heavy bitter waters of the Dead Sea, 1312 feet below the level of the Mediterranean, is a scene of indescribable desolation and solitude, encompassed by desert sands, and bleak, stony, salt hills. Thus there is a difference of level of 983 feet in little more than 60 miles, which makes the course of the river Jordan very rapid. The water of the Dead Sea is so acrid from the large proportion of saline matter it contains, that it irritates the skin: it is more buoyant, and has a greater proportion of salt,[2] than any that is known except the small lake of Eltonsk east of the Volga.

[1] The Lake of Perugia or Thrasymene is an exception.

[2] It would appear to be completely saturated with salt, if it be true, as stated by the American expedition under Lieut. Lynch, that the sounding lead brought up crystals of salt from its bottom in several parts. The extreme saltness of the Dead Sea appears to be owing to the saliferous strata

Though extensive sheets of water exist in many parts of Asia Minor, especially in Bithynia, yet the characteristic feature of the country, and of all the table-land of western Asia and the adjacent steppes, is the number and magnitude of the saline lakes. A region of salt lakes and marshes extends at least 200 miles along the northern foot of the Taurus range, on a very elevated part of the table-land of Anatolia. There are also many detached lakes, some exceedingly saline. Fish cannot live in the lake of Toozla; it is shallow, and subject to excessive evaporation. Neither can any animal exist in the Lake of Shabee or Urmiah, on the confines of Persia and Armenia, 300 miles in circumference: its water is perfectly clear, and contains a fourth part of its weight of saline matter. These lakes are fed by springs, rain, and melted snow, and, having no emissaries, the surplus water is carried off by evaporation.

It is possible that the volcanic soil of the table-land may be the cause of this exuberance of salt water. Lake Van, a sheet of salt water 240 miles in circumference, is separated from the equally salt lake Urmiah only by a low range of hills: and there are many pieces of fresh water in that neighbourhood, possibly in similar hollows.

Persia is singularly destitute of water; the Lake of Zurrah, on the frontiers of Afghanistan, having an area of 18 square miles, is the only piece of water on the western part of the table-land of Iran.

It is evident from the saline nature of the soil, and the shells it contains, that the plains round the Caspian, the Lake Aral, and the steppes, even to the Ural Mountains, had once formed part of the Black Sea; 57,000 square miles of that country are depressed below the level of the ocean—a depression which extends northwards beyond the town of Saratov, 300 miles distant from the Caspian. The surface of the Caspian itself, which is 82 feet below the level of the ocean, is its lowest part, and has an area of 140,000 square miles, nearly equal to the area of Great Britain and Ireland. In Europe alone it drains an extent of 850,000 square miles, receiving the Volga, the Ural, and other great rivers on the north. It has no tide, and its navigation is dangerous from heavy gales, especially from the south-east, which drive the water miles over the land: a vessel was stranded 46 miles inland from the shore. It is 3000 feet deep in some parts, but is shallower to the east, where it contains several islands, and where it is bounded by impassable swamps many miles broad. The Lake of Eltonsk, on the steppe east of the Volga, has an area of 130 square miles, and furnishes two-thirds of the

---

which form its banks, especially towards the south, where true pillars of salt, as stated in the Bible, were found projecting from the sandstone beds, by the American surveyors.

salt consumed in Russia. Its water yields 29·13 per cent. of saline matter, and from this circumstance is more buoyant than any that is known.[1]

The Lake of Aral, which is shallow, is 117 feet higher than the Caspian, and has an area of 23,300 square miles; it has its name from the number of small islands at its southern end, Aral signifying "island" in the Tartar language. Neither the Caspian nor the Lake of Aral have any outlets, though they receive large rivers; they are salt, and, in common with all the lakes in Persia, they are decreasing in extent, and becoming more salt, the quantity of water supplied by tributaries being less than that lost by evaporation. Most of the rivers that are tributary to the Lake of Aral are diminished by canals, that carry off water for irrigation; for that reason a very small portion of the waters of the Oxus reaches the lake. Besides, the Russian rivers yield less water than formerly from the progress of cultivation. The small mountain-lake, Sir-i-Kol, in the high table-land of Pamer, from whence the Oxus flows, is 15,630 feet above the sea; consequently there is a difference of level between it and the Dead Sea of nearly 17,000 feet.

The small number of lakes in the Himalaya is one of the peculiarities of these mountains. The Lake of Wuler, in the valley of Cashmere, is the only one of any magnitude; it is but 10 miles in

---

[1] The water of the Dead Sea contains 26·41 per cent. of saline ingredients, one of which is chloride of magnesium. The water of Lake Eltonsk contains a small quantity of chloride of calcium.

[An analysis of the water of the Dead Sea, by James C. Booth and Alexander Muckle of Philadelphia, is given in Lieut. Lynch's official report of the United States' Expedition to the Dead Sea and River Jordan. It is as follows:

Specific gravity at 60° F	1·22742
Chloride of magnesium	145·8971
" calcium	31·0746
" sodium	78·5537
" potassium	6·5860
Bromide of potassium	1.3741
Sulphate of lime	0·7012
	264·1867
Water	735·8133
	1000
Total amount of solid matter found by experiment	267·000

The great specific gravity of this water does not indicate saturation, because it is still capable of holding much chloride of sodium, and of course still more chloride of magnesium in solution. Crystals of chloride of sodium were found at a depth of 116 fathoms, which shows "that the water of the Dead Sea is very unequally charged with its constituents, and that no safe inference can be drawn from an analysis of surface water, and still less of any specimen" from an unknown depth.—*Official Report.*]

length and 4½ in breadth, but seems to be the residue of one that had filled the whole valley at some early period. There are many great lakes, both fresh and salt, on the table-land; the annular form of Lake Palte, at the northern base of the Himalaya, as represented on maps, is unexampled; the sacred lakes of Manasarowar and of Rakas Tal, in Great Tibet, occupy a space of about 400 square miles, in the centre of the Himalaya, between the gigantic peaks of Gurla on the south and of Kailas on the north; it is from the westernmost of these lakes (which communicate with each other), the Cho Lagan of the Tibetians, that the Sutlej rises, at an elevation of 15,250 feet above the level of the sea. These remarkable lakes mark the point from around which all the great rivers rising in the Himalaya have their origin. Tibet is full of lakes, many of which contain borax in solution. As most of the great lakes on the table-land are in the Chinese territories, strangers have not had access to them; the Koko-nor and Lake Lop seem to be very large; the latter is said to have a surface of 2187 square miles, and there are others not inferior to it in the north. The lakes in the Altaï are beautiful, larger and more numerous than in any other mountain-chain. They are at different elevations on the terraces by which the table-land descends to the flats of Siberia, and are, owing to geological phenomena, essentially different from those which have produced the Caspian and other steppe lakes. They seem to have been hollows formed where the axes of the different branches of the chain cross, and are most numerous and deepest in the eastern Altaï. Baikal, the largest mountain-lake, supposed to owe its origin to the sinking of the ground during an earthquake, has an area of 14,800 square miles, nearly equal to the half of Scotland. It lies buried in the form of a crescent, amid lofty granite mountains, which constitute the edge of the table-land to the south, ending in the desert of the Great Gobi, and in the north-west they gird the shore so closely that they dip into the water in many places; 160 rivers and streams fall into this salt lake, which drains a country probably twice the size of Britain. The river Angara, which runs deep and strong through a crevice at its eastern end, is its principal outlet, and is supposed to carry off but a small proportion of its water. Its surface is 1793 feet above the sea-level, and its climate is as severe as it is in Europe 10° farther north; yet the lake does not freeze till the middle of December, possibly from its depth, being unfathomable in some places with a line of 600 feet.

Two hundred and eighty years before the Christian era, the large fresh-water lake of Oitz, in Japan, was formed in one night, by a prodigious sinking of the ground, at the same time that one of the highest and most active volcanoes in that country rose from the depths of the earth.

Very extensive lakes occur in Africa; there appears to be a great number on the low lands on the east coast, in which many of the rivers from the edge of the table-land terminate. Among others there is the Bahr Assal (salt lake), 25 miles west of Tadjurra, in the country through which the Hawash flows, which has a depression of 570 feet below the level of the ocean, according to the measurement of Lieutenant Cristopher. Notwithstanding the arid soil of the southern table-land, it contains the fresh-water lake of N'yassi or Zambeze, one of the largest, being some hundred miles long; and, though narrow in proportion, it cannot be crossed in a boat of the country in less than three days, resting at night on an island, of which there are many. It lies between 300 and 400 miles west from the Mozambique Channel, and begins 200 miles north of the town of Tete, which is situate on the river Zambeze, from whence it extends from south-west to north-west, possibly to within a degree or two of the equator. It receives the drainage of the country to the south-east. In latitude 20° 20′ and east longitude 22° nearly, is situated the recently discovered Lake of Ngami, which has hitherto been but imperfectly explored. It is supposed to be more than 70 miles long. A large river, the Zouga, leaves it at its eastern extremity, where it is 200 yards wide; it is supposed to join the Limpopo. The elevation of the lake, as determined by its discoverers, Messrs. Livingstone, Oswell, and Murray, is 2825 feet above the sea.[1] No one knows what there may be in the unexplored regions of the Ethiopian desert; but Abyssinia has the large and beautiful lake of Dembia, situate in a spacious plain — the granary of the country — so high above the sea that spring is perpetual, though within the tropics. There are many other lakes in this great projecting promontory so full of rivers, mountains, and forests; but the lowlands of Soudan and the country lying along the base of the northern declivity of the table-land is the region of African lakes, of which the Tchad, almost the size of an inland sea, is in the very centre of the continent. Its extent, and the size of its basin, are unknown; it receives many affluents from the high lands called the Mountains of the Moon, certainly all those that flow from them east of Bornou, and it is supposed but not proved to be drained by the Tchadda, a principal tributary of the Niger. Other lakes of less magnitude are known to exist in these regions, and there are probably many more that are unknown. Salt-water lakes are numerous on the northern boundaries of the great lowland deserts, and many fine sheets of fresh water are found in the valleys and flat terraces of the Great and Little Atlas.

Fresh-water lakes are characteristic of the higher latitudes of both continents, but those in the old continent sink into insignificance in comparison with the number and extent of those in the new. In-

[1] Journal of Geographical Society of London, Vol. xx. p. 143.

deed a very large portion of North America is covered with fresh water; the five principal lakes — Superior, Huron, Michigan, Erie, and Ontario—with some of their dependants, probably cover an area of 94,000 square miles; that of Lake Superior alone, 32,000, which is only 1800 square miles less than the whole of England. The American lakes contain more than half the amount of fresh water on the globe. The altitude of these lakes shows the slope of the continent; the absolute elevation of Lake Superior is 672 feet; Lake Huron is 30 feet lower; Lake Erie 32 feet lower than the Huron; and Lake Ontario is 331 feet below the level of Erie. The river Niagara, which unites the two last lakes, is 33½ miles long, and in that distance it descends 66 feet; it falls in rapids through 55 feet of that height in the last half-mile, but the upper part of its course is navigable. The height of the cascade of Niagara is 162 feet on the American side of the central island, and 1125 feet wide. On the Canadian side the fall is 149 feet high, and 2100 feet wide—the most magnificent sheet of falling water known, though many are higher. The river St. Lawrence, which drains the whole, slopes 234 feet between the bottom of the cascade and the sea. The bed of Lake Superior is 300 feet, and that of Ontario 268 feet, below the surface of the Atlantic, affording another instance of deep indentation in the solid matter of the globe. Some lakes are decreasing in magnitude, though the contrary seems to be the case in America; between the years 1825 and 1838, Ontario rose nearly seven feet: and, according to the American engineers, Lake Erie had gained several feet in the same time. Lake Huron is said to be the focus of peculiar electrical phenomena, as thunder is constantly heard in one of its bays. The lakes north of this group are innumerable; the whole country, to the Arctic Ocean, is covered with sheets of water which emit rivers and streams. Lake Winnipeg, Reindeer Lake, Slave Lake, and some others, may be regarded as the chief members of separate groups or basins, each embracing a wide extent of country almost unknown. There are also many lakes on each side of the Rocky Mountains; and in Mexico there are six or seven lakes of considerable size, though not to be compared with those in North America.

There are many sheets of water in Central America, though only one is of any magnitude, the Lake of Nicaragua, in the province of that name, about 100 miles from the sea, which communicates with the Gulf of Mexico by the River San Juan.

In Central America the Andes are interrupted by plains and mere hills on the Isthmus of Tehuantepec and of Nicaragua, on each side of which there is a series of lakes and rivers, which, aided by canals, might form a water communication between the Atlantic and Pacific oceans. In the former, the line proposed would connect the river Guasacualco, on the Gulf of Mexico, with the Bay of Tehuantepec in the Pacific. In the Isthmus of Nicaragua, the Gulf of San Juan would be connected by the river of that name, and the chain of Lakes of Nicaragua and Leon, with the Bay of Realejo or the Gulf of Fonseca, with the Gulf of Costa Rica. Here the water-shed is only 615 feet above the sea, and of easy excavation, and the lake, situate in an extensive plain, is deep enough for vessels of considerable size.[1]

A range of lakes goes along the eastern base of the Andes, but the greater part of them are mere lagoons or marshes, some very large, which inundate the country to a great extent in the time of the tropical rains. There appears to be a deep hollow in the surface of the earth at the part where Bolivia, Brazil, and Paraguay meet, in which lies the Lake Xarayos, extending on each side of the river Paraguay, but, like many South American lakes, it is not permanent, being alternately inundated and dry, or a marsh. Its inundations cover 36,000 square miles. Salt and fresh water lakes are numerous on the plains of La Plata, and near the Andes in Patagonia, resembling in this respect those in northern latitudes, though on a smaller scale.

In the elevated mountain-valleys and table-lands of the Andes there are many small lakes of the purest blue and green colours, intensely cold, some being near the line of perpetual congelation. They are generally of considerable depth. The great fresh-water lake of Titicaca, however, in the Bolivian Andes, has an area of 2225 square miles, of 60 to a degree, and is more than 120 fathoms deep in many places, surrounded by splendid scenery. Though 12,846 feet above the level of the Pacific, and consequently higher than the Peak of Teneriffe, it contains several species of fish; its shores are cultivated, producing corn, barley, and potatoes; and peopled by a large aboriginal population, inhabiting towns and villages. Numerous vestiges of Peruvian civilization are everywhere to be met with; and in the island from which it derives its name, and where tradition places the origin of the last Inca dynasty, numerous specimens of Peruvian architecture still exist. It receives several rivers from either branch of the Andes, but has only one exit, the river Desaguadero, the waters of which are lost by evaporation and infiltration in the sandy soil through which it flows, and in its terminal lake or marsh of Aullagas.[2]

The limpid transparency of the water in lakes, especially in moun-

[1] The reader is referred to a paper recently published by Capt. R. Fitzroy, in the Journal of the Geographical Society, for a lucid description and review of the different projected canals and routes across the American isthmus, viz., by the Lake Nicaragua, River Guasacualco, Panama, and by Darien.

[2] See Pentland's map of the Lake of Titicaca, 1847, published at the Hyd. Office.

tainous countries, is remarkable; minute objects are visible at the bottom through many fathoms of water. The vivid green tints so often observed in Alpine lakes may be produced by vegetable dyes dissolved in the water, though chemical analysis has not detected them.

Lakes, being the sources of some of the largest rivers, are of great importance for inland navigation as well as for irrigation; while, by their constant evaporation, they maintain the supply of humidity in the atmosphere so essential to vegetation, besides the embellishment a country derives from them.

CHAPTER XXI.

Temperature of the Earth—Temperature of the Air—Radiation—Foci of Maximum Cold — Thermal Equator — Its Temperature, mean and absolute — Isothermal Lines — Continental and Insular Climates — Extreme Climates—Stability of Climate—Decrease of Heat in Altitude—Line of Perpetual Snow—Density of the Atmosphere—The Barometer—Measurement of Heights —Variations in Density and their Causes — Horary Variations — Independent Effect of the dry and aqueous Atmospheres — Mean Height of Barometer in different Latitudes—Depression in the Antarctic Ocean and in Eastern Siberia — Barometric Storms — Polar and Equatorial Currents of Air — Trade Winds — Monsoons — Land and Sea Breezes—Gyration of the Winds in the Extra-Tropical Zones—Winds in Middle European Latitudes—Hurricanes — The Laws of their Motion — Their Effect on the Barometer—How to steer clear of them—The Storm-Wave — Storm-Currents — Arched Squalls — Tornadoes—Whirlwinds — Water Spouts.

THE atmosphere completely envelops the earth to the height of about 50 miles; it bulges at the equator, and is flattened at the poles, in consequence of the diurnal rotation. It is a mixture of water in an invisible state and of air; but the air is not homogeneous; in 100 parts 79 are nitrogen or azote gas, and 21 of oxygen, the source of combustion[1] and animal heat. Besides these, there is a small quantity of carbonic acid gas, varying from 3 to 5 ten-thousandths, which is sufficient to supply all the vegetation on the earth with wood and

[1] [Oxygen is a supporter of combustion, but not the only one. The rapid chemical union of oxygen and a combustible body, accompanied by an extrication of light and heat, is termed combustion.]

leaves, and a very minute proportion of ammoniacal gas.[1] No doubt exhalations of various kinds ascend into the air, such as those which produce miasmata, but they are in quantities too minute to be detected by chemical analysis, so that the atmosphere is found to be of the same composition at all heights above the sea hitherto attained.[2]

The temperature of the earth's surface, and the phenomena of the atmosphere, depend upon the revolution and rotation of the earth, which successively expose all the parts of it, and the air which surrounds it, to a perpetual variation of the gravitating forces of the two great luminaries, and to annual and diurnal vicissitudes of solar heat. Atmospheric phenomena are consequently periodical and connected with one another, and their harmony, and the regularity of the laws which govern them, become the more evident in proportion as the mean values of their vicissitudes are determined from simultaneous observations made over widely extended tracts of the globe. The fickleness of the wind and weather is proverbial, but, as the same quantity of heat is annually received from the sun, and annually radiated into space, it follows that all climates on the earth are stable, and that their changes, like the perturbations of the planets, are limited and accomplished in fixed cycles, whose periods are still in many instances unknown. It is possible, however, that the earth and air may be affected by secular variations of temperature during the progress of the solar system through space, or from periodical changes in the sun's light and heat, similar to those which take place in many of the fixed stars. The secular variation in the moon's mean distance will no doubt alter the amount of her attractive force, though probably by a quantity inappreciable in the aërial tides; at all events variations arising from such circumstances could only become perceptible after many ages.

From experiments made by M. Peltier it appears that, if the absolute quantity of heat annually received by the earth were equally dispersed over its surface, it would, in the course of a year, melt a stratum of ice 46 feet deep covering the whole globe. It is evident

[1] The recent researches of Mr. Ville show that the quantity of ammoniacal vapour in the air is so very minute as to exercise no influence, as was supposed by Liebig, on vegetation. M. Lewy has lately found that in some situations, as at Santa Fé de Bogotà, the proportion of carbonic acid gas varies with the seasons: thus in eleven months out of the twelve, the air contains the ordinary dose, from $\frac{3}{10000}$ to $\frac{5}{10000}$ of its volume, whereas in September this proportion increases to $\frac{47}{10000}$, a circumstance quite inexplicable in the present state of meteorology.

[2] Professor Schoenbein of Basle attributes the peculiar smell, when bodies are struck by lightning, to a principle existing in the atmosphere, which he calls ozone, liberated by the decomposing action of electricity, and possessing the same electrical characters as bromine, chlorine, and iodine. He ascribes the luminous appearance of the ocean to the action of that principle on the animal matter it contains.

that, if so great a quantity of heat had been continually accumulated in the earth, instead of being radiated into space, it would have been transmitted through the surface to the poles, where it would have melted the ice, and the torrid zone, if not the whole globe, would by this time have been uninhabitable. In fact, every surface absorbs and radiates heat at the same time, and the power of radiation is always equal to the power of absorption, for, under the same circumstances, bodies which become soon warm also cool rapidly, and the earth, as a whole, is under the same law as the bodies at its surface.

Although part of the heat received from the sun in summer is radiated back again, by far the greater part sinks into the earth's surface, and tempers the severity of the winter's cold while passing through the atmosphere into the etherial regions.

The power of the solar rays depends on the manner in which they fall, as may be seen from the difference of climates. The earth is about 3,000,000 of miles nearer to the sun in winter than in summer, but the rays strike the northern hemisphere more obliquely in winter than in the other half of the year.

Sir John Herschel has shown that the elliptical form of the earth's orbit has but a trifling share in producing the variation of temperature corresponding to the difference of seasons. For although in one half of its orbit the earth is nearer the sun than in the other half, its motion is so much more rapid in the former than in the latter, that it is exposed for a shorter time to the sun's influence; thus a compensation takes place, and an equal distribution of light and heat is accorded to both hemispheres.

But on account of the present position of the earth's orbit, the *direct* heating power of the sun in summer is greater by one-fifteenth of its whole intensity in the southern than in the northern hemisphere in equal latitudes and under equal circumstances of exposure; for that reason the sufferings of travellers in the southern deserts are much more intolerable than in the northern. In the account of the exploring expedition into the interior of Australia, Captain Sturt mentions that "the ground was almost a molten surface, and if a match accidentally fell on it, it immediately ignited." Sir John Herschel has observed the temperature of the surface soil in South Africa as high as 159° Fahrenheit.[1]

Diurnal variations of heat are perceptible only to a small distance below the surface of the ground, because the earth is a bad conductor: the annual influence of the sun penetrates much farther. At the equator, where the heat is greatest, it descends deeper than elsewhere, with a diminishing intensity, but there, and everywhere throughout the globe, there is a stratum, at a depth varying from

[1] Sir John Herschel's Outlines of Astronomy, p. 218, 1849.

40 to 100 feet below the surface of the ground, where the temperature never varies, and is nearly the same with the mean temperature of the country over it. This zone, unaffected by the sun's heat from above, or by the internal heat from below, serves as an origin whence the effects of solar heat are estimated on one hand, and the internal temperature of the globe on the other. Below it the heat of the earth increases, as already mentioned, at the rate of one degree of Fahrenheit's thermometer for every 50 or 60 feet of perpendicular depth; were it to continue increasing at that rate, every substance would be in a state of fusion at the depth of 21 miles; hitherto, however, the experiments in mines and Artesian wells, whence the earth's temperature below the constant stratum is ascertained, have not been extended below 1700 feet.[1]

M. de Beaumont has estimated by the theory of Fourier, from the observations of M. Arago, that the quantity of central heat which reaches the surface of the earth is capable, in the course of a year, of melting a shell of ice covering the globe a quarter of an inch thick.[2]

---

[1] The protuberant matter at the earth's equator occasions a nutation in the lunar orbit, and the action of the sun and moon on that protuberant matter produces those inequalities in the earth's rotation known as the Luni-Solar Nutation and Precession: (See Connection of Physical Sciences, sections 5 and 11.) These inequalities have been computed on the hypothesis of the earth being a solid mass. Mr. Hopkins has found that the result would be the same if the earth consisted of a solid shell, enclosing a nucleus of liquid fire, provided the shell were from 800 to 1000 miles in thickness. According to the actual increase of internal heat, the earth must be in fusion at the depth of twenty-one miles, a circumstance equally inconsistent with the preceding result, and with the amount of precession. However, the temperature at which fusion takes place is probably different at different depths on account of the enormous pressure (see Connection of Phys. Sciences, p. 83). Now Mr. Hopkins has recently shown, that if the pressure has no effect in increasing the temperature of fusion, the existing temperature cannot be due to original central heat; but if it does affect it, then, along with the increasing tendency of heat to prevent solidification as the depth increases, there would be an increasing tendency to promote it, by rendering the mass fusible at a higher temperature. According as one or other of these tendencies predominates, different cases occur, consequently the internal state of the globe may be determined by experiments on the effect of high pressure on the temperature of fusion. Were the earth composed of a solid shell filled with fluid matter, the lava would stand at the same height in all volcanoes, which it does not; and the same would happen if the globe had a solid nucleus from high pressure, and a solid crust from refrigeration, with matter between, which is one of the possible cases arising from Mr. Hopkins's investigation. He shows, however, that from various circumstances the solid nucleus and the solid crust may be so united at intervals as to divide the molten matter into basins or seas of lava, which may be at different levels below the surface, a state that agrees better than any other with the phenomena of volcanoes.

[2] Annales des Sciences Géologiques, par M. Rivière, 1842.

The superficial temperature of the earth is great at the equator, it decreases gradually towards the poles, and is an exact mean between the two at the 45th parallel of latitude; but a multitude of causes disturb this law even between the tropics. It is affected chiefly by the unequal distribution of land and water, by the height above the sea, by the nature of the soil, and by vegetation, so that a line drawn on a map through all the places where the mean temperature of the earth is the same would be very far from coinciding with the parallels of latitude, but would approximate more to them near the equator. Between the tropics the temperature of the earth's surface is greater in the interior of continents than on the sea-coasts and islands, and in the interior of Africa it is greater than in any other part of the globe.

Temperature depends upon the property all bodies possess, more or less, of perpetually absorbing and emitting or radiating heat. When the interchange is equal, the temperature of a substance remains the same; but when the radiation exceeds the absorption, it becomes colder, and *vice versâ*. The temperature of the air is certainly raised by the passage of the solar heat through it, because it absorbs one-third of it before reaching the earth, but it is chiefly warmed by heat transmitted and radiated from the earth. The radiation is abundant when the sky is still, clear, and blue, but clouds intercept it; so that a thermometer rises in cloudy weather, and sinks when the air becomes clear and calm; even a slight mist diminishes radiation from the earth, because it returns as much heat as it receives. The temperature of the air is subject to such irregularities from these circumstances, and from the difference in the radiating powers of the bodies at the surface of the globe, that it is necessary to find, by experiment, the mean or average warmth of the day, month, and year, at a great variety of places, in order to have a standard by which the temperature in different parallels of latitude may be compared.

The mean diurnal temperature of the air, at any place, is equal to half the sum of the greatest and least heights of the thermometer during 24 hours, and, as the height of the thermometer is twice in the course of that time equal to the mean temperature of the place of observation, it might seem easy to obtain its value; yet that is not the case, for a small error in observation produces a very great error in such minute quantities, so that accuracy can only be attained from the average of a great number of observations, by which the errors, sometimes in excess and sometimes in defect, neutralize or balance each other. The mean value of quantities is a powerful aid to the imperfections of our nature in arriving at truth in physical inquiries, and in none more than in atmospheric phenomena; almost all the certain knowledge man has acquired with regard to the density and temperature of the air, winds, rain, &c., has been acquired by that method

The mean temperature of any one month at the same place differs from one year to another, but the mean temperature of the whole year remains nearly the same, especially when the average of 10 or 15 years is taken: for although the temperature in any one place may be subject to very great variations, yet it never deviates more than a few degrees from its mean state.[1]

[An illustration of the above statement is annexed: — The differences of mean temperature for the months and for the year are given in the following table, deduced from observations made at Philadelphia during sixty years, from 1790 to 1849 inclusive.[2]

	Mean of Sixty Years.	Range of Mean for Sixty Years.		Difference.
		Max.	Min.	
January	30·46	44	24	20
February	29·18	40	24	16
March	38·24	47	34	13
April	51·42	56	44	12
May	62·94	71	56	15
June	73·12	77	64	13
July	75 58	81	68	13
August	72·10	77	66	11
September	65·67	70	60	10
October	53·99	64	49	15
November	42·67	51	28	23
December	31·56	39	26	13
Annual mean for 60 years	52·04	54	49	5]

The motion of the sun in the ecliptic occasions perpetual variations in the length of the day, and in the direction of his rays with regard to the earth; yet, as the cause is periodic, the mean annual temperature from the sun's motion alone must be constant in each parallel of latitude. For it is evident that the accumulation of heat in the long days in summer, which is but little diminished by radiation during the short nights, is balanced by the small quantity of heat received during the short days of winter and its radiation in the long frosty and clear nights.[3] Were the globe everywhere on a level

[1] The mean of any number of unequal quantities is equal to their sum divided by their number: thus the mean temperature of the air at any place during a year is equal to the sum of the mean temperature of each month divided by 12. This method, however, will only give an approximate value; therefore, to ascertain the mean annual temperature at any place accurately, the mean of a number of years must be taken.

[2] [Dr. Ruschenberger's Report on Meteorology and Epidemics for the year 1851. *Transactions of the College of Physicians of Philadelphia.—New Series*, vol. i. p. 281.]

[3] The warmest time of the day is between two and three in the afternoon; the coldest, shortly before sunrise; but on mountain tops, where there is little radiation from the ground, the time of greatest warmth depends on

with the surface of the sea, and of uniform substance, so as to absorb and radiate heat equally, the mean heat of the sun would be regularly distributed over its surface in zones of equal annual temperature parallel to the equator, and would decrease regularly to each pole. The distribution of heat, however, in the same parallel is very irregular in all latitudes, even between the tropics, from the inequalities in the level and nature of the surface of the earth, so that lines drawn on a map through all places having the same mean annual temperature are nearly parallel to the equator only between the tropics: in all other latitudes they deviate greatly from it, and from one another.[1] Radiation is the principal modifying cause of temperature; hence the heat of the air is most powerfully modified by the ocean, which occupies three times as much of the surface of the globe as the land, and is more uniform in its surface, and also in its radiating power. On the land the difference in the radiating force of the mountains and table-lands from that of the plains — of deserts from grounds covered with rich vegetation — of wet land from dry, are the most general causes of variation: the local causes of irregularity are beyond enumeration.

There are two points in the northern hemisphere, both in the 80th parallel of latitude, where the cold is more intense than in any other part of the globe with which we are acquainted. One north of Canada in 100° W. long. has a mean temperature of —3°·5 of Fahrenheit,; while at the Siberian point, in 95° E. long., the mean temperature of the air is +1°; consequently it is four and a half degrees warmer than that north of Canada—a difference that has an influence even to the equator, where the mean temperature of the air is different in the different longitudes. Sir David Brewster has computed that the mean temperature of the North Pole of the earth's rotation is not under 5° of Fahrenheit, and may be even 17°, supposing the ocean to extend so far; but M. Arago's estimate on the hypothesis of there being land at the North Pole makes the cold much greater, for land increases the cold by abstracting heat from the air in high latitudes, and augments the heat by radiation in low latitudes.

The line of the maximum temperature of the atmosphere, or the atmospheric thermal equator, which cuts the terrestrial equator in the meridians of Otaheite and Singapore, passes through the Pacific

direct rays of the sun, and is therefore a little before noon. The maximum annual temperature occurs about the middle of July in the northern hemisphere, the least is in January, so that the former takes place some time after the summer solstice, because the earth absorbs more heat than it radiates during that interval, and for the contrary reason the greatest cold is some time after the winter solstice; the mean takes place in April and October.

[1] Lines drawn on a map or globe through all places where the mean annual temperature is the same are isothermal lines.

in its southern course, and through the Atlantic in its northern, has a mean temperature of 83°·84 of Fahrenheit. But by the comparison of many observations the mean equatorial temperature of the air is 82°·94 in Asia, 85°·10 in Africa, and 80°·96 in America: thus it appears that tropical Africa is the hottest region on earth. Moreover, the atmosphere in the tropical zone of the Pacific, when free from currents, is two degrees and a quarter warmer than the corresponding zone in the Atlantic, which is 82°·40. Local circumstances increase both heat and cold immensely; in the Nubian Desert, for example, the heat has been 150° of Fahrenheit in the sun, and 130° in the shade. Perhaps the greatest degree of heat on record was that experienced by Captain Griffiths near the Euphrates, where the thermometer stood at 156° in the sun, and 132° in the shade. In December, 1738, at Kiringa, in Siberia, Gmelin the elder experienced cold of 120°; the gentlest breeze would have rendered that cold fatal by the rapid abstraction of heat from the body.—(Dr. Thomson's Introduction to Meteorology.)

On account of the great extent of ocean, the isothermal lines in the southern hemisphere coincide more nearly with the parallels of latitude than in the northern. In the Antarctic Ocean the only flexure is occasioned by the cold of the south polar current, which flows along the western coast of the American continent. In the northern hemisphere the predominance of land and its frequent alternations with water, the prevalence of particular winds, irregularities of the surface, and the difference in the temperature of the points of maximum cold, cause the isothermal lines to deviate more from the parallels of latitude. They make two deep bends northward, one in the Northern Atlantic and another in the north-east of America, and at last they separate into two parts, and encircle the points of maximum cold.

Professor Dove has discovered that, in consequence of the excess of land in the northern hemisphere, and the difference in the effect produced by the sun's heat according as it falls on a solid or liquid surface, there is an annual variation in the aggregate mean temperature at the surface of the earth, whose maximum takes place during the sun's northern declination, and its minimum during its southern.[1]

Places having the same mean annual temperature, often differ materially in climate: in some the winters are mild and the summers

[1] For example, Professor Dove has found that the mean temperature of December, January, and February, at Toronto in Canada, added to the mean temperature of the same months at Hobart Town in Van Diemen's Land, exceeds the sum of the mean temperature of June, July, and August, at the same places, added together, by 22°·7 of Fahrenheit. Similar results, though varying in amount, were obtained for many corresponding places in the two hemispheres, which establishes the law given in the text.

cool, whereas in others the extremes of heat and cold prevail: England is an example of the first; Quebec, St. Petersburg, and the Arctic regions are instances of the second. The solar heat penetrates more abundantly and deeper into the sea than into the land; in winter it preserves a considerable portion of that which it receives in summer, and from its saltness does not freeze so soon as fresh water; hence the ocean is not liable to the same changes of temperature as the land, aud by imparting its heat to the winds it diminishes the severity of the climate on the coasts and in islands, which are never subject to such extremes of heat and cold as are experienced in the interior of continents. The difference between the influence of sea and land is strikingly exemplified in the high latitudes of the two hemispheres. In consequence of the unbounded extent of the ocean in the south, the air is so mild and moist that a rich vegetation covers the ground, while in the corresponding latitudes in the north the country is barren from the excess of land towards the Polar Ocean, which renders the air dry and cold. A superabundance of land in the equatorial regions, on the contrary, raises the temperature, while the sea tempers it.

Professor Dove has shown from a comparison of observations that northern and central Asia have what may be termed a true continental climate both in summer and in winter — that is to say, a hot summer and cold winter; that Europe has a true insular or sea climate in both seasons, the summers being cool and the winters mild; and that in North America the climate is inclined to be continental in winter, and insular in summer. The extremes of temperature in the year are greater in central Asia than in North America, and greater in North America than in Europe, and that difference increases everywhere with the latitude. In Guiana within the tropics the difference between the hottest and coldest months in the year is 2°·2 of Fahrenheit, in the temperate zone it is about 60°, and at Yakutsk in Siberia 114°·4. Even in places which have the same latitude as in northern Asia, compared with others in Europe or North America, the diversity is very great. At Quebec the summers are as warm as those in Paris, and grapes sometimes ripen in the open air, yet the winters are as severe as those in St. Petersburg. In short, lines drawn on a map through places having the same mean summer or winter temperature are neither parallel to one another, to the isothermal or geothermal lines, and they differ still more from the parallels of latitude.[1]

---

[1] In the same manner as isothermal lines are supposed to pass through all parts of the globe where the mean temperature of the air is the same, so the isogeothermal lines are supposed to pass through all places where the mean heat of the ground is the same: the isothermal lines are supposed to be drawn through all places having the same mean summer temperature; and the isochimenal lines pass through all places where the mean winter

Observations tend to prove that all the climates on the earth are, and have remained so from the remotest historical periods, stable; and that their vicissitudes are only oscillations of greater or less extent, which vanish in the mean annual temperature of a sufficient number of years. There may be a succession of cold summers and mild winters, but in some other country the contrary takes place; the distribution of heat may vary from a variety of circumstances, but the absolute quantity gained and lost by the whole earth in the course of a year is invariably the same.[1]

Since the air receives its warmth chiefly from the earth, its temperature diminishes with the height so rapidly, that at a very small elevation the cold becomes excessive, as the perpetual snow on the mountain-tops clearly shows. Besides, as the warm air ascends it expands, and its capacity for heat being increased more becomes latent, which gradually diminishes the sensible heat shown by the thermometer: the decrease is at the rate of a degree of Fahrenheit's thermometer for every 334 feet. By computations founded on the capacity of the air for heat, and absorption of the solar light in the atmosphere, M. Fourier has estimated the temperature of the ethereal regions to be — 50° of Fahrenheit, while M. Pouillet estimates it at — 220° from direct experiments on the radiation of terrestrial heat into a clear blue sky during the night.

The atmosphere, being a heavy and elastic fluid, decreases in density upwards, according to a determinate law, so rapidly, that three-fourths of the whole air it contains are within four miles of the earth, and all the phenomena perceptible to us—as clouds, rain, snow, and thunder—occur within that limit. The air even on the tops of mountains is so rare as to diminish the intensity of sound, to affect respiration, and to occasion a loss of muscular strength in man and animals.[2]

---

temperature is the same. The practice of representing to the eye these lines on a map or terrestrial globe is of the greatest use in following and understanding the complicated phenomena of temperature and magnetism.

[1] According to the researches of M. Arago, the climate of France has not altered since a century before the Christian era, that is, in a period of two thousand years; and M. Dureau de la Malle has arrived at the conclusion that the climate of Italy has not varied from the time of Cato the Censor, who died 147 years before Christ, and the present time, or in 20 centuries, by comparing the times of ripening of different vegetables and plants, the periods of the vintage, and of the harvest, as given in the writings of Varro, Columella, &c., with the ripenings and harvests as they take place at present, and in the same localities.—Dureau de la Malle sur la Climatologie, &c., l'Italie, &c., Paris, 1850, 8vo. It has been established by actual observation of the thermometer, that the climate of central Italy has not varied since the time of Galileo, 220 years ago.

[2] If the heights above the earth increase by equal quantities, as a foot or a mile, the densities of the strata of air, or the heights of the barometer which are proportional to them, will decrease in geometrical progression:

Since the space in the top of the tube of a barometer is a vacuum, the column of mercury is suspended in the tube by the pressure of the atmosphere on the surface of the mercury in the cistern: hence every variation in the density or height of the atmosphere occasions a corresponding rise or fall in the barometric column.[1] The actual mean pressure of the atmosphere at the level of the sea is 15 pounds on the square inch; hence the pressure on the whole earth is enormous.

The decrease in the density of the air affords an accurate method of finding the height of mountains above the level of the sea, which would be very simple, were it not for changes of temperature which alter the density and interfere with the regularity of the law of its decrease. But as the heat of the air, as before stated, diminishes with the height above the earth at the rate of one degree of Fahrenheit's thermometer for every 334 feet, tables are constructed by the aid of which heights may be determined with great accuracy. In consequence also of diminished pressure, water boils at a lower temperature on mountain-tops than at the level of the sea, which affords another method of ascertaining heights.[2] [Lieut. Herndon, U. S. Navy, while descending the Amazon, observed the temperature at which water boiled, in order to ascertain the height at which he was daily above the level of the sea.]

---

for example, if the height of the barometer at the level of the sea be 29·922 inches, it will be 14·961 inches at the height of 18,000 feet, or one-half as great; it will be one-fourth as great at the height of 36,000 feet, one-eighth at the height of 54,000 feet, and so on.

[[1] For a lucid description of the structure and uses of the barometer, the reader is referred to Dr. Lardner's Hand-Books of Natural Philosophy.]

[2] A very ingenious little instrument, called the Aneroid Barometer, has been lately invented in France; which, at the same time that it forms an exact and very portable *weather*-glass in the common acceptation of that term, may be employed with considerable accuracy in ascertaining differences of level. Although not to be compared, as an instrument of precision, with the ordinary mercurial barometer, it is infinitely more portable, and gives with promptitude and accuracy small differences of pressure; it will be found, under proper precautions, and comparison from time to time with the mercurial barometer, a most useful companion to the traveller in mountain districts.

A friend of the author's has recently tested it in the latter respect on some of our railways, and found that observations made with it carefully, will give, on a line of 200 miles in extent (on the Great Western Railway, for instance, between London and Plymouth), the relative levels of the different stations within a very few feet. The observations can be made in a couple of minutes. The gentleman in question writes to us, that he considers the Aneroid Barometer will prove a very useful instrument to the geological and the botanical traveller.

See, for a description of this instrument, a pamphlet recently published at 84, Strand, by Mr. E. J. Dent, on the construction and Uses of the Aneroid Barometer. London, 1849.

By the annual and diurnal revolutions of the earth, each column of air is alternately exposed to the heat and cold of summer and winter, of day and night, and also to variations in the attraction of the sun and moon, which disturb its equilibrium, and produce tides similar to those in the ocean. Those produced by the moon ebb and flow twice during a lunation, and diurnal variations in the barometer, to a very small amount, are also due to the moon's attraction.[1] The annual undulations occasioned by the sun have their greatest altitudes at the equinoxes, and their least at the solstices, and the diurnal variations in the height of the barometer, which accomplish their rise and fall twice in 24 hours, are chiefly due to the effects of temperature on the dry air and moisture of the atmosphere, which, according to Mr. Dove's discoveries, produce independent pressures upon the mercurial column.

[Whenever a liquid passes into an aeriform condition, or, in other words, is converted into vapour, it does so at the expense of the temperature of surrounding objects. Water cannot become vapour without an addition of heat, which becomes *latent;* that is, not appreciable by any instrument or by the senses. When vapour passes to the liquid state, its *latent* heat becomes sensible. Hence, vaporization and condensation are always accompanied by an absorption and an extrication of heat. Dr. Lardner, in his excellent "Hand-Books of Natural Philosophy," states that "as much heat is absorbed in converting a given quantity of water at 212° into steam, as would be sufficient to raise the same quantity of water to the temperature of 1200° when not vaporized." The heat absorbed in vaporization is less as the temperature of the vaporizing liquid is higher. "Thus a given weight of water vaporized at 212° absorbs less heat than would the same quantity vaporized at 180°." Water at a temperature of 50° in passing to a state of vapour, also at 50°, absorbs 1080°; but water at 212° in becoming vapour, also at the temperature of 212°, absorbs 966°, or 114° less.]

A quantity of vapour is continually raised by the heat of the sun from the surface of the globe, which mixes in an invisible state with the dry air or gaseous part of the atmosphere. It is most abundant in the torrid zone, and, like the heat on which it depends, varies with the latitude, the season of the year, the time of the day, the

[1] The moon's orbit is very much elongated, so that her distance from the earth varies considerably, and consequently her attractive force. Moreover her attraction varies with the rotation of the earth, which brings her twice in 24 hours in the meridian of any place, once in the superior and once in the inferior meridian; but her action on the atmosphere is much inferior to that of the heat of the sun. The amplitude of the diurnal variation due to the action of the sun and moon is about 0·1043 of an inch at the equator, and diminishes to 0.015 towards the poles, the change taking place, according to Professor J. Forbes, in 64° 8′ of latitude.

elevation above the sea, and also with the nature of the soil, the land, and the water. There is no chemical combination between the aërial and aqueous atmospheres, they are merely mixed; and the diurnal variations arise from the superposition of two distinct diurnal oscillations, each going through its complete period in 24 hours; one taking place in the aërial atmosphere from the alternate heating and cooling of the air, which produce a flux and reflux over the point of observation; the other arising from the aqueous atmosphere, owing to the alternate production and destruction of vapour by the heat of the day and the cold of the night. The diurnal variations of the vapour have their maximum at or near the hottest hour of the day, and the minimum at or near the coldest, which is exactly the converse of the diurnal variations of the dry air. On the whole there are two maxima and two minima heights of the barometer in the course of the 24 hours from the combinations of these, but in the interior of continents far from water, where the air is very dry, there ought to be one maximum and one minimum during that period according to this theory. That appears to be actually the case in some parts of Asiatic Siberia, at Prague in Europe, at Toronto in Upper Canada, and in some places between the tropics.

Between the tropics, the barometer attains its greatest height at nine or half-past nine in the morning; it then sinks till four in the afternoon, after which it again rises and attains a second maximum at half-past ten or eleven in the evening; it then begins to fall till it reaches a second time its lowest point at four in the morning. The difference in the height is 0·117 of an inch, which gradually decreases north and south. Baron Humboldt mentions that the diurnal variations of the barometric pressure are so regular between the tropics, that the hour of the day may be inferred from the height of the mercury to within fifteen or sixteen minutes, and that it is undisturbed by storm, tempest, rain, or earthquake, both on the coasts and at altitudes 13,000 feet above them. The mean height of the barometer between the tropics at the level of the sea is 30 inches with very little fluctuation, but, owing to the ascending currents of air from the heat of the earth, it is less under the equator than in the temperate zones. It attains a maximum in western Europe between the parallels of 40 and 45°; in the North Atlantic the maximum is about the 30th parallel, and in the southern part of that ocean it is near the Tropic of Capricorn; the amplitude of the oscillations decreases from the tropics to about the 70th parallel, where the diurnal variations cease. They are affected by the seasons, being greatest in summer and least in winter. It appears also that the fluctuations are the reverse on mountain-tops from what they are on the plains, and probably at a certain height they would cease altogether.[1] It is a singular fact, discovered during Sir James

[1] Mr. Pentland has, however, found within the tropics, in the Peru-Bo-

C. Ross's last voyage, that the mean height of the barometer is an inch lower throughout the Antarctic Ocean and at Cape Horn than it is at the Cape of Good Hope or Valparaiso: that difference in the pressure of the atmosphere is probably connected with the perpetual gales off the extremity of South America. M. Erman observed a similar depression not far from the Sea of Okhotsk in eastern Siberia.

Besides the small horary undulations, there are vast waves moving over the oceans and continents in separate and independent systems, being confined to local yet very extensive districts, probably occasioned by long-continued rains or dry weather over wide tracts of country. By numerous barometrical observations made simultaneously in both hemispheres, the courses of several have been traced, some of which take 24, others 36 hours, to accomplish their rise and fall. One especially of these vast barometric waves, many hundreds of miles in breadth, has been traced over the greater part of Europe, and not its breadth only, but also the direction of its front and its velocity, have been clearly ascertained. The course of another wave has been made out from the Cape of Good Hope, through many intermediate stations, to the observatory at Toronto in Canada. Since every undulation has its perfect effect independently of the others, each one is marked by a change in the barometer, and this is beautifully illustrated by curved lines on paper, constructed from a series of observations. The general form of the curve shows the course of the principal wave, while small undulations in its outline mark the maxima and minima of the minor oscillations. Although, like all other waves, these in the atmosphere are but waving forms, in which there is no transfer of air, yet winds arise from them like tide-streams in the ocean, and Sir John Herschel is of opinion that the crossing of two of these vast aërial waves, coming in different directions, may generate at the point of intersection those tremendous revolving storms, or hurricanes, which spread desolation far and wide.

The air expands and becomes lighter with heat, contracts and becomes heavier with cold, and, as there are 82 degrees difference between the equatorial and polar temperature, the light warm air at the equator is constantly ascending to the upper regions of the atmosphere, and flowing north and south to the poles, from whence the cold, heavy air rushes along the surface of the earth to supply its place between the tropics, for the same tendency to restore equili-

---

livian Andes, at elevations between 11,000 and 14,000 feet, the horary oscillations of the barometer as regular, and nearly as extensive, as on the level of the sea in the same latitude; they have also been found to observe the same regularity at still more elevated stations in the Himalaya, although the extent of the oscillation was less, owing possibly to the extra-tropical position of that region.

brium exists in air as in other fluids.[1] These two superficial currents, which have no rotatory motion when they reach and leave the poles, are deflected from their meridional paths by friction from the continually increasing velocity of the earth's rotation, as they approach the tropics; and, as they revolve slower than the corresponding parts of the earth at which they arrive, the bodies on its surface strike against them with the excess of their velocity, so that the wind appears to a person who thinks himself at rest, to blow in a direction contrary to that of the earth's rotation. For that reason the current from the north pole becomes a north-east wind before arriving at the Tropic of Cancer, and that from the south pole becomes a south-east wind before it comes to the Tropic of Capricorn, their limit being about the 28th parallel of latitude on each side of the equator. In fact the difference of temperature puts the air in motion, and the direction of the resulting wind, at every place, depends upon the difference between the rotatory motion of the wind and the rotatory motion of the earth—the whole theory of the winds depends upon these circumstances.

Near the equator the trade-winds, north and south of it, so completely neutralize each other, that far at sea a candle burns without flickering — [that is, when perfectly calm; but there is no spot on the surface of the earth where the air is forever at rest.] This zone of calms and light breezes, known as the *Variables*, which has a breadth of abont five degrees and a half, is subject to heavy rains and violent thunder-storms. On account of the unequal distribution of land and water in the northern and southern hemispheres, the terrestrial equator is not the line of greatest heat, therefore the centre of the zone in question does not coincide with it, but runs along the sixth parallel of north latitude; however, it changes in position and extent with the declination of the sun, but never extends south of the equinoctial line.

Though the trade-winds extend to the 28th degree on each side of the equator, their limits vary considerably in different parts of the ocean, moving two or three degrees to the north or south, according to the position of the sun; and in the Atlantic the north-east trade-

[1] Clouds carried by the upper currents are frequently seen flying in a contrary direction to those nearer to the earth, and volcanic ashes have been carried to a distance of several hundred miles, when the trade winds below were blowing in an opposite direction. In January, 1839, Mr. Pentland collected volcanic ashes at nearly midway between the African and American continents, between lat. 10° and 14° N. lat., in the Atlantic Ocean, and which were evidently brought by the upper currents in a direction opposite to that of the then prevailing E.N.E trade-wind, and probably from some of the active volcanoes of central America. It is a well-known fact, that the constant trade-winds have only a limited vertical extent, and that at a certain elevation, on the top of the Peak of Teneriffe, for instance, the wind blows in an entirely contrary direction from that prevailing at the same time at the level of the sea in the same island.

wind is less steady than the south-east.[1] These perennial winds are known by recent observation to be less uniform in the Pacific than in the Atlantic; they only blow permanently over that portion between the Galápagos Archipelago, off the coast of America and the Marquesas. In the Indian Ocean the south-east trade-wind blows from a few degrees east of Madagascar to the coast of Australia, between 10° and 28° S. lat. The trade-winds are only constant far from land, because continents and islands intercept them, and change their course. On that account the numerous groups of islands westward from the Marquesas change the trade-winds into the periodical monsoons,[2] which are steady currents of air in the Arabian Gulf, the Indian Ocean, and China Sea, arising from diminished atmospheric pressure at each tropic alternately, from the heat of the sun, thereby producing a regular alternation of north and south winds, which, combining with the rotation of the earth on its axis, become a north-east wind in the northern hemisphere, and a south-east in the southern. The former blows from April to October, the latter from October to April; the change is accompanied by heavy rain and violent storms of thunder and lightning. The ascent of the warm air between the tropics occasions a depression of the barometer amounting to the tenth of an inch, which is a measure of the force producing the trade-winds.[3] In both hemispheres there is a regular variation in the mean height of the barometer within the zone in which these great aërial currents flow; it is higher at their polar limits, and decreases with extreme uniformity towards their equatorial boundaries, the difference in both hemispheres being 0·25 of an inch.

The unequal temperature of the land and sea causes sea-breezes which blow towards the land during the day, and land-breezes which blow sea-ward in the night: the former are by much the strongest, for the difference of the temperature of the air over the land and over the sea is greater during the day than in the night; they are not perceptible in the mornings and evenings, because the temperature of the land and water is then nearly the same.

[1] Lieutenant Maury, of the United States' navy, is led to believe that there is a region within the limit of the N.E. trade-winds, in the Atlantic, in which the prevailing winds are from the south and west: this region is somewhat in the shape of a wedge, with its base towards the coast of Africa, between the equator and 10° N. lat., and between the meridians of 10° and 25° W. long. In this space, in which the law of the trade-winds is reversed, there are great atmospheric disturbances, violent squalls, sudden gusts of wind, thunder-storms, heavy rains, baffling airs, and calms.

[2] Monsoon is derived from the Arabic and Malay word Moussin, a season. (Marsden, in Asiatic Researches.

[3] Sir John Herschel has observed, that on account of the upper flow of heated air not being immediately compensated by polar currents the barometer is two-tenths of an inch higher at the tropics than at the equator.

The Harmattan or N.E. wind of Africa is a periodical land breeze, which comes from the deserts of Northern Africa, occasionally carrying the sand 300 miles into the Atlantic; when violent it is said to have carried dust 700 miles; it fell on the deck of the Clyde Indiaman at that distance, on the 19th of January, 1826.[1]

The trade-winds and monsoons are permanent, depending on the apparent motion of the sun; but it is evident from theory that there must be partial winds in all parts of the earth, occasioned by the local circumstances that affect the temperature of the air. Consequently the atmosphere is divided into districts, both over the sea and land, in which the winds have nearly the same vicissitudes from year to year. The regularity is greatest towards the tropics, where the causes of disturbance are fewer. In the higher latitndes it is more difficult to discover any regularity, on account of the greater proportion of land, the difference in its radiating power, and the greater extremes of heat and cold. But even there a degree of uniformity prevails in the succession of the winds; for example, in all places where north and south winds blow alternately, a vane veers through every point of the compass in the transition, and in some places the wind makes several of these gyrations in the course of the year.[2] The south-westerly winds so prevalent in the Atlantic Ocean between the 30th and 60th degrees of north latitude are produced by the upper current being drawn down to supply the superficial current which goes towards the equator, and, as it has a greater rotatory motion than the earth in these latitudes, it produces a south-westerly wind. On this account the average voyage from Liverpool to New York in a sailing vessel is 40 days, while it is only 23 days from New York to Liverpool. For the same reason the average direction of the wind in England, France, Germany, Denmark,

[1] It is not improbable that many of the recorded falls of sand on vessels in the Atlantic, at great distances from land, and supposed to be derived from the coast of Africa, have been volcanic ashes, carried in the upper or counter current of the trade-winds, from the volcanoes of tropical America, in the instance cited by Mr. Pentland, at p. 40.

[2] In the northern hemisphere a north wind sets out with a less rotatory motion than the places have at which it successively arrives, consequently it veers through all the points of the compass from N. to N.E. and E. If a south wind should now spring up, it would gradually veer from S. to S.W. and W., because its rotatory velocity would be greater than that of the places it successively comes to. The combination of the two would cause a vane to veer from E. to S.E. and S.; but the rotation of the earth would now cause the south wind to veer round from S. to S.W. and W.; and should a north wind now arise, its combination with the west wind would bring the vane round from W. to N.W. and N. again. At the Greenwich Observatory the wind makes five gyrations in that direction in the course of a year. In Europe it is the contention of the N.E. and S.W. winds which causes the rotation of the wind, and the principal changes of weather, the S W. being warm and moist, the N.E. cold and dry, except where it comes over the German Ocean.

Sweden, and North America, is some point between south and west. North-westerly winds prevail in the corresponding latitudes of the southern hemisphere from the same cause. In fact, whenever the air has a greater velocity of rotation than the surface of the earth, a wind more or less westerly is produced; and when it has less velocity of rotation than the earth, a wind having an easterly tendency results. Thus there is a perpetual change between the different masses of the atmosphere, the warm air tempering the cold of the higher latitudes, and the cold air mitigating the heat of the lower; it will be shown afterwards that the aërial currents are the bearers of principles on which the life of the animal and vegetable world depends.

Hurricanes are those storms of wind in which the portion of the atmosphere that forms them revolves in a horizontal circuit round a vertical or somewhat inclined axis of rotation, while the axis itself, and consequently the whole storm, is carried forwards along the surface of the globe, so that the direction in which the storm is advancing is quite different from the direction in which the rotatory current may be blowing at any point; the progressive motion may continue for days, while the wind accomplishes many gyrations through all the points of the compass in the same time. In the Atlantic the principal region of hurricanes is to the east of the West India islands, and in the Pacific it lies east of the island of Madagascar; consequently the former is in the northern hemisphere, the latter in the southern; but in every case the storm moves in an elliptical or parabolic curve. The West Indian hurricanes generally have their origin eastward of the Lesser Antilles or Caribbean islands, and the vertex of their path near the tropic of Cancer, or about the exterior limit of the north-east trade wind. As the motion of the storm before it reaches the tropic is in a straight line from S.E. to N.W., and after it has passed the tropic from S.W. to N.E., the bend of the curve is turned towards Florida and the Carolinas. In the South Pacific Ocean the body of the storms moves in an exactly opposite direction. The hurricanes which originate south of the equator, and whose initial path is from N.E. to S.W., turn at the tropic of Capricorn and then tend from N.W. to S.E., so that the bend of the curve is turned towards Madagascar.

The extent and velocity of the Atlantic hurricanes are great; the most rapid move at the rate of 90 miles an hour. The hurricane which took place on the 12th of August, 1830, was traced from the eastward of the Caribbean islands to the banks of Newfoundland, a distance of more than 3000 miles, which it passed over in six days. Although that of the 1st of September, 1821, was not so extensive, its velocity was greater, as it moved at the rate of 30 miles an hour. Small storms are generally more rapid than those of great magnitude. Sometimes they appear to be stationary, sometimes they stop and again proceed on their course, like water-spouts. Hurricanes are occasionally contemporaneous, and so near to one another as to travel in almost parallel tracks. This happened in the China Seas, in October, 1840, when the two storms met at an angle of 47°, and it was supposed that the ship Golconda foundered in that spot with 300 people on board. A hurricane has been split or divided by a mountain into two separate storms, each of which continued its new course, and the gyrations were made with increased violence. This occurred in the gale of the 25th of December, 1821, in the Mediterranean, when the Spanish mountains and the maritime Alps became new centres of motion.

By the friction of the earth the axis of the storm bends a little forward, and the whirling motion begins in the higher regions of the atmosphere before it is felt on the earth: this causes a continual intermixture of the lower and warmer strata of air with those that are higher and colder, producing torrents of rain, and sometimes violent electric explosions.

The rotation as well as the course of the storm is in a different direction in the two hemispheres, though always alike in the same. In the northern the gyratory movement of the wind is from east, through the north, to west, south, and east again; while in the southern hemisphere the rotation about the axis of the storm is in the contrary direction. Hurricanes happen south of the equator between December and April; in the West Indies, between June and October. Rotatory storms frequently occur in the Indian Ocean, and the typhoons of the China Seas are real hurricanes of great violence. Both conform to the laws of such winds in the northern hemisphere. The Atlantic storms probably reach Spain, Portugal, and the coast of Ireland. Two circular storms have passed over Great Britain, and small ones often occur between the Chops of the Channel and Madeira. A true hurricane passed over Ireland and the west coast of England in January, 1839: a strong gale had blown from S.S.E. on the 6th, when about ten in the evening the air became suddenly calm and warm, which was evidently during the passage of the axis of the storm, for soon after the gale was renewed with the utmost violence, but now it was from the S.W. and W.S.W., and on the evening of the 7th was accompanied by snow, thunder, lightning, and intense cold. At Leeds, 70 miles distant from the Irish Sea, and separated from it by a ridge of hills, there was everywhere a saline deposit.

The temperature of winds depends upon the nature of the surface over which they blow: in Europe the coldest and driest wind is from the N. and N.N.E.; in America it is from the N. and N.N.W., because both come from the polar ice, and sweep over extensive tracts of land. The warm and moist winds in Europe are from the

S.W., because they blow over a great extent of ocean, especially on the western side of the continent.

The revolving motion accounts for the sudden and violent changes observed during hurricanes. In consequence of the rotation of the air, the wind blows in opposite directions on each side of the axis of the storm, and the violence of the blast increases from the circumference towards the centre of gyration, but in the centre itself the air is in repose: hence, when the body of the storm passes over a place, the wind begins to blow moderately, and increases to a hurricane as the centre of the whirlwind approaches; then in a moment a dead and awful calm succeeds, suddenly followed by a renewal of the storm in all its violence, but now blowing in a direction diametrically opposite to what it had before; this happened in the island of St. Thomas on the 2nd of August, 1837, where the hurricane increased in violence till half-past seven in the morning, when perfect stillness took place for 40 minutes, after which the storm recommenced in a contrary direction. The breadth of a hurricane is greatly augmented when its path changes its direction in crossing the tropic. In the Atlantic the vortex of one of these tempests has covered an area from 600 to 1000 miles in diameter. The breadth of the lull in the centre varies from 5 to 30 miles: the height is from 1 to 5 miles at most; so that a person might see the strife of the elements from the top of a mountain, such as Teneriffe or Mowna Roa, in a perfect calm, for the upper clouds are frequently seen to be at rast during the hideous turmoil in the lower regions.

The sudden fall of the mercury in the barometer in latitudes habitually visited by hurricanes is a certain indication of a coming tempest. In consequence of the centrifugal force of these rotatory storms, the air becomes rarefied, and, as the atmosphere is disturbed to some distance beyond the actual circle of gyration or the limits of the storm, the barometer often sinks some hours before its arrival; it continues sinking the first half of the hurricane, and again rises during the passage of the latter half, though it does not attain its greatest height till the storm is over. The diminution of atmospheric pressure is greater, and extends over a wider area, in the temperate zones than in the torrid, on account of the sudden expansion of the circle of rotation where the gale crosses the tropic.

As the fall of the barometer gives warning of the approach of a hurricane, so the laws of the storm's motion afford to the seaman knowledge to avoid it. In the northern temperate zone, if the gale begins from the S.E. and veers by S. to W., the ship should steer to the S.E.; but if the gale begins from the N.E. and changes through N. to N.W., the vessel ought to go to the N.W. In the northern part of the torrid zone, if the storm begin from the N.E and veer through E. to S.E., the ship should steer to the N.E.; but if it begin from the N.W. and veer by W. to S.W., the ship should steer to the S.W., because she is on the south-western side of the storm. Since the laws of storms are reversed in the southern hemisphere, the rules for steering vessels are necessarily reversed also.[1]

A heavy swell or storm-wave is peculiarly characteristic of these tempests. In the centre of the hurricane the pressure of the atmosphere is so much diminished by rotation, that the mercury in the barometer falls from one to two, and even two and a half inches. On that account the pressure of the ocean beyond the range of the wind raises the water in the centre of the vortex about two feet above its usual level, and proportionally to the degree of diminished pressure over the whole area of the storm. This mass of water, or storm-wave, is driven bodily along with, or before the tempest, and rolls in upon the land like a huge wall of water. It is similar to the earthquake-wave, and is by no means the heaping up of the water after a long gale. Ships have been swept by it out of docks

---

[1] In all hurricanes hitherto observed, the sinking of the mercury, and the increase of the wind, have been more or less regularly progressive till within three or four hours' sail of the centre of the storm; and in one class they have continued so even to the centre; while in another class, and by far the most terrible, the depression of the mercury has been sudden and excessive when within that distance of the centre, and the violence of the tempest far beyond the average. When a ship is within 50 or 60 miles of the centre, the storm has the mastery, and seamanship is of little avail. Rules for avoiding this calamity, and for managing a ship when involved in a hurricane, are fully explained in the 'Hurricane Guide,' by Wm. Radcliff Birt, published under the sanction of the Admiralty, in 12mo., London, 1850; a little book in which the navigator will find information conveyed in a very intelligible manner on the subject; and in the new edition, 1851, of the 'Sailor's Horn-Book for the Laws of Storms,' by H. Piddington, Esq., President of the Marine Courts of Inquiry at Calcutta. The following approximate table is given by the latter author to serve as a guide till better data shall be obtained: —

Average fall of the barometer per hour.	Distance of a ship from the centre of the storm, in miles.
From 0·020 to 0·060	From 250 to 150
" 0·060 " 0·080	" 150 " 100
" 0·080 " 0·120	" 100 " 80
" 0·120 " 0·150	" 80 " 50

The rate of fall per hour doubles after the storm has lasted six hours, and within three hours of the centre of the hurricane the mercury will fall four times as fast, if it be of the violent class.

Colonel James Capper discovered the rotatory motions of storms, and W. C. Redfield, Esq., of New York, was the first who determined their laws. Colonel Reid, Governor of Barbadoes, Dr. Thom, of the 86th regiment, and Mr. Piddington, of Calcutta, have also written, and added greatly to our knowledge, on the subject; whilst Mr. Birt has united in a very abridged form the practical information collected by the authors who preceded his little essay on Hurricanes.

and rivers, and it has sometimes carried vessels over reefs and banks so as to land them high and dry; this happened to two ships on the coast of the Eastern Andaman islands, in 1844. Coringa, on the Coromandel coast, is particularly subject to inundations from that cause. In 1789 the town and 20,000 inhabitants were destroyed by a succession of these great waves during a hurricane, and as many perished there in 1839.

Besides storm-waves, storm-currents are raised, which revolve with the rotation of the wind, and are of the greatest force near the centre of the vortex.

The rise of the sea by the pressure of the surrounding ocean, and the irresistible fury of the wind, make a tremendous commotion in the centre of the storm, where the sea rises, not in waves, but in pyramidal masses: the noise during its passage resembles the deafening roar of the most tremendous thunder; and in the typhoons in the China seas it is like numberless voices raised to the utmost pitch of screaming. In general there is very little thunder and lightning; sometimes a vivid flash occurs during the passage of the centre, or at the beginning of the storm; yet in Barbadoes the whole atmosphere has been enveloped in an electric cloud.

A thick lurid appearance, with dense masses of cloud in the horizon, ominous and terrible, are the harbingers of the coming tempest. The sun and clouds frequently assume a fiery redness, the whole sky takes a wild and threatening aspect, and the wind rises and falls with a moaning sound, like that heard in old houses on a winter's night: it is akin to the "calling of the sea," a melancholy noise which, in a dead calm, presages a storm on some parts of the English coast.

Those intensely violent gales, of short duration, called *arched squalls*, because they rise from an arch of clouds on the horizon, are not rotatory; they occur in the Straits of Malacca, attended by fierce thunder and lightning and a lurid phosphorescent gleam. The north-western gales in the Bay of Bengal, the tornadoes on the African coast, and the pampéros of the Rio de la Plata, are of the same nature. On an average a strong gale moves at the rate of 40 miles an hour, a storm at about 56, and hurricanes at 90. Deserts, especially those of Africa and Asia, are subject to intensely hot winds of short duration, frequently fatal to exhausted travellers; of these the simoon and sand wind are the most formidable; a red lurid appearance in the atmosphere caused by the quantity of burning sand raised by the wind gives warning of their approach; everything is scorched in their passage, and breathing becomes painful; it is probably owing to the sand wafted by them that these winds are so deleterious, and not to their temperature, since air heated to a much higher degree may be breathed with impunity, as has been proved by Sir Joseph Banks and by Sir Francis Chantrey, in an atmosphere raised to more than 300°. The simoon generally blows only a few hours, but sometimes it continues for two or three days, when it comes in gusts driving clouds of sand — nothing can withstand it. There can be no doubt that unaccountably sudden changes of temperature occasion these formidable winds.

Whirlwinds are frequent in tropical countries, especially in deserts; sometimes several are seen at one time in the Arabian deserts, of all sizes, from a few feet to some hundred yards in diameter. They occur in all kinds of weather, by night as well as by day, and come without the smallest notice, rooting up trees, overwhelming caravans, and throwing down houses; and as they produce water-spouts when they reach the sea, they dismantle and even sink ships. Pillars of sand are often raised by them on the African deserts two or three hundred feet high. In Nubia, Bruce saw eleven advancing towards him with considerable swiftness: it was vain to think of flying where the speed of the swiftest horse could have been of no avail, and that conviction riveted him to the spot. They retreated, leaving him in a state of mind between fear and astonishment, to which he could give no name. Whirlwinds advance with a loud rushing noise, and are frequently attended by electrical explosions. The water-spouts so frequently seen on the ocean originate in adjacent strata of air of different temperatures, running in opposite directions in the upper regions of the atmosphere. They condense the vapour, and give it a whirling motion, so that it descends tapering to the sea below, and causes the surface of the water to ascend in a pointed spiral till it joins that from above, and then it looks like two inverted cones, being thinner in the middle than either above or below. When a water-spout has a progressive motion, the upper and under part must move in the same direction, and with equal velocity — otherwise it breaks, which frequently happens.

CHAPTER XXII.

Evaporation — Distribution of Vapour — Dew—Hoar Frost—Fog—Region of Clouds—Forms of Clouds—Rain—Distribution of Rain—Quantity — Number of rainy Days in different Latitudes—Rainless Districts—Snow Crystals—Line of Perpetual Snow—Limit of Winter Snow on the Plains —Sleet—Hail—Minuteness of the Ultimate Particles of Matter—Their Densities and Forms—Their Action on Light—Colour of Bodies—Colour of the Atmosphere—Its Absorption and Reflection of Light — Mirage — Fog Images—Coronæ and Halos—The Rainbow—Iris in Dewdrops—The Polarization of the Atmosphere — Atmospheric Electricity — Its Variations — Electricity of Fogs and Rain — Inductive Action of the Earth — Lightning—Thunder—Distribution of Thunder-Storms—Back Stroke — St. Elmo's Fire — Phosphorescence — Aurora — Magnetism—Terrestrial Magnetism—The Dip—Magnetic Poles and Equator—Magnetic Intensity —Dynamic Equator—Declination—Magnetic Meridian — Lines of Equal Variation—Horary Variations—Line of Alternate Horary Phenomena—Magnetic Storms—Coincidence of the Lines of equal Magnetic Intensity with Mountain Chains—Diamagnetism.

MOISTURE is evaporated in an invisible form from every part of the land and water, and at all temperatures, even from snow. Mr. Darwin mentions that the snow once entirely disappeared from the Nevado of Aconcagua in Chile, which is 23,910 feet high, from evaporation under a cloudless sky and an excessively dry air. The vapour rises and mixes with the atmosphere; and as its pressure and density diminish with the height above the surface of the earth, in consequence of gravitation, there is absolutely less moisture in the higher than in the lower regions of the air.[1]

Seven-tenths of the atmosphere rests on the ocean; therefore the sea has the greatest influence in modifying climates and supplying the air with moisture. The evaporation is greatest between the tropics, from the excess of heat and the preponderance of the ocean,

[1] The humidity of the air is measured by the Hygrometer, an instrument which shows the rapidity of evaporation at all temperatures; for the rate of evaporation is in proportion to the dryness of the atmosphere, and is nearly in the inverse ratio of the density. When the evaporation is below 15° on the scale of the Hygrometer the air is very damp, when above 70° it is intensely dry. The best mode of determining the quantity of moisture in the air is by the wet bulb thermometer, which shows the temperature at which the atmosphere is saturated with humidity: hence the amount of the latter is easily found in the tables.

and its average quantity decreases from thence to the poles. Over the open sea, in all latitudes, the air is saturated with moisture; and in that over the coasts the quantity is very great, but it diminishes from the coasts to the interior of the continents. In the interior of the United States of North America, in the deserts of Asia, and in the interior of New Holland, the air is continually dry. There is scarcely any evaporation in the deserts of Africa, and the extreme heat, increased by the reverberation of the sand, opposes aqueous precipitation, so this land is doomed to perpetual sterility. The air over the steppes of Siberia is likewise nearly deprived of moisture. The greatest degree of dryness on record is that observed by M. Erman between the valleys of the Irtish and Obi, after a continued south-west wind and a temperature of 74° 7′ of Fahrenheit.

Throughout all the countries in the northern hemisphere where observations have been made on the variations of atmospheric moisture, it appears that the air contains less vapour in January than in any other month of the year, yet at that time there is the greatest dampness to our sensations; while in July the air is driest, and yet, on account of the heat, evaporation is the greatest: the reason is, that the heat in July dissolves the moisture and increases its elasticity or tension so much that it becomes insensible, whereas the cold of winter condenses it and renders it apparent. The proportion of vapour in the air varies with the direction of the wind: in Europe it is greatest in a S.W. wind, and least in a N.E.; the former being part of the equatorial current drawn down to the surface of the globe, comes warm and moist over the Atlantic, while the northern wind blows dry and cold from the pole. When the moisture is abundant and the tension great, which is often the case before rain, the air is very transparent, and distant objects appear nearer, and all their details are distinctly seen: from that circumstance the clearer view of distant mountains and headlands predicts wet weather. Very dry air is also exceedingly transparent, as on the tops of very lofty mountains, and in sandy deserts where the stars are seen to shine with uncommon lustre, and the brighter planets are visible in the daytime. On account of the heat the air between the tropics contains more moisture than elsewhere, and were it not for the amount of evaporation, the warmth there would be greater than it is, for a depression of temperature takes place during evaporation by the absorption of the heat which becomes latent and insensible to the feelings and to the thermometer. The evaporation and consequent absorption of heat may be so rapid as to produce intense cold; upon that principle M. Boutigny froze water and even quicksilver in a red-hot crucible.

The quantity of atmospheric moisture varies also with the hours of the day and night. In early morning the evaporation accumulates near the surface of the ground from the resistance of the air above it, but as the sun rises above the horizon the warm air descends and carries the vapour with it; so that the quantity near the ground is diminished till evening, when, on account of the lowness of the temperature, the ascending currents cease, and the air becomes loaded with vapour and deposits its excess in the shape of dew or hoar-frost. For in the night the earth radiates part of the heat it received during the day through the atmosphere into space, and the temperature of the bodies on its surface sinks below that of the air; and by abstracting part of the heat which holds the humidity of the air in solution, a deposition takes place. The dew-point is the temperature at which vapour is deposited on bodies colder than itself, but before any deposition takes place the air must be saturated with moisture to the temperature of the body upon which the dew is deposited. It is very abundant on the shores of continents, but it is not deposited on small islands in the midst of large seas, because on them the difference between the temperature of the day and night is not sufficiently great. Dr. Dalton has estimated that the quantity of dew that falls in England annually would form a bed of water uniformly spread over the whole kingdom of five inches in depth. If the radiation be great, the dew is frozen and becomes hoar-frost, which is the ice of dew. Cloudy and windy weather is unfavourable for the formation of dew by preventing the free radiation of heat, and actual contact is necessary for its formation, as it is never suspended in the air like fog. Dew falls in calm serene nights, but not on all substances indifferently; it wets them in proportion to their powers of radiation, leaving those dry that radiate feebly or not at all. Dew is most abundant on coasts; in the interior of continents there is very little, except near lakes or rivers. When dew is congealed into hoar-frost it forms beautiful crystals, and the cold which produces it is very hurtful to vegetation, but the slightest covering preserves plants from its effects.

When the atmosphere is so saturated with the vapour of water that it is precipitated in the air itself, a fog is the result, which consists of small globular particles of water. When dew is formed, the earth is colder than the air in contact with it; but the case is exactly the contrary when fogs take place, the moist soil being warmer than the air. In countries where the soil is moist and warm, and the air damp and cold, thick and frequent fogs arise, as in England, where the coasts are washed by a sea of elevated temperature, and the excess of the heat of the Gulf-stream above the cold moist air is the cause of the perpetual fogs in Newfoundland, and on the approach of winter those dense fogs known to the seamen as frost-smoke steam from the Polar Ocean till it is frozen over.

Superior to all these phenomena, and at a considerable height above the earth, the air is very dry, because under ordinary circumstances the vapour ascends in a highly elastic and invisible state till it reaches a stratum of air of lower temperature, and then it is condensed into clouds. The region of clouds is a zone at a height varying from one to four miles above the surface of the earth, which is saturated with moisture. From friction and other causes the currents of air in the lower part of that zone run horizontally on each other; and as they generally differ in moisture, temperature, and velocity, the colder condense the invisible vapour in the warmer, and make it apparent in the form of a cloud, which differs in no respect from a fog, except that one floats high in the air, while the other rests on the ground.

At moderate heights clouds consist of vapour, but at great elevations where the cold is severe they are an assemblage of minute crystals of ice. They assume three primary characters, from whence four subordinate forms are derived. The cirrus, or cat's-tail [mare's tail] of sailors, is the highest; it sometimes resembles a white brush, at other times it consists of horizontal bands of slender silvery filaments. To these all Kämtz's measurements assign a height of 19,500 feet, which is confirmed by their appearance being the same when seen from the tops of mountains or from the plains; consequently they must consist of minute particles of ice or flakes of snow floating in the higher regions of the zone of clouds. The cirri for the most part arrange themselves in parallel bands which converge to opposite points in the horizon, by the effects of perspective, and as they travel in their longitudinal direction they appear to be stationary. In the middle and higher latitudes of the northern hemisphere they tend from south-west to north-east, and at the equator from south to north. It is supposed that their parallel form arises from their being conductors between two foci of electricity, but, whatever the cause of this arrangement may be, it is very extensive; they are supposed by Baron Humboldt and M. Arago to be connected with the aurora. Among these clouds, which occasionally appear like fleecy cotton or wool, halos and parhelia are formed, which often precede a change of weather, announcing rain in summer, in winter frost and snow.

Cumuli or summer-clouds are rounded forms resting on a straight band in the horizon, and resemble mountains covered with snow. They are formed by ascending currents drawing the vapours into the higher regions of the atmosphere; sometimes they rise and cover the whole sky, and in the evening they frequently become more numerous and of deeper tint, presaging storm or rain.

The stratus is the third of the primary characters of clouds: it is a horizontal band, which forms at sunset and vanishes at sunrise. The subordinate varieties of clouds are combinations of these three principal classes.[1] The winds, the great agents in all atmospheric

[1] The four subordinate forms of clouds are the cirro-stratus, composed of little bands of filaments, more compact than the cirrus, forming hori-

changes, carry the vapour to a distance, where it is often condensed on the tops of mountains into clouds which seem to be stationary, but which in reality are only maintained by a constant condensation of fresh vapour, which is carried off, as soon as formed, by the wind, and becomes invisible on entering warmer air.

When two masses of air of different temperature meet, the colder, by abstracting the heat which holds the moisture in solution, causes the particles to coalesce and form drops of water, which fall in the shape of rain by their gravitation. And when two strata of different temperature moving rapidly in contrary directions come into contact, a heavy fall of rain takes place; and as the quantity of aqueous vapour is most abundant in tropical regions, the drops are larger and the rain heavier than elsewhere.

Since heat is the cause of evaporation, rain is very unequally distributed, and with it decreases from the equator to the poles. From the island of Haiti in the Antilles, to Uleaborg in Finland, the annual quantity of rain that falls decreases from 150 inches to 13. It is, however, more abundant in the New World than in the Old; 115 inches fall annually in tropical America, while in the Old World the annual fall is only 76 inches; so also in the temperate zone of the United States the annual quantity is 37 inches, while in the Old Continent it is but 31¾ inches.[1]

Between the tropics the rains follow the course of the sun: when he is north of the equator the rains prevail in the northern tropic; and when he is south of that line, in the southern: hence one period of the year is extremely wet and the other extremely dry; the change taking place near the equinoxes. Nevertheless, in countries situate between the 5th and 10th parallels of latitude, north and south, there are two rainy seasons, and two dry; one occurs when

---

zontal strata, which seem to be numerous thin clouds when in the zenith, and at the horizon a long narrow band. The cumulo-stratus consists of the summer-cloud, like snowy mountains heaped on one another, which at sunrise have a black or bluish tint at the horizon, and pass into the nimbus, or rain-cloud, which has a uniform grey tint, fringed at the edges, — it often becomes a thunder-cloud; and the fourth is the cirro-cumulus, a combination of filaments and heaped-up cumuli or summer-clouds. In 1848 the sky was nearly cloudless and free of fog night and day, at London, for the first eight days of May, and was almost free from clouds till the 15th, a circumstance without parallel on record. — General Quarterly Report of Weather at the Royal Observatory, Greenwich.

[1] Local circumstances have great influence, especially the vicinity of mountains: probably the greatest average annual quantity on record is 302 inches, which falls on the western Ghauts, in 18° N. lat. At Guadaloupe it is 286 inches. In the Silvas of the river Amazon and at Honduras it is said to be excessive. In England the average annual quantity is 32 inches; nearly half that quantity fell in the first six weeks of 1848, which is more than had occurred for 33 years, and probably within a century.— Greenwich Meteor. Register.

the sun passes the zenith in his progress to the nearest tropic, and the other at his return, but in the latter the rains are less violent and of shorter duration. Although the quantity of water which falls between the tropics in a month is greater than that of a whole year in Europe, yet the number of rainy days increases with the latitude, so that there are fewest where the quantity is greatest. Neither does it fall continually during the rainy season between the tropics, for the sky is generally clear at sunrise, it becomes cloudy at ten in the morning, at noon the rain begins to fall, and, after pouring for four or five hours, the clouds vanish at sunset, and not a drop falls in the night, so that a day of uninterrupted rain is very rare.[1]

At sea within the region of the trade-winds it seldom rains,[2] but in the narrow zone between them known as the *Variables*, in both the great oceans, it rains almost continually, attended by violent thunder and lightning.

Throughout the whole region where the monsoons prevail, it is not the sun, directly, but the winds, that regulate the periodical rains. That region extends from the eastern coasts of Africa and Madagascar across the Indian Ocean to the northern districts of Australia, and from the tropic of Capricorn to the face of the Himalaya, the interior of China, and even to Corea, inclusive. In these countries the western coasts are watered during the south-west monsoon, which prevails from April to October; and the eastern coasts are watered during the north-east monsoon, which blows from October to April. For example, the south-west wind condenses the vapour on the summit of the Ghauts, and violent rains fall daily on the coast of Malabar, while on the Coromandel coast the sky is serene. Exactly the contrary takes place during the north-east monsoon; it rains on the coast of Coromandel, while there is fair weather on the Malabar coast, and the table-land of the Deccan partakes of both. In the southern hemisphere the rainy season corresponds with the south-west monsoon, and the dry with the south-eastern.

Between the tropics it rains rarely during the night, and for months together not a drop falls;[3] while in the temperate zone it often rains in the night, and rain falls at all seasons, though more abundantly in some than in others. It seldom rains in summer throughout the north of Africa, Madeira, the southern parts of Spain and Portugal,

[1] At Demerara six inches have been known to fall in 12 hours. The quantity that falls in Italy is sometimes very great; at Rome half the yearly average quantity fell in 15 hours.

[[2] Showers and squalls of rain occur frequently in the regions of both the N.E. and S.E. trade-winds.]

[3] [According to the experience of the writer, derived from cruising several years within the tropics, rain is most frequent at night. It is certainly the case in the Chinese Seas, and in the vicinity of Sumatra and Java.]

Sicily, southern Italy, all Greece, and the north-western part of Asia; but it falls copiously during the other seasons, especially in winter; consequently that extensive region is called the province of winter rains.

The province of autumnal rains includes all Europe south of the Carpathians, western France, the delta of the Rhine, northern and western Scandinavia, and the British isles; throughout these countries more rain falls in autumn than in the other three seasons.

The province of summer rains comprises the eastern parts of France, the Netherlands (with the exception of the delta of the Rhine), the north of Switzerland, all Germany north of the Alps, the Carpathian mountains, Denmark, southern Scandinavia, all central Europe, and the countries beyond the Ural Mountains to the interior of Siberia, where showers are very rare in winter. In some places it rains almost perpetually, as in the island of Sitka, on the north-eastern coast of North America, where the year has sometimes passed with only 40 days of fair weather.

In the southern hemisphere, in Chile and the south-western part of America, winter is the rainy season, while on the eastern side of the Cordilleras, in the interior of that chain, the rains occur in summer. In Tierra del Fuego and the extreme point of the continent the two provinces meet, the periodical precipitation disappears, and it snows and rains throughout the year in torrents. At Cape Horn the quantity of rain which fell in 41 days measured nearly 154 inches. This excessive fall of rain occurs along the whole western shores of Patagonia, from the Straits of Magellan to Cape Tres Montes — a circumstance favoured by the high and rugged coasts, and the incessant westerly winds, which carry the vapour exhaled from the ocean to be precipitated here in the form of rain.

South Africa and Australia resemble each other in their rainy seasons, which in both countries take place in the winter months.

The annual amount of rain at the equator is 95 inches, which falls in 78 or 80 days, giving an average of 1·14 inch daily; while at St. Petersburg the annual amount is 17 inches, which falls in 169 days, the average being little more than the tenth of an inch daily.

The quantity of rain decreases in ascending from the plains to table-lands, especially if these be edged by mountains, because they precipitate the vapour before it arrives at the high plains. On the contrary, the quantity increases in ascending from plains to the tops or slopes of rugged mountains, on account of partial currents of air which condense the moisture into clouds.

The quantity of rain decreases on receding from the coasts into the interior of continents, because more vapour rises from the sea than from the land. The vapour from the Gulf-stream produces a greater quantity of rain and fog in the southern counties of England and Ireland than that which falls in the other parts of the islands.

The number of rainy days depends upon the direction of the wind. In Europe, if the wind always blew from the north-east, it would seldom rain, because it blows over a great extent of continent; whereas it would never cease raining were the wind always to blow from the south-west, because it would come loaded with vapour from the Atlantic. Hence the greatest quantity of rain falls on the west coasts of Great Britain and Ireland, the coast of Scandinavia, the Eastern Alps, and the centre of Portugal; in the two last it depends partly on the height and serrated form of the mountains. In western Europe it rains on twice as many days as in the eastern part: in Ireland there are three times as many rainy days as in Italy or Spain. In fact, on the western side of Ireland, it rains on 208 days out of the 365. In England, France, and the north of Germany, there are from 152 to 155 rainy days in the year; the number decreases towards the interior of the continent, so that in Siberia it only rains on 60 days in the year. Occasionally it rains over a wide extent of country at the same time; on the 2nd of February, 1842, it rained in North America over 1400 miles in length, but the breadth to which it extended was not ascertained. Rain sometimes falls without clouds, from a partial condensation of vapour; Sir James C. Ross mentions a smart shower without clouds which he saw in the South Atlantic on the 20th December, 1839: it continued an hour.

There are enormous tracts of land on which rain never falls, and others where it rains at long intervals and in small quantities. The most extensive rainless district stretches from the borders of Morocco eastward through the desert of Africa, the low coasts of Arabia, Persia, and the desert province of Mekran, in Beloochistan, occupying a space of 80 degrees of longitude and 17 of latitude. The desert of Gobi, on the table-land of Tibet, and part of Mongolia, form another rainless province in the great continent; while, in the New World, the rainless districts are — the table-land of Mexico, part of Guatemala and California, and the western declivity of the Andes of Peru, towards the Pacific; in all occupying a surface equal to 5,500,000 square miles. The whole of the moisture is intercepted by the Andes of Peru; so that rain only occurs on the coast once or twice in a century — to the great terror of the inhabitants when it does fall. South Africa and Australia beyond the tropics, suffer from droughts, which are periodical in Australia; they recur in the countries of the eastern coasts in a period of 12 years, and continue 3 years. The Pampas of South America are also subject to droughts, though they do not appear to be periodical, nor do they continue more than a season.[1]

When the temperature of the air is near the freezing point or

[1] The reader is referred to the chart of the distribution of rain in the Physical Atlas of Alexander Keith Johnston, Esq., where the value of the practice referred to in note p. 274 is shown.

below it, snow falls instead of rain; but the colder the air the less moisture does it contain, consequently the less snow falls, which is the reason of the comparatively small quantity on the high plains of the Himalaya and Andes. Snow sometimes assumes the form of grains; but is generally in regular crystals of great beauty, varying in form according to the degree of cold. Dr. Scoresby, whose voyages in the Polar Seas afforded him constant opportunities of studying them, of which he so diligently availed himself, mentions five principal kinds of snow-crystals, each of which had many varieties, in all amounting to 96. M. Kämtz, however, is of opinion that there are several hundred. The whiteness of the snow is owing to the reflection of light from the minute faces of its crystals, which are like small mirrors.

Snow never falls between the tropics except on the tops of very high mountains. The mean elevation of the line of perpetual snow above the level of the sea in these hot regions is about 15,207 feet, from whence it decreases on both sides, and at last grazes the surface of the earth at the arctic and antarctic circles, subject, however, to various flexures. In the Andes, near Quito, the lowest level has an elevation of 15,795 feet, which is higher than the top of Mont Blanc; from thence it varies very irregularly, both to the north and south. In 18° of N. lat. it descends to 14,772 feet on the mountains of Mexico, while on the south it rises to 18,000 feet in some parts of the western Cordillera of the Bolivian Andes, owing to the extensive radiation from the subjacent plains and valleys. The line is at an altitude of 17,000 feet on the western Cordillera, whence it sinks to 13,800 feet at Copiapo, to 12,780 near Valparaiso; it is only 7960 in the southern prolongation of the Chilian Andes, on the volcano of Antuco, lat. 37° 40′, and 3390 in the Straits of Magellan. In lat. 31° N. the snow-line is at an elevation of 12,981 feet on the southern side of the Himalaya, and at 16,620 feet on the northern side, while Captain Gerard gives from 18,000 to 19,000 as its altitude on the mountains in the middle of the plain of Tartary. On Mont Blanc the line is at the height of 8500 feet, so that mountain is snow-clad for 7000 feet below its summit. In the Pyrenees it is 8184 feet, and at the island of Mageroe it is at 2160 feet, above the Polar Ocean.

In the southern hemisphere snow never falls on the low lands at the level of the sea north of the 48th parallel of latitude, on account of the predominance of water, whereas in the northern hemisphere it falls on the plains much nearer the equator, on account of the excess of land, but its limit is a curved line, on account of the alternations of land and water. In the western part of the great continent the southern limit of the fall of snow on the low lands nearly coincides with the 30th parallel of north latitude, so that it includes all Europe. In the American continent it follows nearly the same line, extending through the southern parts of the United States. In China snow falls at the level of the sea as far south as Canton;[1] on the north-western coast of America, on the contrary, it does not fall at that level till about the 48th degree of N. lat.—these are the two extremes. Although Europe lies within the region of snow, the quantity that falls is very different in different places, increasing greatly from south to north. On an average it snows only one day and a half at Rome in the year, while at Petersburg there are 171 snowy days, but in that city the quantity of rain is to that of snow as 1000 to 384. Snow, by protecting the ground from cold winds, as well as by its slow conducting power, and by preventing radiation, maintains the earth at a higher temperature than it otherwise would have. In Siberia the difference between the temperature of the ground beneath the snow and that of the air above it has amounted to 38° Fahrenheit.

Sleet, which is formed of small particles of rounded hail mixed with rain, falls in squally weather in spring and autumn. True hail, when large, is pear-shaped, and consists of a nucleus of frozen snow coated with ice, and sometimes with alternate layers of snow and ice. Hailstones have often fallen as large as pigeons' and even hens' eggs. The masses and blocks of ice of great size, which have not unfrequently fallen, appear to have been formed of hailstones of large size frozen together. It appears to be formed in the high cold regions of the atmosphere, by the sudden condensation of vapour during the contention of opposing winds, and is intimately connected with electricity, since its fall is generally accompanied with thunder and lightning. Hail-showers are of short duration, exceedingly partial, and extend over a country in long narrow bands: one which took place on the 13th of July, 1788, began in the morning in the south of France, and reached Holland in a few hours, destroying a narrow line of country in its passage.

Local circumstances, no doubt, have a great influence on its formation; it occurs more frequently in countries at a little distance from mountains than in those close to them or farther off, and at all hours, but most frequently at the hottest time of the day, and rarely in the night. In the interior of Europe one-half of the hail-storms take place in summer. Hail is very rare on the tropical plains, and often altogether unknown, though it frequently falls at heights of 1700 or 1800 feet above them, and at still greater elevations, in the Bolivian Andes, for example, above 12,000; and on the table-land of Ethiopia at heights between 6000 and 10,000 feet. If the air

[[1] Snow falls at Canton only at very long intervals of time. The night of the 7th of February, 1836, snow fell "and covered the ground and roofs of the houses with a coat nearly two inches thick. Such occurrences in Canton are very unfrequent, probably not once in a century."—*Chinese Repository*, Vol. iv., p. 487.]

is very cold throughout the greater part of the stratum through which hail falls, it is probably increased in size during its descent; and, on the contrary, large drops of rain which precede a thunder-storm are supposed to be hail melted in its passage through low warm air.

LIGHT.

We know nothing of the size of the ultimate particles of matter, except that they must be inconceivably small, since organized beings possessing life and exercising all its functions have been discovered so minute that a million of them would occupy less space than a grain of sand.

The air is only visible when in mass; the smallest globule of steam tells no more of its atoms than the ocean; the minutest grain of sand magnified appears like the fragment of a rock — no mechanical division can arrive at the indivisible. Although the ultimate atoms are beyond the power of vision, chemical compounds show that the divisibility of matter has a limit, and that the particles have different densities; moreover the cleavage of crystalline substances gives reason to believe that they have different forms.[1] Thus the reasoning power of man has come to the aid of his imperfect sense of vision, so that what were before imaginary things are now real beings with definite weights, and uniting by fixed laws. Though nothing had been known of their size, their effects were evident in the perceptions of sweet and sour, salt and bitter, and in the endless varieties of aroma in the food we eat and the liquors we drink. Moreover, their different densities are evident, as they arise by their buoyancy in the perfume of the rose, or sink by their weight in the heavy odour of mignonette. Every substance on earth is merely a temporary compound of the ultimate atoms, sooner or later to be resolved into its pristine elements, which are again to be combined in other forms, and according to other laws; so that literally there is nothing new under the sun, for there is no evidence of new matter being added to the earth, nor of that which exists being annihilated. Fire, which seems utterly to destroy, only resolves bodies into their elementary parts, to become what they were before, the support of animal or vegetable life, or to form new mineral compounds. It is to the action of these particles on the light of the sun that nature owes all its colours.

When a sunbeam passes through a glass prism[2] an oblong image

---

[1] The reader is referred to the 'Connexion of the Physical Sciences' for an account of Dr. Dalton's theory of definite proportions, and the relative weight of atoms; and to Dr. Daubeny's recently published work on the Atomic Theory. [See also, Graham's Chemistry.]

[2] The reader is referred to the 18th section of the 'Physical Sciences' for reflection, refraction, and absorption of light, and to the 19th section for the constitution of the solar light and colours. [Also, to Lardner's Hand-Books of Natural Philosophy.]

of the sun is formed consisting of colours in the following order—red, orange, yellow, green, blue, indigo, and violet. Sir John Herschel discovered lavender rays beyond the violet, and dark red rays exterior to the red, which are not so easily brought into evidence as the rest.

Even the most transparent substances absorb light; air, water, the purest crystal, stop some of the rays as they pass through them. A portion of the light is also reflected from the surface of all bodies; were it otherwise, they would be invisible. We should be unconscious of the presence and form of material substances beyond our reach except by the reflected rays,—

> "The mist of light from whence they take their form
> Hides what they are."

As the same light does not come to all eyes, each person sees his own rainbow, the same flower by different rays. White substances reflect all the light, black substances absorb all but that which renders them visible, while coloured bodies decompose the light, absorb some of the colours, and reflect or transmit the rest. Thus a violet absorbs all but the violet rays, which it reflects; a red flower only reflects the red and absorbs the rest; a yellow substance absorbs all but the yellow. In the same manner transparent substances, whether solid or fluid, absorb some colours and transmit others: thus an emerald absorbs all but the green, a ruby all but the red; whereas a diamond does not decompose the light, but transmits every ray alike. Very few, however, of the colours, whether transmitted or reflected, are pure, but the substance takes its hue from the colour that predominates.

The atmosphere, where rare, absorbs all the colours of the sun's light except the blue, which is its true colour. In countries where the air is pure, the azure of the sky is deep; it is still more so at great elevations, where the density of the air is less; and its colour is most beautiful as it gradually softens the outlines of the mountains into extreme distance, or blends the sea with the sky. When the sun is near the horizon, the atmosphere, on account of its superior density, absorbs the violet and blue, and leaves the yellow and red rays in excess; that property, together with the refractive power of the aqueous vapour, which is most abundant near the earth's surface, gives the roseate hue to the early morning, and the gold and scarlet tints to the closing day. The blending of these colours with the blue above produces that beautiful vivid green so frequently seen in Italy and other warm countries. The last reflected rays of the setting sun are red, which gives a rose-coloured tint to the Alpine snow, and below the red the shadow of the earth is sometimes cast upon the atmosphere in the form of a deep blue segment, known as the ante-twilight. The air reflects and scatters part of the white solar beams, whence the brightness and cheerfulness of day; were it not for that reflective power, the sun and moon would be like sharply defined balls of fire in the profoundly black vault of the heavens, and dark night would instantly follow sunset. When the sun is 18 degrees below the horizon, the air, at the height of 30 miles, is still dense enough to reflect his rays, and divide the day from the night by the sober shades of twilight.[1]

A considerable portion of the sun's light is absorbed by the atmosphere: the loss increases with the obliquity of incidence and the density of the air. It is diminished 1300 times by the thickness of the air at the horizon, which enables us to look at the sun when setting without being dazzled.[2]

The bending or refraction of the sun's light passing through the atmosphere causes distant objects, as mountains, to appear higher than they are. It increases with the density of the air and the obliquity of incidence, and on that account the sun is seen above the horizon after he is really below it, or has set. During the winter of 1820, which the expedition under Sir Edward Parry passed at Melville Island, in 74° 47′ N. lat., the sun did not rise for 92 days; but in consequence of extraordinary refraction he appeared above the horizon on the 3d of February, which was three days sooner than he ought to have done. Berentz is said to have seen the snn at Nova Zembla on the 20th of January, 1597, fifteen days before he was expected.

The sun and moon often appear distorted at their rising and setting, because the looming, or extraordinary refraction, is greatest in the morning or evening, from the increased density of the air at the surface of the earth by reason of the cold; the distortion of objects is occasioned by the rays of light passing through the strata of air of different densities; from this cause objects are sometimes seen inverted, and three images of the same object occasionally appear, two direct and one inverted.

[" In climates subject to sudden and extreme vicissitudes of tem-

[1] Between the tropics twilight continues from the setting of the sun till he is 16° below the horizon, in middle latitudes until he is 18°, and in the polar regions until he is 20°; then and then only does real night begin: at Edinburgh there is no real night from the 6th of May until the 7th of August, in London there is none from the 21st of May till the 22nd of July, and in Paris there is no true night in the month of June.

[2] The photometer is an instrument invented by Sir John Leslie for measuring the relative intensity of light and its variations, upon the principle that the heat contained in solar light is a measure of the intensity of light; Sir John computed that one-fourth of the light of the sun is absorbed by the atmosphere, and, with regard to obliquity of incidence, that, out of 1000 rays which fall obliquely on the earth, only 378 reach it at the equator, 288 in the latitude of 45°, and 110 at the poles; in England the light measured by the photometer is 65° greater in intensity in summer than in winter.

perature, the strata of air are often affected in an irregular manner as to their density, and consequently as to their refracting power. If it happen that rays proceeding from a distant object directed upwards after passing through a denser be incident upon the surface of a rarer stratum of air, and that the angle of incidence in this case exceeds the limit of transmission, the ray will be reflected downwards; and if it be received by the eye of an observer, an inverted image of the object will be seen at an elevation much greater than the object itself."[1]]

Mirage, [*fata morgana*,] or the delusive appearance of water, so frequent in deserts, is owing to the reflection of light between two strata of air of different densities, occasioned by the radiation of heat from the arid soil. It is very common on the extensive plains in Asia and Africa, and especially in Upper Egypt; villages on small eminences above the plain appear as if they were built on islands in the middle of a lake when the dry sandy ground is heated by the mid-day sun. Sometimes objects appear double, and occasionally several images appear above one another, some direct and some inverted; this is particularly the case in high latitudes, where the Icy Sea cools the stratum of air resting on it.[2]

In the polar regions, or on the tops of mountains, when the sun is in the horizon the shadow of a person is sometimes thrown on an opposite cloud or mist, the head being surrounded by concentric coloured rings or circles, the number varying from one to five. Dr. Scoresby saw four of these rings, on one occasion, round the shadow of his head, as he stood between the sun and a thick low fog: the first ring consisted of concentric bands of white, yellow, red, and purple; the second consisted of concentric bands of blue, green, yellow, red, and purple; the third of green, white, yellowish white, red, and purple; and in the fourth the bands were greenish white, deeper on the edges. Mr. Green, at the height of two miles, saw the shadow of his balloon, surrounded by three coloured rings, on a cloud below. These appearances, called *glories*, or fog-images, and the coronæ or small concentric coloured circles which surround the sun or moon when partly obscured by thin white clouds, are owing to the refraction of the light in the aqueous particles of the cloud or fog. The colours in the concentric bands of the coronæ, however, differ from the foregoing; that nearest the sun is of deep blue, white, and red; the circle exterior to that consists of purple, blue, green, pale yellow, and red; but the series is very rarely complete.

Halos, which surround the sun in large circles, or a complicated combination of circles, are, on the contrary, supposed to be produced by the light falling on minute crystals of ice suspended in the at-

[[1] Dr. Lardner's Hand-Books of Natural Philosophy.]

[2] For the cause of mirage, see the 'Connexion of the Physical Sciences.'

mosphere; they are particularly brilliant and frequent in high latitudes. It is scarcely possible to give an idea of these beautiful and singular objects. Sometimes a large coloured circle surrounds the sun or passes through his centre, which is occasionally touched or cut by segments of others. One seen at St. Petersburg on the 29th of June, 1790, consisted of four coloured circles of different sizes intersecting each other, which were either cut or touched by segments of eight others, and at the points of intersection mock suns or parhelia appeared. The sky is very hazy on these occasions. Mock suns, without circles and halos, are by no means uncommon, and halos are often seen round both sun and moon, but seldom of that complicated kind. They are situate between the observer and the sun, whereas the rainbow is always in that part of the sky opposite the sun, because it is produced by refraction and reflection of the sun's rays in the drops of rain; and when the light is intense and the rain abundant, there are two concentric bows, the prismatic colours of the innermost of which are the most vivid, the violet being within and the red outside: sometimes the inner edge exhibits a repetition of colours in fine fringes, in which red and green predominate. The colours are reversed in the exterior bow, the violet being outside and the red on the inner edge. Besides these two principal and most common bows, supernumerary rainbows occasionally appear within the interior bow, generally green and violet, though there are sometimes more or less perfect repetititions of all the colours.[1] The visible extent of the bow depends upon the altitude of the sun and the position of the spectator. As a line joining the centres of the sun and bow must pass through the eye of the spectator, the altitude of the sun must be less than 45°, and only a portion of the bow can be seen from a plain; but the complete circle may be visible to a person on the top of a high mountain when the sun is low, except the small portion intercepted by his shadow. In squally weather a rainbow is sometimes seen on a blue sky when rain is falling, but it is generally on clouds; it is constantly seen when the sun shines on the fine drops of fountains and cascades, and on the grass in a dewy morning. As the light of the moon is feeble, lunar rainbows are rare, and, for the most part, colourless. In the early morning when the sun throws his slanting beams across the fields, a miniature bow, with all its vivid colours, may be seen in each dewdrop as it hangs on the points of the bending grass.

Light is said to be polarized when, after having been once refracted or reflected, it is rendered incapable of being again refracted or re-

---

[1] In the primary bow the light is twice refracted and once reflected in the rain-drops, while in the external bow it is twice refracted and twice reflected, and as light is lost at each refraction and reflection the interior bow is the brightest. Sir David Brewster has found that the light of the rainbow is polarized.

flected at certain angles. For example, if a crystal of brown tourmaline be cut longitudinally into thin slices, and polished, the light of a candle may be seen through a slice as if it were glass. But if one of these slices be held perpendicularly between the eye and the candle, and a second slice be turned round between the eye and the other plate of tourmaline, the image of the candle will vanish and come into view at every quarter-revolution of the plate, varying through all degrees of brightness down to total or almost total evanescence, and then increasing again by the same degrees as it had decreased. Thus the light, in passing through the first plate of tourmaline, is said to be polarized because it has been rendered incapable of passing through the second piece of tourmaline in certain positions.

A ray of light acquires the same property if it be reflected from a pane of plate glass at an angle of 57 degrees; it is by that rendered incapable of being reflected by another pane of plate glass in certain definite positions, for the image of the light vanishes and reappears alternately at every quarter-revolution of the second pane.

If a thin plate of mica be interposed when the image of the candle has vanished, the darkness will instantly disappear, and a succession of the most gorgeous colours will come into view, varying with every inclination of the mica from the richest reds to the most vivid greens, blues and purples. The most splendid colours arranged in symmetrical forms are exhibited by thin plates of an infinite variety of substances besides mica. They display some of the most beautiful objects in nature, and show differences otherwise inappreciable in the arrangement of the molecules of crystalline bodies.[1]

M. Arago discovered that the light of the sun is polarized by the reflection of the atmosphere, but not equally so on every part of the sky; the polarization is least in the vicinity of the sun, and greatest at 90° from him, for there his light is reflected at an angle of 45°, which is the polarizing angle for air.[2] There are three points in the sky where the light is not polarized: one of these neutral points, discovered by M. Arago, is 18° 03′ above the point diametrically opposite to the sun when he is in the horizon; the second neutral point, discovered by M. Babinet, is 18° 30′ above the sun when he is rising or setting; and the third, discovered by Sir David Brewster, is 15° or 16° below the sun. These points vary with the height of the sun, and the two latter rise and coincide in his centre when he is in the zenith.[3]

---

[1] For phenomena and theory of polarized light, see section 21, 'Connexion of the Physical Sciences.'—[Also, Dr. Lardner's Hand-Books of Natural Philosophy.]

[2] Every substance, whether solid or fluid, has its own polarizing angle.

[3] The reader is referred to a plate in Johnston's Physical Atlas, showing the phenomena of the polarization of the atmosphere.

Now the portion of polarized light sent to the eye from any part of a clear sky is in a plane passing through that point, the eye of the observer, and the centre of the sun. If that point be the north pole of the heavens, it is clear that, as the sun moves in his diurnal course, the plane will move with him, as an hour circle, and may be used as a dial to determine the hour of the day. Professor Wheatstone, by whom that beautiful application of the polarization of the atmosphere has been made, has constructed a clock, of very simple form, which shows the time of day with great accuracy, and which has many advantages over a sundial.

ELECTRICITY.

Electricity pervades the earth, the air, and all substances, without giving any visible sign of its existence when in a latent state, but, when elicited, it exhibits forces capable of producing the most sudden, violent, and irresistible effects. It is roused from its dormant state by every disturbance in the chemical, mechanical, or calorific condition of matter, and then experience shows that bodies in one electric state repel, and in another they attract each other. Probably their mutual attraction and repulsion arise from the redundancy and defect of electricity; in the first case they are said to be positively, in the latter negatively electric.[1] When they have different kinds of electricity they attract each other, and, when not opposed, the electricity coalesces with great rapidity, producing the flash, explosion, and shock, and that with the more violence the greater the tension or pressure of the electricity on the surrounding air which resists its escape. Equilibrium is then restored, and the electricity remains latent till called forth by a new exciting cause. The electrical state of substances is easily disturbed, for, without contact, positive electricity tends to produce negative electricity in a body near it, and *vice versâ:* the latter is then said to be electric by induction.

The electricity of the atmosphere arises from evaporation, and the chemical changes that are in perpetual progress on the globe; no electricity, however, is developed by the evaporation of pure water, but it arises abundantly from water containing matter susceptible of chemical action during the evaporation; consequently the ocean is one of the greatest sources of atmospheric electricity; combustion is another; and a large portion arises from vegetation. The air, when pure, is almost always positively electric; but as the chemical changes on the earth sometimes produce positive and sometimes negative electricity, it is subject to great local variations; a passing cloud or a puff of wind produces a change, and a distant storm renders it negative for the time, but the earth is always in a negative state.

[1] See sections 28 and 29 of the 'Connexion of the Physical Sciences:' on Electricity. [Also, Dr. Lardner's Hand-Books of Natural Philosophy.]

The quantity of electricity varies with the hours of the day and the seasons; it is more powerful in the day than in the night, in winter than in summer, and it diminishes from the equator to the poles. It thunders daily in many places, in others never, as on the east coast of Peru and in the Arctic regions, except where there are violent volcanic explosions, which always generate electricity, as in Iceland. Wherever there are no trees or high objects to conduct it to the ground, the quantity of positive electricity increases with the height above the surface of the earth. Violent thunder-storms take place on the highest summits of the Andes and Himalaya mountains. On the high table-land of Ethiopia they are violent and so frequent, that M. d'Abbadie calculates it thunders fifty-six days out of every hundred. In general thunder-clouds in our latitudes float at the height of from 3000 to 5000 feet above the earth.

Electricity becomes very strong when dew is deposited, and in some cases it is strongly developed in fogs. Mr. Cross found it so powerful on one occasion, that it was dangerous to approach the apparatus for measuring its intensity. A continued succession of explosions lasted nearly five hours, and the stream of fire between the receiving ball and the atmospheric conductor was too vivid to look at. M. Peltier has found that the common fogs arising from the mere condensation of the moisture in the air are neutral, but that others which are produced by exhalations from the earth are sometimes positive, sometimes negative; the subject, however, requires further investigation.

Though in long-continued mild rains there are no traces of electricity, yet when rain or snow falls from the higher regions of the atmosphere, it is more or less developed, sometimes positive, sometimes negative, depending a good deal on the direction of the wind. During a drifting fall of snow, Mr. Cross collected electricity enough to decompose water. The atmosphere being positively electric, negative rain is supposed to arise from the evaporation of the drops in passing through dry air; the vapour carries off the positive electricity and leaves the drop in a negative state—a circumstance which seems to be confirmed by the electricity of cascades, near which there always is more or less negative electricity; the positive flows into the earth, while the other remains united to the drops of the cascade.

The inductive action of the earth upon the clouds and of the different strata of clouds on each other, produces great variations in their electrical state. If rain falls from the lowermost of two strata of positively electrical clouds, the inductive action of the earth renders the under surface positive and the upper negative, and the rain is positive. By-and-bye the under surface of the cloud and the earth become neutral; and after a time the lower cloud becomes charged with negative electricity by the induction of the upper strata, and the rain is then negatively electric. Clouds are very differently charged; grey clouds have negative — red, white, and orange clouds positive electricity; and when clouds differently charged meet, an explosion takes place. When the sky is clear and the air calm and warm, a succession of small white fleecy clouds rising rapidly above the horizon and flying swiftly in the very high regions of the atmosphere, is a certain presage of a thunder-storm.

Electricity of each kind is probably elicited by the friction of currents of air, or masses of clouds moving rapidly in different directions, as in thunder-storms, when small white clouds are seen flying rapidly over the black mass; yet the quick and irregular motion of clouds in storms is probably owing to the strong electrical attraction and repulsion among themselves, though both may be concerned in these hostile encounters. When two clouds differently charged by the sudden condensation of vapour, and driven by contending winds, approach within a certain distance, the thickness of the coating of electricity increases on the two adjacent sides, and when the accumulation becomes so great as to overcome the coercive pressure of the atmosphere between them, a discharge takes place which occasions a flash of lightning. The actual quantity of electricity in any part of a cloud is very small. The intensity of the flash depends upon the extent of surface occupied by the electricity, which acquires its intensity by its instantaneous condensation.

The air, being a non-conductor, does not convey the electricity from the clouds to the earth, but it acquires from them an opposite electricity, and when the tension is very great the force of the electricity becomes irresistible, and an interchange takes place between the clouds and the earth, but the motion of the lightning is so rapid, that it is difficult to ascertain when it goes from the clouds to the earth, or from the earth to the clouds, though there is no doubt it does both: explosions have burst from the ground, and people have been killed by them.

When the air is highly rarefied by heat its coercive power is diminished, so that the electricity escapes from the clouds in the form of diffuse lambent sheets of lightning without thunder or rain, frequently seen in warm summer evenings, sometimes even near the zenith, and quite different from that sheet lightning at the horizon which is in general only the reflection of the forked lightning of a distant storm.[1] When the quantity of electricity developed by the sudden condensation of vapour is very great, the lightning is always forked; its zigzag form is occasioned by the unequal conducting

[1] The author saw a very remarkable instance of lightning without thunder. There were no clouds on the sky except one in the zenith, over which diffused sheets of lambent lightning played for more than an hour without thunder: the cloud did not appear to be so high but that thunder might have been heard had there been any.

power of the air, by which it is sometimes divided into several branches. The author once saw a flash divided into four parallel streams—a very uncommon occurrence. Occasionally in very great storms the lightning sends off lateral branches. It often appears as a globe of fire moving so slowly that it is visible for several seconds, while the flashes of forked lightning do not last the millionth part of a second. Professor Wheatstone, who has measured the veolocity of lightning by experiments of great ingenuity, found that it far surpasses the velocity of light, and would encircle the globe in the twinkling of an eye. This inconceivable velocity is beautifully exemplified in the electric telegraph, [the invention of an American, Professor Morse,] by which the most violent and terrific agent in nature is rendered obedient to man, and conveys his thoughts as rapidly as they are formed. The colour of lightning is generally a dazzling white or blue, though in highly rarefied air it is rose-colour or violet.

The sudden compression of the air during the passage of lightning must convert a great quantity of latent into sensible heat, for heat in a latent or insensible state exists in all bodies independent of their temperature. Heat is absorbed and becomes insensible to the thermometer when solids become liquids, and when liquids are changed to vapour; and it again becomes sensible when vapour is condensed, and when liquids become solid. When water freezes, all the heat that kept it liquid is given out; and when ice melts, it absorbs heat from everything near it. The air is full of heat in a latent state, whatever its temperature may be, but it can be squeezed out by sudden compression so as to kindle tinder. Every aërial wave, every sound, every word spoken must set free an infinitesimal quantity of heat; so everything that tends to rarefy the air must cause it to absorb a proportional quantity.

The rolling noise of thunder is probably owing to the difference between the velocity of lightning and that of sound. Thunder may be regarded as originating in every point of a flash of lightning at the same instant; and as sound takes a considerable time to travel, it will arrive first from the nearest point; and if the flash run in a direct line from a person, the noise will come later and later from the remote points of its path, in a continued roar. Should the direction of the flash be inclined, the succession of sounds will be more rapid and intense; and if the lightning describe a circular course above a person, the sound will arrive at the same instant from every point with a stunning crash.[1]

[1] Sound travels at the rate of 1120 feet in a second in air at the temperature of 62° Fahrenheit; so if that number be multiplied by the number of seconds elapsed between the flash of lightning and the thunder, the result will be the distance in feet at which the stroke took place. A relative of the author's was fishing in the Tweed on a very sultry day, and iay

In passing to the earth, lightning follows the best conductors—metals by preference, then damp substances—which is the reason why men and animals are so often struck. If it meets with a bad conductor, it shivers it to pieces and scatters the fragments to a considerable distance. A powerful flash scatters gunpowder, while a feeble one ignites it; the hardest trees are split and torn to shreds; when a tree is struck, the heat of the flash converts the sap into steam, the expansive force of which shivers the tree. The surface of rocks is vitrified by it; and when it falls on a sandy soil, its course underground is marked by vitrified tubes many feet long.

Thunder-storms occur daily within the region of the Variables, [winds,] which is also the region of storms: in countries under the influence of the monsoons they are tremendous at the changes of these periodical winds; where the trade-winds prevail they are hardly known, though electrical discharges are frequent at their limits. In Greece and Italy there are about 40 thunder-storms annually, which occur in spring and autumn, while north of the Alps they take place in summer. There are about 24 in the year on the coasts of the Atlantic and in Germany, but they are much more frequent among mountains than on plains. In the interior of the old continent they rarely occur in winter, and three-fourths of the number happen in summer. They are of such rare occurrence in high latitudes that in a residence of 6 years in Greenland Sir Charles Geiseke only heard it thunder once.

Some storms arise from the contention of opposite currents in the air; others are occasioned by currents of warm air ascending from the earth, which are suddenly condensed as they enter the upper regions of the atmosphere, and, as this sometimes happens at the hottest hour of the day, these storms are periodical for many successive days, recurring always at the same hour. Sometimes they extend over a great expanse of country, and the lightning darts from all points of the compass. A person may be killed at the distance of 20 miles from the explosion by the *back stroke.* If the two extremities of a highly charged cloud dip towards the earth, they will repel the electricity of the earth, if it be of the same kind with their own, and will attract the other kind; and if a discharge should take place at one end of the cloud, the equilibrium will instantly be restored by a flash from that part of the earth which is under the other, sufficiently strong to destroy life, and it is the most dangerous, though never so strong as the direct stroke.

When thunder-clouds are very low, there is frequently no lightning; the electricity produced by induction is so powerful that it

---

down on the grass to rest: he was astonished to hear repeated peals of thunder, as there was not a cloud to be seen in the sky; two hours afterwards clouds began to rise, and in the afternoon there was a thunderstorm; the sound had been conveyed down the river by the stream.

escapes from pointed objects in the shape of flame without heat, known as St. Elmo's fire. These flames are not unfrequently seen at the topmasts of ships and the extremities of their yards. Bodies between the clouds and earth may be electrized by induction, and their electricity will be seen in the form of flame, as showers of phosphorescent snow.

Phosphorescence is ascribed to electricity; various substances emit light when decaying, as fish and wood. Although many marine animals are phosphorescent, yet the luminous appearance which the sea often assumes is not always to be attributed to them, but probably to the decaying animal matter it contains.

The aurora is decidedly an electrical phenomenon. It generally appears soon after sunset in the form of a luminous arch stretching more or less from east to west, the most elevated point being always in the magnetic meridian of the place of the observer: across the arch the coruscations are rapid, vivid, and of various colours, darting like lightning to the zenith, and at the same time flitting laterally with incessant velocity. The brightness of the rays varies in an instant: they sometimes surpass the splendour of stars of the first magnitude, and often exhibit colours of admirable transparency, blood-red at the base, emerald-green in the middle, and clear yellow towards their extremity. Sometimes one, and sometimes a quick succession of luminous currents run from one end of the arch or bow to the other, so that the rays rapidly increase in brightness: but it is impossible to say whether the coruscations themselves are actually affected by a horizontal motion of translation, or whether the more vivid light is conveyed from ray to ray. The rays occasionally dart far past the zenith, vanish, suddenly reappear, and, being joined by others from the arch, form a magnificent corona or immense dome of light. The segment of the sky below the arch is quite black, as if formed by dense clouds; yet M. Struve is said to have seen stars in it, consequently the blackness must be from contrast. The lower edge of the arch is evenly defined; its upper margin is fringed by the streamers which converge by the effect of perspective to the magnetic poles, that is, to a point in the northern hemisphere 70° below the horizon and 23° west of north from London, and to a point diametrically opposite in the southern. The apparent convergence of the arch is owing to the same cause.

Either the aurora must be high above the earth, or its coruscations must be very extensive, since the same display is visible at places wide asunder. It has frequently been seen in North America and all over the north of Europe at the same time, sometimes even as far south as Italy, yet Sir Edward Parry certainly saw a ray dart from it to the ground near him. M. Struve, Admiral Wrangel, and others who have had many opportunities of seeing the aurora in high latitudes, assign a very moderate elevation to it. The arch probably passes through the magnetic pole; hence in the north of Greenland it lies south of the observer, and Sir Edward Parry saw it to the south in Melville Island, which is in 70° N. lat.; consequently it must appear in the zenith in some places. Dr. Faraday conjectures that the electric equilibrium of the earth is restored by the aurora conveying the electricity from the poles to the equator, for it appears in the high southern latitudes, as well as in the northern; and the Rev. G. Fisher has lately suggested that, as the principal display of the aurora takes place at or near the margin of the polar ice, the electricity may be conveyed by the conducting power of the frozen particles which abound in the air in these latitudes, and which, being rendered fitfully luminous by the passage of the electricity, produce the arch and the ever-varying flashes of the aurora.

The aurora has a powerful influence on the magnetic needle, even in places where the display is not seen. Its vibrations seem to be slower or quicker according as the auroral light is quiescent or in motion, and the disturbances of the compass during the day show that the aurora is not peculiar to the night. Observations have proved that the disturbances of the magnetic needle and the auroral displays were simultaneous at Toronto, in Canada, on 13 days out of 24, the remaining days having been clouded; and contemporaneous observations show that on these thirteen days there were also magnetic disturbances at Prague and at Van Diemen's Land, so that the "occurrence of aurora at Toronto on these occasions may be viewed as a local manifestation connected with magnetic effects, which, whatever may have been their origin, probably prevailed on the same day over the whole surface of the globe."[1] It has been observed that the two kinds of auroral action bear a strong analogy to the two modes of magnetic action discovered by Dr. Faraday, the ordinary auroral beams or streamers being parallel to the magnetic meridian, and the auroral arch at right angles to it.

MAGNETISM.

Magnetism is one of those unseen imponderable existences which, like electricity and heat, are known only by their effects. It is certainly identical with electricity, for, although it never comes naturally into evidence, magnets can be made to exhibit all the pheno mena of electrical machines.

Terrestrial magnetism, which pervades the whole earth, is extremely complicated; it varies both with regard to space and time, and probably depends upon the heat of the sun, upon his motion in the ecliptic, which produces changes of temperature, on galvanic

[1] Colonel Sabine's Notes to the English translation of Humboldt's 'Cosmos,' vol. ii.

currents circulating through the surface of the globe, and possibly on the earth's rotatory motion.

The distribution of terrestrial magnetism is determined by the declination-needle, or mariner's compass, and the dipping-needle; they consist of magnetised needles or bars of steel, so suspended that the declination-needle revolves in a horizontal direction, and the dipping-needle moves in a plane perpendicular to the horizon. The north end of the declination-needle or magnet points to the north, and the south end to the south, and it only remains at rest when in that position. The direction of the needle is the magnetic meridian of the place of observation.

The north end of the dipping-needle bends or dips below the horizon in the northern hemisphere, and the south end bends or dips beneath it in the southern hemisphere, and between the two there is a line which encircles the whole earth, where the dipping-needle remains horizontal. That line, which is the magnetic equator or line of no dip, crosses the terrestrial equator in several places, extending alternately on each side, but never deviating more than twelve degrees from it. The deviation is greater in that part of the Pacific where there are most islands, and it is greatest both to the south and north in traversing the continents of Africa and America; thus it appears that the configuration of the land and water has an influence on terrestrial magnetism. North and south of the magnetic equator the needle dips more and more, till at last it becomes perpendicular to the horizon in two points, or rather linear spaces, known as the north and south magnetic poles, which are quite distinct from the poles of the earth's rotation. One, whose position was determined by Captain Sir James Clark Ross, is in 70° N. lat. and 97° W. long., while that in the southern hemisphere, placed by the same celebrated navigator, from his observations in 1841, in the interior of Victoria Island, is in 75° 5′ S. lat. and 154° 8′ E. long. Lines of equal dip are such as may be drawn on a globe through all those places where the dipping-needle makes the same angle with the horizon. The angle of the dip is not always the same: according to Colonel Sabine, who is the highest authority on this subject, it has been decreasing in the northern hemisphere for the last fifty years, at the rate of three minutes annually; it is also subject to variations of short periods, and it seems to be affected by shocks of earthquakes, even when very distant.

The intensity of the magnetic force is as variable and even more complicated than the other magnetic phenomena: it is measured by the number of vibrations made by the declination-needle in a given time. It is very different in different parts of the earth, but there are four points in which the intensity is greater than anywhere else Two of these are in the northern and two in the southern hemisphere; they neither coincide with the poles of the earth's rotation nor with the magnetic poles, nor are they all of equal intensity.

One of these foci of maximum magnetic intensity is situate in North America, south-west from Hudson's Bay; another is in northern Siberia, in 120° E. long. In the southern hemisphere, one of the points of maximum magnetic intensity is in the South Atlantic in 20° S. lat. and 324° E. long., and the other is situate in 60° S. lat. and 131° 20′ E. long.[1] In consequence of the unequal intensity of the force in these 4 foci, the decrease in magnetic power from them towards the equator is extremely irregular, so that the dynamic equator, which is a line supposed to be drawn through all the points on the earth where the intensity is the least, encircles the globe in a waving line, which neither coincides with the geographical nor magnetic equator; it forms the division between the magnetic intensities in the two hemispheres. Lines drawn on a globe through all the points where the magnetic intensity is the same are so complicated that it is scarcely possible to convey an idea of them in words. They form a series of ovals round each of the foci of maximum force, then a figure of 8 in each hemisphere having a focus and its ovals in each loop, then they open into tortuous lines which encompass the globe, but which become less so as they approach the dynamic equator. The complication is increased by the foci in the two hemispheres being unsymmetrically placed with regard to one another, as well as by the difference in their intensities.

The declination or horizontal needle only remains at rest when in a magnetic meridian, that is when it points to the north and south magnetic poles. The magnetic meridians coincide with the geographical meridians in some places, and in these the magnet points to the true north and south, that is, to the poles of the earth's rotation. But if it be carried successively to different longitudes, it will deviate sometimes to the east, sometimes to the west of the true north. Imaginary lines on the globe, passing through all places where the magnet points to the poles of the earth's rotation, are lines of no variation; and lines passing through all places where the magnet deviates by an equal quantity from the geographical meridians are lines of equal variation;[2] they are also very irregular and form two closed systems or loops—that is, they surround two points,

[1] The foci are all of different intensities; that in the South Atlantic, discovered by M. Erman, has the least intensity of the four, and the other in the southern hemisphere, discovered by Sir James Ross, has the greatest; taking 1 as the unit at the magnetic equator in Peru, their intensities are as 20·071 and 0·706. In the northern hemisphere the American focus is more intense than that in Siberia, which is moving from west to east, while the minor focus in the southern hemisphere is moving from east to west.

[2] A very interesting series of general charts, on which these lines of equal variation are laid down, is on the point of being published by the Hydrographic Office.—January, 1851.

one in northern Siberia and another in the Pacific, nearly in the meridian of the Pitcairn Islands and the Marquesas.[1]

The whole magnetic system is perpetually undergoing secular and periodical changes, which are so irregular and complicated that half a century is sufficient to alter the form and position of all the lines that have been mentioned. The foci of magnetic intensity, and the whole system represented by the magnetic lines, are moving along the two hemispheres in opposite directions; those in the northern hemisphere are going from west to east, and those in the southern from east to west; and as the foci of maximum intensity move with different velocities, the forms as well as the places of the curves are slowly, yet continually, changing. The weaker magnetic focus in the northern hemisphere moved through 50 degrees of longitude in 250 years.

The declination is subject to periodic variations, depending upon the position of the moon, and to annual variations arising from the motion of the sun in the ecliptic, as well as to horary variations corresponding to changes of temperature from the diurnal rotation of the earth.

Throughout the middle latitudes of the northern hemisphere the north end of the magnet has a mean motion from east to west from eight in the morning till half-past one, it then moves to the east till evening, after which it makes another excursion to the west, and returns again to its original position at eight in the morning. The extent of its variation is greater in the day than in the night, in summer than in winter. It decreases from the middle latitudes in Europe, where it is 13 or 14 minutes, to the equator, where it is only 3 or 4; but at the equator the variations are performed with extreme regularity. The horary motions of the south end of the magnet in the southerh hemisphere are accomplished in an exactly opposite direction. Between these two magnetic hemispheres there is a line passing through an infinity of places, and very nearly coinciding with the line of minimum magnetic intensity, where the horary phenomena of both hemispheres are combined, each predominating alternately at opposite seasons. At St. Helena, which is one of the places in question and nearly on the line of minimum intensity, the horary motion of the north end of the magnet corresponds in direction during one-half of the year with the movement in the northern hemisphere, and in the other half of the year the direction at the same hours corresponds with that in the southern

---

[1] The author is indebted to the admirable and profound investigations of Colonel Sabine for almost all she knows on the subject of terrestrial magnetism. In these, and in his notes on the English translation of Humboldt's 'Cosmos,' the reader will find all that is most interesting on the subject. In his own works there are plates of the course of the different magnetic lines mentioned in the text.

hemisphere, the passage from the one to the other being at the equinoxes, when the diurnal variations at the usual hours partake more or less of the characteristics of both on different days.[1]

It thus appears that there are six points on the earth peculiarly remarkable for magnetic phenomena, all of which are distinct from one another, and from the poles of the earth's rotation—namely, two magnetic poles where the dipping needle makes an angle of 90 degrees with the horizon. The magnetic equator corresponds with these in every point of which the angle of the dip is zero: it encircles the earth, and intersects the terrestrial equator, but does not coincide with it. The other four points are the foci of maximum magnetic intensity, and to them the dynamical equator or line of minimum magnetic intensity corresponds, also surrounding the earth in an irregular line, but which coincides with neither the terrestrial nor magnetic equator. Besides these, and either partly or nearly coinciding with the line of minimum intensity, is that line which is supposed to pass through all places where the hoary variations of the magnet partake of the phenomena of each hemisphere alternately.

The earth's magnetism is subject to vast unaccountable commotions or storms of immense extent, which occur at irregular intervals and are of short duration. In 1818 a magnetic storm, shown by a violent agitation of the needle, took place at the same time over 47 degrees of longitude, extending through all the countries from Paris to Kasan; and on the 25th of September, 1841, one of these storms was simultaneously observed at Toronto in North America, at the Cape of Good Hope, Prague in Europe, at Macao in China, and there is reason to believe that it extended to Van Diemen's Land. Similar storms have happened simultaneously in Sicily and at Upsala in Sweden; others of less extent and shorter periods more frequently occur, and are, like the greater storms, not to be attributed to any known cause.

M. Necker de Saussure has traced a marked coincidence between the prevailing direction of the stratified masses of the mountain chains and that of the curves of equal magnetic intensity. The coincidence is perfect in the Ural chain, for there the lines of force tend north and south; and they do not deviate much from the stratification in the great plains of European Russia. There is every reason to believe that a coincidence takes place in the Scandinavian mountains, for a line of equal magnetic intensity passes parallel to

[1] At St. Helena the north end of the needle reaches its eastern extreme in May, June, July, and August, and nearly at the same hours it reaches its western extreme in November, December, January, and February. The passage from one to the other takes place at, or soon after, the equinoxes in March and April, September and October. — Colonel Sabine's Notes to 'Cosmos,' Vol. ii.

the Norwegian coast. In Scotland a line almost coincides with the Grampians; and as it becomes less northerly before reaching Portugal and Spain, it is there also in singular coincidence with the sierras on the table-land; the Pyrenees however form an exception to the law. A magnetic line follows the break of the chain of the Alps with great precision. The intersection of two upheavals makes these mountains alter their direction from S.W. and N.E. to E. nearly, and near to that change the magnetic line takes a similar bend and coincides with the Caucasus, Taurus, Hindoo-Coosh, Himalaya, and Chinese mountains, after which it again tends to the north, and follows the Yablonoi chain to Behring's Straits.

In Africa the lines of equal magnetic force coincide with the Komri, and with the lofty sea-coast range which unites the mountains of Abyssinia with those at the Cape of Good Hope. Throughout North America the lines of equal force coincide with the Alleghanies, and on the coast of the Pacific they take the direction of the Rocky Mountains. In Mexico the stratified rocks are parallel to the mountains of Anahuac, which is the same with the direction of the magnetic curves, and a similar coincidence takes place in the Parima ranges, and in the coast-chain of Venezuela. The Andes and the lines of equal magnetic intensity are completely discordant, for they cross one another; but lines of equal magnetic force stretch from the southern promontories of America and Asia to the mountains of Victoria Land.

There is strong presumptive evidence of the influence of the electric and magnetic currents on the formation and direction of the mountain masses and mineral veins, but their slow persevering action on the ultimate atoms of matter has been placed beyond doubt by the formation of rubies and other gems, as well as various other mineral substances, by voltaic electricity.

The existence of electric currents on the surface of the earth has been deduced from terrestrial magnetism, and from the connection between the diurnal variations of the magnet and the apparent motion of the sun; also from the electro-magnetic properties of metalliferous veins, and from atmospheric electricity, which is continually passing between the air and the earth.

[Professor Faraday has shown that oxygen is magnetic, being attracted towards the pole of a magnet; and that, like other magnetic bodies, it loses and gains in power as its temperature is raised and lowered, and that these changes occur with the range of natural temperature. These properties it carries into the atmosphere.

Oxygen loses its sensible magnetism in almost all gases when it unites with them in *chemical* combination.

A magnetic gas, *mechanically* mixed with any other gas, preserves its magnetism whatever be the density of the mixture; but in the neighbourhood of the poles, there appears to be, to a certain extent, a separation of the gases, which must slightly augment the attraction of the entire mass.

Lieutenant Maury has endeavoured to show a probable connection between the magnetism of the oxygen of the air and the circulation of the atmosphere, between the equatorial and polar regions of the earth.[1]]

Dr. Faraday's brilliant discoveries have changed the received opinions with regard to the magnetic properties of matter. Although all bodies are magnetic, they show that it assumes a totally different form in different substances. For example, if a bar of iron be freely suspended between the poles of an electro-magnet, or very powerful horse-shoe magnet, it will be attracted by both poles, and will rest in the direction between them—that is, on the line of force. But if a bar of bismuth be suspended in the same manner, it will be repelled by both poles, and will assume a direction at right angles to that which the iron took, and thus the same force, whether electric or magnetic, produces opposite effects upon these two metals. Substances affected after the manner of iron are magnetic—those affected after the manner of bismuth are said to be *diamagnetic*. All substances come under one or other of these two classes: the diamagnetic are infinitely more abundant than the magnetic; almost all bodies on earth belong to that class. Many of the metals, acids, oils, sugar, starch, animal matter, flame, and all the gases, whether light or heavy, have the diamagnetic property less or more, but oxygen less than any other, and that is the reason why atmospheric air is the most feebly diamagnetic of all substances at its natural temperature; for when very hot it becomes more diamagnetic, and if extremely cold it takes a place among the magnetic class. Important results with regard to the magnetic state of the globe will undoubtedly be deduced from this new property of matter, and Dr. Faraday's observations on that subject show that he is not without such anticipations.[2]

["Five years ago," says Dr. John Tyndall, in a paper presented to the British Association in 1851, "Faraday established the existence of a force called *diamagnetism*, and from that time to the present, some of the first minds in Germany, France and England, have been devoted to the investigation of this subject. One of the most important aspects of the inquiry is the relation which subsists between magnetism and diamagnetism. Are the laws which govern both forces identical? Will the mathematical expression of the at-

[[1] The laws of magnetism are lucidly explained by Dr. Lardner in his Hand-Books of Natural Philosophy.]

[2] These anticipations appear to be fully verified by Dr. Faraday's important discoveries, which have been just announced, on the effects of solar heat upon the oxygen of the atmosphere as the grand moving cause in magnetic phenomena.—December, 1850.

traction in one case be converted into the expression of the repulsion in the other by a change of sign from positive to negative?

The following are the principal results of Dr. Tyndall's investigation.

1. The repulsion of a diamagnetic substance placed at a fixed distance from the pole of a magnet is governed by the same law as the attraction of a magnetic substance.

2. The entire mass of a magnetic substance is most strongly attracted when the attracting force acts parallel to that line which sets axial when the substance is suspended in the magnetic field; and the entire mass of a diamagnetic substance is most strongly repelled when the repulsion acts parallel to the line which sets equatorial in the magnetic field.

3. The superior attraction and repulsion of the mass in a particular direction is due to the fact, that in this direction the material particles are ranged more closely together than in any other directions; the force exerted being attractive or repulsive, according as the particles are magnetic or diamagnetic. This is a law applicable to matter in general, the phenomena exhibited by crystals in the magnetic field being particular manifestations of the same.]

"When we consider the magnetic condition of the earth as a whole, without reference to its possible relation to the sun, and reflect upon the enormous amount of diamagnetic matter which forms its crust; and when we remember that magnetic curves of a certain amount of force, universal in their presence, are passing through these matters, and keeping them constantly in a state of tension, and therefore of action, we cannot doubt that some great purpose, of utility to the system and to us its inhabitants, is fulfilled by it. If the sun have anything to do with the magnetism of the globe, then it is possible that part of this effect may be due to the action of the light that comes to us from that body; and in that view the air seems most strikingly placed round our sphere, investing it with a transparent diamagnetic, which therefore is permeable to his rays, and at the same time moving with great velocity across them. Such conditions seem to suggest the possibility of magnetism being thence generated."

CHAPTER XXIII.

Vegetation — Nourishment and Growth of Plants — Effects of the different Rays of the Solar Spectrum — Classes — Botanical Districts.

In the present state of the globe a third part only of its surface is occupied by land, and probably not more than a fourth part of that is inhabited by man, but animals and vegetables have a wider range. The greater part of the land is clothed with vegetation and inhabited by quadrupeds, the air is peopled with birds and insects, and the sea teems with living creatures and plants. These organised beings are not scattered promiscuously, but all classes of them have been originally placed in regions suited to their respective wants. Many animals and plants are indigenous only in determinate spots, while a thousand others might have supported them as well, and to many of which they have been transported by man.

Plants extract inorganic substances from the ground which are indispensable to bring them to maturity, but the atmosphere supplies the vegetable creation with the principal part of its food.

The black or brown mould which is so abundant is the produce of decayed vegetables. When the autumnal leaves, the spoil of the summer, fall to the ground, and their vitality is gone, they enter into decomposition, and combining with the oxygen of the atmosphere convert it into an equal volume of carbonic acid gas, which consequently exists abundantly in every good soil, and is the most important part of the nourishment of vegetables. This process is slow, and stops as soon as the air in the soil is exhausted; but the plough, by loosening the earth, and permitting the atmosphere to enter more freely and penetrate deeper into the ground, accelerates the decomposition of the vegetable matter, and consequently the formation of carbonic acid.

In loosening and refining the mould, the common earth-worm is the fellow-labourer with man; it eats earth, and, after extracting the nutritious part, ejects the refuse, which is the finest soil, and may be seen lying in heaps at the mouth of its burrow. So instrumental is this creature in preparing the ground, that it is said there is not a particle of the finer vegetable mould that has not passed through the intestines of a worm: thus the most feeble of living things is employed by Providence to accomplish the most important ends.

The food of the vegetable creation consists of carbon, hydrogen, azote, [nitrogen,] and oxygen — all of which plants obtain entirely from the atmosphere in the form of carbonic acid gas, water, and ammonia.

They imbibe these three substances, and, after having decomposed them, they give the oxygen to the air, and consolidate the carbon, water, and azote into wood, leaves, flowers, and fruit.

The vitality of plants is a chemical process entirely due to the sun's light; it is most active in clear sunshine, feeble in the shade, and nearly suspended in the night, when plants, like animals, have their rest.

The atmosphere contains about one-three-thousandth part of carbonic acid gas, yet that small quantity yields enough of carbon to form the solid mass of all the magnificent forests and herbs that clothe the face of the earth, and the supply of that necessary ingredient in the composition of the atmosphere is maintained by the breathing of animals, by volcanoes, by decomposition of animal and vegetable matter, and by combustion. The green parts of plants constantly imbibe carbonic acid in the day; they decompose it, assimilate the carbon, and return the oxygen pure to the atmosphere. As the chemical action is feeble in the shade and in gloomy weather, only a part of the carbonic acid is decomposed, then both oxygen and carbonic acid are given out by the leaves; but during the darkness of the night a chemical action of a different character takes place, and almost all the carbonic acid is returned unchanged to the atmosphere, together with the moisture which is evaporated from the leaves both night and day. Thus plants give out pure oxygen during the day, and carbonic acid and water during the night.

Since the vivifying action of the sun brings about all these changes, a superabundance of oxygen is exhaled by the tropical vegetation in a clear unclouded sky, where the sun's rays are most energetic, and atmospheric moisture most abundant. In the middle and higher latitudes, on the contrary, under a more feeble sun and a gloomy sky, subject to rain, snow, and frequent atmospheric changes, carbonic acid is given out in greater quantity by the less vigorous vegetation. But here, as with regard to heat and moisture, equilibrium is restored by the winds; the tropical currents carry the excess of oxygen along the upper strata of the atmosphere to higher latitudes, to give breath and heat to men and animals; while the polar currents, rushing along the ground, convey the surplus carbonic acid to feed the tropical forests and jungles. Harmony exists between the animal and vegetable creations; animals consume the oxygen of the atmosphere, which is restored by the exhalation of plants, while plants consume the carbonic acid exhaled by men and animals; the existence of each is thus due to their reciprocal dependence. Few of the great cosmical phenomena have only one end to fulfil, they are the ministers of the manifold designs of Providence.

When a seed is thrown into the ground, the vital principle is developed by heat and moisture, and part of the substance of the seed is formed into roots, which suck up water mixed with carbonic acid from the soil, decompose it, and consolidate the carbon. In this stage of their growth, plants derive their whole sustenance from the ground. As soon, however, as the sugar and mucilage of the seed appear above the ground, in the form of leaves or shoots, they absorb and decompose the carbonic acid of the atmosphere, retain the carbon for their food, give out the oxygen in the day, and pure carbonic acid in the night. In proportion as plants grow, they derive more of their food from the air and less from the soil, till their fruit is ripened, and then the whole of their nourishment is derived from the atmosphere. Trees are fed from the air after their fruit is ripe, till their leaves fall, annuals till they die. Air-plants and several species of cactus and others derive all their food from the atmosphere. It is wonderful that so small a quantity of carbonic acid as exists in the air should suffice to supply the whole vegetation of the world—and still more wonderful that a seed minute enough to be wafted invisibly by a breath of air should be the theatre of all the chemical changes that make it germinate.[1]

Plants absorb water from the ground by their roots; they decompose it, and the hydrogen combines in different proportions with their carbonic acid to form wood, sugar, starch, gum, vegetable oils and acids. As the green parts combine with the oxygen of the air, especially during night, when the function of plants are torpid, it is assimilated on the return of daylight, and assists in forming oils, resins, and acids. The combination of the oxygen of the air with the leaves, and also with the blossom and fruit, during night, appears to be unconnected with the vital process, as it is the same in dead plants. An acid exists in the juice of every plant, generally in combination with an alkali. It must be observed, however, that these different substances are produced at different stages in the growth; for example, starch is formed in the roots, wood, stalk, and seed, but it is converted into sugar as the fruit ripens, and the more starch the sweeter the fruit becomes. Most of these new compounds are formed between the flowering of the plant and the ripening of the fruit, and indeed they furnish the materials for the flowers, fruit, and seed.

Ammonia, the third organic constituent of plants, is the last residue from the decay and putrefaction of animal matter. It is volatilized, and rises into the atmosphere, where it exists as a gas, but in so small a quantity that it is with difficulty detected by chemical analysis; yet, as it is very soluble in water, enough is brought to

[1] The sporules or seeds of the fungi are so minute that M. Freis counted above ten millions in a single plant of the reticularia maxima: they were so subtile that they were like smoke.

the ground by rain to supply the vegetable world. Ammonia enters plants by their roots along with rain-water, and is resolved within them into its constituent elements, hydrogen and nitrogen. The hydrogen aids in forming the wood, acids, and other substances before mentioned; while the nitrogen enters into every part of the plant and forms new compounds; it exists in the blossom and fruit before it is ripe, and in the wood, as albumen; it also forms gluten, which is the nutritious part of wheat, barley, oats, and all other cerealia, as well as of esculent roots, as potatoes, beet-root, &c. Nitrogen exists abundantly in peas, beans, and pulse of every kind;[1] it enters into the composition of most elementary vegetable substances; in short, a plant may grow without ammonia, but it cannot produce seed or fruit; the use of animal manure is to supply plants with this essential article of their food. Thus the decomposition and consolidation of the elementary food of plants, the formation of the green parts, the exhalation of moisture by their leaves, its absorption by their roots, and all the other circumstances of vegetable life, are owing to the illuminating power of the sun. Heat can be supplied artificially in our northern climates, but it is impossible to replace the splendour of a southern sun. His illuminating influence is displayed in a remarkable degree by the cacalia ficoides; its leaves combine with the oxygen of the atmosphere during the night, and are as sour as sorrel in the morning: as the sun rises they gradually lose their oxygen, and are tasteless at noon; by the continued action of light they lose more and more, till towards evening they become bitter. The difference of a clear or cloudy sky has an immense effect on vegetation; the ripening of fruit depends upon the habitual serenity of the sky more than on summer temperature alone.

The blue rays of the solar spectrum have most effect on the germination of seed; the yellow rays, which are the most luminous, on the growing plant. That is on account of the chemical rays, now so well known by their action in Daguerreotype impressions. They are most abundant beyond the visible part of the solar spectrum, and diminish through the violet, blue, and green, to the yellow, where they cease. They penetrate the ground, and have a much greater influence on the germination of seeds than ordinary light or darkness. That invisible principle, together with light, is essential to the formation of the colouring matter of leaves; it is most active in spring, and is in very considerable excess compared with the quantity of light and heat; but as summer advances, the reverse takes place; the calorific radiation, or those hot rays corresponding to the extreme red of the spectrum, which facilitate the

[1] It is very doubtful, from some late researches noticed elsewhere, that the air contains any appreciable quantity of ammoniacal gas, or that it contributes in a material degree to vegetation. See M. de Ville's researches in 'Comptes Rendus.'

flowering and forming of the fruit, become by far the most abundant; and a set of invisible rays, which exist near the point of maximum heat in the solar spectrum, are also most abundant in summer. Mr. Hunt found that the hot rays immediately beyond the visible red destroy the colour of certain leaves; and for that reason the glass of the great palm-house at Kew Gardens is tinged pale yellow-green, to exclude the scorching rays in question, though it is permeable by the other rays of heat, those of light, and the chemical rays.[1]

In spring and summer the oxygen taken in by the green leaves in the night aids in the formation of oils, acids, and the other parts that contain it; but as soon as autumn comes, the vitality or chemical action of vegetables is weakened; and the oxygen, no longer given out in the day, though still taken in during the night, becomes a minister of destruction; it changes the colour of the leaves, and consumes them when they fall. Nitrogen, so essential during the life of plants, also resumes its chemical character when they die, and by its escape hastens their decay.

Although the food which constitutes the mass of plants is derived principally from the water and the gases of the atmosphere, fixed substances are also requisite for their growth and perfection, and these they obtain from the earth by their roots. The inorganic matters are the alkalis, phosphates, silica, sulphur, iron, and others.

It has already been mentioned that vegetable acids are found in the juices of all the families of plants. They generally are in combination with one or other of the alkaline substances, as potash, lime, soda, and magnesia, which are as essential to the existence of plants as the carbonic acid by which these acids are formed: for example, vines have potash; plants used as dyes never give vivid colours without it; all leguminous plants require it, and only grow naturally on ground that contains it. None of the corn tribe can produce perfect seeds unless they have both potash and phosphate of magnesia; nor can they or any of the grasses thrive without silica, which gives the

---

[1] The solar spectrum, or coloured image of the sun, formed by passing a sunbeam through a prism, is composed of a variety of invisible as well as visible rays. The chemical rays are most abundant beyond the violet end of the spectrum, and decrease through the violet, blue, and green, to the yellow, where they cease. The rays of heat are in excess a little beyond the red end, and gradually decrease towards the violet end. Besides these there are two insulated spots at a considerable distance from the red, where the heat is a maximum. Were the rays of heat visible, they would exhibit differences as distinct as the coloured rays, so varied are their properties according to their position in the spectrum. There are also peculiar rays which produce phosphorescence, others whose properties are not quite made out, and probably many undiscovered influences; for time has not yet fully revealed the sublimity of that creation, when God said, "Let there be light—and there was light."

hard coating to straw, to the beard of wheat and barley, to grass, canes, and bamboos; it is even found in solid lumps in the hollows and joints of cane, known in India by the name of tabashir. To bring the cerealia to perfection, it is indispensable that in their growth they should be supplied with carbonic acid for the plant, silica to give it strength and firmness, and nitrogen for the grain.

Phosphoric acid, combined with an earth or alkali, is found in the ashes of all vegetables, and is essential to many. Pulse contain but little of it, and on that account are less nutritious than the cerealia. The family of the cruciferæ, as cabbages, turnips, mustard, &c., contain sulphur in addition to the substances common to the growth of all plants; each particular tribe has its own peculiarities, and requires a combination suited to it. On that account there is often a marked difference in the arborescent vegetation on the same mountain, depending on the nature of the rocks.

The ocean furnishes some of the matters found in plants; the prodigious quantity of sea-water constantly evaporated carries with it salt in a volatilized state, which, dispersed over the land by the wind, supplies the ground with salt and the other ingredients of sea-water. The inorganic matters which enter plants by their roots are carried by the sap to every part of the vegetable system. The roots imbibe all liquids presented to them indiscriminately, but they retain only the substances they require at the various stages of their growth, and throw out such parts as are useless, together with the effete or dead matter remaining after the nutriment has been extracted from it. Plants, like animals, may be poisoned, but the power they have of expelling deleterious substances by their roots generally restores them to health. The feculent matter injures the soil; besides, after a time the ground is drained of the inorganic matter requisite for any one kind of plant: hence the necessity for a change or rotation of crops.

A quantity of heat is set free and also becomes latent in the various transmutations that take place in the interior of plants; so that they, like the animal creation, have a tendency to a temperature of their own, independent of external circumstances.

The quantity of electricity requisite to resolve a grain weight of water into its elementary oxygen and hydrogen is equal to the quantity of atmospheric electricity which is active in a very powerful thunder-storm; hence some idea may be formed of the intense energy exerted by the vegetable creation in the decomposition of the vast mass of water and other matters necessary for its sustenance. But there must be a compensation in the consolidation of the vegetable food, otherwise a tremendous quantity would be in perpetual activity. It is said to be given out from the points of their leaves, so possibly some part of the atmospheric electricity may be ascribed to this cause; but there is reason to believe that electricity, excited by the power of solar light, constitutes the chemical vitality of vegetation.

The colouring matter of flowers is various, if we may judge from the effect which the solar spectrum has upon their expressed juices. The colour is very brilliant on the tops of mountains and in the Arctic lands. Possibly the diminished weight of the air may have some effect, for it can scarcely be supposed that barometrical changes should be entirely without influence on vegetation.

The perfume of flowers and leaves is owing to a volatile oil, which is often carried by the air to a great distance: in hot climates it is most powerful in the morning and evening. The odour of the Humiria has been perceived three miles from the coast of South America; a species of Tetracera sends its perfume as far from the island of Cuba; and the aroma of the Spice Islands is wafted to a considerable distance out to sea. The variety of perfumes is infinite, and shows the innumerable combinations of which a few simple substances are capable, and the extreme minuteness of the particles of matter.

In northern and mean latitudes winter is a time of complete rest to the vegetable world, and in tropical climates the vigour of vegetation is suspended during the dry, hot season, to be resumed at the return of the periodical rains. The periodical phenomena of the appearance of the first leaves, the flowering, ripening of the fruit, and the fall of the leaf, depend upon the annual and diurnal changes of temperature, moisture, electricity, and perhaps on magnetism, and succeed with such perfect harmony and regularity, that, were there a sufficient number of observations, lines might be drawn on a globe passing through all places where the leaves of certain plants appear simultaneously, and also for the other principal phases of vegetation. In places where the same plant flowers on the same day, the fruit may not ripen at the same period in both; it would therefore be interesting to know what relation lines passing through those would have to one another and to the isothermal lines; more especially with regard to the plants indispensable to man, since the periodicity of vegetation affects his whole social condition.[1]

Almost all plants sleep during the night; some show it in their leaves, others in their blossom. The Mimosa tribe not only close their leaves at night, but their foot-stalks droop; in a clover-field not a leaf opens until after sunrise. The common daisy is a familiar instance of a sleeping flower; it shuts up its blossom in the

[1] Professor Quetelet is desirous that the periodical phenomena of vegetation should be observed at a number of places, in order to establish a comparison between the periods at which they take place; and for that purpose he gives a list of the commonest plants, as lilac, laburnum, elder, birch, oak, horse-chestnut, peach, pear, crocus, daisy, &c., which he himself observes annually at Brussels.

evening, and opens its white and crimson-tipped star, the "day's eye," to meet the early beams of the morning sun; and then also "winking mary-buds begin to ope their golden eyes."

The crocus, tulip, convolvulus, and many others, close their blossoms at different hours towards evening, some to open them again, others never. The ivy-leaved lettuce opens at eight in the morning, and closes for ever at four in the afternoon. Some plants seem to be wide awake all night, and to give out their perfume only, or at nightfall. Many of the jessamines are most fragrant during the twilight: the Olea fragrans, the Daphne odorata, and the night-stock reserve their sweetness for the midnight hour, and the night-flowering Cereus turns night into day. It begins to expand its magnificent sweet-scented blossom in the twilight, it is full blown at midnight, and closes, never to open again, with the dawn of day;—these are "the bats and owls of the vegetable kingdom."[1]

Many plants brought from warm to temperate climates have become habituated to their new situation, and flourish as if they were natives of the soil; such as have been accustomed to flower and rest at particular seasons change their habits by degrees, and adapt themselves to the seasons of the country that has adopted them. It is much more difficult to transfer alpine plants to the plains. Whether from a change of atmospheric pressure or mean temperature, all attempts to cultivate them at a lower level generally fail: it is much easier to accustom a plant of the plains to a higher situation.

Plants are propagated by seeds, offsets, cuttings, and buds; hence they, but more especially trees, have myriads of seats of life, a congeries of vital systems acting in concert, but independently of each other, every one of which might become a new plant. In this respect the fir and pine tribe are inferior to deciduous trees, which lose their leaves annually, because they are not easily propagated except by seeds. It has been remarked that all plants that are propagated by buds from a common parent stock have the same duration of life: this has been noticed particularly with regard to some species of apple-trees in England. It appears that all the garden varieties of fruit, whether from buds, layers, or cuttings, wear out after a time; and that seedlings have a great tendency to revert to the original wild character of the plant.

A certain series of transitions takes place throughout the lives of plants, each part being transformed and passing into another; a law

[1] Dandelion opens at five or six in the morning, and shuts at nine in the evening; the goat's-beard wakes at three in the morning, and shuts at five or six in the afternoon. The orange-coloured Escholtzia is so sensitive that it closes during the passage of a cloud. "The marigold that goes to bed wi' the sun, and with him rises weeping," with many more, are instances of the sleep of plants: the gentianella, veronica, and other plants close their blossoms on the approach of rain.

that was first observed by the illustrious poet Göthe. For example, the embryo leaves pass into common leaves, these into bracteæ, the bracteæ into sepals, the sepals into petals, which are transformed into stamens and anthers, and these again pass into ovaries with their styles and stigmas, that are to become the fruit and ultimately the seed of a new plant.

Plants are naturally divided into three classes, differing materially in organization:—The Cryptogamia, whose flowers and seeds are either too minute to be easily visible, or are hidden in some part of the plant, as in fungi, mosses, ferns, and lichens, which are of the least perfect organization. Next to these are the monocotyledonous plants, as grasses and palms, in which the foot-stalks of the old leaves form the outside of the stem; plants of this class have but one seed lobe, which forms one little leaf in their embryo state. Their flowers and fruit are generally referable to some law in which the number 3 prevails, as, for example, the petals and other parts are three in number. The dicotyledonous plants form the third class, which is the most perfect in its organization and by much the most numerous, including the trees of the forest and most of the flowering shrubs and herbs. They increase by coatings from without, as trees, where the growth of each year forms a concentric circle of wood round the pith or centre of the stem; the seeds of these plants have two lobes, which in their embryo state appear first in two little leaves above ground, like most of the European species. The parts of the flowers and fruit of this class generally have some relation to the number 5.

The three botanical classes are distributed in very different proportions in different zones: monocotyledonous plants, such as grasses and palms, are much more rare than the dicotyledonous class. Between the tropics there are four of the latter to one of the grass or palm tribes, in the temperate zones six to one, and in the polar regions only two to one, because mosses and lichens are most abundant in the high latitudes, where dictyledonous plants are comparatively rare. In the temperate zones one-sixth of the plants are annuals, omitting the cryptogamia; in the torrid zone scarcely one plant in twenty is annual, and in the polar regions only one in thirty. The number of ligneous vegetables increases on approaching the equator, yet in North America there are 120 different species of forest-trees, whereas in the same latitudes in Europe there are only 34. The social plants, grasses, heaths, furze, broom, daisies, &c., which cover large tracts, are rare between the tropics, except on the mountains and table-lands and on the llanos of equatorial America.

Equinoctial America has a more extensive and richer vegetation than any other part of the world; Europe has not above half the number of indigenous species of plants; Asia, with its islands, has somewhat less than Europe; Australia, with its islands, in the Pacific, still less; and there are fewer vegetable productions in Africa than in any part of the globe of the same extent.

Since the constitution of the atmosphere is very much the same everywhere, vegetation depends principally on the sun's light, moisture, and the mean annual temperature, and it is also in some degree regulated by the heat of summer in the temperate zones, and also by exposure, for such plants as require warmth are found at a lower level on the north than on the south side of a mountain. Between the tropics, wherever rain does not fall, the soil is burnt up and is as unfruitful as that exposed to the utmost rigour of frost; but where moisture is combined with heat and light, the luxuriance of the vegetation is beyond description. The abundance and violence of the periodical rains combine with the intense light and heat to render the tropical forests and jungles almost impervious from the rankness of the vegetation. This exuberance gradually decreases with the distance from the equator; it also diminishes progressively as the height above the level of the sea increases, so that each height has a corresponding parallel of latitude where the climates and floras are similar, till the perpetual snow on the mountain-tops, and its counterpart in the polar regions, have a vegetation that scarcely rises above the surface of the ground. Hence, in ascending the Himalaya or Andes from the luxuriant plains of the Ganges or Amazon, changes take place in the vegetation analogous to what a traveller would meet with in a journey from the equator to the poles. This law of decrease, though perfectly regular over a wide extent, is perpetually interfered with by local climate and soil. From the combination of various causes, as the distribution of land and water, their different powers of absorption and radiation, together with the form, texture, and clothing of the land, and the prevailing winds, it is found that the isothermal lines, or imaginary lines drawn through places on the surface of the globe which have the same mean annual temperature, do not correspond with the parallels of latitude. Thus, in North America, the climate is much colder than in the corresponding European latitudes. Quebec is in the latitude of Paris, and the country is covered with deep snow four or five months in the year, and it has occurred that a summer has passed there in which not more than 60 days have been free from frost.

In the southern hemisphere, beyond the 34th parallel, the summers are colder and the winters milder than in corresponding latitudes of the northern hemisphere. Neither does the temperature of mountains always vary exactly with their height above the sea; other causes, as prevailing winds, difference of radiation, and geological structure, concur in producing irregularities which have a powerful effect on the vegetable world.

However, no similarity of existing circumstances can account for whole families of plants being confined to one particular country, or even to a very limited district, which, as far as we can judge, might have grown equally well on many others. Latitude, elevation, soil, and climate are but secondary causes in the distribution of the vegetable kingdom, and are totally inadequate to explain why there are numerous distinct botanical districts in the continents and islands, each of which has its own vegetation, whose limits are most decided when they are separated by the ocean, mountain-chains, sandy deserts, salt-plains, or internal seas. Each of these districts is the focus of families and genera, some of which are found nowhere else, and some are common to others, but, with a very few remarkable exceptions, the species of plants in each are entirely different or representative.[1] This does not depend upon the dif-

[1] M. de Candolle established 20 botanical regions, and Professor Schow 20; but Professor Martius, of Munich, has divided the vegetation of the globe into 51 provinces, namely, 5 in Europe, 11 in Africa, 13 in Asia, 3 in New Holland, 4 in North and 8 in South America, besides Central America, the Antilles, the Antarctic Lands, New Zealand, Van Diemen's Land, New Guinea, and Polynesia. To these, other divisions might be added, as the Galapagos, which is so strongly defined.

Baron Humboldt gives the following concise view of the distribution of plants, both as to height and latitude:—

The equatorial zone is the region of palms and bananas.
The tropical zone is the region of tree-ferns and figs.
The subtropical zone, that of myrtles and laurels.
The warm temperate zone, that of evergreen trees.
The cold temperate zone, that of European or deciduous trees.
The subarctic zone, that of pines.
The arctic zone, that of rhododendrons.
The polar zone, that of alpine plants.

Upper Limit of Trees on Mountains.—The upper limit of trees is distinguished by the Escalloniæ, on the Andes of Quito, at the height of 11,500 feet above the level of the sea.

In tropical Mexico the upper limit of trees, at the height of 12,789 feet, is distinguished by the Pinus occidentalis.

In the temperate zone the limit of trees is marked by the Quercus Semicarpifolia, at 11,500 feet, on the south side of the Himalaya, and by the Betula Alba, on the north side, at the height of 14,000 feet; the same birch forms the limit of the Caucasus, at the elevation of 6394 feet. On the Pyrenees and Alps the limit is marked by the Coniferæ or pine tribe: on the Pyreness by the Pinus uncinata, at the height of 10,870 feet; on the south side of the Alps by the larch, at the elevation of 6700 feet; and by the Pinus abies, at 5883 feet on the north.

In Lapland the Betula Alba forms the upper limit of trees, at the height of only 1918 feet.

The upper Limit of Shrubs.—In the Andes of Quito the Bejarias are the shrubs that attain the greatest height, and terminate at 13,420 feet above the sea-level.

The juniper, Salix, and Ribes, or currant tribe, form the upper limit of shrubs on the south side of the Himalaya, at the height of 11,500 feet. The tama, or Genista versicolor, a species of broom, flourishes at the height

ference in latitude, for the vegetation of the United States of North America is totally unlike that of Europe under the same isothermal lines, and even between the tropics the greatest dissimilarity often prevails under different degrees of longitude; consequently the cause of this partial distribution of plants, and that of animals also, which is according to the same law, must be looked for in those early geological periods when the earth first began to be tenanted by the present races of organised beings.

As the land rose at different periods above the ocean, each part, as it emerged from the waves, had probably been clothed with vegetation, and peopled with animals, suited to its position with regard to the equator, and to the climate and condition of the globe then being. And as the conditions and climate were different at each succeeding geological epoch, so each portion of the land, as it emerged from the ocean, would be characterized by its own vegetation and animals, and thus at last there would be many centres of creation, as at this day, all differing more or less from one another, and hence alpine floras must be of older date than those in the plains. The vegetation and faunas of those lands that differed most in age and place would be most dissimilar, while the plants and animals of such as were not far removed from one another in time and place would have correlative forms or family likenesses, yet each would form a distinct province. Thus, in opposite hemispheres, and everywhere at great distances, but under like circumstances, the species are representatives of one another, rarely identical: when, however, the conditions which suit certain species are continuous, identical species are found throughout, either by original creation or by migration. The older forms may have been modified to a certain extent by the succeeding conditions of the globe, but they never could have been changed, since immutability of species is a primordial law of nature. Neither external circumstances, time, nor human art, can change one species into another, though each to a certain extent is capable of accommodating itself to a change of external circumstances, so as to produce varieties even transmissible to their offspring.

The flora of Cashmere and the higher parts of the Himalaya mountains is similar to that of southern Europe, yet the species are representative, not identical. In the plains of Tartary, where from their elevation the degree of cold is not less than in the wastes of Siberia, the vegetation of one might be mistaken for that of the other; the gooseberry, currant, willow, rhubarb, and in some places

---

of 17,000 feet on the north side, and vegetation is prolonged to nearly 18,000 feet.

The Rhododendron forms the upper limit of shrubs on the Caucasus, at 8825 feet; in the Pyrenees it grows to 8312 feet; in the Alps to 7480 feet; and in Lapland it forms the upper limit of shrubs at an elevation of 3000 feet above the Arctic Ocean.

the oak, hazel, cypress, poplar, and birch, grow in both, but they are of different species. The flora near the snow-line on the lofty mountains of Europe, and lower down, has also a perfect family likeness to that in high northern latitudes. In like manner many plants on the higher parts of the Chilian Andes are similar, and even identical, with those in Terra del Fuego; nay, the Arctic flora has a certain resemblance to that of the Antarctic regions, and even occasional identity of species. These remarkable coincidences may be accounted for by the different places having been at an early geological period at the same level above the ocean, and that they continue to retain part of their original flora after their relative positions have been changed. The tops of the Chilian Andes were probably on a level with Terra del Fuego when both were covered with the same vegetation, and in the same manner the lofty plains of Tartary may have acquired their vegetation when they were on the level of southern Siberia.

In the many vicissitudes the surface of the globe has undergone, continents formed at one period were broken up at another into islands and detached masses by inroads of the sea and other causes. Professor E. Forbes has shown that some of the primary floras and faunas have spread widely from their original centres over large portions of the continents before the land was broken up into the form it now has, and thus accounts for the similarity and sometimes identity of the plants and animals of regions now separated by seas,— as, for example, islands, which generally partake of the vegetation and fauna of the continents adjacent to them. Taking for granted the original creation of specific centres of plants and animals, Professor E. Forbes has clearly proved that "the specific identity, to any extent, of the flora and fauna of one area, with those of another, depends on both areas forming, or having formed, part of the same specific centre, or on their having derived their animal and vegetable population by transmission, through migration, over continuous or closely contiguous land, aided, in the case of alpine floras, by transportation on floating masses of ice."

By the preceding laws the limited provinces and dispersion of animal and vegetable life are explained, but the existence of single species in regions very far apart has not yet been accounted for.

Very few of the exogenous or dicotyledonous plants are common to two or more countries far apart: among the few, the Samolus Valerandi, a common English plant, is a native of Australia; the Potentilla tridentata, not found in Britain, except on one hill in Angusshire, is common to Arctic Europe and the mountains of North America; and in the Falkland Islands there are more than 30 flowering plants identical with those in Great Britain.

There are many more instances of wide diffusion among the monocotyledonous plants, especially grasses: the Phleum alpinum of Switzerland grows without the smallest variation at the Straits of Magellan, and Mr. Bunbury met with the European quaking-grass in the interior of Southern Africa north of the Cape of Good Hope;[1] but the cellular or cryptogamous class is most widely diffused—plants not susceptible of cultivation, of little use to man, and of all others the most difficult to transport. The Stricta aurata, found in Cornwall, is a native of the Cape of Good Hope, St. Helena, the West India islands, and Brazil; the Trichomanes brevisetum, long supposed to be peculiar to the British isles, is ascertained to grow in Madeira, South America, &c.; and our eminent botanist, Mr. Robert Brown, found 38 British lichens and 28 British mosses in New Holland, yet in no two parts of the world is the vegetation more dissimilar; and almost all the lichens brought from the southern hemisphere by Sir James Ross, amounting to 200 species, are also inhabitants of the northern hemisphere, and mostly European.

In islands far from continents the number of plants is small, but of these a large proportion occur nowhere else. In St. Helena, of 30 flower-bearing plants, 1 or 2 only are native elsewhere, but in 60 species of cryptogamous plants Dr. Hooker found only 12 peculiar to the island.

Some plants are more particularly confined to certain regions: the species of Cinchona which furnish the Peruvian bark grow along the eastern declivity of the Andes, as far as 18° S. lat.; the cedar of Lebanon is indigenous on that celebrated mountain only; and the disa grandiflora is limited to a very small spot on the top of the Table-mountain at the Cape of Good Hope; but whether these are remnants whose kindred have perished by a change of physical circumstances, or centres only beginning to spread, it is impossible to say.

Plants are dispersed by currents: of 600 plants from the vicinity of the river Zaire on the coast of Africa, 13 are found also on the shores of Guiana and Brazil, evidently carried by the great equatorial current to countries congenial in soil and climate. The seeds of the mimosa scandens, the guilandina Bonduc, and the cachew-nut, are wafted from the West India islands to the coasts of Scotland and Ireland by the Gulf-stream, a climate and soil which do not suit them, therefore they do not grow. Of all the great orders, the species of Leguminosæ are most widely dispersed on coasts, be-

[1] The Aquatic Monocotyledonous plants offer perhaps more striking examples of wide diffusion over the surface of the globe than any others. The Pistia stratiotes is found in India and in many parts of South America; the Lemna trisulca and gibba are found throughout Europe and in Australia. Dr. Weddell found the well known Caulinia fragilis of Europe in South America, but the Chara fœtida is perhaps the most widely distributed plant among the monocotyledonous aquatics.

cause their seeds are not injured by the water. Winds also waft seeds to great distances; birds and quadrupeds, and above all man, are active agents in dispersing plants.

CHAPTER XXIV.

Vegetation of the Great Continent — of the Arctic Islands — And of the Arctic and North Temperate Regions of Europe and Asia.

The southern limit of the polar flora, on the old continent, lies mostly within the Arctic Circle, but stretches along the tops of the Scandinavian mountains, and reappears in the high lands of Scotland, Cumberland, and Ireland, on the summits of the Pyrenees, Alps, and other mountains in southern Europe, as well as on the table-land of eastern Asia, and on the high ridges of the Himalaya. The great European plain to the Ural Mountains, as well as the low lands of England and Ireland, were at one period covered by a sea full of floating ice and icebergs, which made the climate much colder than it now is. At the beginning of that period the Scandinavian range, the other continental mountains, and those in Britain and Ireland, were islands of no great elevation, and were then clothed with the Arctic flora, or a representative of it, which they still retain now that they form the tops of the mountain-chains, and at that time both plants and animals were conveyed from one country to another by the floating ice. It is even probable, from the relations of the fauna and flora, that Greenland, Iceland, and the very high European latitudes, are the residue of a great northern land which had sunk down at the close of the glacial period, for there were many vicissitudes of level during that epoch. At all events it may be presumed that the elevation of the Arctic regions of both continents, if not contemporaneous, was probably not far removed in time. Similarity of circumstances had extended throughout the whole Arctic regions, since there is a remarkable similarity and occasional identity of species of plants and animals in the high latitudes of both continents, which is continued along the tops of their mountain-chains, even in the temperate zones; and there is reason to believe that the relations between the faunas and floras of Northern America, Asia, and Europe, must have been established towards the close of the glacial period.

The flora of Iceland approaches that of Britain, yet only one in four of the British plants are known in Iceland. There are 870 species in Iceland, of which more than half are flower-bearing: this is a greater proportion than is found in Scotland, but there are only 32 of woody texture. This flora is scattered in groups according as the plants like a dry, marshy, volcanic, or marine soil. Many grow close to the hot-springs; some not far from the edge of the basin of the Great Geyser, where every other plant is petrified; and species of Confervæ flourish in a spring said to be almost hot enough to boil an egg. The cerealia cannot be cultivated on account of the severity of the climate, but the Icelanders make bread from metur, a species of wild corn, and also from the bulbous root of Polygonum viviparum; their greatest delicacy is the Angelica archangelica; Iceland moss, used in medicine, is an article of commerce. There are 583 species in the Feroe islands, of which 270 are flowering plants: many thrive there that cannot bear the cold of Iceland.

ARCTIC FLORA OF THE GREAT CONTINENTS.

In the most northern parts of the Arctic lands the year is divided into one long intensely cold night and one bright and fervid day, which quickly brings to maturity the scanty vegetation. Within the limit of perpetual congelation the Palmella nivalis (or red snow of the Arctic voyagers), a very minute red or orange-coloured plant, finds nourishment in the snow itself, the first dawn of vegetable life; it is also found colouring large patches of snow in the Alps and Pyrenees.

Lichens are the first vegetables that appear at the limits of the snow-line, whether in high latitudes or mountain-tops, and they are the first vegetation that takes possession of volcanic lavas and new islands, where they prepare soil for plants of a higher order; they grow on rocks, stones, and trees, in fact on anything that affords them moisture. More than 2400 species are already known; no plants are more widely diffused, and none afford a more striking instance of the arbitrary location of species, as they are of so little direct use to man that they could not have been disseminated by his agency. The same kinds prevail throughout the Arctic regions, and the species common to both hemispheres are very numerous. Some lichens produce brilliant red, orange, and brown dyes; and the tripe de roche, a species of Gyrophora, is a miserable substitute for food, as our intrepid countryman, Sir John Franklin, and his brave companions experienced in their perilous Arctic journey.

Mosses follow lichens on newly formed soil, and they are found everywhere throughout the world in damp situations, but in greatest abundance in temperate climates; 800 species are known, of which a great part inhabit the Arctic regions, constituting a large portion of the vegetation.

In Asiatic Siberia, north of the 60th parallel of latitude, the ground is perpetually frozen at a very small depth below the surface:

a temperature of 70° below zero of Fahrenheit is not uncommon, and in some instances the cold has been 120° below zero. Then it is fatal to animal life, especially if accompanied by wind. In some places trees grow and corn ripens even at 70° of north latitude; but in the most northern parts boundless swamps, varied by lakes both of salt and fresh water, cover wide portions of this desolate country, which is buried under snow nine or ten months in the year. As soon as the snow is melted by the returning sun, these extensive morasses are covered with coarse grass and rushes, while mosses and lichens mixed with dwarf willows clothe the plains; saline plants abound, and whole districts produce Diotis ceratoides.

In Nova Zembla and other places in the far north the vegetation is so stunted that it barely covers the ground, but a much greater variety of minute plants of considerable beauty are crowded together there in a small space than in the Alpine regions of Europe, where the same genera grow. This arises from the weakness of the vegetation; for in the Swiss Alps the same plant frequently occupies a large space, excluding every other, as the dark blue gentian, the violet-coloured pansy, the pink and yellow stone-crops. In the remote north, on the contrary, where vitality is comparatively feeble and the seeds do not ripen, thirty different species may be seen crowded together in a brilliant mass, no one having strength to overcome the rest. In such frozen climates plants may be said to live between the air and the earth, for they scarcely rise above the soil, and their roots creep along the surface, having scarcely power to enter it. All the woody plants, as the betula nana, and reticulated willow, Andromeda tetragona, with a few berry-bearing shrubs, trail along the ground, never rising more than an inch or two above it. The salix lanata, the giant of these Arctic forests, never grows more than five inches above the surface, while its stem, 10 or 12 feet long, lies hidden among the moss, owing shelter to its lowly neighbour.

The chief characteristic of the vegetation of the Arctic regions is the predominance of perennial and cryptogamous plants, and also of the sameness of its nature; but more to the south, where night begins to alternate with day, a difference of species appears with that of longitude as well as of latitude. A beautiful flora of vivid colours adorns these latitudes both in Europe and Asia during their brief but bright and ardent summer, consisting of potentillas, gentians, chickweeds, saxifragas, sedums, ranunculi, spiræas, drabas, artemisias, claytonias, and many more. Such is the power of the sun and the consequent rapdity of vegetation, that these plants spring up, blossom, ripen their seed, and die, in six weeks: in a lower latitude woody plants follow these, as berry-bearing shrubs, the glaucous kalmia, the trailing azalea, and rhododendrons. The Siberian flora differs from that in the same European latitudes by the North American genera Phlox, Mitella, Claytonia, and the predominance of asters, solidagos, spiræas, milk-vetches, wormwood, and the saline plants goosefoot, and saltworts.

Social plants abound in many parts of the northern countries, as grass, heath, furze, and broom; the steppes are an example of this on a very extensive scale. Both in Europe and Asia they are subject to a rigorous winter, with deep snow and chilling blasts of wind; and as the soil generally consists of a coating of vegetable mould over clay, no plants with deep roots thrive upon them; hence the steppes are destitute of trees, and even bushes are rare except in ravines; the grass is thin but nourishing. Hyacinths and some other bulbs, mignonette, asparagus, liquorice, and wormwood, grow in the European steppes; the two last are peculiarly characteristic. The nelumbium speciosum grows in one spot five miles from the town of Astracan, and nowhere else in the wide domains of Russia: the leaves of this beautiful aquatic plant are often two feet broad, and its rose-coloured blossoms are very fragrant. It is also native in India and Tibet, where it is held sacred, as it was formerly in Egypt, where it is said to be extinct: it is one of the many instances of a plant growing in countries far apart.

Each steppe in Siberia has its own peculiar plants; the Peplis and Camphorosma are peculiar to the steppe of the Irtish, and the Amaryllis tartarica abounds in the meadows of eastern Siberia, where vegetation bears a great analogy to that of north-western America; several genera and species are common to both.

Half the plants found by Wormskiold in Kamtchatka are European, with the exception of eight or ten, which are American. Few European trees grow in Asiatic Siberia, notwithstanding the similarity of climate, and most of them disappear towards the rivers Tobol and Irtish.

In Lapland and in the high latitudes of Russia large tracts are covered with birch-trees, but the pine and fir tribes are the principal inhabitants of the north. Prodigious forests of these are spread over the mountains of Norway and Sweden, and in European Russia 200,000,000 acres are clothed with these Coniferæ alone, or occasionally mixed with willows, poplars, and alders. Although soils of pure sand and lime are absolutely barren, yet they generally contain enough of alkali to supply the wants of the fir and pine tribes, which require ten times less than oaks and other deciduous trees.

The Siberian steppes are bounded on the south by great forests of pine, birch, and willow: poplars, elms, and Tartarian maple overhang the upper courses of the noble rivers which flow from the mountains to the Frozen Ocean, and on the banks of the Yenessei the pinus Cembra, or Siberian pine, with edible fruit, grows 120 feet high. The Altaï are covered nearly to their summit with similar forests, but on their greatest heights the stunted larch crawls on the ground, and the flora is like that of northern Siberia: round lake Baikal the pinus Cembra grows nearly to the snow-line.

Forests of black birch are peculiar to Daouria, where there are also apricot and apple trees, and rhododendrons, of which a species grows in thickets on the hills with yellow blossoms. Here and everywhere else throughout this country are found all the species of Caragana, a genus entirely Siberian. Each terrace of the mountains and each steppe on the plains has its peculiar plants, as well as some common to all; perennial plants are more numerous than annuals.

If temperature and climate depended upon latitude alone, all Asia between the 50th and 30th parallels would have a mild climate; but that is far from being the case, on account of the structure of the continent, which consists of the highest table-lands and the lowest plains on the globe.

The table-land of Tibet, where it is not cultivated, has the character of great sterility, and the climate is as unpropitious as the soil; frost, snow, and sleet begin early in September, and continue with little interruption till May; snow, indeed, falls every month in the year. The air is always dry, because in winter moisture falls in the form of snow, and in summer it is quickly evaporated by the intense heat of the sun. The thermometer sometimes rises to 144° of Fahrenheit in the sun, and even in winter his direct rays have great power for an hour or two, so that a variation of 100° in the temperature of the air has occurred in twelve hours. Notwithstanding these disadvantages, there are sheltered spots which produce most of the European grain and fruits, though the natural vegetation bears the Siberian character, but the species are quite distinct. The most common indigenous plants are Tartarian furze and various prickly shrubs resembling it, goose-berries, currants, hyssop, dog-rose, dwarf sow-thistle, equisetum, rhubarb, lucern, and asafœtida, on which the flocks feed. Prangos, an umbelliferous plant with broad leaves and scented blossom, is peculiar to Ladak and other parts of Tibet. Mr. Moorcroft says it is so nutritious, that sheep fed on it become fat in twenty days. There are three species of wheat, three of barley, and two of buckwheat, natives of the lofty table-land, where the sarsinh is the only fruit known to be indigenous. Owing to the rudeness of the climate, trees are not numerous, yet on the lower declivities of some mountains there are aspens, birch, yew, ash, Tartaric oak, various pines, and the Pavia, a species of horse-chesnut. Much of the table-land of Tartary is occupied by the Great Gobi and other deserts of sand, with grassy steppes near the mountains; but of the flora of these regions we know nothing.

FLORA OF BRITAIN AND OF MIDDLE AND SOUTHERN EUROPE.

The British islands afford an excellent illustration of distinct provinces of animals and plants, and also of their migration from other centres. Professor E. Forbes has determined five botanical districts, four of which are restricted to limited provinces, whilst the fifth, which comprehends the great mass of British plants, is everywhere, either alone or mixed with the others. All of these, with a very few doubtful exceptions, have migrated before the British islands were separated from the continent. The first, which is of great antiquity, includes the flora of the mountain districts of the west and south-west of Ireland, and is similar to that in the south of Spain, but the more delicate plants had been killed by the change of climate after the separation of Ireland from the Asturias. The flora in the south of England and the south-east of Ireland is different from that in all other parts of the British Islands; it is intimately related to the vegetation of the Channel Islands and the coast of France opposite to them, yet there are many plants in the Channel Islands which are not indigenous in Britain. In the south-west of England, where the chalk plants prevail, the flora is like that on the adjacent coast of France.

The tops of the Scottish mountains are the focus of a separate flora, which is the same with that in the Scandinavian Alps, and is very numerous. Scotland, Wales, and a part of Ireland received this flora when they were groups of islands in the Glacial Sea. The rare Eriocaulon is found in the Hebrides, in Connemara, and in Northern America, and nowhere else. Some few individuals of this flora grow on the summits of the mountains in Cumberland and Wales. The fifth, of more recent origin than the alpine flora, including all the ordinary flowering plants, as the common daisy and primrose, hairy ladies' smock, upright meadow crowfoot, and the lesser celandine, together with our common trees and shrubs, has migrated from Germany before England was separated from the Continent of Europe by the British Channel. It can be distinctly traced in its progress across the island, but the migration was not completed till after Ireland was separated from England by the Irish Channel, and that is the reason why many of the ordinary English plants, animals, and reptiles are not found in the sister island, for the migration of animals was simultaneous with that of plants, and took place between the last of the tertiary periods and the historical epoch, that of man's creation: it was extended also over a great part of the continent.[1]

[1] The British flora contains at least 3000 species.

Deciduous trees are the chief characteristic of the temperate zone of the old continent, more especially of middle Europe; these thrive best in soil produced by the decay of the primary and ancient volcanic rocks, which furnish abundance of alkali. Oaks, elms, beech, ash, larch, maple, lime, alder, and sycamore, all of which lose their leaves in winter, are the prevailing vegetation, occasionally mixed with fir and pine.

The undergrowth consists of wild apple, cherry, yew, holly, hawthorn, broom, furze, wild rose, honeysuckle, clematis, &c. The most numerous and characteristic herbaceous plants are the umbelliferous class, as carrot and anise, the campanulas, the Cichoraceæ, a family to which lettuce, endive, dandelion, and sow-thistle belong. The cruciform tribe, as wallflower, stock, turnip, cabbage, cress, &c., are so numerous, that they form a distinguishing feature in the botany of middle Europe, to which 45 species of them belong. This family is almost confined to the northern hemisphere, for, of 800 known species, only 100 belong to the southern, the soil of which must contain less sulphur, which is indispensable for these plants.

In the Pyrenees, Alps, and other high lands in Europe, the gradation of botanical forms, from the summit to the foot of the mountains, is similar to that which takes place from the Arctic to the middle latitudes of Europe. The analogy, however, is true only when viewed generally, for many local circumstances of climate and vegetation interpose; and although the similarity of botanical forms is very great between certain zones of altitude and parallels of latitude, the species are for the most part different.

Evergreen trees and shrubs become more frequent in the southern countries of Europe, where about a fourth part of the ligneous vegetation never entirely lose their leaves. The flora consists chiefly of ilex, oak, cypress, hornbeam, sweet chesnut, laurel, laurustinus, the apple tribe, manna or the flowering ash, carob, jujube, juniper, terebinths, lentiscus and pistaccio, which yield resin and mastic, arbutus, myrtle, jessamine (yellow and white), and various pines, as the Pinus maritima, and Pinus Pinea, or stone pine, which forms so picturesque a feature in the landscape of southern Europe. The most prevalent herbaceous plants are Caryophylleæ, as pinks, Stellaria, and arenarias, and also the labiate tribe, mint, thyme, rosemary, lavender, with many others, all remarkable for their aromatic properties, and their love of dry situations. Many of the choicest plants and flowers which adorn the gardens and grounds in northern Europe are indigenous in these warmer countries: the anemone, tulip, mignonette, narcissus, gladiolus, iris, asphodel, amaryllis, carnation, &c. In Spain, Portugal, Sicily, and the other European shores of the Mediterranean, tropical families begin to appear in the arums, plants yielding balsams, oleander, date and palmetto palms, and grasses of the group of Panicum or millet, Cyperaceæ or sedges, Aloe and Cactus. In this zone of transition there are six herbaceous for one woody plant.

FLORA OF TEMPERATE ASIA.

The vegetation of western Asia approaches nearly to that of India at one extremity, and Europe at the other; of 281 genera of plants which grow in Asia Minor and Persia, 109 are European. Syria and Asia Minor form a region of transition, like the other countries on the Mediterranean, where the plants of the temperate and tropical zones are united. We owe many of our best fruits and sweetest flowers to these regions. The cherry, almond, oleander, syringa, locust-tree, &c., come from Asia Minor; the walnut, peach, melon, cucumber, hyacinth, ranunculus, come from Persia; the date-palm, fig, olive, mulberry, and damask-rose, come from Syria; the vine and apricot are Armenian; the latter grows also everywhere in middle and northern Asia. The tropical forms met with in more sheltered places are the sugar-cane, date and palmetto palms, mimosas, acacias, Asclepias gigantea, and arborescent Apocineæ. On the mountains south of the Black Sea, American types appear in rhododendrons and the Azalea pontica, and herbaceous plants are numerous and brilliant in these countries.

The table-land of Persia, though not so high as that of Eastern Asia, resembles it in the quality of the soil, which is chiefly clayey, sandy, or saline, and the climate is very dry; hence vegetation is poor, and consists of thorny bushes, acacias, mimosas, tamarisk, jujube, and asafœtida. Forests of oak cover the mountains of Lusistan, but the date-palm is the only produce of the parched shores of the Arabian Gulf and of the oases on the Persian table-land. In the valleys, which are beautiful, there are clumps of Oriental plane and other trees, hawthorn, tree-roses, and many of the odoriferous shrubs of Arabia Felix.

Afghanistan produces the seedless pomegranate, acacias, date-palms, tamarisks, &c. The vegetation has much the same general character as that of Egypt. The valleys of the Hindoo Coosh are covered with clover, thyme, violets, and many odoriferous plants: the greater part of the trees in the mountains are of European genera, though all the species of plants, both woody and herbaceous, are peculiar. The small leguminous plant from whose leaves and twigs the true indigo dye is extracted grows spontaneously on the lower offsets of the Hindoo Coosh. This dye has been in use in India from the earliest times, but the plant which produces it was not known in England till towards the end of the 16th century.

Since that time it has been cultivated in the West Indies and tropical America, though in that country there is a species which is indigenous.

Hot arid deserts bound India on the west, where the stunted and scorched vegetation consists of tamarisks, thorny acacia, deformed Euphorbiæ, and almost leafless thorny trees, shaggy with long hair, by which they imbibe moisture and carbon from the atmosphere. Indian forms appear near Delhi, in the genera Flacourtia and others, mixed with Syrian plants. East of this transition the vegetation becomes entirely Indian, except on the higher parts of the mountains, where European types prevail.

The Himalaya mountains form a distinct botanical district. Immediately below the snow-line the flora is almost the same with that on the high plains of Tartary, to which may be added rhododendrons and andromedas, and among the herbaceous plants primroses appear. Lower down vast tracts are covered with prostrate bamboos, and European forms become universal, though the species are Indian, as gentians, plantagos, campanulas, and gale. There are extensive forests of Coniferæ, consisting chiefly of Pinus excelsa, Deodora, and Morinda, with many deciduous forest and fruit trees of European genera. A transition from this flora to a tropical vegetation takes place between the altitudes of 9000 and 5000 feet, because the rains of the monsoons begin to be felt in this region, which unites the plants of both. Here the scarlet and other rhododendrons grow luxuriantly; walnuts, and at least 25 species of oak, attain a great size, one of which, the Quercus semicarpifolia, has a clean trunk from 80 to 100 feet high. Geraniums and labiate plants are mixed in sheltered spots with the tropical genera of Scitamineæ, or the ginger tribe; bignonias and balsams, and camellias, grow on the lower part of this region.

It is remarkable that Indian, European, American, and Chinese forms are united in this zone of transition, though the distinctness of species still obtains: the Triosteum, a genus of the honeysuckle tribe, is American; the Abelia, another genus of the same, together with the Camellia and and Tricyrtis, are peculiarly Chinese; the daisy and wild thyme are European. A few of the trees and plants mentioned descend below the altitude of 5000 feet, but they soon disappear on the hot declivities of the mountain, where the Erythrina monosperma and Bombax heptaphyllum are the most common trees, together with the Millingtoniæ, a tribe of large timber-trees, met with everywhere between the Himalaya and 10° N. lat. The shorea robusta, Dalbergia, and Cedrela, a genus allied to mahogany, are the most common trees in the forests of the lower regions of these mountains.

The temperate regions of eastern Asia, including Chinese Tartary, China, and Japan, have a vegetation totally different from that of any other part of the globe similarly situated, and show in a strong point of view the distinct character which vegetation assumes in different longitudes. In Manchouria and the vast mountain-chains that slope from the eastern extremity of the high Tartarian table-land to the fertile plains in China, the forests and flora are generally of European genera, but Asiatic species; in these countries the buckthorn and honeysuckle tribes are so numerous as to give a peculiar character to the vegetation. Mixed with these and with roses are thickets of azaleas covered with blossoms of dazzling brightness and beauty.

The transition zone in this country lies between the 35th and 27th parallels of north latitude, in which the tropical flora is mixed with that of the northern provinces. The prevailing plants on the Chinese low grounds are Glycine, Hydrangea, the camphor laurel, stillingia sebifera, or wax-tree, Clerodendron, Hibiscus rosa-sinensis, thuia orientalis, olea fragrans, the sweet-blossoms of which are mixed with the finer teas to give them flavour; Melia azedarach, or Indian pride, the paper mulberry, and others of the genus, and camellia sasanqua, which covers hills in the province of Kiong-si. The tea-plant, and other species of Camellia, grow in many parts; the finest tea is the produce of a low range of hills from between the 33rd and 25th parallels, an offset from the great chain of Peling. Thea viridis and bohea are possibly only varieties of the same plant; the green tea is strong and hardy, the black a small delicate plant. The quality of the tea depends upon the stage of growth at which it is gathered; early leaves make the best tea, those picked late in the season give a very coarse tea. Bohea grows in the province of Fukian, hyson in Song-lo. Pekoe or pak-ho, which means white down in Chinese, consists of the first downy sprouts or leaf-buds of three-years-old plants. A very costly tea of this kind, never brought to Europe, and known as the tea of the Wells of the Dragon, is used only by persons of the highest rank in China. The true Imperial tea also, called Flos theæ, which is not, as was supposed, the flower-buds, but merely a very superior quality of tea, seldom reaches Europe; that sold under this name is really Chusan tea flavoured with blossoms of olea fragrans.[1] The Chinese keep tea a year before they use it, because fresh tea has an intoxicating quality which produces disturbance of the nervous system. It is a remarkable circumstance that tea and coffee, belonging to different families, natives of different quarters of the globe, should possess the same principle, and it

[1] The plants with which the Chinese give flavour to tea are the olea fragrans, Chloranthus inconspicuus, gardenia florida, aglaia, odorata, mogorium sambac, vitex spicata, camellia sasanqua, camellia odorifera, illicium anisatum, magnolia yulan, rosa indica odoratissima, turmeric, oil of Bixa orellana, and the root of the Florentine iris.

The principles of caffeine and theine are in all respects identical.

is not less remarkable that their application to the same use should have been so early discovered by man.

The tea-plant grows naturally in Japan and upper Assam; it is hardy, and possesses great power of adaptation to climate. It has lately been cultivated in Brazil,[1] in Provence, and in Algiers, but at an expense which renders it unprofitable. Tea comes to Europe almost exclusively from China, but the plant thrives so well in the north-western provinces of India that the English will ultimately compete with the Chinese in producing it, especially for the consumption of Tibet. Tea was first brought to Europe by the Dutch in 1610; a small quantity came to England in 1666, and now the annual consumption of tea in Great Britain is upwards of fifty millions of pounds.[2]

The climate of Japan is milder than its latitude would indicate, owing to the influence of the surrounding ocean. European forms prevail in the high lands, as they do generally throughout the mountains of Asia and the Indian Archipelago, with the difference of species, as Abies, Cembra, Strobus, and Larix. The Japanese flora is similar to the Chinese, and there are 30 American plants, besides others of Indian and tropical climates. These islands, nevertheless, have their own peculiar flora, distinct in its nature; as the Sophora, Kerria, Aucuba, Mespilus, and pyrus japonica, rhus vernix, illicium anisatum, or the anise-tree, daphne odorata, the soap-tree, various species of the Calycanthus tribe, the custard-apple, the Khair mimosa, which yields the catechu, the litchi, the sweet orange, the cycas revoluta, a plant resembling a dwarf palm, with various other fruits. Many tropical plants mingle with the vegetation of the cocoa-nut and fan palms.

Thus the vegetation in Japan and China is widely different from that in the countries bordering the Mediterranean, though between the same parallels of latitude. In the tropical regions of Asia, where heat and moisture are excessive, the influence of latitude vanishes altogether, and the peculiarities of the vegetation in different longitudes become more evident.

---

[1] [The produce of tea in Brazil is almost sufficient to meet the demand; there is no tea imported into that country direct from China. Only a small quantity from the latter country is brought by American and European traders, which is used chiefly by the foreign population.]

[2] Davis on China.

CHAPTER XXV.

Flora of Tropical Asia — Of the Indian Archipelago, India, and Arabia.

TROPICAL Asia is divided by nature into three distinct botanical regions: the Malayan peninsula, with the Indian Archipelago; India, south of the Himalaya, with the island of Ceylon; and the Arabian peninsula. The two first have strong points of resemblance, though their floras are peculiar.

FLORA OF THE INDO-CHINESE PENINSULA AND THE INDIAN ARCHIPELAGO.

Many of the vegetable productions of the peninsula beyond the Ganges are the same with those of India, mixed with plants of the Indian Archipelago, so that this country is a region of transition, though it has a splendid vegetation of innumerable native productions, dyes of the most vivid hues, spices, medicinal plants, and many with the sweetest perfume. The soil in many places yields three crops in the year: the fruits of India, and most of those of China, come to perfection in the low lands. The Arang forms an exception to the extreme beauty of the multitude of palms which adorn the Malayan peninsula; though it is eminently characteristic of that country, it is an ugly plant, covered with black fibres like horsehair, sufficiently strong to make cordage. It is cultivated for the sugar and wine made from its juice. Teak is plentiful; almost all that is used in Bengal comes from the Birman empire, though it is less durable than that of the Malabar coast. The Hopea odorata is so large that a canoe is made of a single trunk; the Gordonia integrifolia is held in such veneration that every Birman house has a beam of it.

There are seven species of native oak in the forests; the mimosa catechu, which furnishes the terra japonica used in medicine; the trees which produce varnish and stick-lac; the glyphyria nitida, a myrtle, the leaves of which are used as tea in Bencoolen, called by the natives the tree of long life. The coasts are wooded by the heritiera robusta, a large tree which thrives within reach of the tide; bamboos with stems a foot and a half in diameter grow in dense thickets in the low lands. The Palmyra palm and the borassus flabelliformis grow in extensive groves in the valley of the Irrawaddy: it is a magnificent tree, often 100 feet high, remarkable for its gigantic leaves, one of which would shelter 12 men.

The anomalous plants the Zamias and Cycadeæ, somewhat like a palm with large pinnated leaves, but of a different family, are found here and in tropical India; those in America are of a different species. Orchideæ and tree-ferns are innumerable in the woody districts of the peninsula.

The vegetation of the Indian Archipelago is gorgeous beyond description; although in many instances it bears a strong analogy to that of the Malayan peninsula, tropical India, and Ceylon, still it is in an eminent degree peculiar. The height of the mountains causes variety in the temperature sufficient to admit of the growth of dammar pines, oaks, rhododendrons, magnolias, valerians, honeysuckles, bilberries, gentians, oleasters, and other European orders of woody and herbaceous plants; yet there is not one species in common.

Palm-trees are more abundant in these islands than in any other part of the world, especially in the Sunda group, the origin of many, a few of which are now widely spread over the eastern countries. Three species of Areca, attaining a height of from 40 to 50 and more feet, are cultivated in all the hot parts of India; and caroyta Urens, the fruit of which is acrid, yet it yields wine and sugar, are all native. The attempt is vain to specify the multitudes of these graceful trees which form so characteristic a feature in the vegetation of these tropical islands, where a rich moist soil with intense heat brings them to such perfection. It has been observed that monocotyledonous plants are generally more plentiful in islands than on continents, and also that they extend farther into the southern than into the northern hemisphere, which may be accounted for by the moist and mild climate of the former.

Jungle and dense pestilential woods entirely cover the smaller islands and the plains of the larger; the coasts are lined with thickets of mangroves, a matted vegetation of forest-trees, bamboos, and coarse grass, entwined with climbing and creeping plants, and overgrown by orchideous parasites in myriads: the gutta-percha is also a native of these alluvial tracts. The forest-trees of the Indian Archipelago are almost unknown; teak and many of the continental trees grow there, but the greater number are peculiarly their own. The naturalist Rumphius had a cabinet inlaid with 400 kinds of wood, the produce of Amboyna and the Molucca islands.

Sumatra, Java, [Borneo,] and the adjacent islands are the region of the dryobalanops camphora, in the stems of which solid lumps of a remarkable and costly kind of camphor are found. All the trees of that order, and of several others, are peculiar to these islands, and 78 species of trees and shrubs of the Melastomaceous tribe grow there and in continental India. There are thickets of the sword-leaved vaquois-tree and of the Pandanus or screw-pine, a plant resembling the anana, with a blossom like that of a bulrush very odoriferous, and in some species edible.

This is the region of spices, which are very limited in their distribution: the myristica moschata (the nutmeg and mace-plant) is confined to the Banda Islands, but it is said to have been discovered lately in New Guinea. The Amboyna and the Molucca groups are the focus of the caryophyllus aromaticus, a myrtle, the buds of which are known as cloves. Various species of cinnamon and cassia, both of the laurel tribe, together with varieties of pepper, different from those in India and Ceylon, grow in this archipelago. All the pepper-plants require great heat: they are rare in Africa, but plentiful in America and the Indian Archipelago; the common black pepper is peculiar to the hottest parts of Asia, extending only a few degrees on each side of the equator. In 1842 more than 30,000,000 pounds weight of pepper were produced in Sumatra alone. Some of the most excellent fruits are indigenous here only, as the dourio, the ayer ayer, loquat, the choapa of Molucca, peculiar kinds of orange, lemon, and citron, with others known only by name elsewhere. Those common to the continent of India are the jambrose, rose-apple, jack, various species of bread-fruit, mango, mangosteen, and the banana.

Here the nettle tribe assumes the most pernicious character, as the upas-tree of Java, one of the most deadly vegetable poisons: and even the plants resembling our common nettle are so acrid that the sting of one in Java occasions not only pain but illness, which lasts for days. A nettle in the island of Timor, called by the natives the "Devil's leaf," is so poisonous that it produces long illness and even death. The chelik, a shrub growing in the dense forests, produces a poison even more deadly than the upas. Some of the fig genus, which belongs also to the natural order of nettles, have acrid juices. Trees of the cachew tribe have a milky sap: the fine japan lacquer is made from the juice of the stigmaria verniciflua. Barringtonia and palms are very splendid here, the latter generally of peculiar species and limited in their distribution, as the Nipa. No country is richer in club-mosses and orchideous plants, which overrun the trees in thousands in the deep dark mountain forests, choked by huge creeping plants, an undergrowth of gigantic grasses, through which not a ray of light penetrates.

Sir Stamford Raffles describes the vegetation of Java as "fearful." In these forests the air is heavy, charged with dank and deadly vapours, never agitated by a breath of wind; the soil, of the deepest black vegetable mould, always moist and clammy, stimulated by the fervid heat of a tropical sun, produces trees whose stems are of a spongy texture from their rapid growth, loaded with parasites, particularly the orchideous tribes, of which no less than 300 species are peculiar to that island. Tree-ferns are in the proportion of one to twenty of the other plants, and form a large portion of the vegetation of Java and all these islands; and there are above 200 tropical species of club-mosses growing to the length of 3 feet, whereas in cold countries they creep on the ground.

The Rafflesias, of which there are four species, are the most singular productions of this archipelago. The most extraordinary one is common to Java and Sumatra, where it was discovered by Dr. Arnold, and therefore is called Rafflesia Arnoldi. It is a parasitical plant, with buds the size of an ordinary cabbage, and the flower, which smells of carrion, is of a brick-red colour, 3½ feet in diameter: that found by Mr. Arnold weighed 15 pounds, and the cup in its centre could contain 12 pints of liquid.

According to Sir Stamford Raffles there are six distinct climates in Java, from the top of the mountains to the sea, each having an extensive indigenous vegetation. No other country can show an equal abundance and variety of native fruit and esculent vegetables. There are 100 varieties of rice, and of fragrant flowers, shrubs, and ornamental trees the number is infinite. Abundant as the Orchideæ are in Java, Ceylon, and the Birmese empire, these countries possess very few that are common to all, so local is their distribution. Ferns are more plentiful in this archipelago than elsewhere; tree-ferns are found chiefly between or near the tropics, in airless damp places.

INDIAN FLORA.

The plains of Hindostan are so completely sheltered from the Siberian blasts by the high table-lands of Tartary and the Himalaya mountains, that the vegetation at the foot of that range already assumes a tropical character. In the jungles and lower ridges of the fertile valley of Nepal, and on the dark and airless recesses of the Silhet forests, arborescent ferns and orchideous plants are found in profusion, scarcely surpassed even in the islands of the Indian Archipelago—indeed the marshy Tariyane is full of them. Sekein is an extremely rich botanical country. Numerous beautiful species of rhododendrons were discovered by Dr. Hooker, between 5000 and 10,000; and the Arctic vegetation between 10,000 and 17,000 feet is also very rich. In the Khasaya country, south of Assam, at the eastern extremity of Bengal, the vegetation is extremely abundant and varied; oaks abound. The lowest ranges of the Himalaya, the pestilential swamp of the Tariyane, the alluvial ridges of the hills that bound it on the south, and many parts of the plains of the Ganges, are covered with primeval forests, which produce whole orders of large timber-trees, frequently overrun with parasitical loranths.

The native fruits of India are many: the orange tribe is almost all of Indian origin, though some of the species are now widely spread over the warmer parts of the other continents and the more distant countries of Asia. Two or three species are peculiar to Madagascar; one is found in the forests of the Essequibo, and another in Brazil, which are the only exceptions known. The limonia laureola grows on the tops of the high Asiatic mountains, which are covered with snow several months in the year; and the wampee, a fruit much esteemed in China and the Indian Archipelago, is produced by a species of this order. The vine grows wild in the forests; plantain, banana, jambrose, guava, mango, mangosteen, date, areca, palmyra, cocoa-nut, and gameto palms are all Indian, also the gourd family. The Scitamineæ, or ginger tribe, are so numerous, that they form a distinguishing and beautiful feature of Indian botany: they produce ginger, cardamons, and turmeric. The flowers peculiar to India are brilliant in colours, but generally without odour, except the rose and some jessamines.

The greater part of the trees and plants mentioned belong also to tropical India, where vegetation is still more luxuriant; a large portion of that magnificent country, containing 1,000,000 square miles, has been cultivated time immemorial, although vast tracts still remain in a state of nature. Those extensive mountain-chains which traverse and surround the Deccan are rich in primeval forests of stupendous growth with dense underwood. The most remarkable of these trees are the Indian cotton-tree and the Dombeya, which is of the same order; that which produces the Trincomalee wood, used for building boats at Madras; the red-wood tree, peculiar to the Coramandel coast; the satin-wood, the superb butea frondosa, the agallochum tribe, which yields the odorous wood of aloes mentioned in Scripture, the melaleuca leucadendron and the melaleuca cajepute, from which the oil is prepared. The dragon's-blood tree is a native of India, though not exclusively, as some of the best specimens grow [in Brazil and] in Madagascar, where it is planted for hedges. Sandal-wood and dragon's-blood are obtained from the Pterocarpus sandalinus and draco; the sappan-tree gives a purple dye: these are all of the leguminous or bean tribe, of which there are 452 Indian species: ebony grows in these tropical regions, in Mauritius, and on the south coast of Africa.

Trees of the fig tribe are among the most remarkable vegetable productions of India for gigantic size and peculiarity of form, which renders them valuable in a hot climate from the shade which their broad-spreading tops afford. Some throw off shoots from their branches, which take root on reaching the ground, and, after increasing in girth with wonderful rapidity, produce branches which also descend to form new roots, and this process is continued till a forest is formed round the parent tree. Mr. Reinwardt saw in the island of Simao a large wood of the Ficus Benjamina which sprang from one stem. The Ficus Indica, or Banyan tree, is another instance of this wide-spreading growth; it is found in the islands, but is in greatest perfection around the villages in the Circar mountains:

there is a tree of it on the banks of the Nerbudda, in the province of Guzerat, with 350 main stems, occupying an area of 2000 feet in circumference, independent of its branches, which extend much farther. The camphor genus is mostly Indian, as well as many more of the laurel tribe of great size. The banana is the most generally useful tree in this country; its fruit is food, its leaves are applied to many domestic purposes, and flax fit for making muslin is obtained from its stem. Cotton is a hairy covering of the seeds of several species of the mallow tribe which grow spontaneously in tropical Asia, Africa, and America; it is, however, cultivated in many countries beyond these limits. That grown in China and the United States of America is an herbaceous annual from 18 inches to 2 feet high; there are also cotton-trees, native and cultivated, in India, China, Africa, and America. Herodotus mentions cotton garments 445 years before the Christian era, and the Mexicans and Peruvians manufactured cotton cloth before the discovery of America.

["The most important plants of the family of MALVACÆ are the *cotton-trees*, the fruit of which furnishes the textile (weaveable) material, known under the name of *cotton*. Many species of this genus are known; one called *herbaceous cotton*, varies much in its appearance; sometimes it is an herbaceous annual plant growing scarcely beyond eighteen or twenty inches in height; at other times a shrub from four to six feet high, the stem of which is ligneous and perennial at the lower part. This cotton-tree grows in Egypt, Syria, and India, and is also cultivated in Sicily. The *arborescent cotton-tree* was originally from India; it is now cultivated in Brazil and Peru, and constitutes one of the most important products of the United States: it grows to the height of from fifteen to twenty feet. The leaves of these plants are alternate, petiolate, and divided into five digitate lobes; the flowers borne upon peduncles in the axils of the upper leaves, are yellowish, or purplish. The fruit is an egg-shaped capsule, divided into from two to five cells, each of which contains several seeds; the cotton is found surrounding these seeds."[1]

Herbaceous cotton grows from four to six feet high, and produces two crops annually; the first in eight months after sowing the seed; the second within four months after the first; and the produce of each plant is reckoned at about one pound weight.

According to the census of 1850, there were 1094 manufactories of cotton in the United States, which manufactured 641,240 bales of cotton into sheetings, calicoes, yarns, &c.]

Palms, the most stately and graceful of the vegetable productions of tropical regions, are abundant in India, in forests, in groups, and in single trees. Some species grow near to the limit of perpetual snow, some 900 feet above the sea, others in valleys and on the

[[1] Ruschenberger's Elements of Natural History.—Botany.]

shores of the continent and islands. They decrease in number and variety as the latitude increases, and terminate at Nice, in 44° N. lat., their limit in the great continent. The leaves of some are of gigantic size, and all are beautiful, varying in height from the slender Calamus rotang, 130 feet high, to the Chamærops hnmilis, not more than 15 or 20. Different species yield wine, oil, wax, flour, sugar, thread, and rope; weapons and utensils are made by their stems and leaves; they serve for the construction of houses; the cocoa-nut palm gives food and drink; sago is made from all except the Areca catechu, the fruit of which, the betel-nut, is used by the natives for its intoxicating quality.

Though palms in general are very limited in their distribution, a few species are very widely spread; for example, the cocoa-nut palm, which grows spontaneously on the southern coasts of the Indo-Chinese peninsula and the Sunda Islands, from whence it has been carried to all the intertropical regions of the globe, where it has been extensively cultivated from its usefulness. So luxuriant is its growth in Ceylon that in one year nearly 3,000,000 of nuts were exported; in parts of that island, on the Malabar and Coromandel coasts, and in some districts in Bengal, the Borassus flabelliformis supplies its place.

The island of Ceylon, which may be regarded as the southermost extremity of the Indian peninsula, is very mountainous, and rivals the islands of the Indian Archipelago in luxuriance of vegetable productions, and in some respects bears a strong resemblance to them. The species of laurel, the bark of which is cinnamon, is indigenous, and one of the principal sources of the revenue of Ceylon. The taleput leaves of a species of palm are of such enormous size, that they are applied to many uses by the Cingalese: in ancient times strips of the leaf were written upon with a sharp style, and served as books. The sandal-wood of Ceylon is of a different species from that of the South Sea islands, and its perfume more esteemed. Indigo is indigenous, and so is the choya, whose roots give a scarlet dye. The mountains produce a great variety of beautiful woods used in cabinet work. It is a remarkable circumstance in the distribution of plants that the orchideæ are not very numerous in this island.

ARABIAN VEGETATION.

The third division of the tropical flora of Asia is the Arabian, which differs widely from the other two, and is chiefly marked by trees yielding balsams. Oceans of barren sand extend to the south, from Syria through the greater part of Arabia, varied only by occasional oases in those spots where a spring of water has reached the surface; there the prevalent vegetation consists of the grasses holcus and panicum dicotomum growing under the shade of the date-palm; mimosas and stunted prickly bushes appear here and there in the sand. There is verdure on the mountains, and along some of the coasts, especially in the province of Yemen, which has a flora of its own. The keura odorifera, a superb tree, with agreeable perfume, eight species of figs, the three species of amyris gileadensis, or balm of Gilead, opobalsamum also yielding balsam, and the kataf, from which myrrh is supposed to come, are peculiar to Arabia. Frankincense is said to be the produce of the boswellia serrata; and there are many species of Acacia, among others the acacia arabica, which produces gum arabic. The arak and tamarind trees connect the botany of Arabia with that of the West Indies, while it is connected with that of the Cape of Good Hope by stapelias, mesembryanthemums, and liliaceous flowers. The character of Arabian vegetation, like that of other dry hot climates, consists in its odoriferous plants and flowers.

Arabia produces coffee, which, however, is not indigenous, but is supposed to have come from the table-land of Ethiopia, and to have its name from the province of Kaffa, where it forms dense forests. It was introduced into Arabia in the end of the fifteenth century, and grows luxuriantly in Arabia Felix, where the coffee is of the highest flavour. Most of that now used is the progeny of plants raised from seed and brought from Mocha to the Botanic Garden at Amsterdam in 1690, by Van Hoorn, Governor of Batavia. A plant was sent to Louis XIV., in 1714, by the Magistrates of Amsterdam —it was from this plant that the first coffee-plants were introduced in 1717 into the West India islands. A year afterwards the Dutch introduced coffee-trees into Surinam, from whence they spread rapidly over the warm parts of America and the West India islands. Many thousands of people are now employed in its cultivation there, in Demerara, Java, Manilla, the isle of Bourbon, and other places. 6,300,000 pounds of coffee beans were imported into Great Britain in 1849, and 30,000 tons of shipping were employed in its transport across the Indian and Atlantic Oceans. Coffee was not known till many centuries after the introduction of sugar. The first coffee-house was opened in London in 1652, and the first in France, at Marseilles, 1671.

["The trunk of the *Coffee-tree*—Coffea Arabica—is cylindrical, and rises to from fifteen to twenty feet high; its branches are somewhat knotty; its leaves are lanceolate, shining, and of a deep green; its flowers are white and almost sessile; and its fruit is fleshy, ovoid berries, which are at first green, then red, and finally black; each berry encloses two fleshy nuts, each containing a seed convex outwardly and flat within, and marked on the flat side by a longitudinal groove. This shrub ordinarily flowers twice a year, but there is scarcely an interval between these periods, so that it is always loaded with flowers and fruit; the latter generally ripens four months after inflorescence, and must be gathered with care according to its state of maturity."[1] The tree bears fruit at the age of between two and three years.

By the Arabians, who brought it, Niebuhr states, from Abyssinia to Yemen, coffee has been cultivated for ages in the hilly range of Jabal, in a healthy temperate climate, watered by frequent rains, and abounding in wells and water-tanks. A combination of circumstances seems to favour the cultivation of coffee in Arabia, which can hardly be attained elsewhere. Frequent rains, and a pure and cloudless sky causing an almost uninterrupted flood of light, communicate an excessive stimulus to all the functions of vegetation, and are causes of the perfect elaboration of those delicate principles on which the aroma of the coffee is dependent.

The seed consists chiefly of albumen and a peculiar or proximate principle termed *caféine*, whose ultimate constituents are identical with those of the proximate principle of tea, *thëine*, as well as those of *paraguäine*, the active principle of *matè*, or Paraguay tea—*Ilex paraguensis.*

The commercial importance of this plant renders its history interesting.

The consumption of coffee in Europe, in the year 1848, by the average of various authorities, was 400 millions of pounds; and in the United States and British America, by estimate, 150 millions of pounds, making the total consumption of those countries alone, 550,000,000 pounds.

The consumption of coffee, it is estimated, increases in Europe at the rate of 2½ per cent., and in the United States at the rate of 7½ per cent. per annum.

The quantity of coffee produced in several countries in five different years, has been carefully estimated by comparing various authorities, and is stated in millions of pounds as follows:—

	1841.	1843.	1848.	1851.	1852.
Brazil	160	174	270	300	300
Java	112	125	140	100	120
St. Domingo	25	38	35	40	30
Cuba and Porto Rico	56	50	50	30	25
British West Indies	12	10	12	7	5
French and Dutch West Indies	6	7	5	2	2
Sumatra	12	15	10	10	8
Mocha, &c	10	8	6	5	3
Ceylon	10	15	25	25	30
Laguayra, &c	25	30	30	25	20
Costa Rica	2	3	5	5	5
	430	471	587	559	548

[1] [Ruschenberger's Elements of Natural History.—Botany.]

The quantity of coffee produced has increased 118 millions of pounds since 1841.[1]

The annual average consumption of coffee to the population of the United States is estimated at 6¼ pounds per head.]

CHAPTER XXVI.

African Flora — Flora of Australia, New Zealand, Norfolk Island, and of Polynesia.

The northern coast of Africa, and the range of the Atlas generally, may be regarded as a zone of transition, where the plants of southern Europe are mingled with those peculiar to the country; half the plants of northern Africa are also found in the other countries on the shores of the Mediterranean. Of 60 trees and 248 shrubs which grow there, 100 only are peculiar to Africa, and about 18 of these belong to its tropical flora. There are about six times as many herbaceous plants as there are trees and shrubs; and in the Atlas mountains, as in other chains, the perennial plants are much more numerous than annuals. Evergreens predominate, and are the same as those on the other shores of the Mediterranean. The pomegranate, the locust-tree, the oleander, and the palmetto abound; and the cistus tribe give a distinct character to the flora. The sandarach, or thuia articulata, peculiar to the northern side of the Atlas mountains and to Cyrenaica, yields close-grained hard timber, used for the ceiling of mosques, and is supposed to be the shittim wood of Scripture. The Atlas produces seven or eight species of oak, various pines, especially the pinus maritima, and forests of the Aleppo pine in Algiers. The sweet-scented arborescent heath and Erica scoparia are native here, also in the Canary Islands and the Azores, where the tribe of house-leeks characterises the botany. There are 534 phanerogamous plants, or such as have the parts of fructification evident, in the Canary Islands; of these 310 are indigenous, the rest African; the pinus canariensis is peculiar, and also the Dracænæ, which grow in perfection here. The stem of a dracæna draco, at the Villa Oratava in Teneriffe, measures 46 feet in circumference at the base of the tree, which is 75 feet high. It is known to have been an object of great antiquity in the year 1402, and is still alive, bearing blossoms and fruits. If it be not an instance of the partial location of plants, there must have been intercourse between India and the Canary Islands in very ancient times

[1] [See Hunt's Merchant's Magazine for 1852.]

Plants with bluish green succulent leaves are characteristic of tropical Africa and its islands; and though the group of the Canaries has plants in common with Spain, Portugal, Africa, and the Azores, yet there are many species, and even genera, which are found in them only; and the height of the mountains causes much variety in the vegetation.

On the continent south of the Atlas a great change of soil and climate takes place; the drought on the borders of the desert is so extensive that no trees can resist it, rain hardly ever falls, and the scorching blasts from the south speedily dry up any moisture that may exist; yet, in consequence of what descends from the mountains, the date-palm forms large forests along their base, which supply the inhabitants with food, and give shelter to crops which could not otherwise grow. The date-palm, each tree of which yields from 150 to 160 pounds weight of fruit, grows naturally, and is also cultivated, through northern Africa.[1] It has been carried to the Canary Islands, Arabia, the Persian Gulf, and to Nice, the most northern limit of the palm-tribe. Stunted plants are the only produce of the desert, yet large tracts are covered with the pennisetum dichotomum, a harsh prickly grass, which, together with the alhagi maurorum, is the food of Camels.

The plants peculiar to Egypt are, acacias, mimosas, cassias, tamarisks, and nymphæa Lotus, the blue Lotus, the Papyrus, from which probably the first substance used for writing upon was made, and has left its name to that we now use: also the zizyphus or jujub, various mesembryanthemums, and most of the plants of Barbary grow here. The date-palm is not found higher on the Nile than Thebes, where it gives place to the doom-palm, or Cucifera Thebaica, peculiar to this district, and singular as being the only palm that has a branch stem.

The eastern side of equatorial Africa is less known than the western, but the floras of the two countries, under the same latitude, have little affinity: on the eastern side the Rubiaceæ, the Euphorbiæ, a race peculiarly African, and the Malvaceæ, are most frequent. The genus Danais of the coffee tribe distinguishes the vegetation of Abyssinia, also the Dombeya, a species of vine, various jessamines, a beautiful species of honeysuckle; and Bruce says a caper-tree grows to the height of the elm, with white blossoms, and fruit as large as a peach. The daroo, or ficus sycomorus, and the arak-tree, are native. The kollquall, or euphorbia antiquorum, grows 40 feet high on the plain of Baharnagach, in the form of an elegant branched candelabrum, covered with scented fruit. The kantuffa, or thorny

[1] The best dates are those grown near Tozzer in the Beled el Jerid, in lat. 34° N., a region which, like that of Jericho, also celebrated for its dates, has an extremely warm climate, supposed to be owing to its depression below the sea level.

shrub, is so great a nuisance from its spines, that even animals avoid it. The erythrina abyssinica bears a poisonous red bean with a black spot, used by the Shangella and other tribes for ages as a weight for gold and by the women as necklaces. Mr. Rochet d'Hericourt has lately brought some seeds of new grain from Shoa, that are likely to be a valuable addition to European cerealia.

The vegetation of tropical Africa on the west is known only along the coast, where some affinity with that of India may be observed. It consists of 573 species of flower-bearing plants, and is distinguished by a remarkable uniformity, not only in orders and genera, but even in species, from the 16th degree of N. lat. to the river Congo in 6° S. lat. The most prevalent are the grasses and bean tribes, the Cyperaceæ Rubiaceæ, and the Compositæ. The Adansonia, or baobab of Senegal, is one of the most extraordinary vegetable productions; the stem is sometimes 34 feet in diameter, though the tree is rarely more than 50 or 60 feet high; it covers the sandy plains so entirely with its umbrella-shaped top, that a forest of these trees presents a compact surface, which at some distance seems to be a green field. Cape Verde has its name from the numbers that conceal the barren soil under their spreading tops; some of them are very old, and, with the dragon-tree at Teneriffe, are supposed to be the most ancient vegetable inhabitants of the earth. The pandanus candelabrum, instead of growing crowded together in masses like the baobab, stands solitary on the equatorial plains, with its lofty forked branches ending in tufts of long stiff leaves. Numerous sedges, of which the Papyrus is the most remarkable, give a character to this region, and cover boundless plains, waving in the wind like cornfields, while other places are overgrown by forests of gigantic grasses with branching stems.

A rich vegetation, consisting of impenetrable thickets of mangrove, the poisonous machineel, and many large trees, cover the deltas of the rivers, and even grow so far into the water that their trunks are coated with shell-fish; but the pestilential exhalations render it almost certain death to botanize in this luxuriance of nature.

Various kinds of the soap or sapodilla trees are peculiar to Africa; the butter-tree of the enterprising but unfortunate Mungo Park, the star-apple, the cream-fruit, the custard-apple, and the water-vine, are plentiful in Senegal and Sierra Leone. The ibraculea is peculiarly African; its seeds are used to sweeten brackish water. The safu and bread-fruit of Polynesia are represented here by the musanga, a large tree of the nettle tribe, the fruit of which has the flavour of the hazel-nut. A few palms have very local habitations, as the elais guineaensis, or palm-oil plant, found only on that coast. That graceful tribe is less varied in species in equatorial Africa than in the other continents. It appears that a great part of the flora of this portion of Africa is of foreign origin.

The flora of South Africa differs entirely from that of the northern and tropical zones, and as widely from that of every other country, with the exception of Australia and some parts of Chile. The soil of the table-land at the Cape of Good Hope, stretching to an unknown distance, and of the Karoo plains and valleys between the mountains, is sometimes gravelly, but more frequently is composed of sand and clay; in summer it is dry and parched, and most of its rivers are dried up; it bears but a few stunted shrubs, some succulent plants and mimosas, along the margin of the river-courses. The sudden effect of rain on the parched ground is like magic: it is recalled to life, and in a short time is decked with a beautiful and peculiar vegetation, comprehending, more than any other country, numerous and distinctly defined foci of genera and species.

Twelve thousand species of plants have been collected in the colony of the Cape in an extent of country about equal to Germany. Of these, heaths and proteas are two very conspicuous tribes; there are 300 species of the former, and 200 of the latter, both of which have nearly the same limited range, though Mr. Banbury found two heaths, and the protea cynaroides, the most splendid of the family (bearing a flower the size of a man's hat), on the hills round Graham's Town, in the eastern part of the colony. These two tribes of plants are so limited that there is not one of either to be seen north of the mountains which bound the Great Karoo, and by much the greatest number of them grew within 100 miles of Cape Town; indeed at the distance of only 40 miles the prevailing Proteaceæ are different from those at the Cape. The leucadendron argenteum, or silver-tree, which forms groves at the back of the table-mountains, is confined to the peninsula of the Cape. The beautiful disa grandiflora is found only in one particular place on the top of the table-mountain.

The dry sand of the west coast and the country northward, through many degrees of latitude, is the native habitation of stapelias, succulent plants with square leafless stems, and flowers like star-fish, with the smell of carrion. A great portion of the eastern frontier of the Cape colony and the adjacent districts is covered with extensive thickets of a strong succulent and thorny vegetation, called by the natives the bush: similar thickets occur again far to the west, on the banks of the river Gauritz. The most common plants of the bush are aloes of many species, all exceedingly fleshy and some beautiful: the great red-flowering arborescent aloe, and some others, make a conspicuous figure in the eastern part of the colony. Other characteristic plants of the eastern districts are the spek-boem, or portulacaria afra, schotia speciosa, and the great succulent euphorbias, which grow into real trees 40 feet high, branching like a candelabrum, entirely leafless, prickly, and with a very acrid juice. The euphorbia meloformis, three feet in diameter, lies on the ground, to which it is attached by slender fibrous roots, and is confined to the mountains of Graaf Reynet. Euphorbias, in the Old World, correspond with the Cactus tribe, which belong exclusively to the New. The Zamia, a singular plant, having the appearance of a dwarf-palm without any real similarity of structure, belongs to the eastern districts, especially to the great tract of bush on the Caffir frontier.

Various species of Acacia are indigenous and much circumscribed in their location: the acacia horrida, or the white-thorned acacia, is very common in the eastern districts and in Caffirland. The acacia cafra is strictly eastern, growing along the margins of rivers, to which it is a great ornament. The acacia detinens, or hook-thorn, is almost peculiar to Zand valley.

It appears from the instances mentioned that the vegetation in the eastern districts of the colony differs from that on the western, yet many plants are generally diffused of orders and genera found only in this part of Africa:—Nearly all the 300 species of the fleshy succulent tribe of mesembryanthemum, or Hottentot's fig; a great many beautiful species of the Oxalis, or wood-sorrel tribe; every species of Gladiolus, with the exception of that in the corn-fields in Italy and France; ixias innumerable, one with petals of apple-green colour; geraniums, especially the genus Pelargonium, or stork's bill, almost peculiar to this locality; many varieties of Gnaphalium and Xeranthemum; the brilliant Strelitzia; 133 species of the house-leek tribe, all fleshy, attached to the soil by a strong wiry root, and nourished more or less from the atmosphere: Diosmas are widely scattered in great variety; shrubbery Boragineæ, with flowers of vivid colours, and Orchideæ with large and showy blossoms. The leguminous plants and the Cruciferæ of the Cape are peculiar; indeed all the vegetation has a distinct character, and both genera and species are confined within narrower limits than anywhere else, without any apparent cause to account for a dispersion so arbitrary.

Notwithstanding the peculiarity of character with which the botany of the Cape is so distinctly marked, it is connected with that of very remote countries by particular plants; for example, of the seven species of bramble which grow at the Cape, one is the common English bramble, or blackberry. The affinity with New Holland is greater: in portions of the two countries in the same latitude there are several genera and species that are identical: Proteaceæ are common to both, so are several genera of Irideæ, Leguminosæ, Ficoideæ, Myrtaceæ, Diosmeæ, and some others. The botany of the Cape is connected with that of India, and even that of South America, by a few congeners.

The vegetation of Madagascar, though similar in many respects to the floras of India and Africa, nevertheless is its own: the Brexiaceæ and Chlenaceæ are orders found nowhere else; there are spe cies of Bignonia, Cycadeæ, and Zamias, a few of the mangosteen tribe, and in the mountains some heaths. The hydrogeton fenestralis is a singular aquatic plant, with leaves like the dried skeletons of leaves, having no green fleshy substance, and the tanghinia veneniflua, which produces a poison so deadly that its seeds are used to execute criminals, and one seed is sufficient.

Some genera and species are common and peculiar to Madagascar, the Isle of Bourbon, and Mauritius; yet of the 161 known genera in Madagascar only 54 grow on the other two islands. The three islands are rich in ferns. The Pandanus, or screw-pine genus, abounds in Bourbon and the Mauritius, where it covers the sandy plains, sending off strong aërial roots from the stem, which strike into the ground and protect the plant from the violent winds. Of 290 genera in Bourbon and Mauritius, 196 also grow in India, though the species are different: there is also some resemblance to the vegetation of South Africa, and there is a solitary genus in common with America.

Eight or ten degrees north of Madagascar lies the group of the Seychelles Islands, in which are groves of the peculiar palm which bears the double cocoa-nut, or coco de mer, the growth of these islands only. Its gigantic leaves are employed in the construction of houses, and other parts of the plant are applied to various domestic purposes.

FLORA OF AUSTRALIA.

The interior of the Australian continent is so little known, that the flora which has come under observation is confined to a short distance from the coast; but it is of so strange and unexampled a character, that it might easily be mistaken for the production of another planet. Many entire orders of plants are known only in Australia, and the genera and species of others that grow elsewhere assume new and singular forms. Evergreens, with hard narrow leaves of a sombre, melancholy hue, are prevalent, and there are whole shadowless forests of leafless trees; the foot-stalks, dilated and set edgewise on the stem, supply their place, and perform the functions of nutrition; their altered position gives them a singular appearance. Plants in other countries have glands on the under side of the leaves, but in Australia there are glands on both sides of these substitutes for leaves, which make them dull and lustreless, and the changes of the seasons have no influence on the unvarying olive-green of the Australian forests; even the grasses are distinguished from the gramineæ of other countries by a remarkable rigidity. Torres Straits, in the north, only 50 miles broad, separates this dry, sombre vegetation from the luxuriant jungle-clad shores of New Guinea, where deep and dark forests are rich in more than the usual tropical exuberance — a more complete and sudden change can hardly be imagined.

The peculiarly Australian vegetation is in the southern part of the continent of New Holland distributed in distinct foci in the same latitude, a circumstance of which the Proteaceæ afford a remarkable instance. Nearly one-half of the known species of these beautiful shrubs grow in the parallel of Port Jackson, from which they decrease in number both to the south and the north. In that latitude, however, there are twice as many species on the eastern side of the continent as there are on the western, and four times as many as in the centre. Although the Proteaceæ at both extremities of the continent have all the characters peculiar to Australia, yet those on the eastern coast resemble the South American species, while those on the western side have a resemblance to African forms, and are confined to the same latitudes.

Species of this family are numerous in Van Diemen's Land; where they thrive at the elevation of 3500 feet, and also on the plains. The myrtle tribe form a conspicuous feature in Australian vegetation, particularly the genera Eucalyptus, Melaleuca, Beaufortia, and others, with splendid blossoms — white, purple, yellow, crimson: 100 species of the Eucalypti, most of them large trees, grow in Australia; they form great forests in the colony of Port Jacksón. The leafless acacias, of which there are 93 species, are a prominent feature in the Australian landscape. The leaves, except in very young plants, are merely foliaceous foot-stalks, presenting their margin towards the stem; yet these and the Eucalypti form the densest shade of any trees in the country. The genus Casuarina, with its strange-jointed drooping branches, called the marsh oak, holds a conspicuous place; it is chiefly confined to the principal parallel of this vegetation, and produces excellent timber; it grows also in the Malayan peninsula and South Sea islands. The oxleya xanthoxylon, or yellow wood, one of the mahogany tribe, grows to great size; and the podocarpus aspleniifolia forms a new genus of the cone-bearing tree. Some of the nettle tribe grow 15 or even 20 feet high. The Epacrideæ, with scarlet, rose, and white blossoms, supply the place of, and very much resemble, heaths, which do not exist here. The purple flowering Tremandreæ; the yellow-flowering Dilleniaceæ; the doryanthes excelsa, the most splendid of the lily tribe, 24 feet high, with a brilliant crimson blossom; the Banksia, the most Australian of all the Proteaceæ; with Zamias of new species, are all conspicuous in the vegetation of Port Jackson.

There is a change on the north-eastern coast of Australia. The castanospermum australe is so plentiful that it furnishes the principal food of the natives; a caper-tree of grotesque form, having the colossal dimensions of the Senegal baobab, and extraordinary trees of the fig genus, characterize this region. It sometimes occurs, when the seeds of these fig-trees are deposited by birds on the iron-bark-tree, or euacalyptus resinifera, that they vegetate and enclose the trunk of the tree entirely with their roots, whence they send off enormous lateral branches, which so completely envelop the tree, that at last its top alone is visible in the centre of the fig-tree, at the height of 70 or 80 feet. The Pandanus genus flourishes within the influence of the sea-air. There are only six species of palms, equally local in their habitations as elsewhere, not one of which grows on the west side of the continent. The Araucaria excelsa, or Norfolk Island pine, produces the best timber of any tree in this part of Australia: it, or others of the same genus, extends from the parallel of 29° on the east coast towards the equator, and grows over an area of 900 square miles, including New Norfolk, New Caledonia, and other islands, some of which have no other timber-tree: they are supposed to exist only within the influence of the sea. The Asphodeleæ abound and extend to the southern extremity of Van Diemen's Land.

The south-western districts of Australia exhibit another focus of vegetation, less rich in species than that of Port Jackson, but not less peculiar. The kingia australis, or grass-tree, rises solitary on the sandy plains, with bare blackened trunks as if scathed by lightning, occasioned by the fires of the natives, and tufts of long grassy leaves at their extremities; Banksias, particularly the kind called wild honeysuckle, are numerous; the Stylidium, whose blossoms are even more irritable than the leaves of the sensitive mimosa, and plants with dry, everlasting blossoms, characterize the flora of these districts. The greater part of the southern vegetation vanishes on the northern coasts of the continent, and what remains is mingled with the cabbage-palm, various species of the nutmeg tribe, sandal-wood, and other Malayan forms—a circumstance that may hereafter be of importance to our colonists.

Orchideæ, chiefly terrestrial, are in great variety in the extra-tropical regions of Australia, and the grasses amount to one-fourth of the monocotyledonous plants. Reeds of gigantic size form forests in the marshes, and kangaroo grass covers the plains.

Beautiful and varied as the flora is, Australia is by no means luxuriant in vegetation. There is little appearance of verdure, the foliage is poor, the forests often shadeless, and the grass thin; but in many valleys of the mountains, and even on some parts of the plains, the vegetation is vigorous. It is not the least remarkable circumstance in this extraordinary flora, that, with the exception of a few berries, there is no edible fruit, grain, or vegetable indigenous either in Australia or Van Diemen's Land.

The plants of Australia prevail in every part of Van Diemen's Land; yet the coldness of the climate and the height of the mountains permit genera of the northern hemisphere to be mixed with the vegetation of the country. Butter-cups, anemones, and polygonums of peculiar species grow on the mountain-tops, together with Proteaceæ and other Australian plants. The plains glow with the warm golden flowers of the black wattle, a mimosa, emblematic of the island, and with the equally bright and orange blossom of the gorse, which perfumes the whole atmosphere. Only one tree-fern grows in this country; it rises 20 feet to the base of the fronds, which spread into an elegant top, producing a shadow gloomy as night-fall, and there are 150 species of orchis. The southern extremities both of Australia and Van Diemen's Land are characterized by the prevalence of evergreen plants: but the trees here, as well as in the other parts of the southern hemisphere, do not shed their leaves periodically as with us.

The botany of New Zealand appears to be intimately allied to that of Australia, South America, and South Africa, but chiefly to that of the first. Noble trees form impenetrable forests, 60 of which yield the finest timber, and many are of kinds to which we have nothing similar. Here there are no representatives of our oak, birch, or willow, but five species of beech and ten of Coniferæ have been discovered that are peculiar to the country. They are all alpine, and only descend to the level of the sea in the southern parts of the island. The Coniferæ of the southern hemisphere are more local than in the northern; of the ten species peculiar to New Zealand it is not certain that more than two or three are found in the middle island, or that any of them grow south of the 40th parallel. The Kauri pine, or dammara australis, is indigenous in all the three islands; but it is the only cone-bearing tree in the North Island, where it grows in hilly situations near the sea, shooting up with a clean stem 60 or 90 feet, sometimes 30 feet in diameter, with a spreading but thin top, and generally has a quantity of transparent yellow resin imbedded at its base. This fine tree does not grow beyond the 38th degree of S. lat. The metrosideros tomentosa, with rich crimson blossoms, is one of the greatest ornaments of the forests, and the metrosideros robusta is the most singular. It grows to a very great size, and sends shoots from its trunk and branches to the ground, which become so massive that they support the old stem, which to all appearance loses its vitality; it is in fact an enormous epiphyte, growing to, and not from, the ground. Many of the smaller trees are of the laurel tribe, with poisonous berries. Besides, there is a cabbage-palm, the areca sapida, elder, the fuchsia excorticata, and other shrubs. This country is probably the southern limit of the orchideous plants that grow on trees. Before New Zealand was colonized, the natives lived chiefly on the roots of the edible fern, pteris esculenta, with which the country is densely covered, mixed with a shrub that grows like a cypress, and the tea-plant, which is a kind of myrtle whose berries afford an intoxicating liquor. More than 140 species of fern are natives of these islands, some of which are arborescent and 40 feet high; the country is chiefly covered with these and with the New Zealand flax, phormium tenax, which grows abundantly both on the mountains and plains. The vegetation is so vigorous on these volcanic islands that it grows richly on the banks of hot springs, and even in water too hot to be touched.

In Norfolk Island 152 species of plants are already known, and many, no doubt, are yet to be discovered. The Cape gooseberry or physalis edulis, the guava-tree, pepper, white and swamp oak, iron, blood-wood, and lemon trees, are native; also the bread-fruit tree, which blossoms, but does not bear fruit. The araucaria excelsa and some palms are indigenous, and there are three times as many ferns as of all the other plants together.

The multitude of islands of Polynesia constitute a botanical region apart from all others, though it is but little varied, and characterized principally by the number of syngenesious plants with arborescent forms and tree-ferns. In continental India and the tropical parts of Australia the proportion of ferns to conspicuously flowering plants is as 1 to 26, while on the Polynesian islands it is 1 to 4, and perhaps even as 1 to 3.[2]

The cocoa-nut palm and the pandanus are common to all the islands, but the latter thrives only when exposed to the sea-air. This archipelago produces tacca pinnatifida, which yields arrow-root; the morus papyrifera, whose bark is manufactured into paper; and one of the Dracæna tribe, from which an intoxicating liquor is made. Fifty varieties of the bread-fruit tree are indigenous, which produce three or four crops annually. It is most abundant in the Friendly, Society, and Caroline groups, from whence it has been taken to America, where it thrives in very low latitudes. The Sandwich group is peculiar in the number of Goodenias and Lobelias; while the Coral islands, whose flora is entirely borrowed, rarely have two species belonging to the same genus; the fragrant suriana and sweet-scented Tournfortia are among their scanty vegetation.

The two species of banana-trees which are natives of southern Asia have been introduced at an unknown and probably early period into the Polynesian islands, and all tropical countries in the eastern and western hemispheres. Syria is their northern limit, where the Musa paradisaica grows to 34° N. lat. The sweet fruit of these trees produces, on the same extent of ground, 44 times as much nutriment as the potato, and 133 times more than wheat.

St. Helena, the Sandwich group, New Zealand, Juan Fernandez, and above all the Galapagos islands, are more peculiar in their floras than any other tracts of their size. The Galapagos archipelago consists of 10 principal islands lying immediately under the equator,

[1] Dr. Mantel. [2] Dr. J. D. Hooker

600 miles from the coast of America. They are entirely volcanic, and contain 2000 extinct craters. The vegetation is so peculiar that, of 180 plants which have been collected, 100 are found nowhere else; of 21 species of Compositæ all but one are new, and belong to 10 genera, 8 of which are confined to these islands exclusively.

This flora has no analogy to that of Polynesia, but it bears a double relation to the flora of South America. The plants peculiar to the Galapagos islands are for the most part allied to those on the cooler part of the continent or in high lands, while the others are the same with those that abound in the hot damp intertropical regions of the continent. The greatest number of peculiar plants grow on the tops of the islands where the sea vapour is condensed, and many of them are confined to some one islet of the group. Though this flora is singular, it is poor compared with that of the Sandwich group, or the Cape de Verde Islands.[1]

CHAPTER XXVII.

American Vegetation—Flora of North, Central, and South America—Antarctic Flora—Origin and Distribution of the Cerealia—Ages of Trees—Marine Vegetation.

FROM similarity of physical circumstances the arctic flora of America bears a strong resemblance to that of the northern regions of Europe and Asia. This botanical district comprises Greenland, and extends considerably to the south of the arctic circle, especially at the eastern and western ends of the continent, where it reaches the 60th parallel of N. lat., and even more; it is continued along the tops of the Rocky Mountains almost to Mexico, and it re-appears on the White Mountains and a few other parts of the Alleghanies.

Greenland has a much more arctic flora than Iceland; the valleys are entirely covered with mosses and marsh-plants, and the gloomy rocks are cased in sombre lichens that grow under the snow, and the grasses on the pasture-grounds that line the fiords are nearly four times less varied than those of Iceland. In some sheltered spots the service-tree bears fruit, and birches grow to the height of a few feet: but ligneous plants in general trail on the ground.

---

[1] The Euphorbia and Borreria are the distinguishing features of the low grounds in the Galapagos islands; while the Scleria, croton, and Cordia mark the high grounds. Compositæ and Campanulaceæ distinguish St. Helena and Juan Fernandez. The prevailing plants in the Sandwich group are the Goodeniaceæ and Lobeliaceæ; and in New Zealand ferns and club-mosses prevail, almost to the exclusion of the grasses.—Dr. J. D. HOOKER.

The arctic flora of America has much the same character with that of Europe and Asia: many species are common to all; still more are representative, but there is a difference in the vegetation at the two extremities of the continent; there are 30 species in the east and 20 in the west end which grow nowhere else. The sameness of character changes with the barren treeless lands at the verge of the arctic region, and the distribution of plants varies both with the latitude and the longitude. Taking a broad view of the botanical districts of North America, there are two woody regions, one on the eastern, the other on the western side of the continent, separated by a region of prairies where grasses and herbaceous plants predominate. The vegetation of these three parts, so dissimilar, varies with the latitude, but not after the same law as in Europe, for the winter is much colder and the summer warmer on the eastern coasts of America than on the western coast of Europe, owing in a great measure to the prevalence of westerly winds which bring cold and damp to our shores.

Boundless forests of black and white spruce, with an undergrowth of reindeer moss, cover the country south of the arctic region, which are afterwards mixed with other trees; gooseberries, strawberries, currants, and some other plants thrive there. There are vast forests in Canada of pines, oak, ash, hickory, red beech, birch, the lofty Canadian poplar, sometimes 100 feet high and 36 feet in circumference, and sugar-maple; the prevailing plants are Kalmias, azaleas, and asters, the former vernal, the latter autumnal; solidagos and asters are the most characteristic plants of this region.

The splendour of the North American flora is displayed in the United States; the American sycamore, chestnut, black walnut, hickory, white cedar, wild cherry, red birch, locust-tree, tulip-tree, or Liriodendron, the glory of American forests, liquid-ambar, oak, ash, pine-trees of many species, grow luxuriantly, with an undergrowth of Rhododendrons, Azaleas, Andromedas, Gerardias, Calycanthus, Hydrangea, and many more of woody texture, with an infinite variety of herbaceous and climbing plants.

The vegetation is different on the two sides of the Alleghany mountains; the locust-tree, Canadian poplar, Hibiscus, and Hydrangea, are most common on the west side; the American chestnut and Kalmias are so numerous on the Atlantic side as to give a distinctive character to the flora; here, too, aquatic plants are more frequent; among these the Sarracènia or side-saddle flower, singular in form, with leaves like pitchers covered with a lid, half full of water.

The autumnal tints of the forests in the middle States are beautiful and of endless variety; the dark leaves of the evergreen pine, the red foliage of the maple, the yellow beech, the scarlet oak, and purple Nyssa, with all their intermediate tints, ever changing with the light and distance, produce an effect at sunset that would astonish the native of a country with a more sober-coloured flora under a more cloudy sky.

In Virginia, Kentucky, and the southern States the vegetation assumes a different aspect, though many plants of more northern districts are mixed with it. Trees and shrubs here are remarkable for broad shining leaves and splendid blossoms, as the Gleditschia, Catalpa, Hibiscus, and all the family of Magnolias, which are natives of the country, excepting a very few found in Asia and the Indian islands. They are the distinguishing feature of the flora from Virginia to the Gulf of Mexico, and from the Atlantic to the Rocky Mountains: the magnolia grandiflora and the tulip-tree are the most splendid specimens of this race of plants; the latter is often 120 feet high. The long-leaved pitch-pine, one of the most picturesque of trees, covers an arid soil on the coast of the Atlantic of 60,000 square miles. The swamps so common in the southern States are clothed with gigantic deciduous cypress, the aquatic oak, swamp hickory, with the magnificent nelumbium luteum and other aquatics; and among the innumerable herbaceous plants the singular dionæa muscipula, or American [Venus'] fly-trap: the trap is formed by two opposite lobes of the leaf, covered with spines, and so irritable, that they instantly close upon the insect that lights upon them. This Magnolia region corresponds in latitude with the southern shores of the Mediterranean, but the climate is hotter and more humid, in consequence of which there is a considerable number of Mexican plants. A few dwarf-palms appear among the Magnolias, and the forests in Florida and Alabama are covered with Tillandsia usneoides, an air-plant, which hangs from the boughs.[1]

Ten or twelve species of grass cover the extensive prairies or steppes of the valley of the Mississippi. The forms of the Tartarian steppes appear to the north in the Centaurea, Artemisia, Astragali; but the Dahlias, Œnotheras, with many more, are their own. The Helianthus and Coreopsis, mixed with some European genera, mark the middle regions; and in the south, towards the Rocky Mountains, Clarkia and Bartonia are mixed with the Mexican genera of Cactus and Yucca. The western forest is less extensive and less varied than the eastern, but the trees are larger. This flora in high latitudes is but little known: the thuia gigantea on the Rocky Mountains and the coast of the Pacific is 200 feet high. Claytonias and currants, with plants of northern Asia, are found here. [In his narrative of the United States Exploring Expedition, Commander Wilkes, U. S. Navy, gives a plate representing a portion of a dense forest near Astoria, on the Columbia river. "The largest tree of the sketch was thirty-nine feet six inches in circumference,

---

[1] Of 2981 species of flower-bearing plants in the United States of North America, there are 385 found also in northern and temperate Europe.

eight feet above the ground, and had a bark eleven inches thick. The height could not be ascertained, but it was thought to be upwards of 250 feet, and the tree was perfectly straight." — Vol. v. p. 116.]

Farther west the pinus Lambertiana is another specimen of the stupendous trees of this flora; seven species of pine are indigenous in California, some of which have measured 200 and even 300 feet high, and eighty in circumference. Captain Sir Edward Belcher, in his 'Voyage on the Pacific,' mentions having measured an oak 27 feet in circumference, and another 18 feet girth at the height of 60 feet from the ground, before the branches began to spread. This is the native soil of the currant-bushes with red and yellow blossoms, of many varieties of lupins, pæonies, poppies, and other herbaceous plants so ornamental in our gardens.

There are 332 genera of plants peculiar to North America, exclusive of Mexico, but no family of any great extent has yet been discovered there. About 160 large trees yield excellent timber; the wood of the pine-trees of the eastern forests is of inferior quality to that grown on the other side of the continent, and both appear to be less valuable than the pine-wood of Europe, which is best when produced in a cold climate. The Pinus Cembra and the Pinus uncinata are the most esteemed in the Old World.

The native fruits of North America are mostly of the nut-kind, and there are many of these, to which may be added the Florida orange, the Chicasa plum, the Papaw, the Banana,[1] the red mulberry, and the plumlike fruit of the Persimon. There are seven species of wild grapes, but good wine has not hitherto been produced. Although America has contributed so much to the ornament of our pleasure-grounds and gardens, yet there are comparatively few North American plants which have become an object of extensive cultivation, while America has borrowed largely from other parts of the globe; the grapes cultivated in North America are European. Tobacco,[2] Indian corn, and many others of the utmost commercial value are strangers to the soil, having been introduced by the earliest inhabitants from Mexico and South America, which have contributed much more to general utility.

[[1] The Banana, of which there are six or seven species, is an herbaceous plant, belonging to the torrid zone almost exclusively. By the Christians of the East, it is supposed to be the tree of good and evil, with the fruit of which Eve was tempted in the garden of Eden.]

[[2] According to the census, the tobacco crop of the United States, in the year 1850, amounted to 199,532,494 pounds. The annual crop, on an average of ten years, ending in 1851, is 132,010 hogsheads, valued at $7,834,076.]

FLORA OF MEXICO AND THE WEST INDIES.

Mexico itself unites the vegetation of North and South America, though it resembles that of the latter more nearly. Whole provinces on the table-land and mountains produce alpine plants, oaks, chestnuts, and pines spontaneously. The Cheirostemon, or hand-tree, so named from the resemblance its stigma bears to the human hand, grows here, and also in the Guatimala forests.

The low lands of Mexico and Central America have a very rich flora, consisting of many orders and genera peculiar to them, and species without number, a great portion of which is unknown. The Hymenea Courbaril, from which the copal of Mexico is obtained, logwood, mahogany, and many other large trees, valuable for their timber, grow in the forests; sugar-cane, tobacco, indigo, American aloe, yam, capsicum, and yucca are indigenous in Mexico and Central America. It is the native region of the Malastomas, of which 620 species are known; almost all the pepper tribe, the Passifloræ, the ornament and pride of tropical America and the West Indian islands, begin to be numerous in these regions. The pine-apple is entirely American, growing in the woods and savannahs: it has been carried to the West Indies, to the East Indies, and China, and is naturalised in all. This country has also produced the cherimoya, said to be the most excellent of fruits. All the Vanilla that is used in Europe comes from the states of Vera Cruz and Oaxaca, on the eastern slopes of the Cordillera of Anahuac in Mexico. It is native throughout tropical America, growing in hot, damp, shady places. Hot arid tracts are covered with the Cactus tribe, a family of Central America and Mexico, which is more widely dispersed than the anana: some species bear a considerable degree of cold. They are social plants, inhabiting sandy plains in thickets, and of many species: their forms are various and their blossoms beautiful. A few occur at a considerable distance from the tropics, to the north and the south. The night-flowering Cereus grows in all its beauty in the arid parts of Chile, filling the night air with its perfume. The cactus opuntia grows in the Rocky Mountains; and Sir George Back found a small island in the Lake of the Woods covered with it. This species has been brought to Europe and now grows a common weed on the borders of the Mediterranean. In Mexico the cochineal insect was collected from the cactus coccinellifer long before the Spanish conquest. There are large fields of American aloe, from which a fermented liquor called pulque, and also an ardent spirit, are made. The ancient Mexicans made their hemp from this plant, and also their paper. The forests of Panama contain at least 97 different kinds of trees, which grow luxuriantly in a climate where the torrents of rain are so favourable to vegetation, and so unfavourable to life that the tainted air is deadly even to animals.

The sugar-cane is a native of both continents; Columbus found it wild in many parts of America; the sweet cane is mentioned by the Prophets, and it has grown time immemorial on the coasts of China and in the islands of the Pacific. Its culture ranges throughout the torrid zone and to latitudes where the mean temperature is not under 64° of Fahrenheit. It grows on the plains of Nepaul at an absolute elevation of 4800 feet, and at the height of from 3500 to 5100 feet in the Cordillera of New Grenada. It is now scarcely cultivated in the southern provinces of New Spain, where it was introduced by the Spaniards, but it is extensively raised in Guiana, Brazil, the West India islands, the Mauritius, Bourbon, Bengal, Siam, Java, the Philippine Islands, and China.

[The following is a statement of the quantity of sugar produced in all countries in the year 1851, taken from Hunt's Merchants' Magazine.

	Tons.
Cuba and Porto Rico (2000 lbs. to the ton)	375,000
European beet-root	160,000
British West Indian	153,000
United States (including Maple sugar)	145,000
Brazil	117,000
Java	100,000
Bengal	78,000
French colonies	60,000
Mauritius	55,000
Manila, Siam, &c.	30,000
Dutch and Danish colonies	22,500
Total	1,295,500]

Maize or Indian corn is believed to have come originally from Mexico and South America. It is an annual requiring only summer heat; its limit is 50° N. in the American continent, and 47° N. in Europe; it ripens at an elevation of 7600 feet in low latitudes, and in the lower Pyrenees at the height of 3289 feet.[1]

The flora of each West Indian island is similar to that of the continent opposite to it. The Myrtus pimento, producing allspice, is common in the hills; custard-apple, Guava, the Alligator pear, and Tobacco are indigenous; the cabbage-palm grows to the height of 150 feet; the Palma-real of Cuba is the most majestic of that noble family; and in Barbadoes there still exists a tree, but wearing out rapidly, which has given the island its name.

[1] In the island of Titicaca in Peru-Bolivia, Mr. Pentland has seen a variety of maize ripen as high as 12,800 feet.

FLORA OF TROPICAL AMERICA.

Although the flora of tropical America is better explored than that of Asia or Africa, there must still be thousands of plants of which we have no knowledge; and those which have come under observation are so varied and so numerous, that it is not possible to convey an idea of the peculiarities of this vegetation, or of the extent and richness of its woodlands. The upper Orinoco flows for some hundred miles chiefly through forests; and the silvas of the [country watered by the] Amazon are six times the size of France. In these the trees are colossal, and the vegetation so matted together by underwood, creeping and parasitical plants, that the sun's rays can scarcely penetrate the dense foliage.

These extensive forests are by no means uniform; they differ on each side of the equator, though climate and other circumstances are the same. Venezuela, Guiana, the banks of the Amazon, and Brazil, are each the centre of a peculiar flora. So partial is this splendid vegetation, that almost each vegetation of the great rivers has a flora of its own; particular families of plants are so restricted in their localities, and predominate so exclusively where they occur, that they change the appearance of the forest. Thus, from the prevalence of the orders Laurineæ, Sapotaceæ, and others, which have leathery, shining, and entire leaves, the forests through which the Rio Negro, Cassiquiare, and Atabapo flow, differ in aspect from those in the other affluents of the Amazon. Even the grassy llanos, so uniform in appearance, have their centres of vegetation; and only agree with the pampas of Buenos Ayres in being covered with grass and herbs. In these tropical regions the flora varies with the altitude also. On the Andes, almost at the limit of vegetation, the ground is covered with purple, azure, and scarlet Gentians, Drabas, Alchemillas, and many other brilliantly coloured alpine plants. This zone is followed by thickets of coriaceous-leaved plants, in perpetual bloom and verdure; and then come the forest-trees. Arborescent ferns ascend to 7000 feet; the coffee-trees and palms to 5000; and neither indigo nor cocoa can be cultivated lower than 2000. The tree yielding cocoa, of which chocolate is made, grows wild in Guiana, Mexico, in the inland forests of Peru and Bolivia, and on the coast of the Caraccas; it is now cultivated in Central and South America, and in the Philippine islands, where it was introduced by the Spaniards. The seeds of its fruit, which is like a cucumber, are the cacao or chocolate bean.

Many parts of the coast of Venezuela and Guiana are rendered pestilential by the effluvia of the Mangrove, Avicennia, and the Manchineel, one of the Euphorbia family, consisting of 562 species in tropical America, all having milky juice, deleterious in the greater number. The well-known poison Ourari is prepared by the Indians of Guiana from the fruit and bark of the Strychnos toxicaria, than which nature has probably produced no plant more deadly. This Ourari (or Wourali) is a creeping plant which yields the deadly juice, the powerful effect of which was proved by Mr. Waterton's experiments.

The Cinchona, or true bark-tree, grows only on the Cordillera of the Andes.[1] Some of its medicinal qualities are found in other parts of different genera in Guiana, as the Cusparia carony, which produces the Angostura bark. The Sapindus sapomaria, or soap-tree, is used by the natives for washing. Capsicum, Vanilla, the incense-plant, the dipteryx odorata, whose fruit is the tonquin-bean, and the cassava or mandioc, are natives of the country. There are two kinds of mandioc, a shrub whose fleshy roots yield a farina eaten by the natives of Spanish America and Brazil: the root of one is harmless, but the other (the Mandioca brava of the Brazilians) contains a poisonous milky juice, the effects of which are removed by the washing or pressure of the pulp. It grows to about 30° on each side of the equator, and to 3200 feet above the sea-level. An acre of mandioc is said to yield as much nourishment as six acres of wheat.

Arrow-root is native in South America; it has been transported to the West Indies and Ceylon. The flour is the produce of the root. The plant is said to owe its name to the belief of its being an antidote to the poison of the arrows of the Indians. The cow-tree, almost confined to the coast of the mountains of Venezuela, yields such abundance of nutritious milky juice like that of a cow, that it is preserved in gourds. The chocolate plant, or cacao-tree, fruits of the most excellent flavour, plants yielding balsam, resin, and gum, are numerous in the tropical regions. There the laurel-tribe assumes the character of majestic trees; some are so rich in oil, that it gushes from a wound in the bark. One of these laurels produces the essential oil which dissolves caoutchouc, or Indian rubber, used in rendering cloth waterproof.

Palms are the most numerous and the most beautiful of all the trees in these countries. There are 90 species of them; and they are so local that a change takes place every 50 miles. They are the greatest ornament of the upper Orinoco.

The llanos of Venezuela and Guiana are covered with high grass,

---

[1] Dr. Weddell, a very distinguished English botanist, employed by the French Government, who has recently returned from an exploration of the districts of the Andes which furnish the Peruvian bark of commerce, has discovered several new species of Cinchona, the total number of which, according to his beautiful monography, now amounts to 21. — (Weddell, Histoire Naturelle des Quinquinas, 1 Vol. folio, avec 34 planches, Paris, 1849.)

mixed with lilies and other bulbous flowers, sensitive mimosas and palms constantly varying in species.

No language can describe the glory of the forests of the Amazon and Brazil, the endless variety of form, the contrasts of colour and size: there even the largest trees bear brilliant blossoms; scarlet, purple, blue, rose-colour, and golden yellow, are blended with every possible shade of green. Majestic trees, as the bombax ceiba (or silk-cotton tree), the dark-leaved mora with its white blossoms, the fig, cachew, and mimosa tribes, which are here of unwonted dimensions, and a thousand other giants of the forest, are contrasted with the graceful palm, the delicate acacia, reeds of 100 feet high, grasses of 40, and tree-ferns in myriads. Passifloræ and slender creepers twine round the lower plants, while others as thick as cables climb the lofty trees, drop again to the ground, rise anew and stretch from bough to bough, wreathed with their own leaves and flowers, yet intermixed with the vividly coloured blossoms of the Orchideæ. An impenetrable and everlasting vegetation covers the ground; decay and death are concealed by the exuberance of life; the trees are loaded with parasites while alive — they become masses of living plants when they die.

One twenty-ninth part of the flowering plants of the Brazilian forests are of the coffee tribe, and the rose-colouring and yellow-flowering bignonias are among their greatest ornaments, where all is grace and beauty. Thousands of herbs and trees must still be undescribed where each stream has its own vegetation. The palm-trees are the glory of lhe forest: 81 species of these plants are natives of the intertropical parts of Brazil alone; they are of all sizes, from such as have hardly any stem to those that rise 130 feet.[1] In those parts of Brazil less favoured by nature the forest consists of stunted deciduous trees, and the boundless plains have grasses, interspersed with myrtles and other shrubs.[2]

The forests on the banks of the Paraguay and Vermejo are almost as rich as those of the tropics. Noble trees furnish timber and fruit; the algaroba, a kind of acacia, produces clusters of a bean, of which the Indians make bread, and also a strong fermented liquor; the palm and immense forests of the Copernicia cerifera grow there; and the yerbamaté, the leaves and twigs of which are universally used as tea in South America, and were in use before the Spanish conquest. It is a species of holly, [*Ilex paraguensis*,] with leaves three inches long.

---

[1] Professor Martius, of Munich, in his great work on Palms, has described 500, accompanied with excellent coloured plates. It is supposed that the number of species throughout the world amounts to 1000.

[2] There are innumerable points of analogy between the vegetation of the Brazils, equinoctial Africa, and India; but the number of species common to these three continents is very small.

The sandy deserts towards the mountains are the land of the cactus in all their varieties. Some larger species of cactus give a light and durable timber for building; and the cochineal insect, which feeds on one of them, is a valuable article of commerce.

Glass, clover, and European and African thistles, which have been introduced, with a solitary Ombu at wide intervals, are the unvarying features of the pampas; and thorny stunted bushes, characteristic to all deserts, are the only vegetation of the Patagonian shingle. But on the mountain valleys in the far south may be seen the winter's-bark, arbutus, new species of beech-trees, stunted berberries and Misodendron, which latter is a singular kind of parasitical plant.

Large forests of Araucaria imbricata grow on the sides of the Andes of Chile and Patagonia. This tall and handsome pine, with cones the size of a child's head, supplies the natives with a great part of their food. It is said that the fruits of one large tree will maintain eighteen persons for a year.

Nothing grows under these great forests; and when accidentally burnt down in the mountainous parts of Patagonia, they never rise again, but the ground they grew on is soon covered with an impenetrable brushwood of other plants. In Chile the violently stinging Loasa appears first in these burnt places, bushes grow afterwards, and then comes a tree-grass, 18 feet high, of which the Indians make their huts. The new vegetation that follows the burning of primeval forests is quite unaccountable. The ancient and undisturbed forests of Pennsylvania have no undergrowth, and when burnt down they are succeeded by a thick growth of rhododendrons.

The southern coasts of Chile are very barren, and all plants existing there, even the herbaceous, have a tendency to assume a hard knotty texture. The stem of the wild potato, which is indigenous in Chile, becomes woody and bristly as it grows old. It is a native of the sea-strand, and is never found naturally more than 400 feet above it. In its wild state the root is small and bitter; it is one of many instances of the influence of cultivation in rendering unpromising plants useful to man.

It was cultivated in America at the time of its discovery, and is so now, at the height of from 9800 to 13,000 feet above the sea on the Andes, and as high as 4800 feet on the Swiss Alps; it does not succeed on the plains in hot countries, nor farther north than Iceland. It had been introduced into Europe by the Spaniards before the time of Sir Walter Raleigh; he brought it to England from Virginia in 1586.

Coca, the Erythroxylon Coca of botanists, is a native of the tropical valleys on the eastern declivity of the Andes of Peru and Bolivia, where it is extensively cultivated for its leaf, of which the tree furnishes 3 or 4 crops annually; the cocoa-leaf, which possesses nutritive qualities, is chewed by the aborigines mixed with an alkaline substance: it allays hunger, and enables the Indian to undergo great fatigue without any other nourishment for days together; it is an article of great trade, being universally used by the aboriginal population of the Andes, and absolutely indispensable in the more laborious professions, such as that of the miner.

Between the southern parallels of 38° and 45° Chile is covered with extensive forests. Stately trees of many kinds, having smooth and brightly-coloured trunks bound together by parasitical plants; large and elegant ferns are numerous, and arborescent grasses entwine the trees to the height of 20 or 30 feet; palm-trees grow to the 37th parallel of latitude, which appears to be their southern limit.

Although the flora, at an elevation of 9000 feet on the Chilian Andes, is almost identical with that about the Straits of Magellan, yet the climate is so mild in some valleys, that of Antuco, for example, that the vegetation is semi-tropical. In it broad-leaved and bright-coloured plants, and the most fragrant and brilliant Orchideæ, are mixed with the usual alpine genera. Dr. Poeppig says, that whatever South Africa or New Holland can boast of in beauty, in variety of form, or brilliancy of colour, is rivalled by the flora in the highest zone in this part of the Andes, even up to the region of perpetual snow; and, indeed, it bears a strong analogy to the vegetation of both these countries.[1]

The Andes so completely check the migration of plants, that almost throughout their whole length there is no mingling of the floras on their east and west sides, except at the Isthmus of Panama, where the mahogany-tree crosses from the Atlantic to the Pacific side, and in the same way many of the plants on the lands on the east are brought to the west, and spread to California on one side, and as far as the dry plains of Peru on the other.[2]

The humidity or dryness of the prevailing winds makes an immense difference in the character of the countries on each side of the Andes. Within the southern tropic the trade-winds come loaded with vapour from the Atlantic, which is partly precipitated by the mountains of Brazil, and supplies the noble forests of that country

[1] The natural history of Chile in all its departments, and especially in its botany, has been well illustrated in the 'Historia Naturel de Chile,' by M. Claude Gaye, a French naturalist of very varied talent who resided many years there, employed by the Chilian Government in writing its political and natural history. This beautiful work, which is now on the eve of its completion, has been published in Spanish at Paris, under the patronage of and supported by the President of the Republic; and whilst it reflects great credit on its author and the authorities of the prosperous state it is destined to illustrate, is well worthy of imitation by the other Spanish American Republics, where so little has been hitherto effected of a similar nature.

[2] Dr. J. D. Hooker.

with never-ceasing moisture, while the remainder is condensed by the Andes, so that on their eastern side there is an exuberant vegetation, while on the western declivities and in the space which separates them from the Pacific they are almost barren, and on the plains and in the valleys of Peru, where rain very seldom falls, completely so, except where artificial irrigation is employed. Even on the eastern side of these mountains the richness of the vegetation gradually disappears with the increasing height, till at an elevation of about 15,000 feet arborescent plants vanish, and alpine races, of the most vivid beauty, succeed; which in their turn give place to the grasses at the height of 16,100 feet. Above these, in the dreary plains of Bonbon, and other lands of the same altitude, even the thinly-scattered mosses are sickly; and at a height exceeding 20,000 feet the snow-lichen forms the last show of vegetable life on the rocky peaks projecting from the snow; confirming the observatiou of Don [Juan de] Ulloa, that the produce of the soil is the thermometer of the Peruvian Andes.

ANTARCTIC FLORA.

Kerguelen's Land and Terra del Fuego are the northern boundary of the antarctic lands, which are scattered round the south pole at immense distances from one another. On these the vegetation decreases as the latitude increases, till at length utter desolatation prevails; not a lichen covers the dreary storm-beaten rocks; and, with the exception of a few microscopic marine plants, not a seaweed lives in the gelid waves. In the arctic regions, on the contrary, no land has yet been discovered that is entirely destitute of vegetable life. This remarkable difference does not so much depend on a greater degree of cold in winter as on the want of warmth in summer. In the high northern latitudes the power of the summer sun is so great as to melt the pitch between the planks of the vessels; while in corresponding southern latitudes Fahrenheit's thermometer does not rise above 14° at noon at a season corresponding to our August. The perpetual snow comes to a much lower latitude in the southern lands than it does in the north. Sandwich Land, in a latitude corresponding to that of the north of Scotland, is perpetually covered with many fathoms of snow. A single species of grass, the Aira antarctica, is the only flowering plant in the South Shetland group which are no less ice-bound; and Cockburn Island, which forms a part of it, in the 60th parallel, contains the last vestiges of vegetation; while the Shetlands in our hemisphere, in an equally high latitude to the north of Scotland, are inhabited and cultivated: nay, South Georgia, in a latitude similar to that of Yorkshire, is always clad in frozen snow, and only produces some mosses, lichens, and wild burnet; while Iceland, 10 degrees nearer the pole, has 870 species, more than half of which are flower-bearing.

The forest-covered islands of Terra del Fuego are only 360 miles from the desolate South Shetland group. Such is the difference that a few degrees of latitude can produce in these antarctic regions, combined with an equable climate and excessive humidity. The prevalence of evergreen plants is the most characteristic feature in the Fuegian flora. Densely entangled forests of winter's bark, and two species of Beech-trees, grow from the shore to a considerable height on the mountains. Of these the Fagus betuloides, which never loses its brownish-green leaves, prevails almost to the exclusion of the ever-green winter's bark and the deciduous beech, which is very beautiful. There are dwarf species of arbutus, the Myrtus nummularia, which is used instead of tea, besides berberry, currant, and fuchsia; peculiar species of Ranunculi, Calceolarias, Caryophylleæ, cruciform plants and violets. Wild celery and scurvy-grass are the only edible plants; and a bright yellow fungus, which grows on the beech-trees, forms a great part of the food of the natives. There is a great number of plants in Terra del Fuego, either identical with those in Great Britain, or representatives of them, than exists in any other country in the southern hemisphere. The sea-pink, or thrift, the common sloewort, a primula farinosa, and at least 30 other flowering plants, with almost all the lichens, 48 mosses, and many other plants of the cryptogamous kinds, are identically the same, while the number of genera common to both countries is still greater, and, though unknown in the intermediate latitudes, reappear here. Hermite Island, west from Cape Horn, is a forest-land, covered with winter's bark and the Fuegian beeches; and is the most southern spot on earth on which arborescent vegetation is found. An alpine flora, many of the species of European genera, grows on the mountains, succeeded higher up by the mosses and lichens. Mosses are exceedingly plentiful throughout Fuego; but they abound in Hermite Island more than in any other country, and are of singular and beautiful kinds.

Although the Falkland Islands are in a lower latitude than Terra del Fuego, not a tree is to be seen. The Veronica elliptica, resembling a myrtle, which is extremely rare, and confined to West Falkland, is the only large shrub. A white-flowering plant like the Aster, about four feet high, is common; while a bramble, a crowberry, and a myrtle, bearing no resemblance, however, to the European species, trail on the ground, and afford edible fruit. The bog balsam, or Bolax globaria, and grasses, form the only conspicuous feature in the botany of these islands; and together with rushes and the dactylis cæspitosa, or tussack grass, cover them, almost to the exclusion of other plants. The Bolax grows in tufted hemispherical masses, of a yellow-green colour, and very firm substance, often four feet high, and as many in diameter, from whence a strong-smelling resinous substance exudes perceptibly at a distance. This plant has umbelliferous flowers, and belongs to the carrot order, but forms an alpine and antarctic genus quite peculiar.

The Tussack grass is the most useful and the most singular plant in this flora. It covers all the small islands of the group, like a forest of miniature palm-trees, and thrives best on the shores exposed to the spray of the sea. Each Tussack is an isolated plant, occupying about two square yards of ground. It forms a hillock of matted roots, rising straight and solitary out of the soil, often six feet high and four or five in diameter; from the top of which it throws out a thick grassy foliage of blades, six feet long, drooping on all sides, and forming with the leaves of the adjacent plants an arch over the ground beneath, which yields shelter to sea-lions, penguins, and petrels. Cattle are exceedingly fond of this grass, which yields annually a much greater supply of excellent fodder than the same extent of ground would do either of common grass or clover. Both the tussack grass[1] and the bolax are found, though sparingly, in Terra del Fuego; indeed, the vegetation of the Falkland Islands consists chiefly of the mountain plants of that country, and of those that grow on the arid plains of Patagonia; but it is kept close to the ground by the fierceness of the terrific gales that sweep over these antarctic islands. Peculiar species of European genera are found here, as a calceolaria, wood-sorrel, and a yellow violet; while the shepherd's purse, cardamine hirsuta, and the primula farinosa, appear to be identical with those at home. In all, there are scarcely 120 flowering plants, including grasses. Ferns and mosses are few, but lichens are in great variety and abundance, among which many are identical with those in Britain.

In the same hemisphere, far, far removed from the Falkland group, the Auckland Islands lie in the boisterous ocean south of New Zealand. They are covered with dense and all but impenetrable thickets of stunted trees, or rather shrubs, about 20 or 30 feet high, gnarled by gales from a stormy sea. There is nothing analogous to these shrubs in the northern hemisphere; but the veronica elliptica, a native of Terra del Fuego and New Zealand, is one of them. Fifteen species of ferns find shelter under these trees, and their fallen trunks are covered with mosses and lichens. Eighty flowering plants were found during the stay of the discovery ships, of which 56 are new; and half of the whole number are peculiar to this group and to Campbell's Island. Some of the most beautiful flowers grow on the mountains, others are mixed with the ferns in the forests. A beautiful plant was discovered, like a purple aster, a veronica, with large spikes of ultramarine colour; a white

[1] The cultivation of this useful plant has been recently introduced into some of the western islands of Scotland, especially Lewis, by the praiseworthy efforts of its proprietor, Sir James Matheson, M.P.

one, with a perfume like jessamine; a sweet-smelling alpine Hierochloe; and in some of the valleys the fragrant and bright-yellow blossoms of a species of asphodel were so abundant that the ground looked like a carpet of gold. A singular plant grows on the seashore, having bunches of green waxy blossoms the size of a child's head. There are also antarctic species of European genera, as beautiful red and white gentians, geraniums, &c. The vegetation is characterized by an exuberance of the finer flowering plants, and an absence of grasses and sedges; but the landscape, though picturesque, has a sombre aspect, from the prevalence of brownish-leaved plants of the myrtle tribe.

Campbell's Island lies 120 miles to the south of the Auckland group, and is much smaller, but from the more varied form of its surface it is supposed to produce as many species of plants. During the two days the discovery ships, under the command of Sir James Ross, remained there, between 200 and 300 were collected; of these 66 were flowering plants, 14 were peculiar to the country. Many of the Auckland Island plants were found here, yet a great change had taken place; 34 species had disappeared and were replaced by 20 new, all peculiar to Campbell's Island alone, and some were found that hitherto had been supposed to belong to Antarctic America only. In the Auckland group only one-seventh of the plants are common to other Antarctic lands, whilst in Campbell's Island a fourth are natives of other longitudes in the Antarctic Ocean. The flora of Campbell's Island and the Auckland group is so intimately allied to that of New Zealand, that it may be regarded as the continuation of the latter, under an Antarctic character, though destitute of the beech and pine trees. There is a considerable number of Fuegian plants in the islands under consideration, though 4000 miles distant; and whenever their flora differs in the smaller plants from that of New Zealand, it approximates to that of Antarctic America: but the trees and shrubs are entirely dissimilar. The relation between this vegetation and that of the northern regions is but slight. The Auckland group and Campbell's Island are in a latitude corresponding to that of England, yet only three indigenous plants of our island have been found in them, namely, the Cardamine hirsuta, Montia, and Callitriche. This is the utmost southern limit of tree-ferns.

Perhaps no spot in either hemisphere, at the same distance from the Pole, is more barren than Kerguelen Land: lying in a remote part of the South Polar Ocean. Only 18 species of flowering plants were found there, which is less than the number in Melville Island, in the Arctic Seas, and three times less than the number even in Spitzbergen. The whole known vegetation of these islands only amounts to 150, including sea-weeds. The Pringlea, a kind of cabbage, acceptable to those who have been long at sea, is peculiar to the island, and grass, together with a plant similar to the Bolax of the Falkland Islands, covers large tracts. About 20 mosses, lichens, &c., only are found in these islands, but many of the others are also native in the European Alps and north polar regions. It is a very remarkable circumstance in the distribution of plants, that there should be so much analogy between the floras of places so far apart as Kerguelen Land, the groups south from New Zealand, the Falkland Islands, South Georgia, and Terra del Fuego.

ORIGIN AND DISTRIBUTION OF CEREALIA.

The plants which the earth produces spontaneously are thus confined within certain districts, and few of them would survive a change of circumstances; nevertheless Providence has endowed those most essential to man with a greater flexibility of structure, so that the limits of their production can be extended by culture beyond what have been assigned to them by nature. The grasses yielding the grains are especially favoured in this respect, though their extension depends upon the knowledge and industry of man; no grain will be cultivated where it can be procured from a foreign market at less expense; so that with regard to useful plants there is an artificial as well as a natural boundary. The cultivation of plants in gardens and hot-houses is entirely artificial and depends on luxury and fashion.

Tartary and Persia are presumed to have been the original countries of wheat, rye, and oats; but these grains have been so long in use that it is impossible to trace their origin with certainty. Barley grows spontaneously in Tartary and Sicily, probably of different species. Those plants which produce the grains must have had a more extended location than any other, and they can endure the greatest extremes of heat and cold. In high northern latitudes wheat is protected from the inclemency of winter by sowing it in spring, or if sown in autumn a coat of snow defends it: the polar limit is the isothermal line of 57° 2′, and wheat will not form seed within the tropics, except at a considerable height above the sea. In America the northern limit is unknown, the country being uninhabited; but at Cumberland House, in the very middle of the continent, one of the stations of the Hudson's Bay Company, in 54° N. lat., there are fields of wheat, barley, and maize. Wheat thrives luxuriantly in Chile and Peru,[1] and at elevations of 8500 and 10,000 feet above the sea. It even produces grain on the banks of the Lake Titicaca, in the Peru-Bolivian Andes, at the height of 12,900

[[1] There is not wheat enough produced in Peru for the consumption of the country; wheat is imported from Chile, and flour from the United States.]

feet in sheltered situations, and good crops of barley are raised in that elevated region.

Barley bears cold better than any other grain, yet neither it nor any other will grow in Iceland. It is successfully cultivated in the Feroe Islands, near Cape North, the extreme point of Norway, near Archangel on the White Sea, and in Central Siberia to between 58° and 59° N. lat.

Rye is only cultivated where the soil is very poor, and agriculture little understood; yet a third of the population of Europe lives on rye bread, chiefly inhabitants of the middle and especially of the northern parts: its limit is about the 67th parallel of N. latitude.

Oats are scarcely known in middle and southern Europe; in the north they are extensively cultivated to the 65th degree of N. latitude.

Rice is the food of a greater number of human beings than any other grain: it has been cultivated from such high antiquity that all traces of its origin are lost. It contains a greater proportion of nutritious matter than any of the Cerealia, but, since it requires excessive moisture, and a temperature of 73° 4′ at least, its cultivation is limited to countries between the equator and the 45th parallel.

Indian corn and millet are much cultivated in Europe south of the 45th and 47th parallels, and forms an important article of food in France, Italy, Africa, India, and America. Buckwheat is extensively cultivated in Northern Europe and Siberia, and the table-lands of central Asia; it is a native of Asia, from whence it was brought into Europe in the 15th century.

The Cerealia afford one of the most remarkable examples of numberless varieties arising from the seed of one species. In Ceylon alone there are 160 varieties of rice, and at least 30 of panicum. The endless varieties which may be raised from the seed of one plant is more conspicuous in the flower-garden: the rose affords above 1400; the varieties of the pansy, calceolaria, tulip, auricula, and primrose are without end, and often differ so much from the parent plant that it seems almost impossible they should have had a common origin: it seems difficult to believe that red cabbage, cauliflower, and many others should have sprung from the sea-kale or Brassica oleracea, so totally dissimilar from any of them, with its bitter sea-green curly leaves. Fashion changes so much with regard to plants, that it is scarcely possible to form even an approximation to the number known to be in cultivation: new plants are introduced from a foreign country, and are apt to take the place of some of the older, which are neglected and sometimes lost; of 120,000 plants which are known to exist on the earth, not more than 15,000 are believed to be in cultivation.

It is supposed that plants capable of bearing a great range of temperature would exist through longer geological periods than those more limited in their endurance of vicissitudes of heat and cold, and it appears that in many instances at least the existence of varieties depends on the life of the plant from whence they originated; the actual duration of individuals is a subject which has not been sufficiently studied, though the progress of physiological botany has given the means of doing so without destroying the plant.

Since forest-trees increase by coatings from without, the growth of each year forming a concentric circle of wood round the pith or centre of the stem, the age of a tree may be ascertained by counting the number of rings in a transverse section of the trunk, each ring representing a year. Moreover, the progress of the growth is known by comparing the breadth of the rings, which are broader in a favourable than in an unfavourable season, though this may depend also in some measure on the quality of the soil which the roots have come to in their downward growth. If the number of concentric rings in a transverse section has shown the age of a tree, and its girth has been ascertained by measurement, an approximation to the age of any other tree of the same kind still growing, under similar circumstances, may be determined by comparison. In this way the age of many remarkable trees has been ascertained. The yew attains a greater age than any other tree in Europe. According to M. Decandolle this tree increases in girth the twelfth part of an inch in a year during the first 150 years, and rather less in the next hundred, the increase probably decreasing progressively. By that estimate a yew at Fountain's Abbey was reckoned to be 1214 years old; one at Crowhurst, in Surrey, was 1400 years old when measured by Evelyn; it has been shown by the same method that a yew at Fortingal, in Scotland, was between 2500 and 2600 years old; and one at Braburn, in Kent, must have been 3000 years old: these are the veterans of European vegetation.[1]

The cypress rivals the yew in longevity, and may perhaps surpass it. There is a cypress in the palace garden at Granada which had been celebrated in the time of the Moors, and was still known, in the year 1776, as Cipres della Regina Sultana, because a sultana met with Abencerrages under its shade. M. Alphonse Decandolle estimates a deciduous cypress in the churchyard of Santa Maria de Tecla, near Oaxaca in Mexico, to be 6000 years old, Zuccarina 3572, and Dr. Lindley only 870. Oaks come next in order: they are supposed to live 1500 or 1600 years. One in Welbeck Lane, mentioned by Evelyn, was computed to be 1400 years old. Chestnut-trees are known to live 900 years; lime-trees have attained 500 or

[1] It is worthy of remark that the trees which in our temperate latitudes attain the greatest age, belong to the family of the Coniferæ, which have furnished the most ancient vegetable remains imbedded in the strata which form the earth's surface, the oldest fossil plants of the Devonian and Carboniferous series belonging to trees nearly allied to Araucaria.

600 years in France; and birches are supposed to be equally durable. Some of the smaller and less conspicuous European plants perhaps rival these giants of the forest in age; heaths, and the alpine willow, which covers the ground with its leaves, although it is really a subterranean tree spreading to a vast distance, are long lived. Ivy is another example of this; there is one near Montpellier, six feet in girth, which must be 485 years old. A lichen was watched for forty years without the appearance of change.

The antiquity of these European vegetables sinks into insignificance when compared with the celebrated Baobab, or Adansonia digitata, in Senegal: taking as a measure the number of concentric rings counted on a transverse incision made for the purpose in the trunk of that enormous tree, it was calculated to be 5150 years old;[1] yet Baron Humboldt considers a cypress in the garden of Chapultepec to be still older; it had already reached a great age when Montezuma was on the throne of Mexico, in 1520. These two trees are probably the most aged organized beings on the face of the earth. Eight olive-trees on the Mount of Olives are supposed to be 800 years old; it is at least certain that they existed prior to the taking of Jerusalem by the Turks. There is some doubt as to the age of the largest cedar on Lebanon; it is nine feet in diameter, and has probably existed 800 or 900 years.

The age of palms and other monocotyledonous plants is ascertained by a comparison of their height with the time which each kind takes to grow. M. Decandolle thus estimates that the Cocos oleracea, or cabbage-palms, may live 600 or 700 years, while the cocoa-nut palm lives from 80 to 330 years.

Mr. Babbage has made an approximation to the age of peat-mosses from the concentric rings of the trees found in them.

---

[1] Doubts have been expressed by some eminent botanists regarding the great age of the Adansonia digitata: the opinion given in the text is that of one of the most eminent physiological botanists of the age, Decandolle, who says, "the baobab is the most celebrated instance of extreme longevity which has hitherto been noticed with any degree of accuracy; in its own country it bears a name which signifies one thousand years, and, contrary to what is usual, this name expresses what in reality is short of the truth." Adanson has noticed one in the Cape Verd Islands which had been observed by two English travellers, three centuries earlier; he found within its trunk the inscription they had engraved covered over by 300 woody layers, and was thus enabled to estimate the bulk this enormous plant had increased in three centuries; it was on such data that Decandolle formed his opinion, which has been adopted by Humboldt and other eminent naturalists, and which we see no reason to differ from. — See, for a very learned view of the contrary opinion, the Gardener's Chronicle for 1849, p. 340. [Also, see Humboldt's Aspects of Nature.]

MARINE VEGETATION.

A vegetable world lies hid beneath the surface of the ocean, altogether unlike that on land, and existing under circumstances totally different with regard to light, heat, and pressure, yet sustained by the same means. Carbonic acid is as essential, and metallic oxides are as indispensable, to marine vegetation as they are to land-plants. Sea-water contains a minute proportion of carbonic acid gas,[1] and something more than a twelve-thousandth part of its weight of carbonate of lime, yet that is sufficient to supply all the shell-fish and coral-insects in the sea with materials for their habitations, as well as food for vegetation. Marine plants are more expert chemists than we are, for the water of the ocean contains rather less than a millionth part of its weight of iodine, which they collect in quantities impossible for us to obtain otherwise than from their ashes.

Sea-weeds fix their roots to anything—to stone, wood, and to other sea-weeds; they must therefore derive all their nourishment from the water, and the air it contains; and the vital force or chemical energy by which they decompose and assimilate the substances fit for their maintenance is the sun's light.

Marine plants, which are very numerous, consist of two groups—a jointed kind, which include the Confervæ, or plants having a thread-like form; and a jointless kind, to which belong dulse, laver, the kinds used for making kelp, iodine, vegetable glue, that in the Indian Archipelago, of which Hirundo esculenta, a species of swallow, make their edible nests, and all the gigantic species which grow in submarine forests, or float like green meadows in the open sea. Flower-bearing sea-weeds are very limited in their range, which depends upon the depth of water and the nature of the coasts; but the cryptogamic kinds are widely dispersed—some species are even found in every climate from pole to pole. No doubt the currents at the surface, and the stratum of uniform temperature lower down, are the highways by which these cosmopolites travel.[2]

There are fewer vegetable provinces in the seas than on shore, because the temperature is more uniform, and the dispersion of the plants is not so much interfered with by the various causes which disturb it on land.[3]

---

[1] M. Laurens has found $\frac{1}{3000}$ part of this gas in the water of the Mediterranean.

[2] The cosmopolite ulvæ are the Enteromorpha, Codium, &c.

[3] Dr. J. D. Hooker has divided the marine vegetation into ten provinces:—the Northern Ocean, from the Pole to the 60th parallel of north latitude;—the North Atlantic, between the 60th and 40th parallels, which is the province of the delessariæ and focus proper;—the Mediterranean, which is a sub-region of the warmer temperate zone of the Atlantic, lying be-

Marine vegetation varies both horizontally, and vertically with the depth, and it seems to be a general law throughout the ocean that the light of the sun and vegetation cease together; it consequently depends on the power of the sun and the transparency of the water; so different kinds of sea-weeds affect different depths, where the weight of the water, the quantity of light and heat suit them best. One great marine zone lies between the high and low water marks, and varies in species with the nature of the coasts, but exhibits similar phenomena throughout the northern hemisphere. In the British seas, where, with two exceptions, the whole flora is cryptogamic,[1] this zone does not extend deeper than 30 fathoms, but is divided into two distinct provinces, one to the south and another to the north. The former includes the southern and eastern coasts of England, the southern and western coasts of Ireland, and both the channels; while the northern flora is confined to the Scottish seas and the adjacent coasts of England and Ireland. The second British zone begins at low-water mark, and extends below it to a depth from 7 to 15 fathoms. It contains the great tangle sea-weeds, growing in miniature forests, mixed with fuci, and is the abode of a host of animals. A coral-like sea-weed is the last plant of this zone, and the lowest in these seas, where it does not extend below the depth of 60 fathoms, but in the Mediterranean it is found at 70 or 80 fathoms, and is the lowest plant in that sea. The same law prevails in the Bay of Biscay, where one set of sea-weeds is never found lower than 20 feet below the surface; another only in the zone between the depths of 5 and 30 feet; and another between 15 and 35 feet. In these two last zones they are most numerous; at a greater depth the kinds continue to vary, but their numbers decrease. The seeds of each kind float at the depth most genial to the future plant: they must therefore be of different weights. The distribution in the Egean Sea was found by Professor E. Forbes to be perfectly similar, only that the vegetation is different, and extends to a greater depth in the Mediterranean than in more northern seas.[2] He also observed

tween the 40th and 23d northern parallels;—the tropical Atlantic, in which sargassum, rhodomelia, corallinia, and siphinea abound; — the antarctic American region, from Chile to Cape Horn;—the Falkland Islands;—and the whole circumpolar ocean south of the 50th southern parallel; — the Australian and New Zealand province, which is very peculiar, being characterized, among other generic forms, by cystoseiriæ and fuceæ; — the Indian Ocean and the Red Sea;—and the last, which comprises the Japan and China Seas. There are several undetermined botanical marine provinces in the Pacific and elsewhere.

[1] The British flowering sea-weeds are the Zostera and Zanichellia.

[2] The vegetation at different depths in the Egean Sea is as distinctly marked as that at different heights on the declivity of a mountain. The coast plants are the padina pavonia and dictyota dichotoma. A greater depth is characterized by the vividly green and elegant fronds of the

that sea-weeds growing near the surface are more limited in their distribution than those that grow lower down, and that with regard to vegetation depth corresponds with latitude, as height does on land. Thus the flora at great depths, in warm seas, is represented by kindred forms in higher latitudes. There is every reason to believe that the same laws of distribution prevail throughout the ocean and every sea.

Sea-weeds adhere firmly to the rocks before their fructification, but they are easily detached afterwards, which accounts for some of the vast fields of floating weeds; but others, of gigantic size and wide distribution, are supposed to grow unattached in the water itself. There are permanent bands of sea-weed in the British Channel and in the North Sea, of the kind called fucus filum, which grow abundantly on the western coasts of the Channel; they lie in the direction of the currents, in beds 15 or 20 miles long, and not more than 600 feet wide. These bands must oscillate with the tides between two corresponding zones of rest, one at the turn of the flood, and the other at the turn of the ebb. It is doubtful whether the fucus natans or sargassum bacciferum grows on rocks at the bottom of the Atlantic, between the parallels of 40° north and south of the equator, and, when detached, is drifted uniformly to particular spots which never vary, or whether it is propagated and grows in the water; but the mass of that plant, west of the Azores, occupies an area equal to that of France, and has not changed its place since the time of Columbus. Fields of the same kind cover the sea near the Bahama Islands and other places, and two new species of it were discovered in the Antarctic seas.

The macrocystis pyrifera and the laminaria radiata are the most remarkable of marine plants for their gigantic size and the extent of their range. They were met with on the Antarctic coasts two degrees nearer the south pole than any other vegetable production, forming, with one remarkable exception, the utmost limit of vegetable life in the south polar seas. The macrycostis pyrifera exists in vast detached masses, like green meadows, in every latitude from the south polar ocean to the 45th degree N. lat. in the Atlantic, and to the shores of California in the Pacific, where there are fields

caulerpa prolifera, probably the prasium of the ancients; associated with it are the curious sponge-like codium bursa, and four or five others. The codium flabelliforme, and the rare and curious vegetable net called microdictyon umbilicatum, characterize depths of 30 fathoms. The Dictyomenia, with stiff purple corkscrew-like fronds, and some others, go as low as 50 fathoms, beyond which no flexible sea-weeds have been found. The coral-like millepora polymorpha take their place, and range to the depth of 100 fathoms, beyond which there is no trace of vegetable life, unless some of the minute and microscopic infusorial bodies living there be regarded as plants.—'Travels in Lycia,' by Lieutenant Spratt, R.N., and Professor E. Forbes.

of it so impenetrable, that it has saved vessels driven by the heavy swell towards that shore from shipwreck. It is never seen where the temperature of the water is at the freezing-point, and is the largest of the vegetable tribe, being occasionally 300 or 400 feet long. The Laminaria abounds off the Cape of Good Hope and in the Antarctic Ocean. These two species form great part of a band of sea-weed which girds Kerguelen Land so densely, that a boat can scarcely be pulled through it; they are found in great abundance on the coasts of the Falkland group, and also in vast fields in the open sea, hundreds of miles from any land: had it ever grown on the distant shores, it must have taken ages to travel so far, drifted by the wind, currents, and the sand of the seas. The red, green, and purple lavers of Great Britain are found on the coasts of the Falkland Islands; and, though some of the northern sea-weeds are not met with in the intervening warm seas, they reappear here. The Lessonia is the most remarkable marine plant in this group of islands. Its stems, much thicker than a man's leg, and from 8 to 10 feet long, fix themselves by clasping fibres to the rocks above high-water mark. Many branches shoot upwards from these stems, from which long leaves droop into the water like willows. There are immense submarine forests off Patagonia and Terra del Fuego, attached to the rocks at the bottom. These plants are so strong and buoyant, that they bring up large masses of stone; and, as they grow slanting, and stretch along the surface of the sea, they are sometimes 300 feet long. The quantity of living creatures which inhabit these marine forests and the parasitical weeds attached to them is inconceivable; they absolutely teem with life. Of the species of marine plants which are strictly antarctic, including those in the seas of Van Diemen's Land and New Zealand, Dr. Hooker has identified one-fifth with the British Algæ.

The high latitudes of the Antarctic Ocean are not so destitute of vegetation as was at first believed. Most minute objects, altogether invisible to the naked eye, except in mass, and which were taken for silicious shelled animalcules of the infusoria kind, prove to be vegetable. They are a species of Diatomaceæ, which, from their multitudes, give the sea a pale ochreous brown colour. They increase in numbers with the latitude, up to the highest point yet attained by man, and no doubt afford food to many of the minute animals in the antarctic seas. Genera and species of this plant exist in every sea from Victoria Land to Spitzbergen. It is one of the remarkable instances of a great end being effected by small means; for the death of this antarctic vegetation is forming a submarine bank between the 76th and 78th parallels of south latitude, and from the 165th to the 160th western meridian.

Great patches of Confervæ are occasionally met with in the open seas. Bands several miles long, of a reddish-brown species, like chopped hay, occur off Bahia, on the coast of Brazil; the same plant is said to have given the name to the Red Sea; and different species are common in the South Pacific Ocean.

CHAPTER XXVIII.

Distribution of Insects — [Geographical Distribution of Animals.]

[To form a general idea of the animal kingdom, it is not enough to know the principal phenomena by which life is manifest in animate creatures, and to have studied the structure of their bodies, and the mechanism of their functions; we must also look at the manner in which animals are distributed over the face of the earth, and endeavour to appreciate the influence which the different circumstances in which they are placed may exercise over them.

When we look at the manner of distribution of animals on the globe, we are struck at first with the difference of the media they inhabit. Some, as every body knows, always live under water and quickly die when withdrawn from it; others can only exist in the air and almost immediately perish when submerged. Some in fact are destined to inhabit the waters, and others to live upon the land; and when we compare aquatic and terrestrial animals, in their physiological and anatomical relations, we find, at least in part, the causes of the differences in their mode of existence.

In studying respiration, it may be observed, there is a constant relation between the intensity of this function and vital energy. Animals consume in a given time a quantity of oxygen, increasing in proportion to the activity of their motions and rapidity of their nutrition: now, they can obtain this oxygen only from the fluids surrounding them; in a gallon of air there are about 84 cubic inches of this vivifying principle, while in a gallon of water we ordinarily find only about five cubic inches. It is evident then that the degree of activity in the respiratory function, indispensable to the exercise of the faculties belonging to superior animals, must be of more easy attainment in air than in water, and on account of this difference alone, the creatures highest in the animal series cannot dwell in water. We comprehend, indeed, that an animal which, in order to exist, must appropriate a considerable quantity of oxygen every instant, does not find it in sufficient quantity when plunged into water, and therefore perishes of asphyxia. But at first sight, it is not easy to explain why an aquatic animal cannot continue to live when taken from the water and placed in the air, for then we supply it with a fluid richer in oxygen than that, the vivifying action of which was sufficient for all its wants. There are, however, various circumstances which, to a certain degree, explain this phenomenon. Physics teach us that a body carefully weighed in air and in water, is lighter in the last than in the first, and that, to sustain it in equilibrium, there is then only required a weight equal to its weight in air, less that of the bulk of water it displaces. Hence it follows that animals whose tissues are too soft to sustain themselves in air, and are compressed to such an extent as to become unfit to perform their functions in the organism, can nevertheless live very well in water, where these same tissues, being not much more dense than the surrounding fluid, are required to possess only a feeble power of resistance to preserve their forms and to prevent the several parts of the body from falling together on each other. This consideration alone is sufficient to show us why gelatinous animals, such as infusoriæ or medusæ, are necessarily inhabitants of the water; for, when we observe one of these delicate creatures while still in this fluid, we perceive that all the parts, even the most slender tissues, are sustained in their proper position and float easily in the surrounding medium; but the moment they are withdrawn, their body is almost entirely effaced, offering to the eye only a confused and shapeless mass. The influence of the density of the surrounding medium upon the mechanical play of these instruments of life is also felt in animals of a more perfect structure, in which, however, respiration is still carried on by means of ramified membranous appendages, resembling diminutive shrub-branches or plumes. For example, in annelidans, or even in fishes, the branchiæ or gills are composed of flexible filaments, which easily sustain themselves in water, and therefore permit the respirable fluid to reach and renew itself at all points of their surface; but, in air, these same membranous filaments are in a measure effaced by their own weight, falling one on another, and, in this way, exclude the oxygen from the greater part of the respiratory apparatus. It results that this function is then embarrassed, and the animal may die of asphyxia in the air, although it found in water all it required for free respiration. To convince ourselves of the importance of these variations in the physical state of organs placed in air or in water, it is only necessary to be reminded of what is seen in dissecting-rooms: an anatomist desirous of studying the structure of a very delicate part, would succeed very indifferently if he made his dissection in air; but by placing the subject of investigation in water, he much more easily succeeds in distinguishing all the parts; because these parts, sustained in a measure by this liquid, then preserve their natural relations just as if they were of a consistent and stiffer tissue. Another circumstance which influences the possibility of living in air or in water is the evaporation which always takes place from the surface of organized bodies placed in the air, but which cannot take place in water. A certain degree of desiccation causes all organic tissues to lose their distinguishing physical properties, and we find that losses by evaporation always produce death in animals when they exceed certain limits. It follows that creatures whose organization is not calculated to preserve them against the injurious effects of evaporation, can only live in water and quickly perish in air. Now the animal economy is equal to this exigence only when it possesses a very complicated structure. In fact, if an active respiration be requisite, the respiratory surface must be deeply lodged in some internal cavity where the air can be renewed only in proportion as it is required for the support of life. To secure this renovation, the respiratory apparatus must be furnished with proper motive organs; to prevent the desiccation or drying of any portion of the surface of the body, the diffusion of the liquids to the different parts of the body must be easily carried on, and there must be an active circulation, or the surface must be invested by a tunic or covering that is scarcely permeable. This is so true, that even in fishes, in which the circulation is very complete, although slowly carried on, and the capillary net-work not very dense, death speedily takes place in consequence of desiccation of a part of the body, of the posterior portion, for example, even when this portion alone is exposed to the air, while the rest of the animal remains under water.

We may add, too, that in water, feeding may be effected with less perfect instruments of prehension than in air, where the transportation of the food required by the animal is more difficult. In all its most essential relations, life is in a manner, more easily maintained in the midst of the waters than on the surface of the earth; in the atmosphere it demands more perfect and more complicated physiological instruments: the water is the natural element of animals lowest in the zoological series; and if the productions of the creation have succeeded each other in the same order as the transitory states through which every animal passes, during the period of its development, we may conclude that animate creatures first appeared in the midst of the waters, a conclusion in accordance with the observations of geologists and the text of the Scriptures.

In this manner the physiologist can account for the division of animals between the two geological elements of the globe, water and earth; but these fundamental differences are not the only ones observed in the geographical distribution of animate creatures. If a naturalist familiar with the fauna[1] of his own country, visit distant regions, he sees, as he advances, that the land becomes inhabited by

[1] *Fauna*, from the Latin, *faunus*, the name of a rural deity among the Romans. The animals of all kinds peculiar to a country constitute its *Fauna*.

animals new to his eyes; then these species disappear, in their turn to give place to species equally unknown.

If, after leaving France, for example, he land in the south of Africa, he will find there only a small number of animals similar to those he saw in Europe, and he will remark especially the Elephant, with big ears; the Hippopotamus; the Rhinoceros, with two horns; the Giraffe; innumerable herds of Antelopes; the Zebra; the Cape Buffalo, the widened base of whose horns covers the front; the black-maned Lion; the Chimpanzee, which of all animals most resembles man; the Cynocephalus, or dog-faced Monkey; Vultures of particular species; a multitude of birds of brilliant plumage, strangers to Europe; insects, also different from those of the north; for example, the fatal Termite, which lives in numerous societies, and builds, in common, its habitation of earth, which is very curious in its arrangement and of considerable height.

If our zoologist leave the Cape of Good Hope, and penetrate into the interior of the great island of Madagascar, he will there find a different fauna. He will see none of the large quadrupeds he met in Africa; in place of the family of monkeys, he will find other mammals equally well formed for climbing trees, but more resembling the carnaria, designated by naturalists under the name of *makis;* he will meet the *ai-ai*, or sloth, a most singular animal, which appears to be a sort of object of veneration among the inhabitants, and partakes of the nature of both monkey and squirrel; Tenrecs (a kind of hedge-hog), small insecti'vorous mammals, which have spiny backs like hedge-hogs, but do not roll themselves in a ball; the Cameleon, with forked nose, and many curious reptiles not found elsewhere, as well as insects not less characteristic of that region.

Still pursuing his route and arriving in India, our traveller sees an elephant different from that of Africa; oxen, bears, rhinoceros, antelopes, stags, different from those of Africa and Europe; the ourang-outang, and a multitude of other monkeys peculiar to those countries; the royal tiger, the argus, the peacock, pheasants, and an almost innumerable host of birds, reptiles, and insects, unknown elsewhere.

If he now visit New Holland, all will be there again new to him, and the aspect of this fauna will appear to him still more strange than the various zoological populations he has passed in review. He will no longer meet with species analogous to our oxen, horses, bears, and large carnaria; large-sized quadrupeds are almost entirely wanting; he will find kangaroos, flying-phalangers, and the ornithorynchus.

Finally, if our traveller, to get back to his own country, traverses the vast continent of America, he will discover a fauna analogous to that of the Old World, but composed almost entirely of different species; he will there see monkeys with a prehensile tail, large carnaria similar to our lions and tigers, bisons, llamas, armadillos; birds, reptiles, and insects, equally remarkable, and equally new to him.

Differences not less great in the species of animals peculiar to different regions of the globe, are observed, when, instead of confining our observations to the inhabitants of the land, we examine the myriads of animated creatures that dwell in the midst of the waters. Passing from the coasts of Europe to the Indian Ocean, and from the latter into the American seas, we meet with fishes, mollusks, crustaceans, and zoophytes, peculiar to each of these regions. This limitation or colonization of species, whether aquatic or terrestrial, is so marked, that a slightly experienced naturalist cannot mistake, even at first sight, the original localities of zoological collections that may have been gathered in one or the other of the great geographical divisions of the globe, and submitted to his examination. The fauna of each of these divisions is peculiar to it, and may be easily characterized by the presence of certain more or less remarkable species.

Naturalists have formed many theories to account for this mode of distribution of animals over the surface of the globe; but, in the present state of science, it is impossible to give a satisfactory explanation, without admitting that, in the beginning, the different species had their origin in the different regions where they are found, and that by degrees they afterwards spread afar and occupied a more or less considerable portion of the surface of the earth. In short, the presence of a particular animal within narrow limits on the earth, necessarily supposes, when this animal is found nowhere else, that it had its origin on this spot, or that it immigrated there from a more or less remote region, and that subsequently it was entirely destroyed where its race commenced, that is, exactly at the place where, according to every probability, all circumstances most favourable to its existence were found in combination. There is nothing strongly in favour of this last hypothesis, and it is repugnant to common sense to believe that, in the beginning, the same country saw the birth of the horse, the giraffe, bison, and kangaroo, for instance, but that these animals left it afterwards, without leaving any trace of their passage, to colonize, one on the steppes of central Asia, another in the interior of Africa, a third in the New World, and another again in the great islands of Australia. It is much more natural to suppose that every species was placed, from the beginning, by the Author of all things, in the region where it was destined permanently to live, and that by extending from a certain number of these distinct centres of creation, different animals have spread throughout those portions of the globe now forming the domain of each kind. In the present condition of the earth, it is impossible to recognise all those zoological centres: for we can conceive the possibility of exchanges so multiplied between two regions, the faunæ of which were primitively distinct, that they present species common to both, and nothing now points out to the eyes of the naturalist their original separation; but when a country is inhabited by a considerable number of species which are not seen elsewhere, even where local circumstances are most similar, we are warranted in the supposition that this region was the theatre of a peculiar zoological creation, and we must regard it as a distinct region.

What the naturalist should ask, is, not how different portions of the earth have come now to be inhabited by different species, but how animals could be so far extended over the surface of the globe, and how nature placed variable limits to this dissemination according to species. The latter question especially presents itself to the mind when we consider the unequal extent now occupied by this or that group of animated creatures: for example, the ourang-outang is confined to the island of Borneo and the neighbouring lands; the musk-ox is colonized in the most northern part of America, and the llama in the elevated regions of Peru and Chile, while the wild-duck is seen everywhere, from Lapland to the Cape of Good Hope, and from the United States to China and Japan.

The circumstances which favour the dissemination of species are of two kinds: the one pertains to the animal itself, and the other is foreign to it. Among the first is the development of the locomotive power, all things being equal in other respects; the species which live attached to the earth, or which possess only imperfect instruments of locomotion, occupy a very limited extent of the earth's surface, compared to those species whose moving powers are rapid and energetic: among terrestrial animals, birds present us with most examples of cosmopolite species, and, among aquatic animals, the cetaceans, and fishes. Reptiles, on the contrary, are restricted to narrow limits, and the same is true of most molusks and crustaceans. The instinct possessed by certain animals to change their climate periodically, also contributes to the dissemination of species; and this instinct exists in a great number of these creatures.

Among the circumstances foreign to the animal, and in a measure accidental, we place first the influence of man; and to illustrate this point, a few examples will suffice. The horse is originally from the steppes of Central Asia, and, at the time of the discovery of America, no animal of this species existed in the New World; the Spaniards carried it with them there not more than three centuries back, and now, not only do the inhabitants of this vast continent, from Hudson's Bay to Terra del Fuego, possess horses in abundance, but these animals have become wild, and are found in almost countless herds. The same is true of the domestic ox: carried from the Old to the New World, they have multiplied there to such an extent that in some parts of South America they are actively hunted for their hides only, for the manufacture of leather. The dog has been everywhere the companion of man, and we could instance a great many animals that have become cosmopolite by following us; the rat, which appears to be originally from America, overran Europe in the middle ages, and is now met with even on the islands of Oceanica.

In some cases, animals have been able to break through natural barriers, seemingly insurmountable, and spread themselves over a more or less considerable portion of the surface of the globe, by the assistance of circumstances whose importance at first sight seems very trifling, such as the movement of a fragment of ice or wood, often carried to considerable distances by currents: nothing is more common than to meet at sea, hundreds of miles from land, fucus floating on the surface of the water and serving as a resting-place for small crustaceans incapable of transporting themselves, by swimming, far from the shores where they were born. The great maritime current, the gulf-stream, commencing in the Gulf of Mexico, coasts North America to Newfoundland, then directs its course to Iceland, Ireland, and returns towards the Azores, often bearing to the coasts of Europe, trunks of trees which were conveyed by the waters of the Mississippi from the most interior parts of the New World, to the sea; it frequently happens that these masses of wood are perforated by the larvæ of insects, and they may afford attachment to the eggs of mollusks, and of fishes, &c. Finally, even birds contribute to the dispersion of living creatures over the surface of the globe, and that, too, in a most singular manner: frequently they do not digest the eggs they swallow, but, evacuating them at places far from where they were picked up, carry to great distances the germs of races unknown till then in the countries where they were deposited.

Notwithstanding all these means of transportation and other circumstances favouring the dissemination of species, there are very few animals that are really cosmopolites, the most of these creatures being colonized within limited regions. That such should be the case, we can comprehend if we study the circumstances which may oppose their progress. But this study is far from furnishing us a satisfactory explanation of all the cases of limited circumspection of a species, and it is often impossible to divine why certain animals remain restricted to a locality, when nothing seems to oppose their propagation in neighbouring situations.

Whatever may be the reason, the obstacles to the geographical distribution of species are sometimes mechanical, and at others, physiological; among the first are seas and chains of lofty mountains. To terrestrial animals seas of much extent are in general an impassable barrier, and we perceive, all things being equal, the mixture of two distinct faunæ is always most intimate in proportion as the regions to which they belong are, geographically, most approximated, or in communication with each other, by intermediate lands. The Atlantic Ocean prevents species peculiar to tropical America, from extending to Africa, Europe, or Asia; and the fauna of the New World is entirely distinct from that of the old continent, except in the highest latitudes, towards the north pole. But there the land of the two continents is approximated, America being separated from Asia only by Behring's Straits, and is connected to Europe by Greenland and Iceland: on this account zoological exchanges can be more easily effected, and we find there species common to both worlds; for example, the white bear, the reindeer, the castor, the ermine, the bald eagle, &c. Chains of lofty mountains also constitute natural barriers, which arrest the dispersion of species, and prevent the admixture of faunæ, proper to neighbouring zoological regions. For instance, the opposite declivities of the Cordillera of the Andes are inhabited by species which are for the most part different; the insects of the Brazilian side, for example, are almost all distinct from those found in Peru and New Grenada.

The dispersion of marine animals living near coasts is prevented in the same manner by the geographical configuration of the earth; but here it is sometimes a continuation of a long chain of land, and sometimes a vast extent of open sea, which opposes the dissemination of species. Thus most animals of the Mediterranean are also found in the European portion of the Atlantic, but they do not extend to the seas of India, from which the Mediterranean is separated by the isthmus of Suez, nor can they traverse the ocean to gain the shores of the New World.

The physiological circumstances which tend to limit the different faunæ are more numerous; and without doubt, the first in consideration is the unequal temperature of different regions of the earth. There are species which can bear an intense cold and tropical heat equally well; man and the dog, for example; but there are others which, in this respect, are less favoured by nature, and which do not flourish, or even cannot exist, except under the influence of a determined temperature. For instance, monkeys, which thrive in tropical regions, almost always die of phthisis, when exposed to the cold and humidity of our climate; while the reindeer, formed to support the rigours of the long and severe winter of Lapland, suffers from the warmth of St. Petersburgh, and generally succumbs to the influence of a temperate climate. Hence it is that, in a great number of cases, the difference of climate is alone sufficient to arrest species in their march from high latitudes towards the equator, or from the equatorial regions towards the poles. The influence of temperature, on the animal economy, also explains why certain species remain within a chain of mountains, without being able to extend beyond it to analogous localities. We know, in fact, that temperature de creases in proportion to the elevation of the land, and consequently, animals that live at considerable heights cannot descend on to the low plains, to reach other mountains, without traversing countries in which the temperature is much higher than that of their ordinary dwelling. The llama, for example, abounds on the pastures of Peru and Chile, situated at a height of from twelve to fifteen thousand feet above the level of the sea, extending southwards to the extremity of Patagonia, but is not seen either in Brazil or Mexico, because it cannot reach those countries without descending to regions too warm for its constitution.

The nature of the vegetation, and of the previously existing fauna, in a region of the globe, also exerts an influence on its invasion by exotic species. Thus, the dispersion of the silk-worm is limited by the disappearance of the mulberry, beyond a certain degree of latitude; the cochineal cannot spread beyond the region in which the cactus grows; and the large carnaria, except those that live on fishes, cannot exist in the polar regions, where vegetable productions are too poor to nourish any considerable number of herbivorous quadrupeds.

It would be easy to multiply examples of these necessary relations between the existence of an animal species, in a particular place, and the existence of certain climatic, phytological, or zoological conditions; but our limits do not permit these details, and the considerations we have already presented, appear to be sufficient to give an idea of the manner in which nature has effected the dissemination of animal species, on different parts of the earth's surface; and, to attain the end we proposed to ourselves in commencing the subject, it only remains for us to glance at the results brought about by the different circumstances we have just mentioned, that is, the present state of the geographical distribution of animated creatures.

When we compare with each other the different regions of the globe, in respect to their zoological population, we are at first struck by the extreme inequality remarked in the number of species. In one country we find a great diversity in the form and structure of the animals composing its fauna, while in another place, there is great uniformity in this respect; and it is easy to perceive a certain relation existing between the different degrees of zoological richness, and the more or less considerable elevation of temperature. In fact, the number of species, both marine and terrestrial, augments, in general, as we descend from the poles towards the equator. The most remote lands of the polar regions offer little to the observation of the traveller but some insects, and in the glacial seas the fishes and mollusks are but little varied; in temperate climates the fauna becomes more numerous in species; but it is in tropical regions that nature has displayed the greatest prodigality in this respect, and the zoologist cannot behold without astonishment the endless diversity of animals that he there finds assembled.

It is also remarked that there is a singular coincidence between the elevation of temperature in different zoological regions, and the degree of organic perfection of the animals which inhabit them. Those animals which most nearly resemble man, and also those in the great zoological divisions which possess the most complicated organization, and the most developed faculties, live in the warmest climates, while in the polar regions we meet with creatures occupying a low rank in the zoological series. Monkeys, for example, are confined to the warm parts of the two continents; the same is true of parrots among birds, of crocodiles and tortoises among reptiles, and of land-crabs among crustaceans, all of them the most perfect animals of their respective classes.

It is also in warm countries that we find animals the most remarkable for the beauty of their colours, their size, and the strangeness of their forms.

Indeed there seems to exist a certain relation between the climate and the tendency of nature to produce this or that animal form. We observe a very great resemblance between most animals inhabiting the extreme northern and southern regions; the faunæ of the temperate regions of Europe, Asia, and North America, are very analogous in their general aspect, and in the tropical regions of the two worlds similar forms predominate. It is not identical species that we meet in distinct and nearly isothermal regions, but species more or less approximating to each other, which seem to be the representatives of one and the same type. For example, the monkeys of India and of Central Africa are represented in tropical America by other monkeys easily distinguishable from the first; the lion, tiger, and panther, of the old continent, correspond to the cougar, jaguar, and ounce, of the New World. The mountains of Europe, Asia, and North America, nourish bears of distinct species, but differing very little from each other. Seals abound especially in the neighbourhood of the polar circles; and if we seek the proofs of this tendency, not among the highest classes of the animal kingdom, but among the inferior creatures, they will be found not less evident: cray-fishes, for example, appear to be confined to the temperate regions of the globe, and are found throughout Europe, in a species common to European streams; in the South of Russia, there is a different species; in North America, there are two species, distinct from the preceding; in Chile, there is a fourth species; in the south of New Holland, a fifth; in Madagascar, a sixth; and at the Cape of Good Hope, a seventh.

A comparison of the faunæ peculiar to the different zoological regions of the globe leads to other results for which it is more difficult to account; when we examine successively the assemblage of species inhabiting Asia, Africa, and America, we remark that the fauna of the New World is characterized by inferiority, a fact which did not escape the celebrated Buffon. In a word, there are no mammals existing now in the New World as large as those of the old; it is true, we find, in America, a considerable number of monkeys, but among them there is none equal to the ourang-outang, or chimpanzee; the rodentia and edentata abound most, which, of all ordinary mammals, are the least intelligent. Finally, in America, we find opossums, animals belonging to an inferior type of ordinary mammals, which have no representative, neither in Europe, nor Asia, nor Africa. If we pass from the New World to the still newer region of Australia, we shall there see a fauna whose inferiority is still more decided, for there the class of mammals is scarcely represented by the Marsupials and Monotremata.

As to the limitation of the different zoological regions into which the globe is divided, and the composition of the faunæ proper to each, we cannot treat without exceeding our limits; but we regret this less, because, in the present state of science, these questions are far from being settled.—*Ruschenberger—Elements of Nat. Hist.*]

DISTRIBUTION OF INSECTS.[1]

Nearly one hundred and twenty thousand insects are known: some with wings, others without; some are aquatic, others are

[1] The great division of the animal kingdom of the *Articulata*, to which insects belong, consists of the following four classes; the three first breathing air by air-vessels (trachea) or air-pouches.

1. *Insects.*—Head distinct; three divisions of the body, viz. head, thorax, and abdomen; three pairs of legs, and wings in general.
2. *Myriapoda.*—Head distinct; 24 or more pairs of legs, no wings.
3. *Arachnida* or Spiders.—Four pairs of legs, head and thorax united, no antennæ or feelers, on wings.
4. *Crustacea*, as Crabs, Lobsters, &c.—Respiration by means of branchiæ or gills, and in general aquatic, five or seven pairs of legs.

Insects properly so called are divided into eight families:—

1. Coleoptera or Beetles, &c., which have four wings, two hard or wing covers called Elytra, and two soft or membranaceous used for flying, and folded under the latter; Cantharides, the Egyptian Scarabæus, and the Ladybird belong to this family, which numbers upwards of thirty thousand species.
2. Orthoptera, which have also four wings, but the wing covers are like parchment: the imperfect insect, instead of resembling a grub as in the coleoptera, only differs from the perfect one by the want of wings; such are the locust, grasshopper, cricket, &c.
3. Neuroptera, with four pairs of transparent or membranaceous wings, body soft, and in general elongated, as Dragon-flies, May-flies, Fourmileons, &c.
4. Hymenoptera: 4 membranaceous wings; instead of being provided with jaws for grinding or mastication, have a proboscis by means of

aquatic only in the first stage of their existence, and many are parasitical. Some land insects are carnivorous, others feed on vegetables; some of the carnivorous tribe live on dead, others on living animals, but they are not half so numerous as those that live on vegetables. Some change as they are developed; in their first stage they eat animal food, and vegetables when they arrive at their perfect state.

Insects maintain the balance among the species of the vegetable creation by preventing the tendency that plants have to encroach on one another. The stronger would extirpate the weaker, and the larger would destroy the smaller, were they not checked by insects which live on vegetables. On the other hand, many plants would be extirpated by insects were these not devoured by other insects and spiders.[1]

Of the 8000 or 9000 British insects, the greater part are carnivorous, and therefore keep the others within due bounds.

---

which they suck their nourishment; in most of this family the female has a sting. The humblebee, wasp, ichneumon, are examples of this family.

5. Lepidoptera, having four wings covered with minute scales or feathers, whence their name; they derive their food by a proboscis, and their first state of development is that of a caterpillar: such are butterflies, sphinxes, moths, &c.
6. Hemiptera or halfwings: in general four wings, but the upper pair are only in their four parts membranaceous or transparent; hence their name of half wing: some are entirely deprived of them, as the common bug and flea; they have a kind of beak, instead of proboscis as in the three preceding families. The cicada, the wood-bug, common bug, flea, &c., belong to the Hemiptera.
7. Rhipiptera, having also a single pair of wings folding longitudinally in a fan shape: only two small genera constitute this family, the stylops and zenos.
8. Diptera, with a single pair of wings, the mouth entirely adapted for suction, with a long retractile proboscis: the common fly is one of the most abundant species of this family.

Some entomologists have recently added two families to the class of insects, the Parasiticæ and the Thysanoura: the first live on the bodies of other animals; to one of its commonest genera belongs the human louse. [See, Entomology, in Ruschenberger's Elements of Natural History.]

[1] Perhaps one of the greatest checks on the propagation of insect life is from insects themselves, many species depositing their eggs on the larvæ of others, which in their development destroy the animal on which they have been deposited; that most destructive insect to the vine, the Pyralis Vitis, is a very remarkable instance of this, some dozen species of insects depositing their eggs on it in its incomplete state, thus keeping down the number of one of the greatest plagues in wine-producing countries. A celebrated entomologist is of opinion that of insects destroyed in Europe by other animals, indeed by all causes—one half owe their destruction to other insects.

Insects increase in kinds and in numbers from the poles to the equator: in a residence of 11 months in Melville Island, Sir Edward Parry found only 6 species, because lichens and mosses do not afford nourishment for the insect tribes, though it is probable that every other kind of plant gives food and shelter to more than one species; it is even said that 40 different insects are quartered upon the common nettle.

The increase of insects from the poles to the equator does not take place at the same rate everywhere. The polar regions and New Holland have very few specifically and individually; they are more abundant in Northern Africa, Chile, and in the plains west of the Brazils; North America has fewer species than Europe in the same latitude, and Asia has few varieties of species in proportion to its size; Caffraria, the African and Indian islands, possess nearly the same number of species; but by far the richest of all, both in species and numbers, are central and intertropical America. Beetles are an exception to the law of increase towards the equator, as they are infinitely more numerous in species in the temperate regions of the northern hemisphere than in tropical countries. The location of insects depends upon that of the plants which yield their food; and, as almost each plant is peopled with inhabitants peculiar to itself, insects are distributed over the earth in the same manner as vegetables; the groups, consequently, are often confined within narrow limits, and it is extraordinary that, notwithstanding their powers of locomotion, they often remain within a particular compass, though the plants, and all other circumstances in their immediate vicinity, appear equally favourable for their habitation.

The insects of eastern Asia and China are different from those in Europe and Africa; those in the United States differ specifically from the British, though they often approach very near; and in South America the equinoctial districts of New Grenada and Peru have distinct groups from those in Guiana; in fact, under the same parallel of latitude, countries similar in soil, climate, and all other circumstances, present the most striking differences in their insect tribes, even in those that live on animal substances.

Though insects are distributed in certain limited groups, yet most of the families have representatives in all the great regions of the globe, and some identical species are inhabitants of countries far from one another. The vanessa cardui, or "Painted Lady Butterfly," is found in all the four quarters of the globe and in Australia; and one, which never could have been conveyed by man, is native in southern Europe, the coast of Barbary, and Chile.[1] It is evident from these circumstances that not only each group, but also

[1] Some doubts have been raised whether this species is identically the same in the widely extended *habitat* described in the text.

each particular species, must have been originally created in the places they now inhabit.

Mountain-chains are a complete barrier to insects, even more so than rivers; not only lofty mountains like the Andes divide the kinds, but they are even different on the two sides of the Col de Tende in the Alps. Each soil has kinds peculiar to itself, whether dry or moist, cultivated or wild, meadow or forest. Stagnant water and marshes are generally full of them; some live in water, some run on its surface, and every water-plant affords food and shelter to many different kinds. The east wind seems to have considerable effect in bringing the insect or in developing the eggs of certain species; for example, the aphis, known as the blight in our country, lodges in myriads on plants, and shrivels up their leaves after a continued east wind. They are almost as destructive as the locust, and sometimes darken the air by their numbers. Caterpillars are also very destructive; the caterpillar of the Y moth would soon ruin the vegetation of a country were it not a prey to some other. Insects sometimes multiply suddenly to an enormous extent, and decrease as rapidly and as unaccountably.

Temperature, by its influence on vegetation, has an indirect effect on the insects that are to feed upon plants, and extremes of heat and cold have more influence on their locality than the mean annual temperature. Thus in the polar regions the mosquito tribes are more numerous and more annoying than in temperate countries, because they pass their early stages of existence in water, which shelters them, and the short but hot summer is genial to their brief span of life.

In some instances height produces the same effect in the distribution of insect life as difference of latitude. The parnassius Apollo, a butterfly native in the plains of Sweden, is also found in the Alps, the Pyrenees, and a closely allied species in the Himalaya. The parnassius smyntheus, true to the habitat of the genus, has recently been found on the Rocky Mountains of North America. Some insects require several years to arrive at their full development; they lie buried in the ground in the form of grubs: the cockchafer takes 3 years to reach its perfect state, and some American species require a much longer time. [For example, the *Cicada septendecem*, or seventeen-year locust, whose appearance about Philadelphia was first recorded in May 1715; and since that date, "punctually in the same month, every seventeenth year, now certainly for nearly one hundred and fifty years, has this extraordinary insect been known to make its visit. No causes have affected it during that period, not even so far as relates to the month in which it appears."[1] This

[1] John Cassin—in the "Proceedings of the Academy of Natural Sciences of Philadelphia," for September 1851.

insect passes seventeen years in the earth, near the roots of fruit and other trees, and lives a few weeks in the air, only long enough to provide for the continuation of the species by depositing its eggs, beneath the tender bark of the smaller branches or twigs of plants. Very soon after the animal emerges from the egg, it plunges into the earth to assume its state of torpidity for seventeen years.

This animal appeared in and around Philadelphia, in May 1851; but in other localities at different dates; in Ohio it was present in 1846, and will appear there again in 1863, and not before, and it will recur at Philadelphia in 1868.]

Insects do not attain their perfect state till the plants they are to feed upon are ready for them. Hence in cold and temperate climates their appearance is simultaneous with vegetation; and as the rainy and dry seasons within the tropics correspond to our winter and summer, insects appear there after the rains and vanish in the heat; the rains, if too violent, destroy them; and in countries where that occurs there are two periods in the year in which they are most abundant—one before and one after the rains. It is also observed in Europe that insects decrease in the heat of summer and become more numerous in autumn: the heat is thought to throw some into a state of torpor, but the greater number perish.

It is not known that any insect depends entirely upon only one species of plant for its existence, or whether it may not have recourse to congeners should its habitual plant perish. When particular species of plants of the same family occur in places widely apart, insects of the same genus will be found on them, so that the existence of the plant may often be inferred from that of the insect, and in several instances the converse.

When a plant is taken from one country to another in which it has no congeners, it is not attacked by the insects of the country: thus our cabbages and carrots in Cayenne are not injured by the insects of that country, and the tulip-tree and other magnolias are not molested by our insects; but if a plant has congeners in its new country, the insect inhabitants will soon find their way to the stranger.

The common fly is one of the most universal of insects, yet it was unknown in some of the South Sea islands till it was carried there from Europe by ships, where it has now become a plague.

Mosquitoes and culices [gnats] are spread over the world more generally than any other tribe: they are the torment of men and animals from the poles to the equator, by night and by day; the species are numerous and their location partial. In the arctic regions the Culex pipiens, which passes two-thirds of its existence in water, swarms during the summer in myriads: the lake Myvatr, in Iceland, has its name from the legions of these tormentors that cover its surface. They are less numerous in central Europe, though one species of mosquito, the simulium columbaschense, which is very small, appears in such clouds in parts of Hungary, especially the Bannat of Temeswar, that it is not possible to breathe without swallowing many: even cattle and children have died from them. In Lapland there is a plague of the same kind. Of all places on earth the Orinoco and other great rivers of tropical America are the most obnoxious to this plague. The account given by Baron Humboldt is really fearful: at no season of the year, at no hour of the day or night, can rest be found; whole districts in the upper Orinoco are deserted on account of these insects. Different species follow one another with such precision, that the time of day or night may be known accurately from their humming noise, and from the different sensations of pain which the different poisons produce. The only respite is the interval of a few minutes between the departure of one gang and the arrival of their successors, for the species do not mix. On some parts of the Orinoco, the air is one dense cloud of poisonous insects to the height of 20 feet. It is singular that they do not infest rivers that have dark water, and each clear stream is peopled with its own kinds; though ravenous for blood, they can live without it, as they are found where no animals exist.

In Brazil the quantity of insects is so great in the woods, that their noise is often heard in a ship anchored some distance from the shore.

Various genera of btterflies and moths are very limited in their habitations, others are dispersed over the world, but the species are almost always different. Bees and wasps are equally universal, yet each country has its own. The common honey-bee is the European insect most directly useful to man; it was introduced into North America not many years ago, and is now spread over the new continent: and is naturalized over Australia and New Zealand. European bees, of which there are many species, generally have stings; the Australian bee, like a black fly, is without one; and in Brazil there are 30 species of stingless bees.

Fire-flies are mostly tropical, yet there are four species in Europe; in South America there are three species, and so brilliant that their pale green light is seen at the distance of 200 paces: a Scolopendra, or centipede in Asia, is as luminous as the glow-worm, and one in France is so occasionally.

The silk-worm came originally from China, and the cochineal insect is a native of tropical America: there are many species of it in other countries. The coccus lacca is Indian, the coccus ilicis lives in Southern Europe, and there is one in Poland, but neither of these have been cultivated.

Scorpions under various forms are in all warm climates; 2 or 3 species are peculiar to Europe, but they are small in comparison with those in tropical countries: one in Brazil is six inches long. As in mosquitoes, the poison of the same species is more active in some situations than in others. At Cumana the sting of the scorpion is little feared, while that of the same species in Carthagena causes loss of speech for many days.

Ants are universally distributed, but of different kinds. Near great rivers they build their nests above the line of the annual inundations. The insects called white ants, belonging to a different genus and family, are so destructive in South America, that Baron Humboldt says there is not a manuscript in that country a hundred years old. [This assertion is probably inaccurate: the public library at Lima, Peru, contains more than one MS. of greater age.]

There are upwards of 1200 species of spiders and their allies known; each country has its own, varying in size, colour, and habits, from the huge bird-catching spider of South America to the almost invisible European gossamer floating in the air on its silvery thread. Many of this ferocious family are aquatic; and spiders, with some other insects, are said to be the first inhabitants of new islands.

The migration of insects is one of the most curious circumstances relating to them: they sometimes appear in great flights in places where they never were seen before, and they continue their course with perseverance which nothing can check. This has been observed in the migration of crawling insects: caterpillars have attempted to cross a stream. Countries near deserts are most exposed to the invasion of locusts, which deposit their eggs in the sand, and when the young are hatched by the sun's heat they emerge from the ground without wings; but as soon as they attain maturity, they obey the impulse of the first wind, and fly, under the guidance of a leader, in a mass, whose front keeps a straight line, so dense that it forms a cloud in the air, and the sound of their wings is like the murmur of the distant sea. They take immense flights, crossing the Mozambique Channel from Africa to Madagascar, which is 120 miles broad: they come from Barbary to Italy, and a few have been seen in Scotland. Even the wandering tribes of locusts differ in species in different deserts, following the universal law of organized nature. Insects not habitually migratory, sometimes migrate in great flocks. In 1847, lady-birds or coccinellæ and the bean aphis arrived in immense numbers at Ramsgate and Margate from the continent in fine calm weather, and a mass of the Vanessa cardui flew over a district in a column from 10 to 15 yards wide, for 2 hours successively. Why these butterflies should simultaneously take wing in a flock is unaccountable, for had it been for want of food they would probably have separated in quest of it. In 1847 the cabbage butterfly came in clouds from the coast of France to England. Dragon-flies migrate in a similar manner. Professor Ehrenberg has discovered a new world of creatures in the Infusoria, so minute that they are invisible to the naked eye. He found them in fog, rain, and snow, in the ocean, in stagnant water, in animal and vegetable juices, in volcanic ashes and pumice, in opal, in the minute dust that sometimes falls on the ocean; and he detected 18 species 20 feet below the surface of the ground in peat-bog, which was full of microscopic live animals: they exist in ice, and are not killed by boiling water. While inquiring into the causes of the cholera which prevailed at Berlin in 1848, M. Ehrenberg discovered 400 species of living microscopic animalcules in different strata of the atmosphere, so that the air is analogous in the distribution of its inhabitants to the ocean, which has marine animals peculiar to different depths. This lowest order of animal life is much more abundant than any other, and new species are found every day. Magnified, some of them seem to consist of a transparent vesicle, and some have a tail: they move with great rapidity, and show a certain instinct by avoiding obstacles in their course: others have silicious shells. Language, and even imagination, fails in the attempt to describe the inconceivable myriads of these invisible inhabitants of the ocean, the air, and the earth: they no doubt become the prey of larger creatures, and perhaps carnivorous insects may have recourse to them when other prey is wanting.

CHAPTER XXIX.

Distribution of Marine Animals in general—Fishes—the Marine Mammalia —Phocæ, Dolphins, and Whales.

Before Sir James Ross's voyage to the antarctic regions, the profound and dark abysses of the ocean were supposed to be entirely destitute of animal life; now it may be presumed that no part of it is uninhabited, since during that expedition live creatures were fished up from a depth of 6000 feet. But as most of the larger fish usually frequent shallow water near the coasts, deep seas must form barriers as impassable to the greater number of them as mountains do to land animals. The polar, the equatorial ocean, and the inland seas have each their own particular inhabitants; almost all the species and some of the genera of the marine creation are different in the two hemispheres, and even in each particular sea; and under similar circumstances the species are for the most part representative, though not the same. Identity of species, however, does occur, even at the two extremities of the globe, for living animals were brought up from the profound depths of the Antarctic Ocean which Sir James Ross recognised to be the very same species which he had often met with in the Arctic seas. "The only way they could have got from the oné pole to the other must have been through the tropics; but the temperature of the sea in these regions is such that they could not exist in it unless at a depth of nearly 2000 fathoms. At that depth they might pass from the Arctic to the Antarctic Ocean without a variation of 5 degrees of temperature; whilst any land animal, at the most favourable season, must experience a difference of 50 degrees, and, if in winter, no less than 150 degrees of Fahrenheit's thermometer;"—a strong presumption that marine creatures can exist at the depth and under the enormous pressure of 12,000 feet of water. The stratum of constant temperature in the ocean may indeed afford the means of migration from pole to pole to those which live in shallower water, as they would only have to descend to a depth of 7200 feet at the equator. The great currents, no doubt, offer paths for fish without any sudden change of temperature: the inhabitants of the Antarctic Sea may come to the coasts of Chile and Peru by the cold stream that flows along them from the South Polar Ocean, and on the contrary, tropical fish may travel by the Gulf-stream to the middle and high latitudes in the Atlantic, but few will leave either one or other to inhabit the adjacent seas, on account of the difference of heat. Nevertheless, quantities of medusæ or sea-nettles are brought by the Gulf-stream to feed the whales at the Azores, though the whales themselves seldom enter the stream, on account of its warmth.

The form and nature of the coasts have great influence on the distribution of fishes; when they are uniformly of the same geological structure, so as to afford the same food and shelter, the fish are similar. Their distribution is also determined by climate, the depth of the sea, the nature of the bottom, and the influx of fresh water.

The ocean, the most varied and most wonderful part of the creation, absolutely teems with life: "things innumerable, both great and small, are there." The forms are not to be numbered even of those within our reach; yet, numerous as they are, few have been found exempt from the laws of geographical distribution.

The discoloured portions of the ocean generally owe the tints they assume to myriads of insects. In the Arctic seas, where the water is pure transparent ultramarine colour, parts of 20 or 30 square miles, 1500 feet deep, are green and turbid from the quantity of minute animalcules. Captain Scoresby calculated that it would require 80,000 persons, working unceasingly from the creation of man to the present day, to count the number of insects contained in 2 miles of the green water. What, then, must be the amount of animal life in the polar regions, where one-fourth part of the Greenland Sea, for 10 degrees of latitude, consists of that water! These animalcules are of the medusa tribe, or of others of the family of zoophytes. Some medusæ are very large, floating like a mass of jelly; and although apparently carried at random by the waves, each species has its definite location, and its peculiar organs of locomotion. One species comes in spring from the Greenland seas to the coast of Holland; and Baron Humboldt met with an immense shoal of them in the Atlantic, migrating at a rapid rate.

Dr. Pœppig mentions a stratum of red water near Cape Pilares, 24 miles long and 7 broad, which seen from the mast-head appeared dark-red, but on proceeding it became a brilliant purple, and the wake of the vessel was rose-colour. The water was perfectly transparent, but small red dots could be discerned moving in spiral lines. The vermilion sea off California is no doubt owing to a similar cause, as Mr. Darwin found red and chocolate-coloured water which had been before observed by Ulloa on the coast of Chile over spaces of several square miles full of microscopic animalcules, darting about in every direction, and sometimes exploding. Infusoria are not confined to fresh water; the bottom of the sea swarms with them. Silicious-coated infusoria are found in the mud of the coral islands under the equator; and 68 species were discovered in the mud in Erebus Bay, near the Antarctic pole. These minute forms of organised life, invisible to the naked eye, are intensely and extensively developed in both of the polar oceans, and serve for food to the higher orders of fish in latitudes beyond the limits of the larger vegetation, though they themselves probably live on the microscopic plant already mentioned, which abounds in all seas. Some are peculiar to each of the polar seas, and a few are distributed extensively throughout the ocean.

The enormous prodigality of animal life supplies the place of vegetation, so scanty in the ocean in comparison with that which clothes the land, and which probably would be insufficient for the supply of the marine creation, were the deficiency not made up by the superabundant land vegetation and insects carried to the sea by rivers. The fish that live on sea-weed must bear a smaller proportion to those that are predaceous than the herbivorous land animals do to the carnivorous. Fish certainly are most voracious; none are without their enemies; they prey and are preyed upon; and there are two which devour even the live coral, hard as its coating is; nor does the coat of mail of shell-fish protect them. Whatever the proportion may be which predatory fish bear to herbivorous, the quantity of both must be enormous, for, besides the infusoria, the great forests of fuci and sea-weed are everywhere a mass of infinitely varied forms of being, either parasitical, feeding on them, seeking shelter among them, or in pursuit of others.

The observations of Professor E. Forbes in the Egean Sea show that depth has great influence in the geographical distribution of marine animals. From the surface to the depth of 230 fathoms there are eight distinct regions in that sea, each of which has its own vegetation and inhabitants. The number of shell-fish or Mollusca and other marine animals is greater specifically and individually between the surface and the depth of 2 fathoms than in all the regions below taken together, and both decrease downwards to the depth of 105 fathoms; between which and the depth of 230 only eight shells were found; and animal life ceases in that part of the Mediterranean at 300 fathoms. The changes in the different zones are not abrupt; some of the creatures of an under region always appear before those of the region above vanish; and although there are a few species the same in some of the eight zones, only two are common to all. Those near the surface have forms and colours more resembling those of the inhabitants of southern latitudes, while those lower down are more analogous to the animals of northern seas; so that in the sea depth corresponds with latitude, as height does on land. Moreover, the extent of the geographical distribution of any species is proportional to the depth at which it lives; consequently, those living near the surface are less widely dispersed than those inhabiting deep water. Professor Forbes also discovered several shells living in the Mediterranean that have hitherto only been known as fossils of the tertiary strata; and also that the species least abundant as fossils are most numerous alive. These important observations, it is true, were confined to the Mediterranean; but analogous results have been obtained in the Bay of Biscay and in the British seas. There are four zones of depth in our seas, each of which has its own inhabitants, consisting of shell-fish, crustaceæ, corallines, and other marine creatures. The first zone lies between high and low water marks, consequently it is shallow in some places and 30 feet deep in others. In all parts of the northern hemisphere it presents the same phenomena; but the animals vary with the nature of the coast, according as it is of rock, gravel, sand, or mud. In the British seas the animals of this littoral or coast zone are distributed in three groups that differ decidedly from one another, though many are common to all. One occupies the seas on the southern shores of our islands and both channels; a middle group has its centre in the Irish seas; and the third is confined to the Scottish seas, and the adjacent coasts of England and Ireland. The second zone extends from the low-water mark to a depth below it of from 7 to 15 fathoms, and is crowded with animals living on and among the sea-weeds, as radiated animals, shell-fish, and many zoophytes. In the third zone, which is below that of vegetable life, marine animals are more numerous and of greater variety than in any other. It is particularly distinguished by arborescent creatures, that seem to take the place of plants, carnivorous mollusca, together with large and peculiar radiata. It ranges from the depth of 15 to 50 fathoms. The last zone is the region of stronger corals, peculiar mollusca, and of others that only inhabit deep water. This zone extends to the depth of 100 fathoms or more.

Except in the Antarctic seas, the superior zone of Mollusca is the only one of which anything is known in the great oceans, which have numerous special provinces. Many, like the Harp, are tropical; others, as the Nautilus and the pearl-oyster, are nearly so; the latter (*Meleagrina Margaritifera*) abounds throughout the Persian Gulf and on the coasts of Borneo and Ceylon, and is supposed to produce the finest pearls. There are others in the Caribbean Sea, and in the Pacific, and especially in the Bay of Panamá, but whether the species are the same is not well ascertained. Some shells are exceedingly limited in their distribution, as the Haliotis gigantea, which is peculiar to the seas around Van Diemen's Land.

According to Sir Charles Lyell, nearly all the species of molluscous animals in the seas of the two temperate zones are distinct, yet the united species when compared with each other have a strong analogy of type; both differ widely from those in the tropical and arctic oceans; and, under the same latitude, species vary with the longitude. The east and west coasts of tropical America have only one shell-fish in common; and those of both differ from the Mollusca in the islands of the Pacific and the Galapagos Archipelago, which form a distinct region. Notwithstanding the many definite marine provinces, the same species are occasionally found in regions widely separated. A few of the shell-fish of the Galapagos Archipelago are analogous with those of the Philippine islands, though so far apart. The east coast of America, which is poor in mollusca, has a number, however, in common with the coasts of Europe.

The Cypræa moneta lives in the Mediterranean, the seas of South Africa, the Mauritius, the East Indies, China, and the South Pacific even as far as Tahiti; and the Janthina fragilis, the animal of which is of a beautiful violet-colour, floats on the surface in every tropical and temperate sea. Mollusca have a greater power of locomotion than is generally believed. Some migrate in their state of larva, being furnished with lobes which enable them to swim freely. The larva of the scalop is capable of migrating to distant regions; the Argonauta spreads its sail and swims along the surface.

The numerous species of Zoophytes which construct the extensive coral banks and atolls are chiefly confined to the tropical seas of Polynesia, the East and West Indies: the family is represented by a very few species in our seas, and in the Mediterranean they are smaller and different generally from those in the torrid zone.

Fishes[1] properly so called, advance in the water by means of their

[1] The skeletons of fishes are composed either of bone or cartilage, hence Cuvier's division of the finny tribe into osseous or bony, and cartilaginous

flexible bodies; the fins and tail serve chiefly to balance them and direct their motion. These larger and more active inhabitants of the waters obey the same laws with the rest of the creation, though the provinces are in some instances very extensive. Dr. Richardson observes that there is one vast province in the Pacific, extending 42 degrees on each side of the equator, between the meridians including Australia, New Zealand, the Malay Archipelago, China, and Japan, in which the genera are the same; but at its extremities the Arctic and Antarctic genera are mingled with the tropical forms. Many species however which abound in the Indian Ocean range as far north as Japan, from which circumstance it is presumed that a current sets in that direction. The middle portion of this province is vastly extended in longitude, for very many species of the Red Sea, the eastern coast of Africa, and the Mauritius range to the Indian and China Seas, to those of northern Australia and all Polynesia; so in this immense belt, which embraces three-fourths of the circumference of the globe and 60 degrees of latitude, the fish are very nearly alike, the continuous chains of islands in the Pacific being favourable to their dispersion. Few of the Pacific fish enter the Atlantic;[1] and from the depth and want of islands in it the great bulk of species is different on its two sides. North of the 44th parallel however the number common to both shores increases. The salmon of America is identical with that of the British isles, the coasts of Norway and Sweden; the cod-fish is identical, as well as several others of the same family. The Cottus or bullhead genus are also the same on both sides of the North Atlantic, and they increase in numbers and variety on approaching the Arctic seas. The same occurs in the northern Pacific, though the generic forms differ

fishes. The fins are formed of spines or rays of bone united more or less by a thin web or membrane; some are hard and others soft; the bony fishes are subdivided into hard finned or acanthopterygians, as the perch, sea-bream, mullet, mackerel, &c., and the soft-finned or malacopterygians, as the salmon, herring, pike, carp, cod, flat-fish, eels, &c. The cartilaginous fishes, or Chondropterygians, include the families of the sturgeon, shark, ray, or skate, lamprey, sun-fish, diodons, &c.

M. Agassiz has more recently divided fishes into four great orders according to the nature of their scales; the first includes sharks, rays, &c., which are covered with solid plates of enamel; the second sturgeon and siluri, which are partially covered with the same; third perch, &c., covered with toothed scales; fourth, salmon, mullet, &c., covered with simple thin plates.

[1] The Notochanthus and Macrourus are deep-water fish in the Arctic regions; they also inhabit the seas of New Zealand. The Pacific fish that enter the Atlantic are some of the mackerel tribe, sharks, and lophobranches. The genera most prevalent in the southern hemisphere are the Notothemia, Borichthys, and Harpagifer. The same species of these genera are found in the seas of the Falkland Islands, Cape Horn, the Auckland Islands, and Kerguelen's Land.—Dr. Richardson.

from those in the Atlantic. From the near approach of the American and Asiatic coasts at Behring's Straits, the fish on both sides are nearly alike, down to the Sea of Okhotsk on one side and to Admiralty Inlet on the other. The sea of Japan and the neighbouring coasts of China are frequented by fishes having northern forms, which are there mingled with many species common to the temperate and warm parts of the ocean. Species of the genus Gadus or Cod reappear in the southern seas very like those of the northern; and two very remarkable Greenland genera, which inhabit deeper water and are seldom taken except when thrown up by a storm, have been discovered on the coasts of New Zealand and South Australia, where the fish differ but little from those in the seas of Van Diemen's Land. Several genera are peculiar to the southern hemisphere, and range throughout the whole circle of the high latitudes. The sharks of the China seas are for the most part identical with those of Australia: the cartilaginous fish to which they belong have a much wider range than those which have been under consideration.

The British islands lie between two great provinces of fishes — one to the south, the other to the north — from each of which we have occasionally visitors. The centre of the first is on the coasts of the Spanish peninsula, extending into the Mediterranean; that on the north has its centre about the Shetland Islands: but the group peculiarly British, and found nowhere else, has its focus in the Irish Sea. It is, however, mixed with fish from the seas bounding the western shores of central Europe, which forms a distinct group.

The Prince of Canino has shown that there are 853 European species of fishes, of which 210 live in fresh water, 643 are marine, and 60 of these go up rivers to spawn. 444 of the marine fish inhabit the Mediterranean, 216 are found off the British coasts, and 171 are peculiar to the Scandinavian seas; so that the Mediterranean is richest in variety of species. In it there are peculiar sharks, sword-fish, dolphins, anchovies, and six species of scomber or mackerel, one of the largest of which, indeed of all edible fish, is the tunny, for which fisheries are established on the southern coasts of France, in Sardinia, Elba, the Straits of Messina, and the Adriatic. Four of the species are found nowhere else but in the Mediterranean. Rays of numerous species are particularly characteristic of the Mediterranean, especially the Torpedos, which have the power of giving an electric shock, and even the electric spark. The Mediterranean has two or three American species, 41 in common with Madeira, one in common with the Red Sea, and a very few seem to be Indian. Some of these fish have probably entered the Mediterranean before it was separated from the Red Sea by the Isthmus of Suez; but geological changes have had very great influence on the distribution of fishes everywhere. Taking salt and fresh water fishes together, there are 100 species common to Italy and Britain; and although the communication with the Black Sea is so direct, there are only 27 common to it and the Mediterranean; but the Black Sea forms a district by itself, having its own peculiar Ichthyology; and the fishes of the Caspian Sea differ entirely from those in every other part of the globe. The island of Madeira, solitary amid a great expanse of ocean, has many species. They amount in number to half of those in Britain; and nearly as many are common to Britain and Madeira as to that island and the Mediterranean; so that many of our fish have a wide range in the Atlantic; and in return we have occasional visits from the tunny, torpedo, pilot-fishes, and various sharks. The Mediterranean certainly surpasses the British and Scandinavian seas in variety, though it is far inferior to either in the quantity or quality of useful fish. Cod, turbot, brill, haddock, ling, herring, and many more, are better in northern seas than elsewhere, and several exist there only.

The greater number of fish used by man as food frequent shoal water. The coast of Holland, our own shores, and other parts of the North Sea where the water is shallow, teem with a never-ending supply of excellent fish of many kinds.

Vast numbers are gregarious and migratory. Cod arrive in the shallow parts of the coast of Norway in February, in shoals many yards deep, and so closely crowded together that the sounding-lead can hardly pass between them: 16,000,000 have been caught in one place in a few weeks. In April they return to the ocean. Herrings come in astonishing quantities in winter.

The principal cod fisheries are on the banks of Newfoundland and the Dogger-bank. They, like all animals, frequent the places to which they have been accustomed. Herrings come to the same places for a series of years, and then desert them, perhaps from having exhausted the food. Pilchards, sardinias, mackerel, and many others may be mentioned among the gregarious and migratory fish.

Sharks like deep water. They are found of different species in all tropical and temperate seas; and, although always dangerous, they are more ferocious in some places than in others, even in the same species.

Most lakes have fish of peculiar species, as the lake Baikal. The fishes of the great interalpine Lake of Titicaca amount to 7 or 8 species, and belong to genera only found in the higher regions of the Andes. In the North American lakes there is a thick-scaled fish, bearing some analogy to those of the early geological eras: there are five species of perch in the North American waters, one of which is the same as that of Europe; and the Gillaroo trout, which is remarkable in having a highly muscular stomach or gizzard, is found in Ireland only. Pike and salmon are the only species of fresh-water fish common to Europe and North America; the pike is however unknown west of the Rocky Mountains. The common salmon does not exist beyond 45° of N. lat. on the eastern coast of America, and it is probably confined within similar limits on the eastern coast of Asia. It is said to be an inhabitant of all the northern parts of the Old World from the entrance of the Bay of Biscay to North Cape, and along the arctic shores of Asia and Kamtchatka to the Sea of Okhotsk, including the Baltic, White Sea, Gulf of Kara, and other inlets. Other kinds of the Salmon tribe are plentiful in the estuaries of Kamtchatka and on the opposite coast of America down to Oregon, but apparently they do not extend to China. Salmon go up rivers to spawn, and make extraordinary leaps over impediments of rocks or walls, in order to reach the suitable places for depositing their eggs. Forty-four fish inhabit the British lakes and rivers, and 50 those of Scandinavia, of the very best kinds. The fresh-water fish of northern climates are more esteemed in a culinary point of view than those of the southern.

Each tropical river has its own species of fish. The fresh-water fish of China agree with those of India in generic forms, but not in species;[1] and those of the Cape of Good Hope and South America differ from those in India and China. Sea-fish, in immense quantities, frequent the estuaries of rivers everywhere. The mouth of the Mississippi is full of them; and the quantity at the mouth of the Don, in the Sea of Azof, is prodigious.

There are some singular analogies between the inhabitants of the sea and those of the land. Many of the Medusæ, two corallines, the Physalia, or Portuguese man-of-war sailors, sting like a nettle when touched. A cuttle-fish, at the Cape de Verde islands, changes colour like a chameleon, assuming the tint of the ground under it. Herrings, pilchards, and many other fish, as well as sea insects, are luminous. The medusa tribe, the species of which are numerous, have also the faculty of emitting light in a high degree. In warm climates, especially, the sea seems to be on fire, and the wake of a ship is like a vivid flame. Probably fish that go below the depths to which the light of the sun penetrates are endowed with this faculty; and shoals of luminous insects have been seen at a considerable depth below the surface of the water. The glow-worm, certain beetles, and fire-flies, shine with the same pale-green light. The fishes that live at great depths resemble owls, and other nocturnal birds, in having large eyes. The tails of some of the skate family, especially one found in the rivers of Guiana, [as well as

[1] The Chinese fresh-water fish are cyprinidæ, ophicephali, and siluridæ —genera which agree closely with those in India, though the species are different.

those found on the coast of New Jersey,] are armed with two or three serrated prongs, three inches long, with which they defend themselves by lashing their tails, inflicting wounds which, in hot climates, are often followed by violent inflammation and pain, and have hence been improperly considered as venomous. But among the terrestrial animals there is nothing analogous to the property of the gymnotus electricus of the South American lakes, or of the silurus electricus of the African rivers, and the different species of the torpedo of the Mediterranean, which possess the faculty of giving an electric shock by means of a very beautiful organic Voltaic apparatus with which they are provided.

The marine mammalia, which, as their name indicates, suckle their young, form two distinct families—the Phocæ or seals, and the Cetacea or whales and porpoises: whilst fish breathe by means of gills, which separate the air dissolved in the water, the marine mammalia possess lungs and breathe like the terrestrial quadrupeds; they are obliged to come to the surface from time to time to inhale the air.

The first family consists of the Seal tribe, and is most abundant in the polar regions of both hemispheres; they are carnivorous, live exclusively on fish, and are seldom found at a great distance from the land or ice islands. To this division belong the common seal and the Walrus in our northern hemisphere; whilst the genus Otaria or sea-lion, and its congeners of many species, and which attain in general a greater size, are only found in high southern latitudes.

The family of Cetacea consists of three great genera: the Manatus and Dugong, which live in or near the estuaries of tropical rivers, are herbivorous; the Dolphins or Porpoises, which are carnivorous, provided with long jaws and numerous teeth, and are found in almost every latitude and in every sea; and the whales, which, unprovided with cutting teeth, are furnished with whalebone inserted in the upper jaw, the extreme filaments of which are destined as a kind of net to catch the minute marine animals which form their food. The marine Cetacea breathe by an opening in the centre of the head, called, in whales, the blower, corresponding to the nose of terrestrial quadrupeds, and which also serves to expel the water taken into the mouth with the food, in the form of jets, which in the whale tribe varies in height and form according to the species.

The favourite haunts of the seal tribe are the polar oceans and desert islands in high latitudes, where they bask in hundreds on the sunny shores during the brief summer of these inhospitable regions, and become an easy prey to man, who has nearly extirpated the race in many places. A million are annually killed in the South Atlantic alone. Seven species are natives of the Arctic, Atlantic, and Polar Oceans; the Greenland seal, the bearded or great seal, and the phoca leporina are found also in the high latitudes of the Northern Pacific.

The phoca oceanica is only met with in the White Sea and the sea at Nova Zembla, and the phoca sagura on the coast of Newfoundland. The sea-lion is to be found on all the coasts of the South Pacific, but their principal gathering is on the island of St. George, one of the Pruibiloff group, in lat. 56° N. The common seal is 6 or 7 feet long, with a face like that of a dog, and a large intelligent eye. It is easily tamed, and in the Orkney islands it is so much domesticated that it follows its master, and helps him to catch fish. This seal migrates in herds from Greenland twice in the year, and returns again to its former haunts; they probably come to the coasts of Europe and the British islands at the time of their migrations; it may be considered, however, a constant inhabitant of our northern shores. Some of the seal tribe have a very wide range, as the fur species, Arctocephalus ursinus, of the Falkland islands, which at one time frequented the southern coasts of Australia in great numbers, but they and three other species have now become scarce from the indiscriminate slaughter of old and young. Sir James Ross found some of the islands in the Antarctic seas overrun with the sea-elephant, phoca elephantina. The Walrus, a grim-looking creature, with tusks 2 feet long, bent downwards, and its nose covered with strong transparent bristles, has a body like that of a seal, 20 feet long, with a coat of short grey or yellow hair. It sleeps on the floating ice, feeds on sea-weed and marine animals, and never leaves the Arctic seas.

The Manati and Dugong form the first group of the family of the Cetacea; they are exclusively herbivorous, and live near the mouths of the great tropical rivers. The Lamantin, a species of Manatus, is found in the Amazon and Orinoco, and another in some rivers of Western Africa. In the former, where it is known as the sea-cow, its body is round like a wine-bag, and sometimes attains a length of 12 or 15 feet; it browses in herds on the herbage at the bottom of streams; and when attacked, the mother defends her young at the sacrifice of her own life. The Dugong is an inhabitant of the eastern archipelago, and of the shallow parts of the Indian Ocean, where it also feeds on sea-weed; it is more a marine animal than the Lamantin, as it is scarcely ever seen in fresh water. The dugong is so harmless and tame as to allow itself to be handled. When it suckles its young it sits upright, which has given rise to the fable of the Mermaid. This animal, like the Lamantin, will sacrifice its life for its young, and is hence, among the Malays, held as the type of maternal affection. The animal called the Manatus septentrionalis, which frequents the Arctic seas, is very little known, and probably not one of the herbivorous Cetacea.

The second group or genus of the Cetacea consists of those of predatory habits: they live on fish, and consequently have sharp and numerous teeth, such as Porpoises, Dolphins, and spermaceti whales or Cachalots; they have, like all the animals of this family, spouting nostrils in the upper part of the head.[1] The common porpoise is seen spouting and tumbling on the surface of all the seas of Europe; shoals of them go in pursuit of herrings and mackerel, and even swim up the rivers in chase of salmon. They have more the form of fish than the seal tribe, and have a dorsal fin. The several species of Dolphins, so remarkable for their voracity and for the swiftness of their motions, owing to the symmetry of their form and the width of their horizontally-placed tail, are seen in almost every latitude. The white dolphin, eaten by the Icelanders, is 18 feet long, and migrates from the Atlantic to Greenland in the end of November. The Grampus, Delphinus Orca, nearly allied to the killer of the South Sea whalers, is fierce and voracious, often 20 feet long, roams in numerous shoals, preying upon the larger fish, and even attacking the whale. The Grind or black dolphin has been known to run ashore in hundreds in the bays of Feroe, Orkney, and Shetland. This seems to be the same or nearly allied to the black fish which was met with in vast numbers by Sir James Ross in the Antarctic seas: they had so little fear, that they darted below the ship on one side and came up at the other. The delphinus peronii, or white porpoise of the southern whalers, is a rare and elegant species of dolphin which chiefly inhabits the high southern latitudes, but has been seen near the equator in the Pacific. [It is not unfrequent near Macao in the China Sea.] They are about six feet long, the hinder part of the head, the back, and the flukes of their tail are black, and all the rest of the purest white. The Narwhal or sea-unicorn (Monodon monoceros) has no teeth, but a tusk of fine ivory wreathed with spiral grooves extending 8 or 10 feet straight from the head; in general there is only one tusk, but there are always the rudiments of another, and occasionally both grow to an equal length. The old narwhals are white with blackish spots, the young are dark-coloured. This singular creature, which is about 16 feet long without the tusks, swims with great swiftness. Dr. Scoresby has seen 15 or 20 at a time playing round his ship in the Arctic seas, and crossing their long tusks in all directions as if they were fencing; they are found in all parts of the Northern Ocean.

The spermaceti whale, the Cachalot or physeter macrocephalus, belonging to the family of the predaceous spouters, is one of the most formidable inhabitants of the deep. Its average size is 60 feet long and 40 feet in circumference; its head, equal to a third of its length, is extremely thick and blunt in front, with a throat wide enough to swallow a man. The proportionally small swimming

[1] The carnivorous Cetacea, with two remarkable exceptions, inhabit the ocean—the Delphinus Inca, of the Upper Amazon and its tributaries; and the D. Gangeticus, of the Ganges.

paws or pectoral fins are at a short distance behind the head, and the tail, which is a horizontal triangle 6 or 7 feet long and 19 feet wide, with a notch between the flukes, is the chief organ of progressive motion and defence. It has a hump of fat on its back, is of a dark colour, but with a very smooth clean skin. These sperm whales have two nasal apertures on the top of their head, through which they throw at each expiration a succession of jets like smoke, at intervals of 15 or 20 minutes, after which they toss their tails high in the air and go head foremost to vast depths, where they remain for a considerable time, and then return again to the surface to breathe. The jet or spout is from 6 to 8 feet high, and consists of water mixed with air, expired from the lungs. [It is a jet of vapour, and not of water, as erroneously stated.] This whale has sperm-oil and spermaceti in every part of its body, but the latter is chiefly in a vast reservoir in its head, which makes it very buoyant: ambergris is sometimes found in the inside of the body, supposed to be produced by disease. These huge monsters, occasionally 75 feet long, go in great herds, or schools, as the whalers call them, of 500 or 600. Females with their young, and two or three old males, generally form one company, and the young males another, while the old males feed and hunt singly. The sperm whales swim gracefully and equably, with the upper part of the head above the water; but when a troop of them play on the surface, some of these uncouth and gigantic creatures leap with the agility of a salmon several feet into the air, and fall down again heavily with a tremendous crash and noise like a cannon, driving the water up in lofty columns capped with foam. The fishery of the sperm whale is attended with danger; not only the wounded animal, but its companions who come to its aid, sometimes fight desperately, killing the whalers and tossing them into the air with a sweep of their tremendous tails, or biting a boat in two. In 1820, the American whaler Essex was wrecked in the Pacific by a sperm whale; it first gave the ship so violent a blow that it broke off part of the keel, then, retreating to a distance, it rushed furiously, and with its enormous head beat in a portion of the planks, and the people had just time to save themselves in the boats when the vessel filled. They often lie and listen when suspicious of mischief. No part of the aqueous globe, except the Arctic seas, is free from their visits; they have been seen in the Mediterranean and the Adriatic, in the British Channel, and even in the estuary of the Thames, but their chief resort is the deepest parts of the warmer seas within or near the tropics, and in the Antarctic Ocean, where they feed on floating molluscæ, such as sepiæ or cuttle-fish, &c.

The third and last genus of the Cetacea are whalebone whales, such as the Greenland whale and Rorquals. Instead of teeth, the upper jaws of these animals are furnished with plates and filaments of whalebone, which are moveable, and are adapted to retain, as in a net, the medusæ and other small marine animals that are the food of these colossal inhabitants of the deep. The common Greenland species, Balæna Mysticetus, was formerly much more numerous, but it is now chiefly confined to the very high northern latitudes; however, should it be the same with the whale found in such multitudes in shallow water on the coasts of the Pacific and in the Antarctic Ocean by Sir James Ross, it must have a very wide range, but it is more probable that each polar region has its own species. The Greenland whale is from 65 to 70 feet long, but they are so much persecuted that they probably never live long enough to attain their full growth. The head is very large, but the opening of the throat is so narrow that it can only swallow small animals. It has two spouts or nostrils, through which it throws jets like puffs of smoke yards high. It only remains two or three minutes on the surface to breathe, and then goes under water for five or six. The back and tail are velvet-black, shaded in some places into grey, the rest is white; some are piebald. The capture of this whale is often attended with much cruelty, from their affection for their young; indeed the custom of killing the calf in order to capture the mother has ruined the fishery in several places, especially in the New Zealand and Australian seas.

Rorquals are also whale-bone whales, differing from the common whale in the more elongated form of the head. One species is from 80 to 100 feet long, the largest of marine animals. This whale travels to lower latitudes in pursuit of herrings and other fish. It had been caught on the coast of Norway as early as the year 890, and probably long before. The first northern navigators were not attracted by the whale as an object of commerce, but stumbled upon it in their search for a north-west passage to the Pacific. The hump-backed whale, Balæna gibbosa, a rorqual 30 or 40 feet long, is met with in small herds in the intertropical and southern regions of the Pacific and the Atlantic; it is seldom molested by the whalers, and is very dangerous for boats, from the habit it has of leaping and rising suddenly to the surface. None of the senses of the whale tribe are very acute; the whalebone whales alone have the sense of smelling more acute than others, and, although the sperm whale is immediately aware of a companion being harpooned at a great distance, they do not hear well in air, and none appear to have any voice.[1]

---

[1] Captain Scoresby's 'Arctic Voyages.'

CHAPTER XXX.

Distribution of Reptiles — Frogs and Toads — Snakes, Saurians, and Tortoises.

REPTILES, more than any other class of animals, show the partial distribution of animated beings, because, being unable to travel to any great distance, they have remained in the places wherein they were originally stationed; and as they inhabit deserts, forests, and uncultivated ground, they have not been disturbed by man, who has only destroyed some individuals, but has not diminished the number of species, which is probably the same now as it ever was. Of the mammalia few hybernate, or fall into a torpid state in winter, such as the bear, marmot, dormouse, &c. Their fat supplies the carbon consumed by the oxygen during their feeble and imperceptible respiration, and is wasted by the time the warm weather returns, which rouses them from their lethargy, thin and attenuated. But reptiles, being colder blooded, bury themselves in the ground, and hybernate during the winter in cold and temperate climates. In hot countries, they fall into a state of torpor during the dry season, so that they have no occasion to wander either on account of temperature or want of sustenance; and the few that do migrate in quest of food always return to their old haunts. As the blood of reptiles, from the peculiarity of their circulation, receives only a small part of the oxygen they inhale, little heat and strength are generated, and they are for the most part sluggish in their motions, which, however, are more varied than in quadrupeds; but as some reptiles, such as tortoises and lizards, breathe more frequently than others, there are consequently great differences in their energy and sensibility.

The order of Reptiles is divided by naturalists into four classes, commencing in the ascending order:—1. Batracians or frogs, toads and salamanders; 2. Ophidians or Serpents; 3. Saurians, lizards, chameleons, crocodiles; and 4. Emydians or tortoises, and turtle. With very few exceptions they are oviparous; they partake of both terrestrial and aquatic forms, and many are amphibious: they all increase in numbers towards the equator, and few live in cold climates; but they can endure a cold winter better than a cool summer. Frogs and salamanders inhabit the banks of the M'Kenzie river in North America, where the mean temperature is between 7° and 8° of Fahrenheit; the thermometer in winter even sinks to 90° below the freezing point. The southern limit of reptiles, so far as it is known, is in 50° S. lat., where a frog was found on the banks of the river Santa Cruz.

The number of species of reptiles in the torrid zone is at least double that in the temperate; Australia has fewer than Europe, and of all places in the Old World, Java is perhaps the richest in reptiles. America possesses more than half of all the species known, the maximum being in Brazil, but every one of them is peculiar to that continent alone.

The Batracians approach nearest to the nature of fishes, and form a link between land and water animals. As tadpoles they have tails and no feet, but when full-grown they generally acquire feet and lose their tails. Besides, in that early stage they are aquatic and breathe by gills, like fishes; but in a state of maturity they breathe by lungs like quadrupeds, though some of the genera always retain their gills and tails, and some never acquire feet. These animals have the power of retarding and accelerating their respiration without stopping the circulation of their blood, so that they can resist heat and cold to a certain degree—a power most remarkable in the salamander, which forms part of this class, so varied in appearance and nature. Some, as toads and frogs, imbibe a quantity of water, which is evaporated through the pores of the skin more or less quickly, and serves to keep them at the temperature of the medium they live in.

The group of toads and frogs consists of four families, which have four feet, but without tails; namely, frogs, hylas or rainettes, toads, and pipæ. Frogs, which are amphibious, have no nails on their toes, and their hind legs are longer than the fore, and webbed, consequently better fitted for swimming and jumping, which they do by leaps. There are above 50 species, so that they are more numerous and more varied than any other reptile. Of the hyla or tree-frog there are sixty species, all of the most vivid and brilliant tints, and several colours are frequently united on the same animal. They mostly live on high trees, and their feet have little cushions at the points of their toes, forming a kind of sucker, by means of which they can squeeze out the air from under their feet, and, by the pressure of the atmosphere, they adhere firmly to the under side of the smoothest leaf, exactly on the same principle by which flies walk on the ceiling of a room. The Bufo, or Toad, is the ugliest of the race; many are hideous, with swollen bodies, wart-like excrescences, and obtuse toes. They seldom go into water, but frequent marshy damp places, and only crawl, whereas the frog and hyla leap. They are much fewer than either of the other two families; only 30 species are known. The Pipæ are also toads of a still more disgusting form, and are distinguished from their congeners by not having an extensile tongue. All these reptiles produce a noise, which is exceedingly varied; they croak in concert, following a leader, and when he is tired another takes his place. One of the North American frogs croaks in bands; one band begins, another answers, and a third replies, till the noise is heard at a great distance; a pause then takes place, after which the croaking is renewed. Mr. Darwin mentions a little musical hyla at Rio de Janeiro, which croaks a kind of harmony in different notes.

Toads and frogs are found in almost all parts of the earth, though very unequally and partially distributed. America has more than all the other countries taken together, and Europe the fewest. Six species of frogs, one rainette, and two toads are European; and all, except four of the frogs, are also found in Asia and Africa. The rana temporaria lives at the height of 7700 feet in the Pyrenees, and near the snow-line on the Alps.

The law of circumscribed distribution is strongly marked in Asia; for of ten species of frogs peculiar to that continent, three only are in the mainland, two are confined to Japan, and, of the five that are Javanese, one is also common to Amboina, and the other four to Bengal. The eight species of rainettes, or tree-frogs, are still more limited in their domicile; five of them are in Java only, and one in Japan; and the hyla viridis is in Asia Minor. There are nine species of toad peculiar to Asia.

None of these reptiles exist in the Galapagos Archipelago, nor in any of the innumerable islands in Oceanica: there are very few in Australia, and these peculiar. In Africa there are eight species of frogs, two or three rainettes, and two toads. One of the two species of pipa, more horrid in appearance than any toad, is very common at the Cape of Good Hope.

The great extent of marshes, rivers, and forests, together with the heat of the climate, make America the very home of reptiles of this kind, and there they grow to a greater size than anywhere else: 23 species of frogs, 27 species of tree-frogs or rainettes, and 21 of toads are indigenous in that continent, not one of which is the same with any of those in the Old World; and most of those in South America are different from those in the northern part of the continent. All these reptiles have abodes, with fixed demarcations, often of small extent. The pipa, or toad of Surinam, is the most horrid of the tribe; the bufo agua of Brazil, 10 or 12 inches long, and the rana pipiens or bull-frog of Carolina, are the largest.

The second family of this class of reptiles have tails and feet, as the salamanders, which are very like lizards in their general form, having a long round or flattened tail and four feet. Some are terrestrial, others aquatic; the former are known as salamanders or newts, the latter as tritons. Both are met with in Europe, but the greater number are American. The amphibious genera of Amphiuma Menopoma and Syren, possessing both lungs and gills, are American; the latter are peculiar to the marshes and rice-grounds of Carolina, and the Oxolotl is only found in the Lake of Mexico The Proteus anguinus, of a light flesh-colour, has four little feet and a flat tail, and has been only found in the dark subterraneous caverns in Carniola.

The third group of this order of Batracians are the Cæciliæ, of which there are only eight species, all inhabitants of the warm parts of Asia, Africa, and America. They have a cylindrical body, without feet or neck, and move exactly as the serpent, so they seem to form the link between these reptiles and the class of frogs and toads.

There are serpents in all hot and temperate countries, but they abound most in intertropical regions. Java contains 56 species, which is a greater number comparatively than any other country, while in Borneo not one has been found. Those in Japan are peculiar. Wherever snakes exist, there also are some of the venomous kinds, but they are fewer specifically and individually than is generally supposed. Of 263 species, only 57 are venomous, or about one in five, although that proportion is not everywhere the same. In sterile, open countries, the proportion of venomous snakes is greater than in those that are covered with vegetation. Thus, in Australia, seven out of ten species are poisonous; and in Africa, one of every two or three individuals is noxious. In general, however, the number of harmless individuals is 20 times as great as the number of the poisonous.

The three great families of venomous serpents are the colubriform or adder-shaped snakes, the triangular-headed snakes, and sea-serpents.

The adder-formed snakes are divided into three genera, the Elaps, which are slender like a cord, with a small head and of brilliant colours. There are four species in South America, of which two are confined to Guiana, and one to Surinam, while the other is found everywhere from Brazil to Carolina. There is only one in Africa, three in Australia, and the rest are in limited districts in tropical Asia, especially in Sumatra and Java; and an entire genus is found only in India, and the islands of Ceylon and Java. The hooded snakes (or Cobra Capello) are the best known of this family, especially the spectacled or dancing snake of the Indian jugglers, which is common everywhere from Malabar to Sumatra, and two other species are only found in Sumatra and Java. The three or four African species are chiefly met with at the Cape of Good Hope and on the Gold Coast; but the most celebrated is that generally known as the Egyptian asp, which has been tamed by magicians of ancient and modern times, and is frequently figured on the Egyptian monuments; it derives some of its celebrity from Cleopatra's death. Two of the family inhabit Australia, one of which is spectacled, like, but of a different species from, that of India.

The second venomous family consists of the triangular-headed serpents, rattlesnakes, and vipers. The first are of a hideous aspect, a large head, broad at the base like a heart, a wide mouth, with their hooked poisonous fangs strongly developed. They quietly watch their prey till it is within reach, then dart upon it, and inflict the deadly wound in a moment; the yellow viper of the French West India islands, the Trigonocephalus lanceolatus, being amongst the most dangerous snakes in existence. One species in the Old World is to be met with everywhere from Ceylon to the Philippine Islands; one is a native in Sumatra, Timor, and Celebes; the rest are narrowly limited in their abode; two are confined to Java alone. Ceylon, Sumatra, Japan, and Tartary, have each a species of these serpents peculiar to itself.

The rattlesnakes are all American — two in the warm districts of North America, and two in the intertropical parts of South America. One of the latter, however, has a hard horn at the end of its tail, instead of a rattle, and sometimes grows to the length of 10 feet, being, with the Trigonocephalus, the longest of the venomous snakes.[1]

Vipers extend farther north than any other of the noxious tribe: two are Asiatic, though one is also common to Africa, which, however, has four peculiar to itself; and the only venomous serpents in Europe are three species of viper, one of which is also spread over the neighbouring parts of Asia and Africa. The common viper inhabits all central Europe and temperate Asia, even to Lake Baikal, in the Altaï Mountains: it is also found in England and Sweden and the north of France, but does not pass the Alps, beyond which it is replaced by one frequenting dry soils, in the south-east of Europe, and is met with in Styria, Greece, Dalmatia, and Sicily; and the aspic viper, which lives on rocky ground, inhabits France between the Seine and the Pyrenees, Switzerland, Italy, and Sicily.

There are six families of innocuous serpents, consisting of numerous species. Four of the families are terrestrial; their species are very limited in their domicile, the greater number being confined to some of the islands of the Indian Archipelago, Ceylon, or to circumscribed districts in tropical Asia, Africa, and America. Nine or ten species are European, some of which are also found in Asia and Africa.

Tree-serpents of various genera and numerous species live only in the great tropical forests of Asia and America, especially in the latter. They are long and slender, the head for the most part ending in a sharp point, and generally green, though there are some of brighter colours; many of these serpents are fierce, though not venomous; some feed on birds, which they watch hanging by the tail from a bough.

[1] [A "Catalogue of North American Reptiles in the Museum of the Smithsonian Institution, by S. F. Baird and C. Girard," published in January, 1853, contains descriptions of 119 species of 35 genera of serpents, among which are named 12 species of two genera of the family of rattlesnakes.]

In all temperate and warm countries abounding in lakes and rivers, fresh-water snakes are numerous; some live in the water, but they mostly inhabit the banks near it; they are excellent swimmers, and may be seen crossing lakes in shoals. America is particularly rich in them: there are several in Asia, but they are rare in Africa, and none have been yet discovered in Australia.

The genus Boa is peculiarly American. The boa constrictor, generally from 9 to 15 feet long, lives in the great tropical forests of South America, where it often watches its prey hanging from the boughs of trees. Two of smaller size have similar habits, and two are aquatic, one of which is sometimes 20 feet long, and another 6 feet; the latter inhabits banks of the rivers, from the Amazon to Surinam; and a species is found at the foot of the Andes of Quito, as high as 3000 feet.

Pythons are the largest snakes of the eastern world, where they represent the boas of the western; one species, which sometimes attains the length of 20 feet, is spread from the western coast of Africa, throughout tropical Asia, to Java and China. Another, only 14 feet long, is confined to Malacca and some of the Sunda Islands. Two others are found only in the islands of Timor and Saparua, and one in Australia. There are only two species of Acrochordi, which, like boas and pythons, twist themselves round their victims and crush them to death: one aquatic, peculiar to Java, the other is a land snake, found everywhere from India to New Guinea.

The West Indian islands possess the snakes of North and South America and some peculiar to themselves; the snakes of central America are little known.

All the seven species of sea-snakes are very venomous, and more ferocious than any other. They frequent the Indian Ocean in shoals from Malabar to the Philippine Islands and the Bay of Bengal; they never enter fresh water.[1]

[1] The existence of creatures in the ocean resembling enormous serpents had been announced at different times for more than a century, but was never authentically established. Accounts of such monsters having been seen in the northern seas, in the fiords of Norway and Sweden, had been given to the world by Egede and Pantoppidan: by the latter more on hearsay evidence than from his own observation. But as in every instance the pretended Sea-Serpent was represented to possess either the blow-holes of a Cetaceous animal or the head and mane of a seal, it was evident the credulity of our Scandinavian neighbours had converted some well-known creatures into incomprehensible marine monsters. The same may be said of the sea-serpent represented to have been stranded on one of the Orkney Islands in 1808, of which a part of the skeleton is preserved in the Museum of the College of Surgeons, and which, when examined by the naturalist, proved to belong to a large species of shark; and of that fallen in with off the coast of Halifax in 1833, by some British officers engaged on a fishing expedition. The existence of the Sea-Serpent was looked upon therefore as one of those creations of that imaginative credulity so frequently enter-

Saurians have representatives in every warm and temperate climate. The crocodile, from its size and ferocity, claims the first place. There are three genera of this family, all amphibious, living in rivers or in their estuaries: the crocodile, properly speaking, common to the old and new continents;[1] the Alligator or Caiman, peculiar to America; and the Gavial, which comes nearer to the form of certain fossil crocodiles than any other, is limited to the Ganges and other great rivers of India. The various species of crocodiles are confined to local habitations: three are Asiatic; two African, one of which is only found about Sierra Leone; two are peculiar to Madagascar; in America there are two species of crocodiles and five of alligators. The American crocodiles inhabit the estuaries of great rivers; a species is to be met with which ascends as high as 3000 feet at the base of the Andes of Quito.[2]

---

tained by ignorant seafaring persons, and had ceased to attract any attention except occasionally by an allusion to it in some Transatlantic newspaper; when it was again revived in an official report addressed to the Lords of the Admiralty by Captain M'Quhae, of Her Majesty's ship Dædalus, who states that, on the 6th of August, 1848, being in lat. 24° 44′ S., long. 9° 22′ E., consequently not far from the south-western coast of Africa, he descried in broad daylight, and at a short distance, an animal with the head of a serpent and at least 60 feet long, passing his ship to the south-westward at the rate of 15 miles an hour. Professor Owen, after a careful consideration of all the details given of this strange apparition, has shown that the animal seen by the officers of the Dædalus was probably a large species of southern seal, of the genus Otaria. The Otaria is longer in proportion than our Arctic seals, and its fore flappers being situated farther back, the neck of the animal appears longer, and is generally, in the act of swimming, raised out of the water, as seen and represented by Captain M'Quhae in his drawing. Professor Owen supposes that this seal had been carried from its usual haunts in or near the Antarctic circle on an iceberg, which having melted away in these middle latitudes, the animal was obliged to find its way back by its locomotive powers; an opinion rendered the more likely, when we consider that it was making for the nearest land where such animals are known to live, Gough's Island and Tristan d'Acunha, from which it was distant about 1500 miles, or 4 days' journey at the rate and in the direction it is represented by Captain M'Quhae to have been progressing when seen from his ship. This statement of the appearance therefore of the sea-serpent in 1848 adds nothing towards confirming the existence of such monsters; whilst it shows how easy it is for even well-informed persons to raise up imaginary beings out of animals well known to the naturalist.

[1] One of the most celebrated species of this division is the crocodile of the Nile, which probably is to be met with in the upper branch of that river, the Bahr-el-Abiad, as high as 4000 feet above the level of the sea. Immense numbers of this animal, of every size and age, are found embalmed in the catacombs of ancient Egypt, which are perfectly identical with the existing species, offering another proof of the important fact first announced by Cuvier, from his examination of the mummies of the ibis, &c., &c., that no animal, in its wild state, had presented the least change within the most remote historical period.

[2] Mr. Pentland informs me that alligators are found in some of the rivers of Bolivia at a much greater elevation.

The alligators of the Mississippi, and of the rivers and marshes of Carolina, are more ferocious than those of South America, attacking men and animals; they only prey in the night; while in the water, like all their congeners, they cannot swallow their food, but they drown the animal they have caught, hide it under water till it becomes putrid, and then bring it on shore to eat it. Locality has considerable influence on the nature and habits of these animals; in one spot they are very dangerous, while in another, at no great distance, they are cowardly. Alligators are rarely more than 15 feet long, and are seen in large herds basking on the banks of rivers: their cry is like the roar of a bull; in a storm they bellow loudly, and are said to be much afraid of some of the porpoise family that ascend the great American rivers. The female watches her eggs and her young for months, never losing sight of them; but the male devours many of the latter when they go into the water. All animals of this class are covered with scales; those of the crocodile family are hard, horny, often osseous, and impenetrable.

Lizards are chiefly distinguished from crocodiles by having a long, thin, forked tongue like that of the viper; by their rapid motions, smaller size, and by some peculiarities of form.

The monitors, which are entirely confined to the old continent, have the tail compressed, laterally, which enables them to swim rapidly; and they are furnished with strong sharp teeth. Many species inhabit Africa and India, especially the Indian Archipelago: the terrestrial crocodile of Herodotus is common on the deserts which surround Egypt; and an aquatic species in the Nile, which devours the crocodile's eggs, is often represented on the ancient Egyptian monuments.

Another group of the monitor family is peculiarly American: some of the species inhabiting the marshes in Guiana are 6 feet long.[1]

Lizards are very common; more than 63 species are European, of which 17 inhabit Italy, and one lives on the Alps at an elevation of 4500 feet; the Iguanians, which differ from them only in the form of the tongue, are so numerous in genera and species, that it would be vain to attempt to follow all their ramifications, which are nevertheless distributed according to the same laws with other creatures but the dragons, only found in India, are too singular to be passed over. The dragon is in fact a lizard with wings of skin;

[1] Animals of a gigantic size, and allied to the lizard family, formerly inhabited the latitudes of Britain. A monster (the Mosasaurus) much surpassing the largest living crocodile is found in our Sussex chalk-beds; and another allied to the Iguana, the iguanodon of Mantell, is of frequent occurrence in the strata upon which the chalk reposes in the Weald of Sussex, the Isle of Wight, &c. Some bones of the iguanodon would indicate an animal more than 50 feet long.

which are spread along its sides and attached to its fore and hind feet, like those of the flying-squirrel, and though they do not enable it to fly, they act like a parachute when the animal leaps from bough to bough in pursuit of insects. Nocturnal lizards of many species inhabit the hot countries of both continents; they are not unlike salamanders, but they have sharp claws, which they can draw in and conceal like those of a cat, and seize their prey. One of these species, the Gecko, climbs on walls in all the countries round the Mediterranean. Chameleons are frequent in Northern Africa; and several species inhabit different districts and islands in Asia; the only European species is found in Spain; it is also common in Northern Africa.

The Anolis, which lives on trees, replaces the chameleon in the hot regions of South America and in the Antilles, having the property common to chameleons of changing its colour, but it is a more nimble and beautiful animal. In Australia, where everything is anomalous, there is a lizard with a leaf-shaped tail.

Scinks resemble serpents in form, but with four very short feet and sharp nails on their claws; they burrow in the sands of Africa and Arabia: there is a species of gigantic black and yellow scink in New Holland, and those in the islands of the Indian Archipelago are green, with blue tails.

Two anomalous saurians of the genus Amblyrhinchus were discovered by Mr. Darwin in the Galapagos Archipelago. One found only in the central islands is terrestrial, and in many places it has undermined the ground with its burrows; the other is the only lizard known that lives on sea-weed and inhabits the sea; it is about four feet long, and hideously ugly, with feet partially webbed and a tail compressed laterally. It basks on the beach, and in its marine habits and food it resembles, on a small scale, the huge monsters of a former creation.

Tortoises are covered with a shell or buckler, but their head, legs, and tail are free, covered with a wrinkled skin, and the animal can draw them into the shell when alarmed. The head is sometimes defended by a regular shield, and the jaws, instead of teeth, have a horny case. The upper buckler is rounded, and formed of eight parts of plates symmetrically disposed, and often very beautiful; the under shell is flat, and consists of four pair of bones and one in the centre. One family of tortoises is terrestrial, two others are amphibious, one of which lives in fresh water, and the other in tropical and warm seas.

There are more land tortoises in Africa than in all the rest of the world, both specifically and individually. They abound also in the Sunda Islands, in the United States of America, South America, and especially Brazil. There are three European species, of which the common tortoise (Testudo Græca), which is found in all the countries round the Mediterranean, is the largest, attaining as much as a foot in length: it lives on insects and vegetables, and burrows in the ground in winter. Some of the East Indian species are enormously large, above three feet long, and remarkable for the beautiful distribution of their colours; certain species are peculiar to Brazil, one to Demerara, and one to North America; but perhaps the largest known species is that of the Galapagos Islands, the Testudo Indica, which attains 500 or 600 pounds in weight.

There are two genera of fresh-water tortoises that live in ponds and stagnant pools and lakes, the Emys and Chelys. The first is very numerous in species in America; there are no less than forty-six peculiar to its two continents: six have been found in Africa, three in Europe, eighteen in Asia and its islands, and only two in Australia. The emys caspia, in Asia Minor, follows a leader, and plunges into the water when alarmed. The Chelys, furnished with a kind of proboscis, is found in the stagnant waters of South America.

The Potamians Trionyx, or fresh-water turtle, live in the great rivers and lakes in warm countries; there are two species peculiar to North America; they are very large and voracious, devouring birds, reptiles, and young crocodiles, and often are a prey to older crocodiles. One is peculiar to the Nile, where it devours the eggs and young of the crocodile; one to the Euphrates and Tigris; there are four species in the Ganges, which are frequently seen eating the human bodies that are thrown into the sacred stream; one of these animals often weighs 250 pounds. The starred trionyx is found in the rivers of Java only, and another species is common to the rivers of Borneo and Sumatra.

The Chelonians, or sea-turtles, live in the seas of the torrid and temperate zones, as far as the 50th parallel of latitude, some living on sea-weed, and others on small marine animals. Distinct species are found in different parts of the ocean. The green turtle, of which there are many varieties, inhabits the Atlantic within the tropics; they may be seen eating sea-weed at the bottom of the water along the coasts: they repair at certain seasons from distances of many hundred miles in great herds to the mouths of rivers to deposit their eggs in the sand. This turtle is often six or seven feet long, and weighs 600 or 700 pounds; it is much esteemed for food, but the shell is of little value.

The hawk's-bill turtle, which yields the tortoise-shell, is caught in different parts of the Indian Ocean, among the Molucca Islands, and on the north-western coast of New Guinea. It is also found in the western hemisphere off Haiti and the Caiman Islands, but the shell is less valuable than that from the east. There are two species in the Mediterranean, which are only valued for their oil. A very peculiar turtle, with a leathery or coriaceous covering (*Testudo cori-*

acea), has been sometimes caught on our English coasts, weighing as much as 800 pounds: it is the species supposed to have been used by the ancients in the construction of their lyre, and has hence been called Lyre Turtle by the French.

With respect to the whole class of reptiles it may be observed, that not one species is common to the Old and New World, and few are common to North and South America; those in Australia are altogether peculiar; and, as far as is at present known, with the exception of the Marianne Islands, there are neither toads, frogs, nor snakes in any of the Polynesian islands, though the Indian Archipelago abounds in them; neither are they found in Terra del Fuego, in the Straits of Magellan, nor in the Falkland Islands.

Five species of reptiles only appear to have existed in Ireland before its geological separation from England — a lizard, a frog, a toad, and two tritons.

CHAPTER XXXI.

Distribution of Birds in the Arctic Regions — In Europe, Asia, Africa, America, and the Antarctic Regions.

More than 7000 species of birds are known, which, according to the most received system, are arranged in seven natural orders or groups, namely, Birds of prey—or vultures, falcons, owls; Perching birds, by much the most numerous, and which comprise the songsters; Climbers, as parrots, woodpeckers, cuckoos, &c.; Pigeons, Gallinaceous birds, including our domestic fowls, partridges, grouse, pheasants, ostriches; Waders, as snipes, herons, curlews, &c.; Web-footed birds, as ducks, petrels, albatrosses, gulls, &c.[1] Next to tropical America, Asia is richest in species: the greatest number of birds of prey inhabit Europe and America, which last surpasses every country in the number and beauty of species.

There is great similarity in the birds of the northern parts of the old and new continents, and some are identical. Towards the south, the forms differ more and more, till in the tropical and southern temperate zones of Asia, Africa, and America, they become entirely different, whole families and genera often being stationary within very narrow limits. Some, however, are almost universally distributed, especially birds of prey, waders, and sea-fowl.

The bald buzzard is to be met with in every country from Europe

[1] See the arrangement in the very beautiful work on Birds by G. R Gray, Esq.

to Australia; the Chinese gosshawk inhabits the American continent, and every station between China and the west coast of Europe; the peregrine falcon lives in Europe, America, and Australia; and Flamingos of different species fish in almost every tropical river, and on the borders of the lakes of the Andes to the height of 13,000 feet. Many sea-birds also are widely spread; the wagel-gull is at home in the northern and southern oceans. Captain Beechey's ship was accompanied by pintadoes, or Cape pigeons, during a voyage of 500 miles, in the Pacific; and even the common house-sparrow is as much at home in the villages of Bengal as it is in Britain. Many more instances might be given, but they do not interfere with the general law of special distribution.

Birds migrate to very great distances in search of food, passing the winter in one country and the summer in another, many breeding in both. In cold climates insects die or hybernate during winter; between the tropics they either perish or sleep in the dry season: so that, in both cases, insect-eating birds are compelled to migrate. When the ground is covered with snow, the want of seeds forces those kinds whose food is vegetable, to seek it elsewhere; and in tropical countries the annual inundations of the rivers regulate the migrations of birds that feed on fish.

Some migrate singly, some in groups, others in flocks of thousands; and in some instances, the old and the young birds go separately. Those that fly in company generally have a leader, and such as fly in smaller numbers observe a certain order. Wild swans fly in the form of a wedge, wild geese in a line. Some birds are silent in their flight, others utter constant cries, especially those that migrate during night, to keep the flock together, as herons, goat-suckers, and rails.

Birds of passage in confinement show the most insurmountable uneasiness when the time of migration draws near. The Canadian duck rushes impetuously to the north at the usual period of summer flight. Redbreasts, Goldfinches, and Orioles, brought from Canada to the United States when young, dart northwards, as if guided by the compass, as soon as they are set at liberty. Birds return to the same place year after year. Storks and swallows take possession of their former nests, and the times of their departure are exact even to a day. Various European birds spend the winter in Asia and Africa; while many natives of these countries come to central Europe in summer.

The birds of passage in America are more numerous, both in species and individuals, than in any other country. Ducks, geese, and pigeons migrate in myriads from the severity of the northern winters; and when there is a failure of grain in the south, different families of birds go to the north. The Virginian partridge crosses the Delaware and goes to Pennsylvania when grain is scarce in New

Jersey; but it is so heavy on the wing, that many fall into the river, and end the journey by swimming.

The same thing happens to the wild turkey, which is caught by hundreds as it arrives wet on the banks of the Ohio, Missouri, and Mississippi. These birds are not fitted for long flight by their structure, their bones have fewer of those air-cells which give buoyancy to the feathered tribes. The number of air-cells is greatest in birds that have to sustain a continued and rapid flight; probably the extremes are to be met with in the swift and the ostrich—the one ever on the wing, the other never. The strength of the ostrich is in the muscles of its legs; while the muscles on the breast of the swift weigh more than all the rest of the body; hence it flies at the rate of 100 miles an hour. The wild duck and the wild pigeon fly between 400 and 500 miles in a day. The Stork and some other migratory birds do not halt till the end of their journey. Many sea-fowl are never seen to rest; and all the eagles, vultures, and hawks are birds of strong flight, and capable of sustaining themselves at heights beyond the reach of less buoyant creatures.

DISTRIBUTION OF ARCTIC AND EUROPEAN BIRDS.

The birds of Europe and North America are better known than those of any part of the globe. New species are constantly discovered in Asia, Africa, and South America; and extensive regions in the East are yet unexplored.

There are 540 species of birds in Europe, many of which are distributed over Asia and Africa, without any apparent variation; and 100 of our European species are also in North America. Of these 39 are land-birds, 28 waders, and 62 water-fowl; among which are most of the marine birds of northern Europe, which, like all sea-fowl, have a wider range.

More than three-fourths of the species, and a much larger proportion of individuals, of the birds of Greenland, Iceland, and Faroe, are more or less aquatic, and many of the remainder are only occasional visitors. Of the few small birds, the greater number are British; but many that reside constantly in Britain are migratory in Iceland and Faroe, and all the small birds leave Greenland in winter. The aquila albicilla, or fishing eagle, is the largest bird of these northern islands; it feeds on salmon and trout, and builds its nest on the boldest crags. The jer-falcon, or falco islandicus, though native, is rare even in Iceland. The snowy owl lives near the glaciers in the interior of Greenland, and is sometimes seen in Orkney Particular kinds of grouse are peculiar to high latitudes, as the ptarmigan or white grouse. The Columba œnas lives on all the rocky coasts of Europe, and it is also an American bird. The crow family are inhabitants of every part of the globe. The common crow is very generally distributed; the carrion-crow and jackdaw are all over Europe and North America. The Royston crow is the only one of the genus seen within the Arctic circle, and but a summer visitor. The Magpie is met with everywhere in Europe. The Jay, one of the most beautiful birds of its tribe, is found in Europe, North America, and China; the Raven everywhere, from Greenland to the Cape of Good Hope, and from Hudson's Bay to Mexico; it is capable of enduring the extremes of heat and cold, and is larger, stronger, and more ravenous in the Arctic islands than anywhere else. It is said to destroy lambs, and to drive the eider ducks from their nests to take their eggs or young: they unite in flocks to chase intruding birds from their abode.

Waders are more numerous than land-birds in the Arctic regions. The snipe and the golden plover are mere visitors; and the oyster-catcher remains all the year in Iceland, where it makes its nest near streams, and wages war with the crow tribe. The heron, curlew, plover, and most of the other waders, emigrate; sand-pipers and the water-ousel remain all the year round.

Web-footed birds, being clothed with down and oily feathers, are best able to resist the cold of a polar climate. The cygnus musicus, or whistling swan, is the largest migratory bird of Europe or America. It is 5 feet long from the tip of the bill to the end of the tail, and 8 feet from tip to tip of the wings: its plumage is pure white, tinged orange or yellow on the head. Some of them winter in Iceland; and in the long Arctic night their song is heard, as they pass in flocks: it is like the notes of a violin. Various species of the duck tribe live in the far north, in prodigious multitudes. The mallard, supposed to be the origin of our tame duck, is everywhere in the Arctic lands. There are two species of eider-duck: the king duck, or somateria spectabilis, one of these, is widely dispersed over the islands and coasts of the North Atlantic, and all the Arctic regions in Europe and America. In Europe its most southern building-place is the Farne Islands on the coast of Northumberland; in America it never goes south of New York. It lives in the open sea in winter, and resorts to the coast when the grass begins to grow. The duck makes her nest of sea-weed, lined with down from her breast. The islanders take the eggs and down twice in the season; but they do not kill the old birds, because the down of a dead duck is of little value, having lost its elasticity. The third time the drake repairs the nest with down from his breast: the birds are allowed to hatch their brood; and, as soon as the young can feed themselves, they are taken out to sea by the mother. They attain their full growth in 4 years, and then measure 2 feet from tip to tip of the wing. The same couple has been known to frequent a nest 20 years, and the Icelanders think the eider-duck lives to 100.

The Cormorant, which lives on fish, is universal in the northern seas, and is scarcely ever eaten by the natives. It sits singly, or sometimes in flocks, on the rocks, watching the fish with its keen eye: it plunges after them, and pursues them for three or four minutes under water. Auks are very numerous, especially the razor-billed auk, or penguin; but the great auk, which is incapable of flight from its small wings, is now nearly extinct in the Arctic seas. Terns, or sea-swallows, are seen everywhere in these seas, skimming along the surface of the water, catching mollusca and small fish. Gulls of many species, and in countless numbers, are inhabitants of the Arctic regions, whilst in the Antarctic they are represented by the equally numerous genus Procellaria, of which the Mother-Cary's-chicken or stormy petrel is the type. No birds are more widely dispersed than these two genera. They are at home, and brave the storm, in every latitude and in every sea. There are nine or ten species of gulls in the Arctic regions, and the most numerous of these probably are the kittywakes, the young of which cover the rocks in Iceland, packed so close together that 50 are killed at a shot.

The Skua Gull is a bold and rapacious bird, forming a kind of link between gulls and birds of prey. It lives by robbing other birds, and is so audacious that it forces the gulls to disgorge the fish they have swallowed, and has been seen to kill a puffin at a single blow. Its head-quarters are in Faroe, Shetland, and the Hebrides, where it hatches its brood, and attacks animals if they come near them.

A few species of petrel, such as the puffin inhabit the Arctic Ocean, but the South Pacific and the Antarctic seas are the favourite resort of this genus. They take their name from the faculty they have of walking on the water,[1] which they do by the aid of their very large flat webbed feet and widely-extended wings. The stormy petrels, consisting of several distinct species, confounded by sailors under the name of tempest-bird or Mother-Cary's-chicken, are the most widely diffused, about the size of a swallow, and nearly of the same colour as the latter; their flight is rapid; they shelter themselves from the storm in the hollow of a wave, and go to land only at the breeding season.

It is observed that all birds living on islands fly against the wind when they go to sea, so as to have a fair wind when they return home tired. The direction of the prevailing winds, consequently, has great influence on the choice of their abode: for example, the 25 bird-rocks, or Vogel-berg, in Faroe, face the west or north-west; and no bird frequents the cliffs facing the east, though the situation is to all appearance equally good; a preference accounted for by the prevalence of westerly wind in these latitudes.

---

[1] Petrel, from St. Peter.

Most marine birds are gregarious. They build their nests on the same rock, and live in society. Of this a curious instance occurs on the rocks in question. The Vogel-berg lies in a frightful chasm among the cliffs of Westmanshavn in Faroe. The chasm is encompassed by rocks 1000 feet high, and myriads of sea-fowl cluster round the top of the crags; but different kinds have separate habitations; and no race or individual leaves his own quarters, or ventures to intrude upon his neighbours.

Upon some low rocks, scarcely rising above the surface of the water, sits the glossy cormorant; the predatory skuas, on a higher shelf, are anxiously regarded by myriads of kittywakes on nests in crowded rows along the shelving rock above, with nothing visible but the heads of the mothers almost touching one another; the auks and guillemots are seated a stage higher on the narrow shelves, in order as on a parade, with their white breasts facing the sea, and in absolute contact. The puffins form the summit of this feathered pyramid, perched on the highest station, and scarcely discernible from its height, if they did not betray themselves by flying backwards and forwards. Some of these tribes have a watch posted on the look-out for their safety; and such confidence has the flock in his vigilance, that if he is taken the rest are easily caught. When the whole take flight, the ear is stunned by their discordant screams.

The greater part of the marine birds of the Arctic seas are inhabitants also of the northern coasts of the continent of Europe and of the British islands.

Few parts of Europe are richer in birds than Britain, both in species and numbers of individuals; and the larger game is so abundant, that no one thinks of eating nightingales and redbreasts and other small birds, as we see in the south of Europe. Of the 540 species of European birds, 277 are native in our islands. The common grouse, the yellow and pied wagtails, are found nowhere else. It is probable that most of the British birds came from Germany before the separation of our island from the continent, and many of short flight never reached Ireland. The Ptarmigan and Capercailzie came from Norway.

There are five European vultures: the Lemmergeyer of the Alps and Pyrenees builds its nest in the most inaccessible parts of the mountains, and is seldom seen; it lives also on the mountains of Abyssinia and in the steppes of Mongolia. Ten eagles are European; one is peculiar to Sardinia; and few if any of them are common in America: the golden eagle is perhaps one; that beautiful bird, which once gave a characteristic wildness to our Scotch mountains, and the distinguishing feather to the bonnet of our Highland chieftains, is now nearly extirpated. The osprey or fishing eagle of Europe is replaced in America by the bald eagle, and so are some of our numerous hawks; among others the jer or gentil falcon has been so much destroyed, that it is now rare even in Iceland, its native place: there are still a few in Scotland, and several are caught in their migratory flight over the Low Countries and reclaimed by the expert falconers for the now nearly obsolete sport of falconry.

The owl tribe is numerous, and many of them are very handsome. The Bubo maximus, the great horned owl, the largest of nocturnal birds, inhabits the forests of middle and southern Europe; it is rare in France and England, though not uncommon in Ireland and Orkney: in Italy, a small owl is tamed and used as a decoy.

Owls, eagles, and hawks have representatives in every country, but of different species. The two species of European Goatsuckers migrate to Africa in winter; their peculiar cry may be heard on a moonlight night when a large flock takes wing for the journey. Several of our swallows go to Africa: both our kingfishers are African, and only visit us in summer; one, the Alcedo ispida, is a native of Lower Egypt and the shores of the Red Sea. Some of the 7 species of European creeping birds, or certhias, creep on the trunks and branches of trees in search of insects; others pursue their prey clinging to the face of rocks and walls, supported by the stiff elastic feathers of the tail: the Hoopoe, an inhabitant of southern Europe, pursues small reptiles and insects on the ground.

The Fringillidæ or thick-billed birds are by much the most characteristic of Europe; to them belong some of our finest songsters. The sylvias have soft beaks, and feed on insects and worms; the nightingale, the redbreast, the wren, the smallest of European birds, the warblers, white-throat and others are of this family. Thick-billed birds live on seed, as do the common sparrow, the gold and other finches, linnets, buntings, and crossbeaks.

Four species of fly-catchers are peculiar to Europe, and five species of shrikes. Ravens, crows, jays, and magpies abound; the Alpine crow and Nutcracker are found in central Europe only. Compared with America the Starling family is poor, and the woodpecker race still more so, yet we have six species, some of which are very beautiful. There is only one Cuckoo entirely European, the other two kinds only come accidentally, and all are birds of passage. There are four species of the pigeon tribe; the Ringdove frequents the larch forests, and is migratory; the Stockdove also leaves us in October: the Biset or rock pigeon, supposod to be the origin from which the infinite variety of our domestic pigeons has sprung, flies in flocks, and makes its flimsy nest on trees and rocks; it is also found in the Daouria part of the Altaï chain. Of gallinaceous birds there are many; the only native pheasant is in the southern parts of the continent; and the capercailzie, extinct in the British forests, inhabits many parts of Europe; in Scandinavia especially it is plentiful as far as the pine-tree grows, which is nearly to North Cape, and also in the Russian forests. The hazel grouse frequents the pine and aspen forests in northern and central Europe, where the black cock also is plentiful. Five species of grouse and six of partridges afford abundance of game; four of the latter are confined to the southern parts of the continent, and so are the sand and pen-tailed grouse, which form a separate family; the former inhabits the sterile plains of Andalusia and Granada, and the latter the stony uncultivated parts of France, southern Italy, and Sicily. The Ortigis Gibraltarica, a peculiar bird allied to the grouse family, is found in the south of Europe only.

European waders are very numerous, and among them there are specimens of nearly all the known genera; woodcocks, snipes, plovers, and curlews are very abundant, and herons of various species; there are three species of egret or crested heron, and the common heron now assembles on the tops of trees unmolested, since the progress of agriculture has rendered the country unfit for hawking. Several cranes and storks, and two species of ibis, are European: a species of flamingo is met with in the south-eastern parts of the continent, and in the Maremme or marshy plains on the western coasts of Italy. Many of the waders, however, migrate in winter. The stork, so great a favourite in Holland that it is specially protected, is a wanderer; it retires to Asia Minor, and on the return of summer resumes its old nest on a chimney-top, breeding in both countries. About 140 species of European birds either live in the more elevated parts of the Alps, or cross them in their annual migrations. They generally take their flight by the Great St. Bernard, the pass of St. Theodule, the Simplon, and St. Gothard. Europe is particularly rich in web-footed birds; there are three species of wild swans, four of wild geese, and more than 30 of the duck tribe, including the inhabitants of the Arctic seas.

BIRDS OF ASIA AND THE INDIAN ARCHIPELAGO.

European birds are widely spread over Asia; most of the Arctic sea-fowl frequent its northern coasts: between 50 and 60 European birds are also Siberian, and there are above 70 European species in Japan and Corea, which probably are also inhabitants of Siberia and the Altaï Mountains, and several are identical with the birds of North America; so that the same affinity prevails in the feathery tribes of the Arctic regions as in the vegetable productions.

Asia Minor is a country of transition, and many European birds are mixed with those of warmer regions, as the halcyon smyrnensis, a bird with gorgeous plumage, identical with the great Bengal kingfisher, so generally found in India. European birds also inhabit the Caucasus, the shores of the Caspian Sea, and Persia. Moreover, these warmer climates are the winter-quarters of various European species.

In Asia Minor, and especially in Armenia, the number and variety of birds is very great; eagles, vultures, falcons, buzzards, quails, partridges, starlings, herons, storks, cranes, legions of Arctic grebes, swans, wild geese, ducks, and pelicans, are natives of these countries; besides singing-birds, the nightingale, the constant theme of the poet's song, abounds in Persia: hawks are trained for hunting deer in that country, and the Asiatic partridges, or francolins, more vividly coloured than ours, differ also in having beaks fitted for digging up bulbous roots, which is their food in the deserts.

Farther east the types become more Indian; the great peninsulas on each side of the Ganges are the habitations of the most peculiar and the most gorgeous of birds. Many species of Kingfishers of the brightest colouring are found here; the plumage of the fly-catchers has the richest metallic lustre; and the Shrikes, of a sober hue with us, are there decked in the most brilliant colours; the Drongo has a coat of ultramarine, and the Calyptomene has one of emerald green.

The large-beaked climbing-birds are singularly handsome. The small collared parakeet, so easily taught to speak, has inhabited the Indian forests and the banks of the Ganges time out of mind, with a host of congeners of every colour; not one species of these or indeed of the whole parrot tribe is common to Asia, Africa, America, or Australia, nor even to any two of these great continents. They are vividly coloured in India, but the cuckoo tribe rivals them; several genera of these birds exist nowhere else, as the large-beaked Malcahos, the Coucals with their stiff feathers, and the Couroucous or Trogons, dressed in vermilion and gold; the last, however, also inhabit other tropical climates.

Southern Asia is distinguished by the variety of its gallinaceous birds and the gorgeousness of their plumage. To this country we owe our domestic fowls; and two species of peacock are wild in the woods of India and Ceylon. The Polyplectron, the only bird of its genus, and the Tragopons, are Indian; and some of the most brilliant birds of the East are among the pheasant tribe, of which five species are peculiar to China and Tibet. There are various species of the pheasant in the Himalaya, and one whose feathers have a metallic lustre. The gold, the silver, and Reeves' pheasant, the tail-feathers of which are four feet long, belong to China. The Lophophorus refulgens, and some others of that genus, are altogether Indian.

The pigeons also are very splendid in their plumage; they mostly belong to China and Japan; those in the Birman empire are green.

It would be vain to enumerate the fine birds that range in the forests, or fish in the rivers, of the Asiatic continent, yet the birds of the Indian Archipelago far surpass them in splendour of plumage; these islands indeed are the abode of the most gorgeously arrayed birds in existence. Even in Java and Sumatra, though most similar to India in their winged inhabitants, there are many peculiar, especially 12 or 13 species of the climbing tribe, and several of the honey-sucking kind; but the dissimilarity increases with the distance, as in New Guinea and its islands, where the honey-sucking genera are developed in novel forms and sumptuous plumage.

About 35 genera are peculiar to India: 32 genera, with all their numerous species, are found only in the islands of the Indian Archipelago, and several of these are limited to one or two islands. There are the Cassicans, which resemble jays, with plumage of metallic lustre; the only two species of Pirolls, one bright violet, the other of brilliant green; various species of Buceros with large horned beaks, Orioles of vivid colours, the Swallow that builds the edible nest, the numerous and splendid Sylvias, and all the species except one of the Philedons or honey-sucking birds with tongues terminating in a brush. Birds of Paradise of many kinds inhabit New Guinea and the neighbouring Moluccas and Aroo Islands. They are birds of passage, and change their quarters with the monsoon. The King or Royal Bird of Paradise has two long slender filaments from the tail, ending in a curled flat web of emerald green, and the male of the green species has long flowing plumes from the sides of his body, which give him a magnificent appearance. The pigeons are peculiarly beautiful and numerous, but limited in their abode. The Goura, or great crowned pigeon, the largest of its tribe, is an inhabitant of Borneo. Each island has its own species of Lories; many Paroquets and Cockatoos, Couroucous or Trogons, Coucals, and the Barbu, with huge beaks, are peculiar to these islands. Even the partridges have thrown aside their grave colours and assumed the vivid hues of the tropics, as the green and tufted Cryptonyx. But the other gallinaceous birds far surpass them in beauty, as the Argus pheasant and two or three species still more splendid, with a host of others, which Europeans have never seen in their wild state, in the deep jungles and impenetrable forests of these unexplored islands. The Cassowary, a bird akin to the ostrich, without the power of flying, but fleet in its course, has a wide range in the Indian Archipelago and the south-eastern peninsula of Asia, and, though destitute of beauty, is interesting from its peculiar location and the character of the whole race.

AFRICAN BIRDS.

A great number of European birds are also inhabitants of Africa, and many migrate there in winter, yet the birds of this continent are very peculiar and characteristic; those in the north and north-east, and at the Cape of Good Hope, are best known, the greater part of tropical Africa being still unexplored by the naturalist. It may be observed, generally, that the tropical birds differ from those of North Africa, but are, with a few exceptions, the same with those in the southern part of the continent, and the whole of Africa south of the desert differs in species from those of north and western Africa and from Europe. Moreover, there is a strong analogy, though no affinity, between the birds of Africa and America in the same parallels of latitude; there is not a single perching bird common to the two continents, though some birds of prey are found in both.

There are 59 species of birds of prey, of which a few are also European. The Secretary-bird is the most singular of this order: it preys upon serpents at the Cape of Good Hope, in Abyssinia, and other parts of the continent. Africa possesses at least 300 species of the passerine order, of which 10 genera, with all their species, are peculiarly its own. The swallows are more beautiful than ours, especially the Filicoda, with two tail-feathers twice as long as its body. Many kingfishers, the most beautifully coloured of their brilliant race, frequent the lakes and rivers: four species of Hoopoes, one of which visits Europe in summer, are natives; and the honey-birds, the representatives of the humming-birds of South America, are peculiarly African. They abound at the Cape of Good Hope, where the nectaries of Proteas and other plants furnish the saccharine juice for their food. The malurus africanus, and many other singing-birds for the most part unknown elsewhere, inhabit the forests. The canary-bird is confined to the Canary Islands; its song differs in two adjacent districts: there are, however, instances of this among other birds. The Capirote, also indigenous in the Canary Islands, is a finer songster, but it cannot be tamed. Various shrikes are peculiar to Africa, but the species known as the grand Batara is confined to the Azores. There are several birds of the raven tribe, or nearly akin to them: as the lampritornis superba; another with purple wings: the Buphaga, the only bird of its genus; and several species of the Calaos. The Weaving-bird, or ploceus textor, is one of the most remarkable of the granivorous tribe; it weaves its nest with grass and twigs very dexterously; one brought to Europe wove a quantity of thread among the wires of its cage, with great assiduity, into a strong texture. The Widow-bird, the Colious, the blue Bee-eater, and all the fly-catching Touracous, with many species of woodpeckers, are found nowhere else. The parrots and paroquets, which swarm in the tropical forests, from the size of a hen to that of a sparrow, or of peculiar forms. The Trogons, or Couroucous, the most beautiful of the large-beaked climbing-birds, are the same as in Asia; but the Barbu and the four species of Barbicans are altogether African, and so are some of the cuckoos. Among the latter are two species of the Cuculus indicator, so named from indicating where the bees have their nests.

one is peculiar to Abyssinia, the other to the interior at the Cape of Good Hope.

There are at least 13 species of African pigeons; and to Africa we are indebted for the Guinea-fowl, of which there are three or four kinds: it wanders in flocks of hundreds among the brushwood on the banks of rivers and lakes in Numidia and all the tropical regions, and they are even more abundant in Madagascar. Many grouse and partridges are peculiar, especially the Gangas, of which there are five species; some go in coveys, and others traverse the deserts in flocks of many hundreds. The sand-grouse, one of this family, is much more abundant on the arid deserts of north Africa than in Europe; and the partridges in this country are represented by the francolin, which feeds on bulbous roots.

The ostrich takes the wide range of Africa and Arabia; and bustards, also wanderers in the plains, are numerous: the most peculiar are the Houbara and the otis kori, in South Africa, five feet high, and remarkable for the brilliancy of its eye.

Waders of infinite variety inhabit the rivers, lakes, and marshes —woodcocks, snipes, plovers, storks, cranes, herons, and spoonbills. The most peculiar are the Dromes and Marabous, whose feathers form a considerable article of commerce; the cream-coloured plover, the Scopus or Ombrette, the water-treader of Abyssinia, and the Tantalus or Curlew tribe, among which is the Ibis (Ibis Religiosa), held sacred by the ancient Egyptians, so frequently found in mummies in the catacombs, and represented on their monuments, and the recently discovered anomalous bird the Balæniceps Rex, which inhabits the upper branches of the White Nile, where it feeds on fish and lizards.

Swimming-birds are no less numerous: the Bernicla cyanoptera is a goose peculiar to Shoa: the Rhynchops and Pelicans, several of the duck kind or birds allied to them, are found nowhere else. There are 56 genera with all their species entirely African, many of which are confined to limited districts.

BIRDS OF NORTH AMERICA.

Of 500 species of North American birds, about 100 are also found in Europe, the greater number of which are aquatic, and those common to the northern coasts of both continents. The sea-fowl on the North Pacific and Behring's Straits are very much the same with those in the Greenland seas and the North Atlantic, but the great Auk of our northern seas still exists on the North Pacific, and the large white albatross, seldom seen in the North Atlantic, frequents Behring's Straits and the western coasts of North America in immense flocks. It is almost universal in the Pacific as far as the stormy regions near the Antarctic circle. Like the Petrel, it is a bird of the tempest, sailing calmly on its huge wings in the most tremendous gales, and following a ship a whole day without resting on the waves: it is the largest of winged sea-fowls; some measure 14 feet from tip to tip of the wings.

There is no vultnre common to the two continents, but there are eagles, and some other birds of prey, a few of the crow tribe, several waders and web-footed birds which inhabit both: yet the general character of the birds of North America is different from those of Europe: 81 American generic forms and two families are not found in Europe. The humming-birds are altogether American; only four species are in North America; one of these is migratory, and another is common to South America. The Parrot family, distributed with generic differences in almost all tropical countries, has but one representative here, which lives in the forests of the Carolinas. Europe has nothing analogous to these two families. It is singular that a country with so many rivers and lakes should possess only one kingfisher. The woods are filled with many species of creeping birds, and there are 68 peculiar species of sylvias and fly-catchers; among others the todus viridis, which forms a genus by itself. Ravens, Crows, Pies, and Jays abound, and there are many species of Icteridæ or Hangnesters. The finch tribe are very numerous, and there are 16 species of woodpeckers, as might be expected in a country covered with forests. Of pigeons there are eight species, but individually they are innumerable, especially the columba migratoria, which passes over Canada and the northern States in myriads for successive days twice in the year. The poultry-yard is indebted to North America for the domestic turkey, which there ranges wild in its native woods and attains great size. There are no partridges, properly speaking, but the Ortyx, a genus closely allied, represents them, and of 13 American species of grouse, only one probably is European, a family which exists in every country under different forms. The vast expanse of water and marshy ground makes North America the home of innumerable water-fowl and waders. Most of the waders and granivorous birds are migratory; in winter they find no food north of the great lakes, where the ground is frozen upwards of six months in the year. Many pass the winter in California, as Storks and Cranes; wild geese cover acres of ground near the sea, and when they take wing their clang is heard from afar. Even gulls and other northern sea-birds come to the coasts of California, and indeed to the shores of all the north and temperate Pacific.

It may be said generally that, with regard to the web-footed tribe, North America possesses specimens of all the genera of the Old World and many peculiarly its own. The table-land of Mexico has some peculiar forms, and some species of swimming-birds found only in more northern latitudes; but, except the Cotingidæ, there are representatives of every group of North and South America.

BIRDS OF SOUTH AMERICA.

The tenants of the air in South America differ more from those in North America than these latter do from the birds of Europe: there are not more than 50 or 60 species in common. South America has a greater variety of original forms than any other country; more than 138 genera with all their species inhabit that country only; of the passerine family alone there are at least 1000 species, all peculiar to it. The vultures are of different genera from those in Europe; the Condor of the Andes is the largest of these; it frequents the highest pinnacles of the Andes in summer, and builds its nest at the height of 15,000 feet and more above the sea; and Baron Humboldt saw it wheeling in circles at the elevation of 22,000 feet. It inhabits the Andes from the Straits of Magellan to 7° N. lat., but it never crosses the isthmus of Panamá, the Condor of California being a smaller bird. It roams over the plains of Patagonia even to the mouth of the Rio Negro, and at times descends from the Andes in groups to feed on the sea-shore upon dead whales; like all the vulture race, it possesses the faculty of descrying a dead or dying animal from a very great distance. Although the Condor lives principally on dead animals and carrion, it will sometimes attack live animals; its habits are those of our ordinary vulture; much exaggeration has found its way into books as to its size and ferocity; the most remarkable point in its history is the great vertical extent in which it is known to live, from the level of the sea to an elevation of nearly four statute miles. The Vultur papa, or king of the vultures, an inhabitant of the tropical regions, is remarkable for the bright blue and vermilion colour of the head and neck; the black vulture lives in large flocks on the tops of high trees in the silvas of Brazil: another numerous species prey on animals in the llanos or plains. Many other rapacious birds are peculiar to this continent; the burrowing owl, so common in the Pampas of Buenos Ayres, is one of these. The Guacharo bird forms a genus of itself (Steatornis): it is the size of a common fowl, with the form and beak of a bird of prey, and is a singular instance of a nocturnal bird feeding on fruit. It shuns the day, and is found under the natural bridge of Pandi, near Bogota, and in the caverns of Guadeloupe and Trinidad: incredible numbers have taken possession of the dark cavern of Guacharo in the valley of Caripe, where they are killed by thousands every year by the Indians for their fat.

The Troupials represent our Orioles, the Bataras and Becardes our shrikes, while the Tangaras partake of the form both of the Shrike and Pie, which last, with all the rest of the crow family, have various representatives in this country. Swallows, or birds allied to them, are numerous, and many that live on the honeyed juice of flowers, like the humming-bird, so peculiarly characteristic of South America; 300 species of humming-birds, from the size of a wren to that of an humblebee, adorn the tropical regions of Brazil and Guiana. This family, so entirely American, has a range from the Straits of Magellan to the 38th parallel of N. lat.; it may be met with in the forests on the mountains of Orizaba, at an elevation of 11,000 feet above the sea; and some beautiful species of it at still greater heights in the Andes of Bolivia and New Granada. There are only three South American humming-birds which visit the United States, and only a few are permanent in Central America: many of them are migratory; they come in multitudes to northern Chile in summer, and disappear in winter. The climbing-birds, with large bills, are mostly confined to the tropical forests, which swarm with peculiar races of parrots, paroquets, and macaws. It is a remarkable circumstance in the distribution of birds that there should be 40 species of parrots in the torrid zone of America, and only three species on the opposite coast of Africa, though the climate is similar and the vegetation nearly as luxuriant. Parrots range from the Straits of Magellan to the 42nd parallel of N. lat., where the Eider-duck, which is a peculiar Arctic bird, first shows itself. There are whole families of birds in tropical America not to be seen elsewhere: as the vividly-coloured Toucan, with its huge beak; the Araucari, which lives on the fruit of the Araucaria; some peculiar species of the gorgeous Trogons or Couroucous; the Toomatias—birds related to the cuckoo tribe; and the Jacamars, which represent the woodpeckers.

The gallinaceous family is totally different from that of the North American forests; the Guan or Penelope represents our pheasants, the different species of Crax or Alectors the wild turkey, which they equal in size and brilliancy of plumage; whilst the numerous species of Tinamous and cognate genera fill the place of the grouse, quails, and partridges of the old continent. South America furnishes two species of gallinaceous birds of a very peculiar character—the Cariama of Brazil, like to the secretary-bird of the Cape of Good Hope in its form and its habits of destroying reptiles; and the Kamichi, which possesses one or more sharp triangular spurs at the point of each wing, a dreadful instrument of attack and defence, such as is possessed by no other bird to the same extent.

The three-toed or American ostrich, or Struthio Rhea, ranges, like all its congeners, over a wide extent of country. It is found from the silvas of Brazil to the Rio Negro, which bounds the Pampas of Buenos Ayres on the south, and in some of the elevated plains of the Peru-Bolivian Cordilleras; while the Struthio Darwinii roams over the plains of Patagonia to the Straits of Magellan.

The water-fowl and waders in this land of rivers are beyond number; millions of Flamingoes, Spatulas, Cormorants, Herons, fishing falcons, and scissor-beaks, follow the fish as they go up the rivers to spawn; nor are gulls wanting where fish are to be found: a little snow-white heron walks on the back and over the head of the crocodile while it sleeps. The water-fowl are almost all peculiar. Eight or nine genera belonging to the warm climates of the Old World are here under new forms, and the number of specific forms of the same genus is greater than in any other country. The beautiful red Ibis or Tantalus ruber inhabits Cayenne; the Ardea helias, the most beautiful of the heron tribe, from its variegated plumage, is found in the same country.

Ducks migrate in immense flocks, alternately between the Orinoco and the Amazon, on account of the greater supply of fish afforded by the floods of these rivers, which take place at intervals of six months from each other. Between the tropics the vicissitudes of drought and humidity have much influence on the migration of birds, because the supply of their food depends upon these changes.

If anything more were required to show the partial location of birds, the Galapagos Archipelago might be mentioned. Of 26 specimens shot by Mr. Darwin, 25 were peculiar, though bearing a strong resemblance to American types; some birds (the Orpheus or Geospizinæ) were even confined to particular islands. But on this comparatively recent volcanic group, only 500 miles distant from the coast of America, everything is peculiar—birds, plants, reptiles, and fish; and though under the equator, none have brilliant colours.

The coasts of Peru and northern Chile, from their desert nature, are not rich in land-birds, but in southern Chile there are several species of humming-birds, parrots, flamingoes, peculiar ducks and geese; and there commences that inconceivable quantity of sea-fowl that swarm on the seas and coasts of the Antarctic regions. The black sheerwater, or rynchops nigra, has been seen to form a dense mass seven miles long; shags fly in an unbroken line of two miles. Pelicans, terns, petrels, and many others cover the low islands and coasts of the main-land, and those of Terra del Fuego.

In the Antarctic and Southern seas, petrels take the place which gulls occupy in northern latitudes; several species of them inhabit these high southern latitudes in prodigious numbers. Two remarkable species of this genus are found throughout the Southern Ocean —the Giant Petrel (P. gigantea), equal to the albatross in size, and resembling it in its mode of life—it sometimes becomes perfectly white; and the Equinoctial Petrel (P. æquinoctialis), a beautiful bird as large as our domestic fowl, and of a jet-black colour. A flock of what was supposed to be the young of the species known as the Pintado or Cape Pigeon (procellaria capensis) was estimated to have been from six to ten miles long, and two or three miles broad, absolutely darkening the air during the two or three hours they were flying over the Discovery ships. The snowy petrel, a most elegant bird, never leaves the ice, and consequently is seldom seen north of the Antarctic circle in summer. Four species of the southern Penguin (Aptenodytes) inhabit these seas; the A. patachonica, the largest bodied of sea-fowls, is a rare, and, for the most part, solitary bird, lives on the pack-ice, and weighs from 60 to 70 pounds. Two other species are smaller and gregarious; they crowd the snow-clad islands in the high southern latitudes in myriads; every ledge of rock swarms with them, and on the shore of Possession Island, close to Victoria Land, it was difficult to pass through the multitudes. They are fine, bold birds, pecking and snapping with their sharp bills at those who venture among them. They can scarcely walk, and, their wings being mere flappers, they cannot fly; they skim along the sea, and swim rapidly, even under water, resembling more a fish or a seal than a bird in their movements, and the noise they make baffles all description. Two species of albatross breed in the Antarctic islands; a kind of skua gull, which robs their nests; and a goose which, like the eider-duck, makes its nest with the down from its breast. A very curious bird, forming as it were a passage between the gallinaceous birds and waders, the Chionis or Vaginalis alba, is only found near the southern extremity of the American continent: it is of a milky white, and of the size of our domestic pigeon, and often takes refuge on the yards of ships off Cape Horn and Staten Land; it lives chiefly on a small species of cuttle-fish. Few land-birds are met with within the Antarctic circle: there are but seven or eight species in the Auckland Islands, mostly New Zealand birds; among others, the Tooa or Tui, and an olive-coloured creeper, the choristers of the woods. One only was found in Campbell Island.

Many generic forms are the same at the two extremities of the globe, yet with distinct specific differences. Sea-fowls are more excursive than other birds, but even they confine themselves within definite limits, so that the coasts may be known from their winged inhabitants.

AUSTRALIAN BIRDS.

The birds of Australia are in many respects as singular as the quadrupeds and plants of that country: a white falcon is among its birds of prey, a black swan among its water-fowl, and 61 genera are entirely Australian. The passeres are so original, that they have furnished many new genera. The Cassican, a handsome bird of bright colours, approaching somewhat to the crow family, the Choucalcyon, the golden and black Oriole, and one species of Philedon, are peculiarly Australian. The Menura superba, or Lyre-bird, from the resemblance its outspread tail bears to the form of the ancient lyre, is the only bird of its genus, and the only one which approaches the character of the gallinaceous family. Here are many new forms of cuckoos, as the Coucal and the Scythrops, the only bird of its genus. Woodpeckers there are none. The parrots, paroquets, and cockatoos, which live in numerous societies, are all peculiar, especially the black cockatoo, which is found here only; it is not so gregarious, but even more suspicious than the white cockatoos, which have a sentinel to warn them of danger. Chious, with huge bills like the toucan, satin-birds, pigeons and doves of peculiar forms, abound; and the Cereopsis goose is no less peculiar among the web-footed tribe. The desert plains of this great continent are allotted to the Emu, a large struthious bird, like its congener the Cassowary, incapable of flight, and once very plentiful, but now in progress of being extirpated or driven by the colonists to the unexplored regions of the interior.

The Apteryx, a bird of the same family, still lingers in New Zealand, but it is on the verge of extinction, and probably owes its preservation to its nocturnal and burrowing habits. It is one of those anomalous creatures that partakes of the character of several others; its head is in shape something like that of the Ibis, with a long, slender bill, fitted for digging into the ground for worms and grubs; its legs and feet resemble those of the common fowl, with a fourth toe or spur behind, in which it differs from its congeners; and its wings, if wings they can be called, are exceedingly small. In a specimen, whose body measured 19 inches, the wings, stripped of the feathers, were only an inch and a half long, ending in a hard horny claw three inches long. The comparatively small wings are characteristic of the whole family; the Ostrich and Rhea have the largest, which, though unavailing in flight, materially aid their progress in running; the wings of the Emu and Apteryx serve only as weapons of defence: the whole tribe also defend themselves by kicking. No animals have a more remarkable geographical distribution than this family, or show more distinctly the decided limits within which they have originally been placed. These huge birds can neither fly nor swim, consequently they could not have passed through the air or the ocean to distant continents and islands. There are five distinct genera, to each of which very extensive and widely separated countries have been allotted: the Ostrich is spread over Africa, from the Cape of Good Hope to the deserts of Arabia; two species of the Rhea range over the Pampas, the plains of Patagonia, and the elevated plains of Southern Bolivia; the continent of Australia is the abode of the Emu; the Cassowary roves over some of the large islands of the Indian Archipelago; and the Apteryx, as stated above, dwells exclusively in New Zealand. The Dodo, a very large short-winged bird, extirpated by the Dutch navigators, inhabited the Mauritius, and belonged probably to the ostrich tribe. Recent observations of its skeleton have led some naturalists to think it more akin to Trenons, or fruit-eating pigeons. The Solitaire, another species, also allied to the pigeons, lived on the Island of Rodriguez, and the Isle of Bourbon was inhabited by two other species, all of which are extinct.

The remains of a very numerous group of extinct struthious birds have been lately discovered imbedded in the very recent geological deposits of New Zealand. One of its genera, the Dinornis, chiefly found in the north island, has several species, the largest of which, the D. giganteus, attains a height of 11 feet, or double that of the largest ostrich; another, the Palapteryx, almost peculiar to the middle island, upwards of 9 feet. From the geological position in which these bones are found, as well as from their state of conservation, they can scarcely be considered as fossil, although belonging to species which have become extinct. Professor Owen has described no less than six species of Dinornis, and four of Palapteryx; and later discoveries in the colony have added several to these numbers. No better example can be cited, as elucidating the certitude of the deductions of the comparative anatomist, than what led to the first discovery of this extraordinary group of birds. A small portion of a bone, which from its dimensions appeared to belong rather to a quadruped of the size of an ox than to a bird, was submitted to Mr. Owen; he boldly pronounced it, from its structure, to belong to a bird of the ostrich kind — a decision that was soon abundantly confirmed by the discovery not only of the bones of the bird, but of its eggs.

The bones of another extinct bird, perhaps a Nestor, have been found, mixed with those of the Dinornis. It is allied to the curious living genus Strigops, something between an owl and a parrot, but more nearly allied to the latter. There are two living species of the Nestor; one in New Zealand; another, almost extinct, in Philip Island, only five miles in extent, and it is found in no other part of the world. The Notornis, a race supposed to have been extinct, closely allied to the water-hen, of the size of the bustard, had also been an ancient inhabitant of these islands,[1] where birds did and do exist, almost to the entire exclusion of quadrupeds and reptiles: an extinct species of dog, and a rat still existing, are the only land animals which shared in these extensive territories with multitudes of the feathered race.[2]

---

[1] This beautiful bird has just been found living in the Middle Island, at Dusky Bay. Its nearest affinity is with a genus of the Rallidæ or land-rail family.

[2] In some parts of the earth the same conditions which regulated the distribution of the ancient fauna and flora still prevail. The flora of the car-

The ostrich family live on vegetables; the form of those that had their home in New Zealand would lead to the conclusion that they had fed on the roots of the edible fern which abounds in that country; and as no quadruped excepting a rat is indigenous in New Zealand, though 700 miles long, and in many places 90 wide, these birds could have had no enemy but man, the most formidable of all.

The beautiful and sprightly Tui, or parson-bird, native in New Zealand, is jet black with a white tuft on its breast, and so imitative that it can be taught to repeat whole sentences. There are parrots and paroquets, vast numbers of pigeons, fine warblers, many small birds, and a great variety of water-fowl, amongst others a cormorant, which, though web-footed, perches on the trees that overhang the streams and sea, watching for fish; and a snow-white frigate-bird, that pounces on them from a great height in the air. Altogether there are at least 84 species of birds that inhabit these islands.[1]

CHAPTER XXXII.

Distribution of Mammalia.

CARBONIC acid, water, and ammonia contain the elements necessary for the support of animals, as well as of vegetables. They are supplied to the herbivora in the vegetable food, which is converted into animal matter by their vital functions.

Vitality in animals, as in vegetables, is the power they have of assimilating their food, a process independent of volition, since it is

bonaceous epoch is perfectly similar to that of New Zealand, where ferns and club-mosses are so abundant; and the fauna of that ancient period had been representative of that which recently prevailed in these islands, since foot-prints of colossal birds have been discovered in the red sandstone of Connecticut.

The age of reptiles of the Wealden and other secondary periods is representative of the fauna of the Galapagos islands, which chiefly consists of tortoises and creatures of the lizard or crocodile family; and the cycadaceous plants and marsupial animals of the oolite are representative of the flora and fauna of Australia.

The colossal birds which prevailed in New Zealand, almost to the entire exclusion of reptiles and quadrupeds, lasted to a very late period.

[[1] The Academy of Natural Sciences of Philadelphia, contains the most extensive and most beautiful collection of birds in the world. It consists of more than 27,000 specimens, of which upwards of 22,000 specimens are mounted and exhibited. A representative of almost every species of birds known in the world may be seen there.]

carried on during sleep, and is the cause of force. Animals inhale oxygen with the air they breathe; part of the oxygen combines with the carbon contained in the food, and is exhaled in the form of carbonic acid gas. With every effort, with every breath, and with every motion, voluntary or involuntary, at every instant of life, a part of the muscular substance becomes dead, separates from the living part, combines with the remaining portion of inhaled oxygen, and is removed. Food, therefore, is necessary to compensate for the waste, to supply nourishment, and to restore strength to the nerves, on which all vital motion depends; for by the nerves volition acts on living matter. Food would not be sufficient to make up for this waste, and consequent loss of strength, without sleep; during which voluntary motion ceases, and the undisturbed assimilation of the food suffices to restore strength, and to make up for the involuntary motion of breathing, which is also a source of waste.

The perpetual combination of the oxygen of the atmosphere with the carbon of the food, and with the effete substance of the body, is a real combustion, and is supposed to be the cause of animal heat, because heat is constantly given out by the combination of carbon and oxygen; and, without a constant supply of food, the oxygen would soon consume the whole animal, except the bones.

Herbivorous animals inhale oxygen in breathing, and, as vegetable food does not contain so much carbon as animal, they require a greater supply to compensate for the wasting influence of the oxygen; therefore, cattle are constantly eating. But the nutritious parts of vegetables are identical in composition with the chief constituents of the blood; and from blood every part of the animal body is formed.

Carnivorous animals have not pores in the skin, therefore their supply of oxygen is from their respiration only; and as animal food contains a greater quantity of carbon, they do not require to eat so often as animals that feed on vegetables. The restlessness of carnivorous animals when confined in a cage is in some degree owing to the superabundance of carbon in their food. They move about continually to quicken respiration, and by that means procure a supply of oxygen to carry off the redundant carbon.

The quantity of animal heat is in proportion to the amount of the oxygen inspired in equal times. The heat of birds is greater than that of quadrupeds, and in both it is higher than the temperature of amphibious animals and fishes, which have the coldest blood. On these subjects we are indebted to Professor Liebig, who has thrown so much light on the important sciences of animal and vegetable chemistry.

The mammalia consist of nine orders of animals, which differ in appearance and in their nature; but they agree in the one attribute of suckling their young. These orders are — the Quadrumana, animals which can use their fore and hind feet as hands, as monkeys and apes; Cheiroptera, animals with winged arms, as bats; Carnivora, that live on animal food, as the lion, tiger, bear, &c.; Rodentia, or gnawers, as beavers, squirrels, mice; Edentata or toothless animals,[1] as anteaters and armadilloes; Pachydermata, or thick-skinned animals, as the elephant, the horse, hippopotamus, and hog; Ruminantia, animals that chew the cud, as camels, lamas, giraffes, cows, sheep, deer; Marsupialia, possessing a pouch in which the young is received after birth; and Cetaceæ, inhabiting the waters, as whales, dolphins, porpoises, &c.

The animal creation, like the vegetable, varies correspondingly with height and latitude; the changes of species in ascending the Himalaya, for instance, are similar to what a traveller would meet with in his journey from an equatorial to a high latitude. The number of land animals increases from the frigid zones to the equator, but the law is reversed with regard to the marine mammalia, which abound most in high latitudes. Taking a broad view of the distribution of the nine orders of mammalia, it may be observed that the tropical forests are the chief abode of the monkey tribe: Asia is the home of the ape, especially the islands of the Indian Archipelago, as far as the most easterly meridian of Timor, beyond which there are none.

They abound throughout Africa from the Cape of Good Hope to Gibraltar, where the Barbary ape or magot is found: another species of magot inhabits the island of Niphon, the northern limit of monkeys at the eastern extremity of the old continent.

The bats that live on fruits are chiefly met with in tropical and warm climates, especially in the Indian Archipelago; the common bats, which live on insects, and are so numerous in species as to form more than a third of the whole family, are found everywhere except in arctic America. The Vampire is only met with in tropical America. Carnivorous mammalia are distributed all over the globe, though very unequally: in Australia there are only four species, two of which are bats; there are only 13 in South America, and 27 in the Oceanic region; while in the tropical regions of America there are 109, in Africa 130, and in Asia 166 species of carnivora; and so rapid is their increase towards the tropical regions, that there are nearly three times as many in the tropical as in the temperate zones.

With regard to the Gnawers or Rodents, species of the same group frequently have a wide range in the same, or nearly the same, parallels of latitude, but when they are inhabitants of high mountain-ridges they follow the direction of the chain, whatever that may be, and groups confined to high latitudes often appear again at great

[1] Or more properly wanting certain teeth, as the canines or incisors.

elevations in low latitudes. The Edentata are particularly characteristic of South America, where there are three times as many species as there are in Asia, Africa, and Australia taken together. In the three latter countries they only occur at intervals, but in America they extend from the tropic of Cancer to the plains of Patagonia. Thick-skinned animals are very abundant in the old continent, especially in Asia and Africa; they have been introduced into North America by man, but in the southern part of that continent the only indigenous species is the Tapir. The Ruminantia abound all over the temperate and tropical countries of both continents, and three species are found as far as the Arctic regions—there are neither Ruminantia nor Pachydermata in Australia. The Marsupialia are confined to Australia, New Guinea, and America.

The distribution of animals is guided by laws analogous to those which regulate the distribution of plants, insects, fishes, and birds. Each continent, and even different parts of the same continent, are centres of zoological families, which have always existed there, and nowhere else; each group being almost always specifically different from all others.

Food, security, and temperature have little influence, as primary causes, in the distribution of animals. The plains of America are not less fit for rearing oxen than the meadows of Europe; yet the common ox was not found in that continent at the time of its discovery; and with regard to temperature, this animal thrives on the Llanos of Venezuela and the Pampas of Buenos Ayres as well as on the steppes in Europe. The horse is another example: originally a native of the deserts of Tartary, he now roams wild in herds of hundreds of thousands on the grassy plains of America, though unknown in that continent at the time of the Spanish conquest.[1] All animals, however, are not so flexible in their constitutions, for most of them would perish from change of climate. The stations which the different families now occupy must have been allotted to them as each part of the land rose above the ocean; and because they have found in these stations all that was necessary for their existence, many have never wandered from them, notwithstanding their powers of locomotion; while others have migrated, but only within certain bounds.

Instinct leads animals to migrate when they become too numerous: the rat in Kamtchatka, according to Pennant, sets out in spring in great multitudes, and travels 800 miles, swimming over rivers and lakes; and the Lapland marmot or Lemming, a native in the mountains of Kolen, migrates in bands, once or twice in 25 years, to the

[1] There exist, however, remains of an extinct species of horse in several parts of South America, contemporaneous with the mastodons, and gigantic lost Edentata of that continent.

Western Ocean, which they enter and are drowned; other bands go through Swedish Lapland and perish in the Gulf of Bothnia. Thus nature provides a remedy against the over-increase of any one species, and maintains the balance of the whole. A temporary migration for food is not uncommon in animals. The wild ass, a native of the deserts of Great Tartary, in summer feeds to the east and north of the lake of Aral, and in autumn they migrate in thousands to the north of India, and even to Persia.[1] The ruminating animals that dwell in the inaccessible parts of the Himalaya descend to their lower declivities in search of food in winter; and for the same reason the reindeer and musk-ox leave the Arctic snows for a more southern latitude.

The Arctic regions form a district common to Europe, Asia, and America. On this account, the animals inhabiting the northern parts of these continents are sometimes identical, often very similar; in fact, there is no genus of quadrupeds in the Arctic regions that is not found in the three continents, though there are only 27 species common to all, and these are mostly fur-bearing animals. In the temperate zone of Europe and Asia, which forms an uninterrupted region, identity of species is occasionally met with, but for the most part marked by such varieties in size and colour as might be expected to arise from difference of food and climate. The same genera are sometimes found in the intertropical parts of Asia, Africa, and America, but the same species never; much less in the south temperate zones of these continents, where all the animals are different, whether birds, beasts, insects, or reptiles; but in similar climates analogous tribes replace one another.

Europe has no family and no order peculiarly its own, and many of its species are common to other countries; consequently the great zoological districts, where the subject is viewed on a broad scale, are Asia, Africa, Oceanica, America, and Australia; but in each of these there are smaller districts, to which particular genera and families are confined. Yet when the regions are not separated by lofty mountain-chains, acting as barriers, the races are in most cases blended together on the confines between the two districts, so that there is not a sudden change.

EUROPEAN QUADRUPEDS.

The character of the animals of temperate Europe has been more changed by the progress of civilization than that of any other quar-

---

[1] Perhaps no quadruped in the wild state will be found to have so wide a vertical range of habitat as this animal. It is found in the plains of Tartary, in the valley of the Tigris, at a very few feet above the sea-level, and in the most elevated valleys of the Himalaya and in Tibet, at elevations exceeding 15,200 feet.

ter of the globe. Many of its original inhabitants have been extirpated, and new races introduced; but it seems always to have had various animals capable of being domesticated. The wild cattle in the parks of the Duke of Hamilton and the Earl of Tankerville are the only remnants of the ancient inhabitants of the British forests, though they were spread over Europe, and perhaps were the parent stock from which the European cattle of the present time have descended; the Aurochs, a race nearly extinct, and found only in the forests of Lithuania, may also have some claim to having furnished the races of our domestic cattle. Both are supposed to have come from Asia. The Mouflon, which exists in Corsica and Sardinia, is by some supposed to be the parent stock of our domestic sheep. The pig, the goat, the red and fallow-deer have been reclaimed, and also the reindeer, which cannot strictly be called European, since it also inhabits the northern regions of Asia and America. The Cat is European; and altogether eight or ten species of our domestic quadrupeds have sprung from native animals.

A remarkable uniformity prevails in the organization and instincts of each species of animal in its wild state. Many adapt themselves to change of climate: after some generations their habits and organization alter by degrees to suit the new condition in which they are placed, but domestication is the cause of all our tame and useful tribes; by high cultivation and training great changes have been produced in form; and in some instances habits and powers of perception are induced, approaching to reason, which remain hereditary as long as the breed is unchanged.

There are still about 180 wild quadrupeds in Europe: 45 of these are also found in Western Asia, and nine in northern Africa. The most remarkable are the reindeer, elk, red and fallow-deer, the roebuck, glutton, lynx, polecat, several wild cats, the common and black squirrels, the fox, wild boar, wolf, the black and the brown bear, several species of weasels and mice. The otter is common; but the beaver is now found only on the Rhine, the Rhone, the Danube, and some other large rivers; rabbits and hares are numerous; the hedgehog is everywhere; the porcupine in southern Europe only; the chamois and ibex in the Alps and Pyrenees. Many species of these animals are widely distributed over Europe, generally with variations in size and colour. The Chamois of the Alps and Pyrenees, though the same in species, is slightly varied in appearance; and the fox of the most northern parts of Europe is larger than that of Italy, with a thicker fur, and of somewhat different colour.

Some animals never descend below a certain height, as the ibex and chamois, which live on higher ground than any of their order, being usually found between the region of trees and the line of perpetual snow, which is about 8900 feet on the southern, and 8200 on the northern declivities of the Alps. The red deer does not ascend beyond 7000 feet, and the fallow-deer not more than 6000, above the level of the sea: these two, however, descend to the plains, the former never do. The bear, the lynx, and the stoat are sometimes met with nearly at the limit of perpetual snow.

Some European animals are much circumscribed in their locality. The Ichneumon is peculiar to Egypt; the mouflon is confined to Corsica and Sardinia; a species of weasel and bat inhabit Sardinia only; and Sicily has several peculiar bats and mice. There is only one species of monkey in Europe, which lives on the rock of Gibraltar, and is supposed to have been brought from Africa. All the indigenous British quadrupeds now existing, together with the extinct hyæna, tiger, bear, and wolf, whose bones have been found in caverns, are also found in the same state in Germany. Ireland was probably separated by the Irish Channel from England before all the animals had migrated to the latter; so that our squirrel, mole, pole-cat, dormouse, and several smaller quadrupeds, never reached the sister island. Mr. Owen has shown that the Britich horse, ass, hog, the smaller wild ox, the goat, roe, beaver, and many small rodents, are the same species with those which had co-existed with the mammoth or fossil elephant, the great northern hippopotamus, and two kinds of rhinoceros long extinct. So that a part only of the modern tertiary fauna has perished, from whence he infers that the cause of their destruction was not a violent universal catastrophe from which none could escape. The Bos longifrons and the gigantic Elk of the Irish bogs were probably co-existent with man.

ASIATIC QUADRUPEDS.

Asia has a greater number and a greater variety of wild animals than any country, except America, and also a larger proportion of those that are domesticated. Though civilized from the earliest ages, the destruction of the animal creation has not been so great as in Europe, owing to the inaccessible height of the mountains, the extent of the plains and deserts, and, not least, to the impenetrable forests and jungles, which afford them a safe retreat: 288 mammalia are Asiatic, of which 186 are common to it and other countries; these, however, chiefly belong to the temperate zone.

Asia Minor is a district of transition from the fauna of Europe to that of Asia. There the chamois, the bouquetin or ibex, the brown bear, the wolf, fox, hare, and others, are mingled with the hyæna, the Angora goat, which bears a valuable fleece, the Argali or wild sheep, the white squirrel; and even the Bengal royal Tiger is sometimes seen on Mount Ararat, and is not uncommon in Azerbijan and the mountains in Persia.

Arabia is inhabited by the hyæna, panther, jackal, and wolf. Antelopes and monkeys are found in Yemen. Most of these are also indigenous in Persia. The wild ass, or Onagra, a handsome spirited animal of great speed, and so shy that it is scarcely possible to approach it, wanders in herds over the plains and table-lands of Central Asia. It is also found in the Indian desert, and especially in the Run of Cutch—"the wilderness and barren lands are his dwelling"—and in the most elevated regions of Tartary and Tibet, on the shores of the sacred lakes of Manasarowar and Rakasthal, at a height of more than 15,250 feet above the sea.[1]

The table-lands and mountains which divide eastern Asia almost into polar and tropical zones, produce as great a distinction in the character of its indigenous fauna. The severity of the climate in Siberia renders the skins of its numerous fur-bearing animals more valuable. These are reindeer, elks, wolves, the large white bear, that lives among the ice on its Arctic shores, several other bears, the lynx, various kinds of martens and cats, the common, the blue, and the black fox, the ermine, and the sable. The fur of these last is much esteemed, and is only equalled by that of the sea-otter, which inhabits the shores on both sides of the northern Pacific.

Many of the Asiatic species of gnawers are confined to Siberia. The most remarkable of these is the flying squirrel, the Jerboa, which burrows in sandy deserts, on the table-land and elsewhere. The Altaï Mountains teem with wild animals: besides many of those mentioned, we also find here several large stags, bears, some peculiar weasels, the argali, and the wild sheep. The wild goat of the Alps is found in the Sayansk part of the chain; the Glutton and musk-deer in the Baikal; and in Daouria the red-deer and the Antelope Saiga. The Bengal tiger and the Felis Irbis, a species of panther, wander from the Celestial Mountains to the Altaï and into southern Siberia: the Tiger is met with even on the banks of the Obi, and also in China, though in the northern regions it differs considerably, but not specifically, from that of Bengal; thus it can bear a mean annual temperature of from 81° of Fahrenheit to the freezing point. The Tapir, and many of the animals of the Indian Archipelago, are found in the southern provinces of the Chinese empire. The animals of Japan have a strong analogy with those of Europe: many are identical, or slightly varied, as the badger, otter, mole, common fox, marten, and squirrel. On the other hand, a large species of bear in the island of Jesso resembles the grizzly bear in the Rocky Mountains of North America. A chamois in other parts of Japan

[1] It is by no means certain that the wild Ass of the three countries mentioned in the text belongs to the same species. The Kiang of Tibet appears to be the same as the Dziggetai (Equus Hemionus of Pallas), which is met with throughout central Asia; but the species found in the Run of Cutch is of a different colour and form: whilst the one neighs like a horse, the other brays like an ass; in one the striped colour of the zebra family is said to exist in the young, and not in the second.

is similar to the Antelope montana of the same mountains: and other animals native in Japan are the same with those in Sumatra; so that its fauna is a combination of those of very distant regions.

A few animals are peculiar to the high cold plains of the table-land of eastern Asia: the Dziggetai, a very fleet animal, is peculiar to these Tartarian steppes. Two species of antelopes inhabit the plains of Tibet, congregating in immense herds, with sentinels so vigilant that it is scarcely possible to approach them.

The Dzeran, or yellow goat, which is both swift and shy, and the handsome Tartar ox, are natives of these wilds; also the shawl-wool goat and the Manul, from which the Angora cat, so much admired in Persia and Europe, is descended. Most of the animals that live at such heights cannot exist in less elevated and warmer regions, exhibiting a striking instance of the limited distribution of species. Goats and sheep endure best the rarefied air and great cold of high lands: the Cashmere goat and Argali sheep browse on the plains of Tibet at elevations of from 10,000 to 13,000 feet; the Rass, a sheep with spiral horns, lives on the table-lands of Pamer, which are 15,000 feet above the sea; and also the Kutch-gar, a species of sheep which is about the height of a year-old colt, with fine curling horns: they congregate in flocks of many hundreds, and are hunted by the nomade tribes of Kirghis.

The ruminating animals of Asia are more numerous than those of any other part of the world; 64 species are native, and 46 of these exist there only. There are several species of wild oxen; one in the Birmese empire, and on the mountains of north-eastern India, with spiral twisted horns. The buffalo is a native of China, India, Borneo, and the Sunda Islands; it is a large animal, formidable in a wild state, but domesticated throughout the East. It was introduced into Italy in the sixth century, and large herds now graze in the low marshy plains near the sea.

Various kinds of oxen have been domesticated in India from time immemorial: the Zebu or Indian ox, with a hump on the shoulders, has been venerated by the Brahmins for ages; the beautiful white silky tail of the domesticated ox, or Yak, of Tartary, used in the East to drive away flies, was adopted as the Turkish standard; and the common Indian ox differs from all others in the great speed of its course. Some other species of cattle have been tamed, and some are still wild in India, Java, and other Asiatic islands. The Cashmere goat, which bears the shawl-wool, is the most valuable of the endless varieties of goats and sheep of Asia; it is kept in large herds in the great valleys on the northern and southern declivities of the Himalaya, and in the upper regions of Bhotan, where the cold climate is congenial to it.

The Bactrian camel, with two humps, is strong, rough, and hairy, and is said to be found in a wild state in the desert of Shamo: it is the camel of central Asia, north of the Himalaya and Taurus, also of the Crimea and the countries round the Caucasus. The more common or Arabian camel with one hump is a native of Asia, though only known now in a domesticated state: it has been introduced into Africa, Italy, the Canary Islands, and even into the elevated regions of the Peru-Bolivian Andes. The best come from the province of Nejed in Arabia, which, on that account, is called the "mother of Camels." The camel of Oman is remarkable for beauty and swiftness.

Ten species of antelopes and twenty of deer are peculiar to Asia: two species of antelopes have already been mentioned as peculiar to the table-lands, the others are distributed in the Asiatic archipelago. The genuine musk-deer (Moschus moschiferus) inhabits the mountainous countries of central and south-eastern Asia, between China and Tartary, the regions round lake Baikal, the Altaï mountains, Nepaul, Bhotan, Tibet, and the adjacent countries of China and Tonquin.

Asia possesses about ten native species of Pachydermata, including the elephant, horse, and ass, which have been domesticated from the time of the earliest historical records. The horse is supposed to have existed wild in the plains of central Asia, as the dromedary in Arabia; though now they are only known as domestic animals. The Arabian and Persian horses possess acknowledged excellence and beauty, and from these our best European horses are descended; the African horse, which was introduced into Spain by the Moors, is probably of the same race.

The elephant has long been a domestic animal in Asia, though it still roams wild in formidable herds through the forests and jungles at the foot of the Himalaya, in other parts of India, the Indo-Chinese peninsula, and the islands of Sumatra and Ceylon; the hunting elephant is esteemed the most noble. A one-horned rhinoceros is a native of continental Asia.

There are several genera of Asiatic carnivorous animals, of which the royal tiger is the handsomest and the most formidable; its favourite habitation is in the jungles of Hindostan, though it wanders nearly to the limit of perpetual snow in the Himalaya, to the Persian and Armenian mountains, to Siberia and China. Leopards and panthers are common, and there is a nameless variety of the lion in Guzerat; the Cheetah, used in hunting, is the only one of the panthers capable of being tamed. The hyæna is found everywhere, excepting the Birman empire, in which there are neither wolves, hyænas, foxes, nor jackals. There are four species of bears in India; that of Nepaul has valuable fur: the wild boar, hog, and dogs of endless variety, abound.

The edentata have only two representatives in India, which differ from all others except the African, in being covered with imbricated scales. Of these the short-tailed pangolin, or scaly ant-eater, is found throughout the Deccan, Bengal, Nepaul, the southern provinces of China, and Formosa.

The Indian Archipelago and the Indo-Chinese peninsula form a zoological province of a very peculiar nature, being allied to the faunas of India, Australia, and South America, yet having animals exclusively its own. The royal tiger is in great abundance in the peninsula of Malacca, and also the black variety of the panther, leopard, wild cats, multitudes of elephants, the rhinoceros of all three species, the Malayan tapir, many deer, the Babiroussa hog, and another species of that genus. Some groups of the islands have several animals in common, either identical or with slight variations, that are altogether wanting in other islands, which, in their turn, have creatures of their own. Many species are common to the Archipelago and the neighbouring parts of the continent, or even to China, Bengal, Hindostan, and Ceylon. Flying quadrupeds are a distinguishing feature of this archipelago, though some do not absolutely fly, but, by an extension of the skin of their sides to their legs, which serves as a parachute, are enabled to take long leaps and to support themselves in the air. Nocturnal flying squirrels, of several species, are common to the Malayan peninsula and the Sunda islands, especially Java; and three species of flying Lemurs inhabit Sunda, Malacca, and the Pelew Islands. Besides these, there are the frugivorous bats, which really fly, differing from bats in other countries by living exclusively upon vegetable food. The edible Roussette, or Kalong, one of the largest known, appears in flocks of hundreds, and even thousands, in Java, Sumatra, and Banda : the pteropus funereus, another of these large bats, assembles in as great numbers.

A hundred and eighty species of the ape and monkey tribe are entirely Asiatic: monkeys are found only on the coast of India, Cochin-China, and the Sunda Islands: the long-armed apes or Gibbons belong to the Sunda Islands and the Malayan peninsula. The Simayang, a very large ape of Sumatra and Bencoolen, moves about in large troops, following a leader, and makes a howling noise at sunrise and sunset that is heard miles off. Sumatra and Borneo are the peculiar abode of the Orang-outang, a name which in the Malay language signifies the "man of woods;" except perhaps the Chimpanzee of Africa, it approaches nearest to man. It has never spread over the islands it inhabits, though there seems to be nothing to prevent it, but it finds all that is necessary within a limited district. The orang-outang and the long-armed apes have extraordinary muscular strength, and swing from tree to tree by their arms.

The Malays have given the name of orang, or man, to the whole tribe, on account of their intelligence as well as their form.

A two-horned rhinoceros is peculiar to Java, of a different species from the African, also the Felis macrocelis, and a very large bear; there are only two species of squirrels in Java, which is remarkable, as the Sunda Islands are rich in them. The Royal tiger of India and the elephant are found only in Sumatra, and the Babiroussa lives in Borneo; but these two islands have many quadrupeds in common, as a leopard, the one-horned rhinoceros, the black antelope, some graceful miniature creatures of the deer kind, the Tapir, also found in Malacca, besides a wild boar, an inhabitant of all the marshy forests from Borneo to New Guinea. In the larger islands deer abound, some as large as the elk, probably the Hippelaphus of Aristotle.

The Anoa, a ruminating animal about the size of a sheep, a species of antelope, shy and savage, goes in herds in the mountains of Celebes, where many forms of animals strangers to the Sunda Islands begin to appear, as some sorts of phalangers, or pouched quadrupeds. These new forms become more numerous in the Moluccas, which are inhabited by flying phalangers and other pouched animals, with hairless scaly tails. The phalangers are nocturnal, and live on trees. In New Guinea there are Kangaroos, the spotted phalanger, the New Guinea hog, and the Papua dog, said to be the origin of all the native dogs in Australia and Oceanica, wild or tame.

The fauna of the Philippine Islands is analogous to that in the Sunda Islands. They have several quadrupeds in common with India and Ceylon, but there are others which probably are not found in these localities.

AFRICAN QUADRUPEDS.

The opposite extremes of aridity and moisture in the African continent have had great influence in the nature and distribution of its animals; and since by far the greater part consists of plains utterly barren or covered by temporary verdure, and watered by inconstant streams that flow only a few months in the year, fleet animals, fitted to live on arid plains, are far more abundant than those that require rich vegetation and much water. The latter are chiefly confined to the intertropical coasts, and especially to the large jungles and deep forests at the northern declivity of the table-land, where several genera and many species exist that are not found elsewhere. Africa has a fauna in many respects different from that of every other part of the globe; for although about 100 of its quadrupeds are common to other countries, there are 250 species of its own. Several of these animals, especially the larger kinds, are distributed over the whole table-land from the Cape of Good Hope to the highlands of Abyssinia and Senegambia without the smallest variety, and many are slightly modified in colour and size. Ruminating animals are very numerous, though few have been domesticated: of these the ox of Abyssinia and Bornou is remarkable from the extraordinary size of its horns, which are sometimes 2 feet in circumference at the root; and the Galla ox of Abyssinia has horns 4 feet long. There are many African varieties of Buffalo; that at the Cape of Good Hope is a large, fierce animal, wandering in herds in every part of the country, even to Abyssinia: the flesh of the whole race is tainted with the odour of musk. The African sheep and goats, of which there are many varieties, differ from those of other countries; the wool of all is coarse, except that of the Merino sheep, said to have been introduced into Spain by the Moors from Morocco.

No country has produced a ruminating animal similar to the Giraffe, or Camelopard, which ranges widely over South Africa from the northern banks of the Gareep, or Orange river, to the Great Desert; it is also found in Dongola and in Abyssinia. It is a gentle, timid animal, and has been seen in troops of 100. The earliest record we have of it is on the sculptured monuments of the ancient Egyptians, and it is well known that it was brought to Rome to grace the triumphs of a victorious emperor.

Africa may truly be said to be the land of the genus Antelope, which is found in every part of it, where it represents the deer of Europe, Asia, and America. Different species have their peculiar localities, while others are widely dispersed, sometimes with and sometimes without any sensible variety of size or colour. The greater number are inhabitants of the plains, while a few penetrate into the forests. Sixty species have been described, of which at least 26 are found north of the Colony of the Cape of Good Hope and in the adjacent countries. They are of every size, from the pigmy antelope, not larger than a hare, to the Caama, which is as large as an ox. Timidity is the universal character of the race. Most species are gregarious; and the number in a herd is far too great even to guess at. Like all animals that feed in groups, they have sentinels; and they are the easy prey of so many carnivorous animals, that their safety requires the precaution. At the head of their enemies is the lion, who lurks among the tall reeds at the fountain, to seize them when they come to drink. They are graceful in their motions, especially the Spring-Buck, which goes in a compact troop; and in their march there is constantly one which gathers its slender limbs together, and bounds into the air.[1]

Africa has only two species of deer, both belonging to the Atlas: one is the common fallow-deer of Europe.

The 38 species of rodentia, or gnawing quadrupeds, of this con-

[1] The reader will find an interesting enumeration of the South African antelopes in an article in the 179th Number of the 'Quarterly Review,' evidently from the pen of a distinguished naturalist, on Mr. Gordon Cumming's 'Hunter's Life in South Africa.'

tinent, live on the plains, and many of them are leaping animals, as the Jerboa capensis. Squirrels are comparatively rare.

There are some species of the horse peculiar to South Africa; of these the gaily-striped Zebra and the more sober-coloured Quagga wander in troops over the plains, often in company with ostriches. An alliance between creatures differing in nature and habits is not easily accounted for. The two-horned rhinoceros of Africa is different from that of Asia; there are certainly three, and probably five, species of these huge animals peculiar to the table-land. Dr. Smith saw 150 in one day near the 24th parallel of South latitude. The Hippopotamus is exclusively African: multitudes inhabit the lakes and rivers in the tropical and southern parts of the continent; those that inhabit the Nile and Senegal appear to belong to different species. An Elephant differing in species from that of Asia is so numerous, that 200 have been seen in a herd near lake Tchad. They are not domesticated in Africa, and are hunted by the natives for their tusks. The Phacochœre, or Ethiopian hog, and a species of Hyrax, are among the Pachydermata of this country. The monkey is found in all the hot parts of Africa: peculiar genera are allotted to particular districts. Except a few in Asia, the family of Guenons is found in no part of the world but about the Cape of Good Hope, and on or near the coasts of Loando and Guinea, where they swarm.

The species are numerous, and vary much in size and colours; the Cynocephalus, or dog-headed Baboon, with a face like that of a dog, is large, ferocious, and dangerous. A species of these Baboons inhabits Guinea, others the southern parts of the table-land, and one is met with everywhere from Sennaar to Caffraria. A remarkable long-haired species, the Hamadryas, is found in the mountains of Abyssinia, 8000 feet above the sea; the Mandrills, which belong to the same genus, come from the coasts of Guinea. The magot, or Barbary ape, is in North Africa. The African long-haired tailless apes, forming the genus Colobus, are met with in the tropical districts on the west coast; the C. Polycomos, or king of the monkeys, so called by the natives from its beautiful fur and singular head of bushy hair, is met with in the forests about Sierra Leone; another of these is peculiar in the low lands of Gojam Kulla, and Damot. The Chimpanzee, which so nearly approaches the human form, inhabits the forests of South-Western Africa from Cape Negro to the Gambia. Living in troops, like most apes and monkeys, which are eminently gregarious, it is very intelligent and easily tamed. A new species of the African Chimpanzee, equalling in size the Orang-outang, has been recently described by Professor Owen: it is probably the largest of the quadrumana, and by all accounts the most dangerous and ferocious.

Baron Humboldt observes that all apes resembling man have an expression of sadness; that their gaiety diminishes as their intelligence increases.

Africa possesses the cat tribe in great variety and beauty; lions, leopards, and panthers are numerous throughout the continent; servals and viverrine cats inhabit the torrid districts; and the lion of the Atlas has ever been considered the most formidable of carnivorous animals. In no country are foxes so abundant. Various species inhabit Nubia, Abyssinia, and the Cape of Good Hope. A long-eared fox, the Fennec, of Bruce, found from the Cape of Good Hope to Kordofan, is peculiar to Africa. There are also various species of dogs, the hyæna, and the jackal. The hyænas hunt in packs, attack the lion and panther, and end by destroying them.

Two species of Edentata are African—the long-tailed Manis, and the Aard-vark, or earth-hog: the first is covered with scales, the latter with coarse long hair; they burrow in the ground and feed on ants. Great flocks of a large migratory vampire-bat frequent the Slave-coast. Altogether there are 26 species of African bats.

Multitudes of antelopes of various species, lions, leopards, panthers, hyænas, jackals, and some other carnivora, live in the oases of the great northern deserts; jerboas, and endless species of leaping gnawers, rats, and mice burrow in the ground. The dryness of the climate and soil keeps the coats of the animals clean and glossy; and it has been observed that tawny and grey tints are the prevailing colours in the fauna of the North African deserts, not only in the birds and beasts, but in reptiles and insects. In consequence of the continuous desert extending from North Africa through Arabia to Persia and India, many analogous species of animals exist in those countries: in some instances they are the same, or varieties of the same species, as the ass, the dziggettai, antelopes, leopards, panthers, jackals, and hyænas.

The fauna on the eastern side of the great island of Madagascar is in some degree analogous to that of India; on the western side it resembles that of Africa, though, as far as it is known, it seems to be a distinct centre of animal life. It has no ruminating animals; and the monkey tribe is represented by the Lemures, the Galagos, and Indris, which are characteristic of this insular fauna. A frugivorous bat, the size of a common fowl, forms an article of food; and an anomalous animal, the Cheiromys, or Aye-Aye, intermediate between the Quadrumana and the Rodents, has only been found in this island.

AMERICAN QUADRUPEDS.

No species of animal has been yet extirpated in America, which is the richest zoological province, possessing 537 species of mammalia, of which 480 are its own; yet no country has contributed so little to the stock of domestic animals. With the exception of the Llama and Alpaca, and the Turkey, and perhaps some sheep and dogs, America has furnished no animal or bird serviceable to man, while it has received from Europe all its domestic animals and its civilized inhabitants.

Arctic America possesses most of the valuable fur-bearing animals that are found in Siberia; and they were very plentiful till the unsparing destruction of them has driven those yet remaining to the high latitudes, where the hunters that follow them are exposed to great hardships. Nearly 2,000,000 of skins were brought to England in the year 1835, most of which were taken in the forest regions; the barren grounds are inhabited by the Arctic fox, the polar hare, by the brown and the white bear, a formidable animal which often lives on the ice itself. The Reindeer lives on the lichens and mosses of these barren grounds, and wanders to the shores of the Polar Ocean: its southern limit in Europe is the Baltic Sea, in America it is the latitude of Quebec. Some of the fur-bearing quadrupeds of these deserts never pass the 65th degree of N. lat.; the greater number live in the northern forests, as the black bear, raccoon, badger, the ermine, and four or five other members of the weasel tribe, the red fox, the polar and brown lynxes, the beaver, the musquash or musk-rat, of which half a million are killed annually, and the moose-deer, whose northern range ends where the aspen and willows cease to grow. The grizzly bear, the largest and most ferocious of its kind, inhabits the range of the Rocky Mountains to Mexico, as well as the western savannahs. The prairie-wolf, the grey fox, the Virginian hare, live in the prairies; the Wapiti, a large stag, inhabits those on both sides of the Rocky Mountains; and the Prongbuck, an antelope fleeter than the horse, roams throughout the western part of the continent, and migrates in winter to California and Mexico. The musk-ox and shaggy bison are peculiar to North America. The musk-ox travels to Parry's Islands in the Arctic regions, yet it never has been seen in Greenland or on the north-western side of the continent. The shaggy bison goes as far south as the Arkansas, and roams, in herds of thousands, over the prairies of the Mississippi and on both sides of the Rocky Mountains. It seldom wanders farther north than the 60th parallel, the southern limit of the musk-ox. A Marmot known by the name of the Prairie-Dog is universal in the great plains from which it derives its name.

There are at least eight varieties of American dogs, several of which are natives of the far north. The Lagopus, or Isatis, native in Spitzbergen and Greenland, is found in all the Arctic regions of America and Asia, and in some of the Kurile Islands. Dogs are employed to draw sledges in Newfoundland and Canada; and the Esquimaux travel drawn by dogs as well as by reindeer. The dogs are strong and docile. The Esquimaux dogs were mute till they learned to bark from dogs in our discovery ships.

There are 13 species of the ruminating genus in North America, including the Bison, the Musk-ox of the Arctic regions, the big-horned sheep, and the goat of the Rocky Mountains. The horse, now roaming wild in innumerable herds over the plains of South America, was unknown there till the Spanish conquest. The quadrupeds of the temperate zone are distributed in distinct groups: those of the state of New York, consisting of about 40 species, are different from those of the Arctic regions, and also from those of South Carolina and Georgia; while in Texas another assemblage of species prevails. The Raccoon, the Coatimondi, and the Kinkajou are all natives of the southern States.

There are 118 species of rodentia, or gnawing animals, in North America; rats, mice, squirrels, beavers, &c., many of which, especially in the north, appear to be identical with those in the high latitudes of Europe and Asia. The genera of very different latitudes are often representatives, but never identical. Squirrels abound in North America; the grey squirrel is found in thousands.

There are 21 species of the genus Opossum in this continent. Of these the Virginian opossum inhabits the whole extent of America between the great Canadian lakes and Paraguay, and also the West India islands, where it is called the Manicou; and two other animals of that tribe live in Mexico. There is a Porcupine in the United States and Canadian forests which climbs trees. The bats are different from those in Europe, and, excepting two, are very local. In California there are Ounces, Polecats, Bears, and a species of Deer of remarkable size and speed.

The high land of Mexico forms a very decided line of division between the fauna of North and that of South America; yet some North American animals are seen beyond it, particularly two of the bears and an otter, which inhabit the continent from the Icy Ocean to beyond Brazil. On the other hand, the Puma, Jaguar, Opossum, Kinkajou, and Peccari have crossed the barrier from South America to California and the United States.

In the varied and extensive regions of South America there are several centres of a peculiar fauna according as the country is mountainous or level, covered with forest or grass, fertile or desert, but the mammalia are inferior in size to those of the Old World. The largest and most powerful animals of this class are confined to the Old Continent. The South American quadrupeds are on a smaller scale, more feeble and more gentle; many of them, as the toothless groups including the Sloths, are of anomalous and less perfect structure than the rest of the quadruped creation; but the fauna of South America is so local and so peculiar, that the species of five of the terrestrial orders, which are indigenous there, are found nowhere else.

The monkey tribe exist in myriads in the forests of tropical America and Brazil, but they are not seen to the north of the Isthmus of Darien, nor farther south than the Rio de la Plata. They differ widely from those in the Old World, bearing less resemblance to the human race; but they are more gentle and lively, and, notwithstanding their agility, are often victims to birds and beasts of prey.

There are two great American families of monkeys—the Sapajous with prehensile tails, by which they suspend themselves, and swing from bough to bough. Some of these inhabitants of the woods are very noisy, especially the Araguato, a large ape whose bawling is heard a mile off. The Howlers are generally very large, and have a wider range than any of the genus; one species, the mycetus rufimanus, or Beelzebub, ascends the Andes to the height of 11,000 feet. The Cebus, or weepers, which are frequently brought to Europe, belong also to this family; the genus has a greater number of species than any other in the New World, but a very narrow location; they are most abundant in Guiana.

The Saquis, or bushy-tailed monkeys, form the other division of the American monkeys. The fox-monkey sleeps during the day; it frequents the deepest forests from the Orinoco to Paraguay. Squirrel-monkeys inhabit the banks of the Orinoco, and the nocturnal monkeys, with very large eyes, live in Guiana and Brazil. The Marmosets are pretty little animals, easily tamed, especially the Midas leonina, not more than 7 or 8 inches long. Some American monkeys have no thumb on the forefoot, as the Ateles or Spider monkeys; others have a versatile thumb on both their hands and feet; whilst a third kind have no opposable thumb on any of their feet.

The forests are also inhabited by Opossums, a genus of the marsupial tribe, or animals with pouches, in which they carry their young; they are somewhat analogous to those which form the distinguishing feature of the Australian fauna, but of entirely distinct forms and species. Some of these animals are no larger than a rat, and they mostly live on trees. One is aquatic, the Chironectes, resembling a small otter, and appears to be only found in the river Yapock in French Guiana. A species in Surinam carries its young upon its back. All the opossums and the Chironectes of this country have thumbs on their hind feet, opposable to the toes, so that they can grasp; they are, moreover, distinguished from the Australian family by a long prehensile tail, and by greater agility. The numerous tribe of Sapajou monkeys, the Ant-eaters, the Kinkajou, and a species of porcupine, have also grasping tails, a peculiarity of many South American animals.

Five genera and 20 species of the Edentata are characteristic of this continent, and confined to South America: they are the two species of sloths, the Ai and the Unau: several Armadilloes, the

Chlamyphorus, and two Ant-eaters. The animals of these genera have very different habits; the sloths, as their name implies, are the most inactive of animals: they inhabit the forests from the southern limit of Mexico to Rio de Janeiro, and to the height of 3000 feet on the Andes, in the region of Palms and Scitamineæ. Of these the common sloth or Ai ranges from Mexico to Brazil; while the Unau, the larger of the two, is confined to Guiana and Brazil. The Armadillo, in its coat of mail, is in perpetual motion, and can outrun a man in speed. They live on all the plains of South America, as far south as Paraguay and the Pampas of Buenos Ayres. The one-banded armadillo rolls itself up like a ball; the nine-banded species is eaten by the natives; the giant armadillo, 3 feet long, inhabits the forests only. Most species of these animals are nocturnal, and burrow in the earth in the Pampas. The Chlamyphorus is also a burrowing animal, peculiar to the province of Mendoza on the eastern slope of the Chilian Andes; they have the faculty of sitting upright. The great or maned Ant-eater, larger than a Newfoundland dog, with shorter legs, defends itself against the jaguar with its powerful claws; it inhabits the swampy savannahs and damp forests from Columbia to Paraguay, and from the Atlantic to the foot of the Andes; its flesh, like that of some other American animals, has a flavour of musk. The little Ant-eater has a prehensile tail, and lives on trees in the tropical forests, feeding on the larvæ of bees, wasps, honey, and ants; another of similar habits lives in Brazil and Guiana. The cat tribe in South America is beautiful and powerful: the Puma, called the Lion of America, is found both in the mountains and the plains, in great numbers; so different are its habits in different places, that in Chile it is timid and flies from a dog; in Peru it is bold, though it rarely attacks a man. The Jaguar, which inhabits the lower forests, is very abundant, and so ferocious that it has been known to spring upon Indians in a canoe; hunting as it sometimes does in troops, it has been known to destroy the inhabitants of entire Indian villages; it is one of the few South American animals that cross the Isthmus of Panama, being found in California, in the State of Mississippi, and has been seen as far north as Canada; furnishing a remarkable analogy, in its extensive wanderings, with the Royal Tiger of the Old World, which, as we have already seen, is often found amidst the Siberian mountains and steppes.

The Vampire is a very large bat, much dreaded by the natives, because it enters their huts at night; and, though it seldom attacks human beings, it wounds calves and small animals, which sometimes die from the loss of blood. The other South American bats are innocuous.

The only ruminating animals except the deer that existed in South America prior to the Conquest were the four species of the genus Auchenia—the Llama, the Alpaca, the Vicuña, and the Guanaco: the first three are exclusively confined to the colder and more elevated regions of the Peruvian Andes; the last has a wider geographical range, extending to the plains of Patagonia, and even to the southernmost extremity of the continent. The Llama inhabits the high valleys of the Peru-Bolivian Andes, its favourite region being in the valley of the lake of Titicaca: it was the only beast of burthen possessed by the aborigines; hence we find it wherever the Incas carried their conquests and civilization, from the equator to far beyond the southern tropic. It is still extensively employed by the Indian as a beast of burthen, and its wool, though coarse, is used by the natives. Like all domestic animals, it varies in colour: its flesh is black, disagreeable, and ill tasted.

The Alpaca, or Paco, a gentle and handsome animal, although more closely allied to the llama than any of its congeners, appears to be a distinct species: it lives in still more elevated places than the llama, its favourite haunts being on the streams descending from the snowy peaks: it is only found in a domestic state; it is reared for its wool, which is extremely fine, silky, and long, and which now bears a high price, from its introduction into some of our finest woollen tissues. The Vicuña is only found in the wild state in the plains on the Andes, as high as 15,000 feet: the wool is much prized for its fineness. The animal has a shrill whistle; it is easily tamed. The Guanaco, by some naturalists considered erroneously as the parent stock of the llama and alpaca, is also only found in the wild state: it extends to 12° S. lat., is very abundant and in large flocks on the Bolivian and Chilian Andes, and has been seen as far south as the Straits of Magellan. All these animals feed principally on a species of coarse wiry grass called *ichu*.[1]

[1] The attention of the scientific world in France has been recently directed to the advantages that might arise from the naturalization of the Llama tribe in Europe, and especially of its two most useful species, the Llama and the Alpaca. M. I. Geoffroy St. Hilaire, a French zoologist of high standing, ignorant probably of what had been done in Great Britain on the same subject, where the experiment had long since been tried, and with very inadequate success, has presented lately some papers to the Academy of Sciences on this subject. We cannot imagine, even if the naturalization of the Llama on a large scale was possible, what advantage could arise from it to our agriculturists. The wool of the Llama is coarse, and so infinitely inferior to the commonest qualities of sheep's wool, that in its native country it is seldom used for any other purpose than the manufacture of ropes, of a rough carpeting and packing-cloth, and for the coarsest apparel of the poor Indian. As to its use as a beast of burden, whilst the Llama eats as much as the Ass, it does not carry more than a moiety what he can, and cannot travel one half of the same daily distance; besides, the female Llama is useless in this respect. The flesh of the Llama, as above stated, is greatly inferior to that of any of our domestic animals, even of the Italian buffalo.

As to the Alpaca, it is very doubtful if, living as it does in an extremely

Several species of deer are found in the tropical regions of South America, and a remarkable species, with fragile hair like that of the roebuck, the cervus (Andium), as high as 11,000 feet in the Andes.

The Rodentia, or gnawers, of South America, are very numerous; there are 92 in Brazil alone: there are only 8 species of squirrels and 64 species of rats and mice, some of which are very peculiar.

The Agoutis represent our hares in the plains of Patagonia, in Paraguay, &c., and extend as far as Guiana. The tribe of the Cavias, or guinea-pigs, are found in Brazil, and some species in the great table-lands of the Peru-Bolivian Andes; the Echymys, or prickly rat, is an inhabitant of the banks of the Rio de la Plata and Paraguay; the Vizcacha of the Pampas, a burrowing animal, inhabits the great plain of Buenos Ayres; an animal bearing the same

---

dry, elevated, and clear atmosphere, it would ever become accustomed to the damp and variable climate of our northern latitudes, or to that of the great European chains of mountains, the Alps and the Pyrenees, and if it did, that its wool would not be greatly deteriorated. The Vicuña is purely a wild species, and has hitherto resisted all the efforts of the aborigines, the most patient and docile of the human race, to render it prolific, when domesticated in its native country.

It appears, therefore, that the domestication of the several species of Auchenia in Europe would be a costly and useless experiment, on the large scale on which it is proposed to try it; this will appear evident when it is known that in the Peru-Bolivian Andes the llama and alpaca are daily disappearing to make room for the more useful and profitable breed of the common European sheep, whilst, as a beast of burden, the ass is everywhere supplanting it: indeed the experiments recently made on a large scale and at considerable expense by the French Government have proved a complete failure.

Connected with this subject, a very singular fact, and, if well established, a very curious one, has been announced by M. Geoffroy St. Hilaire, on the authority of our countryman, Dr. Weddell, recently returned from South America—that a cross-breed between the Alpaca and the Vicuña had been obtained at Macusani, a village in the Andes south of Cusco, in Peru; and that the mules from this cross-breed were capable of reproducing this newly created species unaltered, the wool of which is represented to be of a very valuable quality. Now, if there exists in zoological science a fact clearly established, it is this: that within historical periods no new species of vertebrate animal has been created — the great zoological law of the immutability of species. The remains of the several wild animals which have been buried for more than 30 centuries in the catacombs of Egypt, and in the ruins of Nineveh, are perfectly identical with those now existing in the most minute details of their anatomical structure. We have examined, in the case referred to, the evidence adduced by Dr. Weddell and adopted by M. Geoffroy St. Hilaire in support of this doctrine, a favourite one of his, and we do not consider it sufficient to shake the conclusion arrived at by all the great zoologists of past times, and by the Cuviers, the Humboldts, and the Owens of our own period, as to the impossibility of the production of a new species of animals by domestication, or the creation of new species in the animal creation.

name, but of a very different species, is frequent in the rocky districts of the Andes, as high as 15,000 feet above the sea; and the beautiful Chinchilla, nearly allied to the latter, whose fur is so highly esteemed, inhabits the same regions, at the same great elevations, in the Andes of South Peru, Bolivia, and Chile: the best fur of the chinchilla is collected in the Bolivian province of Potosi, and in the Chilian province of Copiapo. The largest of all the rodentia, the Cabiai (Myopotamus), inhabits the banks of the great rivers of tropical America, where its habits resemble, according to some travellers, those of the hippopotamus. The Paca, the next in size, is less aquatic in its habits, and lives in the dense forests of Brazil and Paraguay.

It is very remarkable that, in a country which has the most luxuriant vegetation, there should not be one native species of hollow-horned ruminants, as the ox, sheep, goat, or antelope; and it is still more extraordinary that the existing animals of South America, which are so nearly allied to the extinct inhabitants of the same soil, should be so inferior in size not only to them, but even to the living quadrupeds of South Africa, which is comparatively a desert as regards its vegetation. The quantity of vegetation in Britain at any one time exceeds the quantity on an equal area in the interior of Africa very considerably; yet Mr. Darwin has computed that the weight of 10 of the largest South African quadrupeds is 24 times greater than that of the same number of quadrupeds of South America; for in South America there is no animal the size of a cow, so that there is no relation between the bulk of the species and the vegetation of the countries they inhabit.

The largest animals indigenous in the West Indian islands are the Agouti, the Racoon, the Houtias, a native of the forests of Cuba; the Didelphous carnivora and the Kinkajou are common to them and to the continent: the Kinkajou is a solitary instance of a carnivorous animal with a prehensile tail.

AUSTRALIAN QUADRUPEDS.

Australia is not farther separated from the rest of the world by geographical position than by its productions. Its animals are creatures apart, of an entirely unusual type; few in species, and still fewer individually, if the vast extent of country be taken into consideration; and there has not been one large animal discovered There are only 53 species of land quadrupeds in Australia, and there is not a single example of the ruminating or pachydermatous animals, so useful to man, among them. There are no native horses, oxen, or sheep; yet all these thrive and multiply on the grassy steppes of the country, which seem to be so well suited to them.

There are none of the monkey tribe; indeed, they could not exist in a country where there is scarcely any fruit.

Of the species of indigenous quadrupeds, 40 are found nowhere else, and by far the greater number are marsupial, or pouched animals, distinguished from all others by their young being as it were prematurely born and nourished in the pouch till they are able to fare for themselves.[1] Though all the members of this numerous family agree in this circumstance, they are dissimilar in appearance, internal structure, in their teeth and feet, consequently in their habits; two genera live on vegetable food, one set are gnawers, and another toothless. The Kangaroo and the Kangaroo-rat walk on their hind legs, and go by bounds, springing from their strong tail; the kangaroo-rat holds its food in its paws like the squirrel; the phalangers live on trees, and swing by their bushy tail—some burrow in the sand; the flying opossum, or Petaurus, peculiarly an Australian animal, lives at the foot of the Blue Mountains, on the leaves of the Gum-tree; by expanding the skin of its sides as a parachute, it supports itself in the air in its leaps from bough to bough. Several of the genera are nocturnal, a characteristic of many Australian animals.

The pouched tribe vary in size from that of a large dog to a mouse; the kangaroos, which are the largest, are easily domesticated, and are used for food by the natives. Some go in large herds in the mountains, others live in the plains; however, they have become scarce near the British colonies, and, with all other native animals, are likely to be soon extirpated. In Van Diemen's Land they are less persecuted; several species exist there. The kangaroos, of which there are perhaps 40 species, are more widely dispersed than any of the marsupial animals of the Old World. They exist not only in Australia and Tasmania, but also in New Guinea. Some are limited within narrow bounds: the banded kangaroo, the handsomest of his tribe, is found only in the islands of Shark's Bay, on the west coast of Australia. The Wombat is peculiar to Australia, the islands in Bass's Strait, and Van Diemen's Land; to which the two largest carnivorous marsupials peculiarly belong, called by the natives the Tiger Hyæna, and the native Devil; both are nocturnal, predatory, and ferocious. A wild dog living in the woods, whose habits are ferocious, is, with the tiger hyæna, the largest carnivorous animal in Australia.

The gnawing animals are aquatic and very peculiar, but the Eden-

[1] There are 5 tribes, 15 genera, and nearly 150 species of living marsupial animals, amounting to about one-twelfth of all the mammalia. The Opossum and Chironectes are American; the four other families are inhabitants of Australia and the Indian Archipelago. Of the latter the Dasyuridæ and Phalangers are nocturnal: some of the Dasyuridæ and the Wombat burrow in the ground.

tata of New Holland are quite anomalous; of these there are two genera, the Ornithorhynchus, or duck-billed mole, and the Echidna; they are the link that connects the Edentata with the pouched tribe, and mammalia with oviparous animals. The Ornithorhynchus is about 14 inches long, and covered with thick brown fur; its head is similar to that of a quadruped, ending in a bill like that of a duck: it has short furry legs with half-webbed feet, and the hind feet are armed with sharp claws. It inhabits burrows on the banks of rivers, which have two entrances, one above, the other below the level of the water, which it seldom leaves, feeding on insects and seeds in the mud.

The Echidna is similar in its general structure to the ornithorhynchus, but entirely different in external appearance, being covered with quills like the porcupine; it is also a burrowing animal, sleeps during winter, and lives on ants in summer.

A singular analogy exists between Australia and South America in this respect, that the living animals of the two countries are stamped with the type of their ancient geological inhabitants, many of which are gigantic representatives of the now diminutive, in comparison, existing animals; while in England and elsewhere the difference between the existing and extinct generations of beings is most decided. Australia and South America seem still to retain some of those conditions that were peculiar to the most ancient eras. Thus each tribe of the innumerable families that inhabit the earth, the air, and the waters, has a limited sphere. How wonderful the quantity of life that now is, and the myriads of beings that have appeared and vanished![1] Dust has returned to dust through a long succession of ages, and has been continually remoulded into new forms of existence—not an atom has been annihilated; the fate of the vital spark that has animated it, with a vividness sometimes approaching to reason, is one of the deep mysteries of Providence.

---

[1] Sir Charles Lyell estimates the number of existing species of animals and vegetables, independent of the infusoria, to be between one and two millions, which must surely be under the mark, considering the enormous quantity of animal life in the ocean, to the amount of which we have not even an approximation. If the microscopic and infusorial existence be taken into the account, the surface of the globe may be viewed as one mass of animal life—perpetually dying—perpetually renewed. A drop of stagnant water is a world within itself, an epitome of the earth and its successive geological races. A variety of microscopic creatures appear, and die; in a few days a new set succeeds; these vanish and give place to a third set, of different kinds from the preceding; and the débris of all remain at the bottom of the glass. The extinction of these creatures takes place without any apparent cause, unless a greater degree of putrescence of the water be to them what the mighty geological catastrophes were to beings of higher organization—the introduction of the new is not more mysterious in one case than in the other.

CHAPTER XXXIII.

The Distribution, Condition, and future Prospects of the Human Race.

More than 800,000,000 of human beings are scattered over the face of the earth, of all nations and kindreds and tongues, and in all stages of civilization, from a high state of moral and intellectual culture, to savages but little above the animals that contend with them for the dominion of the deserts and forests through which they roam. This vast multitude is divided into nations and tribes, differing in external appearance, character, language, and religion. The manner in which they are distributed, the affinities of structure and language by which they are connected, and the effect that climate, food, and customs may have had in modifying their external forms, or their moral and mental powers, are subjects of much more difficulty than the geographical dispersion of the lower classes of animals, inasmuch as the immortal spirit is the chief agent in all that concerns the human race. The progress of the universal mind in past ages, its present condition, and the future prospects of humanity, rouse the deep sympathies of our nature for the high but mysterious destiny of the myriads of beings yet to come, who, like ourselves, will be subject for a few brief years to the joys and sorrows of this transient state, and fellow-heirs of eternal life hereafter.

[The population of the globe has been stated as follows, and is considered sufficiently accurate for comparison :

Caucasians	419,530,000
Mongolians	406,470,700
Malays	32,500,000
Ethiopians or negro race	69,633,300
American or copper-coloured	10,287,000
	938,421,000

They are classed and enumerated according to their religious belief, as follows :

Pagans	561,821,000
Christians	252,565,000
Mohammedans	120,165,000
Jews	3,930,000
	938,421,000

Sectarian divisions have been stated as follows :

Roman Catholics	134,732,000
Greek Church	56,011,000
Protestants	55,791,000

Monophysites	3,865,000
Armenians	1,799,000
Nestorians	367,000
Christians	252,565,000

Protestant Sects.

Lutherans	24,264,000
Presbyterians and Congregationalists	12,760,000
Episcopalians	14,905,000
Methodists, Baptists, &c	3,862,000
	55,791,000

According to the above statements, only a little more than one-fourth of the entire human population of the earth is under Christian rule.]

Notwithstanding the extreme diversity, physical and mental, in the different races of men, anatomists have found that there are no specific differences—that the hideous Esquimaux, the refined and intellectual Caucasian, the thick-lipped Negro, and the fair blue-eyed Scandinavian, are mere varieties of the same species. The human race forms five great varieties marked by strong distinctive characters. Many nations are included in each; distinguished from one another by different languages, manners, and mental qualities, yet bearing such a resemblance in general physiognomy and appearance as to justify a classification apparently anomalous.

The Caucasian group of nations, which includes the handsomest and most intellectual portion of mankind, inhabit all Europe, except Lapland, Finland, and Hungary; they occupy North Africa, as far as the 20th parallel of north latitude, Arabia, Asia Minor, Persia, the Himalaya to the Brahmapootra, all India between these mountains and the ocean, and the United States of North America. These nations are remarkable for a beautifully-shaped head, regular features, fine hair, and symmetrical form. The Greeks, Georgians, and Circassians are models of perfection in form, especially the last, which is assumed as the type of this class of mankind; of which it is evident that colour is not a characteristic, since they are of all shades, from the fair and florid, to the clear dark brown and almost black. This family of nations has always been, and still is, the most civilized portion of the human race. The inhabitants of Hindostan, the Egyptians, Arabians, Greeks, and Romans, were in ancient times what European nations are now. The cause of this remarkable development of mental power is, no doubt, natural disposition, for the difference in the capabilities of nations seems to be as great as that of individuals. The origin of spontaneous civilization and superiority may generally be traced to the talent of some master-spirit gaining an ascendency over his countrymen. Natural causes have also combined with mental—mildness of climate, fertility of soil;

rivers and inland seas, by affording facility of intercourse, favoured enterprise and commerce; and the double-river systems in Asia brought distant nations together, and softened those hostile antipathies which separate people, multiply languages, and reduce all to barbarism. The genius of this family of nations has led them to profit by these natural advantages; whereas the American Indians are at this day wandering as barbarous hordes in one of the finest countries in the world. An original similarity or even identity of many spoken languages may be adverted to as facilitating communication and mental improvement among the Caucasian variety in very ancient times.

The Mongol-Tartar family forms the second group of nations. They occupy all Asia north of the Persian table-land and of the Himalaya; the whole of eastern Asia from the Brahmapootra to Behring's Straits, together with the Arctic regions of America north of Labrador. This family includes the Tourkomans, Mongol and Tartar tribes, the Chinese, Indo-Chinese, Japanese, the Esquimaux, and the Hungarians, now located in the very heart of Europe. These nations are distinguished by broad skulls and high cheek-bones, small black eyes obliquely set, long black hair, and a yellow or sallow olive complexion; some are good-looking, and many are well-made. A portion of this family is capable of high culture, especially the Chinese, the most civilized nation of eastern Asia, although they never have attained the excellence of the Caucasian group, probably from their exclusive social system, which has separated them from the rest of mankind, and kept them stationary for ages; the peculiarity and difficulty of their language have also tended to insulate them. The Kalmuks, who lead a pastoral wandering life on the steppes of central Asia, and the Esquimaux, have wider domains than any other of this set of nations. The Kalmuks are rather a handsome people, and, like all who lead a savage life, have acute senses of seeing and hearing. The inhabitants of Finland and Lapland are nearly allied to the Esquimaux, who are spread over all the high latitudes of both continents—a diminutive race, equally ugly in face and form.

Malayan nations occupy the Indian Archipelago, New Zealand, Chatham Island, the Society group, and several others of the Polynesian islands, together with the Philippines and Formosa, Mindanao, Gilolo, the high lands of Borneo, Sumbawa, Timor, New Ireland, New Guinea, the continents of Australia and Tasmania. The Australians and the Papuans, who inhabit some of these islands, are the most degraded perhaps of mankind. They are very dark, with lank coarse black hair, flat faces, and obliquely set eyes. Endowed with great activity and ingenuity, they are mild and gentle, and far advanced in the arts of social life, in some places; in others, ferocious and vindictive, daring and predatory; and from their maritime position and skill, they are a migratory race. Several branches of this class of nations had a very early indigenous civilization, with an original literature in peculiar characters of their own.

The Ethiopian nations are widely dispersed; they occupy all Africa south of the Great Desert—half of Madagascar. The distinguishing characters of this group are, a black complexion, black woolly or frizzled hair,[1] thick lips, projecting jaws, high cheek-bones, and large prominent eyes. A great variety, however, exists in this jetty race: some are handsome both in face and figure, especially in Ethiopia; and even in western Africa, where the Negro tribes live, there are groups in which the distinctive characters are less exaggerated. This great family has not yet attained a high place among nations, though by no means incapable of cultivation; part of Ethiopià appears to have made considerable progress in civilization in very ancient times. But the formidable deserts, so extensive in some parts of the continent, and the unwholesome climates in others, have cut off intercourse with civilized nations; and unfortunately, the infamous traffic in slaves, to the disgrace of Christianity, has made the nations of tropical Africa more barbarous than they were before: while, on the contrary, the Foulahs and other tribes who were converts to Mahommedanism 400 years ago, have now large commercial towns, cultivated grounds, and schools.

The American race, who occupy the whole of that continent from 62° N. lat. to the Straits of Magellan, are almost all of a reddish brown or copper colour, with long black hair, deep-set black eyes, aquiline nose, and often of handsome slender forms. In North America they live by hunting, are averse to agriculture, slow in acquiring knowledge, but extremely acute, brave, and fond of war, and, though revengeful, are capable of generosity and gratitude. In South America many are half-civilized, but a greater number are still in a state of complete barbarism. In a family so widely scattered great diversity of character prevails, yet throughout the whole there is a similarity of manners and habits which has resisted all the effects of time and climate.

Each of these five groups of nations, spread over vast regions, is accounted one family; and if they are so by physical structure, they are still more so by language, which expresses the universal mind of a people, modified by external circumstances, of which none have a greater influence than the geographical features of the country they inhabit—an influence that is deepest in the early stages of society. The remnants of ancient poetry in the south of Scotland

[1] Wool is peculiar to quadrupeds, the hair of the negro only resembles it. Both hair and wool [are solid cylinders, which] consist of a transparent tube or sheath containing a white or coloured pith, but the sheath of hair is smooth, whilst that of wool is notched, which gives it the felting property.

partake of the gentle and pastoral character of the country; while Celtic verse, and even the spoken language of the Highlander, are full of poetical images of war and stern mountain scenery. This is particularly to be observed in the noble strains of Homer, and in the heroic poems of the early Hindus, which reflect the lofty and sublime character of eastern scenery.[1] As civilization advances, and man becomes more intellectual, language keeps pace in the progress. New words and new expressions are added, as new ideas occur and new things are invented, till at last language itself becomes a study, is refined and perfected by the introduction of general terms. The improvement in language and the development of the mind have been the same in all nations which have arrived at any degree of refinement, and shows the identity of human nature in every country and climate. The art of printing perpetuates a tongue, and great authors immortalize it; yet language is ever changing to a certain degree, though it never loses traces of its origin. Chaucer and Spenser have become obscure; Shakspeare requires a glossary for the modern reader; and in the few years that the United States of America have existed as an independent nation, the colloquial language has deviated from the mother tongue. When a nation degenerates, it is split by jealousy and war into tribes, each of which in process of time acquires a peculiar idiom, and thus the number of dialects is increased, though they still retain a similarity; whereas when masses of mankind are united into great political bodies, their languages by degrees assimilate to one common tongue, which retains traces of all to the latest ages. The form of the dialects now spoken by some savage tribes, as the North American Indians, bears the marks of a once higher state of civilization.

More than 2000 languages are spoken, but few are independent; some are connected by words having the same meaning, some by grammatical structure, others by both; indeed the permanency of language is so great, that neither ages of conquest, nor mixing with other nations, have obliterated the native idiom of a people. The French, Spanish, and German retain traces of the common language spoken before the Roman conquest, and the Celtic tongue still exists in the British Islands.

By a comparison of their dialects, nations far apart, and differing in every other respect, are discovered to have sprung from a common, though remote origin. Thus all the numerous languages spoken by the American Indians, or red men, are similar in grammatical structure: an intimate analogy exists in the languages of the Esquimaux nations who inhabit the arctic regions of both continents.

[1] Valmiki, the Hindu poet, is supposed to have been contemporary with Homer, if not his predecessor: his great work is the 'Ramayana,' an heroic poem of the highest order, four cantos of which have been translated by Gaspare Corresio, an Italian, in 1843.

Dialects of one tongue are spoken throughout North Africa, as far south as the oasis of Siwah on the east, and the Canary Islands on the west. Another group of cognate idioms is common to the inhabitants of Equatorial Africa, while all the southern part of the continent is inhabited by people whose languages are connected. The monosyllabic speech of the Chinese and Indo-Chinese shows that they are the same people; and all the insular nations of the Pacific derived their dialects from some tribes on the continent of India and the Indian Archipelago. Cognate tongues are spoken by the Tartars, Mandtchoux, Fins, Laplanders, many of the Siberian nations, and by the Hungarians. The Syro-Arabian, or Semitic languages, as the Chaldee, Arabic, and Hebrew, are evidently, from their grammatical construction, of the same origin.

The Persian, Greek, Latin, German, and Celtic tongues are connected by grammatical structure, and words expressive of the same objects and feelings, with the Sanscrit, or sacred language of India; consequently the nations inhabiting that vast extent of country from the mouths of the Ganges to the British Isles, the coast of Scandinavia and Iceland, must have had the same origin. "The words that fall thoughtlessly from our lips in the daily vocations of life are no idle sounds, but magic symbols which preserve for ever the first migrations of the race, and whose antiquity makes Greece and Rome appear but of yesterday."[1]

---

[1] The words which one nation borrows from another do not prove an original connection: it is the "home-bred speech," the words which children learn in early infancy, that show a common origin, such as those of near relationship, of first necessity, as to eat, sleep, walk, &c., the names of the most ordinary natural objects, the numerals, &c. Tribes or families of nations long separated have preserved such words for thousands of years, with a purity that makes them easily to be recognized as having sprung from a common stock. However, nothing can be inferred from a coincidence in the meaning of one or two words common to two languages, but Dr. Thomas Young has computed that if three words were identical in two languages, the odds would be more than ten to one, that in both cases they must have been derived from a common parent tongue; that for six words the chances would be 1700 to 1, and for eight words in common 100,000 to 1; so that in the two latter cases the evidence would be little short of certainty that the languages in question, and consequently the natives who speak them, had a common origin.

But according to the best and more learned school of modern ethnographers, the affinity of languages is not so much to be looked for in the coincidence of words as in the grammatical structure, which is also of remarkable permanency. A similar inflexion of nouns and form of roots prevail trough whole groups or classes of languages which have few words in common, the words, as Klaproth justly remarks, being the material of languages, grammar the fashioning or formation of it. The Syro-Chaldaic, Hebrew, Arabic, and Abyssinian afford a striking example of identity in grammatical structure. In the languages the tenses of the verbs are formed from the third person of the preterite, and in most cases the roots of the

The number of languages spoken from the Ganges to Scandinavia, differing so widely from one another, is a proof of the strength of individual character in nations, which can so powerfully impress its peculiarities on the same mother tongue. In fact every nation, as well as every individual, has its own physical, moral, and intellectual organization, which influences its language and its whole existence.

In the Indo-European nations, which have been dominant for ages, civilization has been progressive, though not without interruptions. Providence has endowed these nations with the richest and most ornamental gifts. Imagination has been liberally granted, and embodied in all that is sublime and beautiful in architecture, sculpture, painting, and poetry. In strength of intellect and speculation, in philosophy, science, laws, and the political principles of society, they have been pre-eminent.

The prevailing races of mankind now inhabiting Europe are the Teutonic, Celtic, and Sclavonian. In the greater part of the continent these races are mixed, but the blood is purely Teutonic throughout Iceland, Scandinavia, round the Gulf of Bothnia, in Denmark, Germany, and the east of England from Portsmouth to the Tyne. Pure Celtic blood is confined to the Basque Provinces in Spain, the south and south-west of France, a part of the Grisons and Switzerland, and some part of Great Britain. The Sclavonian blood is widely dispersed in middle Russia, from the Ural Mountains to the west of the Valdai table-land, and from Novogorod to the lower course of the Don. The three races have been much improved by mixture, in appearance, energy, and versatility of mind.

It is extraordinary that nations should lose their vitality without any apparent cause; throughout the Indian Archipelago there is no longer any one great Malayan nation; in Europe pure Celtic blood has been on the decline for 20 centuries, and even the mixed Celtic

---

verbs, or the consonants they contain, are the same, and follow in the same order; vowels are necessarily used when the words are spoken, but they are constantly omitted when the languages are written and printed, since no letters or characters for vowels exist in the alphabets of the Semitic or Syro-Arabian group, with the exception of the Abyssinian, which has a syllabary but no real alphabet. In religious books however, or difficult passages, where the meaning might be doubtful, signs are occasionally added for the vowels, which are of a comparatively modern date: thus the two words which mean *wrote* and *killed* in the three languages in question are printed c t b and c t l, but when spoken they become c t a b and c t a l in Syro-Chaldaic, c a t a b and c a t a l in Hebrew, c a t a b a and c a t a l a in Arabic, the roots or consonants being the same, and following in the same order in all three. The Hebrew is historically known to have sprung from the Syro-Chaldaic, for Abraham spoke Chaldaic.

In Sanscrit the roots are syllables instead of consonants, and the peculiarity of the Chinese, the Indo-Chinese, and Bhutan languages, is that the words consist of but one syllable, and that each word derives its meaning from its position in a sentence.

variety has not increased in proportion to the Teutonic, although for 2000 years they have been exposed to the same external circumstances.

At present the Teutonic race, including the inhabitants of North America and the British colonies, considerably outnumber the Celtic, though its numbers were far inferior in ancient times. The Teutonic variety has subdued and even exterminated the other varieties in its progress towards the west; it is undoubtedly the most vigorous, both in body and mind, of all mankind, and seems destined to conquer and civilize the whole world. It is a singular fact, whatever the cause may be, that the Celts are invariably Roman Catholic, while the Teutonic population is inclined to Protestanism.

Various other races inhabit Europe, much inferior in numbers to those above mentioned, though occasionally mixed with them, as the Turks, Fins, the Samojedes, who live on the shores of the White Sea and in the north-east of Russia, and the Hungarians, the higher class of which are a fine race of men, and on a par with the most civilized of the European nations.[1] There are many mixed Tartar tribes, chiefly in the south and east of the Russian territories; also Jews and Gipsies, who live among all nations, yet mix with none.[2]

[1] Europe had been inhabited before the arrival of the Asiatic tribes, consequently some of the inhabitants of the more remote regions are probably the aborigines of the country.

[2] European Population.

Pure blood.

Teutonic	52,000,000
Sclavonian	50,000,000
Celtic	12,000,000
Magyar	9,000,000
Fins and Samojedes	3,000,000
Tatar	2,000,000
Jews	2,000,000
Total European population of pure blood	130,000,000

Mixed blood in Europe.

Teutonic Celtic	22,000,000
Teutonic Sclavonian	6,000,000
Teutonic mixed with Walloons in Belgium	1,200,000
Teutonic Northmen in Normandy	1,500,000
Celtic in its different crosses	56,000,000
Sclavonian	6,000,000
Lettons	2,000,000
Turks	4,000,000
Turco-Tatar-Sclavonic in centre, south-east, and east of Russia	2,600,000
Kalmuc, between the rivers Volga and Ural	300,000
The number of people of mixed blood in Europe	101,600,000

The inhabitants of Great Britain are of Celtic and Teutonic origin. The Celtic blood is purest in Cornwall and the Scilly Islands, in Wales, and the Isle of Man: in the highlands of Scotland and the Hebrides, it is more mixed than is generally supposed, as plainly appears from the frequency of red hair and blue eyes. In some parts of Ireland there is pure Celtic blood, but throughout the greater part of that country it is mixed, although the Celtic character predominates; but in Ulster, where the earliest colony settled, the blood is purely Teutonic. In Ireland the difference in the organization of the two races is strongly marked: placed under the same circumstances, the Teutonic part of the population has prospered, which, unfortunately, has not been the case with the Celtic.[1]

The dialects spoken in the Celtic districts are closely allied to the Semitic languages of Asia, aud to one another. The Cornish is worn out, the Manx is nearly so, and the Gaelic is declining fast in the highlands of Scotland.

The Roman invasion had no effect on the Anglo-Saxon or old English, a language of Teutonic origin, but the Normans in ancient times had altered it considerably, and in modern times the English tongue has unfortunately been corrupted by the introduction of

---

The total population of Europe, pure and mixed, amounts to about 232 millions, including 600,000 Gipsies. The Teutonic population in the United States of North America and in the British colonies amounts to 20 millions; so that the total number of people of Teutonic blood is rather more than 100 millions.—Notes accompanying the Ethnographic Map of Europe, by Dr. Gustaf Kombst: 'Phys. Atlas.' By a more recent census the population of Europe amounts to 240,000,000.

[1] POPULATION OF GREAT BRITAIN AND IRELAND.

On an average the pure-blooded population amounts to

Teutonic in England, Scotland, and in the east and north-east of Ireland	10,000,000
Celtic in Cornwall, Wales, the Scottish Highlands, and Ireland	6,000,000
The pure-blooded inhabitants amount to	16,000,000
Mixed blood.	
Mixture in which the Teutonic blood predominates	6,000,000
Mixture in which the Celtic blood predominates	4,000,000
	10,000,000

In all 26,000,000 of inhabitants.

Notes accompanying the Ethnographic Map of Great Britain and Ireland, by Gustaf Kombst: 'Phys. Atlas.'

The fear that Britain may be ruined by over-population may be allayed by considering that we are ignorant of the immense treasures and inexhaustible resources of the natural world — that the ingenuity of man is infinite, and will continually discover new powers and innumerable combinations that will furnish sources of wealth and happiness to millions.

French, Latin, and Latinized words. Scotch spoken throughout the Lowlands of Scotland is a language independent of the English, though of the same stock; it is derived from the low German, the Frisian, Dutch, and Flemish, and differs widely from the Anglo-Saxon.

No circumstance in the natural world is more inexplicable than the diversity of form and colour in the human race. It had already begun in the antediluvian world, for "there were giants in the land in those days." No direct mention is made of colour at that time unless it was the mark set upon Cain, "lest any one finding him should kill him," may allude to it. Perhaps, also, it may be inferred that black people dwelt in Ethiopia, or the land of Cush, which means black in the Hebrew language. At all events, the difference now existing must have arisen after the flood, consequently all must have originated with Noah, whose wife, or the wives of his sons, may have been of different colours, for aught we know.

Many instances have occurred in modern times of albinos and red-haired children having been born of black parents, and these have transmitted their peculiarities to their descendants for several generations; but it may be doubted whether pure-blooded white people have had perfectly black offspring. The varieties are much more likely to have arisen from the effects of climate, food, customs, and civilization upon migratory groups of mankind; and of such, a few instances have occurred in historical times, limited, however, to small numbers and particular spots; but the great mass of nations had received their distinctive characters at a very early period. The permanency of type is one of the most striking circumstances, and proves the length of time necessary to produce a change in national structure and colour. A nation of Ethiopians existed 3450 years ago, which emigrated from a remote country and settled near Egypt, and there must have been black people before the age of Solomon, otherwise he would not have alluded to colour, even poetically. The national appearance of the Ethiopians, Persians, and Jews, has not varied for more than 3000 years, as appears from the ancient Egyptian paintings in the tomb of Rhamses the Great, discovered at Thebes by Belzoni, in which the countenance of the modern Ethiopian and Persian can be readily recognised, and the Jewish features and colour are identical with those of the Israelites daily met with in London. Civilization is supposed to have great influence on colour, having a tendency to make the dark shade more general, and it appears that, in the crossing of two shades, the offspring takes the complexion of the darker, and the form of the fairer. But as there is no instance of a new variety of mankind having been established as a nation since the Christian era, there must either have been a greater energy in the causes of change before that time, or, brief as the span of man on earth has been, a wrong estimate of time antecedent to the Christian period must have made it shorter.[1]

Darkness of complexion has been attributed to the sun's power from the age of Solomon to this day—"*Look not upon me, because I am black, because the sun hath looked upon me*;" and there can be no doubt that, to a certain degree, the opinion is well founded. The invisible rays of the solar beams, which change vegetable colours, and have been employed with such remarkable effect in the Daguerreotype, act upon every substance on which they fall, producing mysterious and wonderful changes in their molecular state—man not excepted.[2]

Other causes must have been combined to occasion all the varieties we now see, otherwise every nation between the tropics would be of the same hue, whereas the sooty Negro inhabits equatorial Africa, the Red man equinoctial America, and both are mixed with fairer tribes. In Asia, the Rohillas, a fair race of Affghan extraction, inhabit the plains north of the Ganges; the Bengalee and the mountaineers of Nepaul are dark, and the Mahrattas are yellow. The complexion of man varies with height and latitude; some of the inhabitants of the Himalaya and Hindoo Koosh are fair, and even a red-haired race is found on the latter. There are fair-haired people with blue eyes in the Ruddhua mountains in Africa. The Kabyles, that inhabit the country behind Tunis and Algiers, are similar in complexion to the nations in high northern latitudes. This correspondence, however, only maintains with regard to the northern hemisphere, for it is a well-known fact that the varieties of the numerous species in the great southern continents are much

---

[1] From the discrepancies in the chronological systems, it is evident that the actual period of man's creation is not accurately known. The Chevalier Bunsen has ascertained from monumental inscriptions, that the successive Egyptian dynasties may be traced back to Meres, 3640 years before the Christian era, and from the high state of civilization during the reign of that prince, proved by the magnificence of the works then executed, he infers that the Egyptians must have existed 500 years previous to their consolidation into one empire by him, which goes back to the received period of man's creation. Compared with geological periods, man is of very recent creation, as appears from the vast extent of uninhabited land, but which would require ages and ages to people, even if the increase of population were as rapid as in the United States of North America. Dr. Pritchard says that the Hebrew chronology has been computed with some approximation to truth up to the arrival of Abraham in Palestine, but that we can never know how many centuries may have elapsed from that event to the time when "the first man of clay received the image of God and the breath of life."

[2] Dark-coloured substances absorb more of the sun's heat than light-coloured ones; therefore, the black skins of the natives of tropical climates absorb more heat than fair skins; but, from some unknown cause, the black skin is protected from a degree of heat that would blister a fair one.

more similar in physical characters to the native races of the torrid zone than any of the aboriginal people of the northern regions. Even supposing that diversity of colour is owing to the sun's rays only, it is scarcely possible to attribute the thick lips, the woolly hair, and the entire difference of form, extending even to the very bones and skull, to anything but a concurrence of circumstances, not omitting the invisible influence of electricity, which pervades every part of the earth and air—and possibly terrestrial magnetism. The rarity of air also affects the structure of the human frame, and even modifies the most important functions of life, for the people who have for centuries inhabited the heights of the Andes have a more capacious chest, and lungs of a larger volume, than other races of men, according to Dr. Prichard.

The flexibility of man's constitution enables him to live in every climate, from the equator to the ever-frozen coasts of Nova Zembla and Spitzbergen, and that chiefly by his capability of bearing the most extreme changes of temperature and diet, which are probably the principal causes of the variety in his form. It has already been mentioned, that oxygen is inhaled with the atmospheric air, and also taken in by the pores in the skin; part of it combines chemically with the carbon of the food, and is expired in the form of carbonic acid gas and water; that chemical action is the cause of vital force and heat in man and animals. The quantity of food must be in exact proportion to the quantity of oxygen inhaled, otherwise disease and loss of strength would be the consequence. Since cold air is incessantly carrying off warmth from the skin, more exercise is requisite in winter than in summer, in cold climates than in warm; consequently more carbon is necessary in the former than in the latter, in order to maintain the chemical action that generates heat and to ward off the destructive effects of the oxygen, which incessantly strives to consume the body. Animal food, wine, and spirits, contain many times more carbon than fruit and vegetables, therefore animal food is much more necessary in a cold than in a hot climate. The Esquimaux, who lives by the chase, and eats 10 or 12 pounds weight of meat and fat in 24 hours, finds it not more than enough to keep up his strength and animal heat, while the indolent inhabitant of Bengal is sufficiently supplied with both by his rice diet. Clothing and warmth make the necessity for exercise and food much less, by diminishing the waste of animal heat. Hunger and cold united soon consume the body, because it loses its power of resisting the action of the oxygen, which consumes part of our substance, when food is wanting. Hence nations inhabiting warm climates have no great merit in being abstemious, nor are those guilty of committing an excess who live more freely in colder countries. The arrangement of Divine Wisdom is to be admired in this as in all other things, for, if man had only been capable of living on vegetable food, he never could have had a permanent residence beyond the latitude where corn ripens. The Esquimaux, and all the inhabitants of the very high latitudes of both continents, live entirely on fish and animal food. What effects the difference of food may have upon the intellect is not known.

A nation or tribe driven by war, or any other cause, from a warm to a cold country, or the contrary, would be forced to change their food both in quality and quantity, which in the lapse of ages might produce an alteration in the external form and internal structure. The probability is still greater, if the entire change that a few years produces in the matter of which the human frame is composed be considered. At every instant during life, with every motion, voluntary and involuntary, with every thought and every exercise of the brain, a portion of our substance becomes dead, separates from the living part, combines with some of the inhaled oxygen, and is removed. By this process it is supposed that the whole body is renewed every 7 years; individuality, therefore, depends on the spirit, which retains its identity during all the changes of its earthly house, and sometimes even acts independently of it. When sleep is restoring exhausted nature, the spirit is often awake and active, crowding the events of years into a few seconds, and, by its unconsciousness of time, anticipates eternity. Every change of food, climate, and mental excitement must have their influence on the reproduction of the mortal frame; and thus a thousand causes may co-operate to alter whole races of mankind placed under new circumstances, time being granted.

The difference between the effects of manual labour and the efforts of the brain appears in the intellectual countenance of the educated man, compared with that of the peasant, though even he is occasionally stamped with nature's own nobility. The most savage people are also the ugliest. Their countenance is deformed by violent unsubdued passions, anxiety, and suffering. Deep sensibility gives a beautiful and varied expression, but every strong emotion is unfavourable to perfect regularity of feature; and of that the Greeks were well aware when they gave that calmness of expression and repose to their unrivalled statues. The refining effects of high culture, and, above all, the Christian religion, by subduing the evil passions, and encouraging the good, are more than anything calculated to improve even the external appearance. The countenance, though perhaps of less regular form, becomes expressive of the amiable and benevolent feelings of the heart, the most captivating of all beauty.[1]

[1] The countenances of the Fuegians brought to England in 1830 by Captain FitzRoy improved greatly in expression by their intercourse with civilized men, but they had not returned to their savage brethren more than a year before their whole appearance was completely changed; the

Thus an infinite assemblage of causes may be assigned as having produced the endless varieties in the human race; but the fact remains an inscrutable mystery. But amidst all the physical vicissitudes man has undergone, the species remains permanent; and let those who think that the difference in the species of animals and vegetables arises from diversity of conditions, consider, that no circumstances whatever can degrade the form of man to that of the monkey, or elevate the monkey to the form of man.

Animals and vegetables, being the sources of man's sustenance, have had the chief influence on his destiny and location, and have induced him to settle in those parts of the world where he could procure them in greatest abundance. Wherever the chase or the spontaneous productions of the earth supply him with food, he is completely savage, and only a degree further advanced where he plants the palm and the banana; where grain is the principal food, industry and intelligence are most perfectly developed, as in the temperate zone. On that account the centres of civilization have generally been determined, not by a hot, but by a genial climate, fertile soil, by the vicinity of the sea-coast or great rivers, affording the means of fishing and transport, which last has been one of the chief causes of the superiority of Europe and Southern Asia. The mineral treasures of the earth have been the means of assembling great masses of men in Siberia and the table-land of the Andes, and have given rise to many great cities, both in the Old and the New World. Nations inhabiting elevated table-lands and high ungenial latitudes have been driven there by war, or obliged to wander from countries where the population exceeded the means of living—a cause of migration to which both language and tradition bear testimony. The belief in a future state, so universal, shown by respect for the dead, has no doubt been transmitted from nation to nation. The American Indians, driven from their hunting-grounds, still make pilgrimages to the tombs of their fathers; and these tribes alone, of all uncivilized mankind, worship the Great Spirit as the invisible God and Father of all — a degree of abstract refinement which could hardly have sprung up spontaneously among a rude people, and which must have been transmitted from races who held the Jewish faith.

It is probable that America had been peopled from Asia before the separation of the continents by Behring's Straits, and there is reason to believe that the location of various races of mankind, now insulated, may have taken place before the separation of the lands by mediterranean seas; whilst others, previously insulated, may be now united by the drying up of inland seas; as those which covered

look of intelligence they had acquired was gone; and when compared with likenesses that had been taken of them when in England, they were not to be recognised as the same persons.

the Sahara Desert, and the great hollow round the Caspian Sea, of which it and the Black Sea are probably the remnants.

M. Boué has observed that mountain chains running nearly east and west establish much more striking differences among nations than those which extend from north to south — a circumstance confirmed by observation through the history of mankind. The Scandinavian Alps have not prevented the countries on both sides from being occupied by people of a common descent; while the feeble barrier of the Cheviot Hills, between England and Scotland, and the moderate elevation of the Highland mountains, have prevented the amalgamation of the Anglo-Saxons and the Celts, even in a period of high civilization. The Franks and Belgians are distinct, though separated by hills of still less elevation. For the same reason the Spaniards and Italians differ far more from their neighbours on the other side of the eastern and western chains, than the Spaniards do from the Portuguese, or the Piedmontese from the Provençals. A similar distinction prevails throughout Asia; and in America, where all the principal chains run north and south, there is but one copper-coloured race throughout the continent, which stretches over more climates than Europe and Africa, or even than Asia and Australia united. It is along chains running north and south that the fusion of languages takes place, and not along those of an easterly and westerly direction. From Poland, for instance, there are intermediate insensible gradations through Germany into France; while in crossing from a German district of the Alps to the valleys of Italy, different tribes and different languages are separated by a single mountain. Even wars and conquest have ever been more easy in one direction than in the other. The difference in the fauna and flora on the two sides of the great table-land and mountains of Asia is a striking illustration of the influence which high lands running east and west have on natural productions, and thus, both directly and indirectly, they affect the distribution of mankind.

The circumstances which thus determine the location of nations, and the fusion or separation of their languages, must, conjointly with moral causes, operate powerfully on their character. The minds of mankind, as well as their fate, are influenced by the soil on which they are born and bred. The natives of elevated countries are attached to their mountains; the Dutch are as much attached to their meadows and canals; and the savage, acquainted only with the discomforts of life, is unhappy when brought among civilized man. Early associations never entirely leave us, however much our position in life may alter; and strong attachments are formed to places which generate in us habits differing from those of other countries.

The Baltic and Mediterranean Seas have had no inconsiderable share in civilizing Europe; one combined with a cold and gloomy climate, the other with a warm and glowing sky, have developed dissimilar characters in the temperament and habits of the surrounding nations, originally dissimilar in race. The charms of climate and the ease with which the necessaries of life are procured were favourable to the development of imagination in the more southern nations, and to an indolent enjoyment of their advantages. In the north, on the contrary, the task imposed upon man was harder, and perhaps more favourable to strength of character. The Dutch owe their industry and perseverance to their unceasing struggle against the encroachments of the ocean; the British are indebted to their insular position for their maritime disposition, and to the smallness of their country and the richness of their mines, for their manufacturing and colonizing habits; the military propensities of the French, to the necessity of maintaining their independence among the surrounding nations, as well as to ambition and the love of fame.

Thus external circumstances materially modify the character of nations, but the original propensities of race are never eradicated, and they are nowhere more prominent than in the progress of the social state in France and England. The vivacity and speculative disposition of the Celt appear in the rapid and violent changes of government and in the succession of theoretical experiments in France; while in Britain the deliberate slowness, prudence, and accurate perceptions of the Teuton are manifest in the gradual improvement and steadiness of their political arrangements. "The prevalent political sentiment of Great Britain is undoubtedly *conservative*, in the best sense of the word, with a powerful undercurrent of *democratic tendencies*. This gives great power and strength to the political and social body of this country, and makes revolutions by physical force almost impossible. It can be said, without assumption or pretension, that the body politic of Britain is in a sounder state of health than any other in Europe; and that those know very little of this country, who, led away by what they see in France, always dream of violent and revolutionary changes in the constitution. Great Britain is the only country in Europe which has had the good fortune to have all her institutions worked out and framed by her in a strictly *organic* manner; that is, in accordance with *organic wants*, which require different *conditions* at different and *successive* stages of national development—and not by *theoretical experiments*, as in many other countries which are still in a state of excitement consequent upon these experiments. The social character of the people of this country, besides the features which they have in common with other nations of Teutonic origin, is, on the whole, domestic, reserved, aristocratic, and exclusive."[1]

---

[1] Johnston's 'Physical Atlas.'

The average age of a nation, or the mean duration of life, has a conside-

In speculating upon the effects of external circumstances, and on the original dispositions of the different races of mankind, the stationary and unchanged condition is a curious phenomenon in the history of nations. The inhabitants of Hindostan have not advanced within the historical period; neither have the Chinese. The Peruvians and Mexicans had arrived at a considerable degree of civilization, at which they became stationary, never having availed themselves of their fine country and noble rivers; and their conquerors, the Spaniards, degenerated into the same apathy with the conquered. The unaccountable gipsies have for ages maintained their peculiarities in all countries; so have the Jews and Armenians, who, by the perseverance with which they have adhered to their language and institutions, have resisted the influence of physical impressions.

The influence of external circumstances on man is not greater than his influence on the material world. He cannot create power, it is true, but he dexterously avails himself of the powers of nature to subdue nature. Air, fire, water, steam, gravitation, his own muscular strength, and that of animals rendered obedient to his will, are the instruments by which he has converted the desert into a garden, drained marshes, cut canals, made roads, turned the course of rivers, cleared away forests in one country, and planted them in another. By these operations he has altered the climate, changed the course of local winds, increased or diminished the quantity of rain, and softened the rigour of the seasons. In the time of Strabo, the cold in France was so intense, that it was thought impossible to ripen grapes north of the Cevennes: the Rhine and the Danube were every winter covered with ice thick enough to bear any weight. Man's influence on vegetation has been immense, but the most important changes had been effected in the antediluvian ages of the world. Cain was a tiller of the ground. The olive, the vine, and the fig-tree have been cultivated time immemorial: wheat, rice, and barley, have been so long in an artificial state, that their origin is unknown; even maize, which is a Mexican plant, was in use among the American tribes before the Spanish conquest; and tobacco was already used by them to allay the pangs of hunger, to which those who depend upon the chase for food must be exposed. Most of the

---

rable influence on the character of a people. The average age of the population of England and Wales is 26 years 7 months. By the census the average age of the population of the United States of North America is 22 years 2 months. In England there are 1365 persons in every 10,000, who have attained 50 years of age, and consequently of experience: while in the United States only 830 in each 10,000 have arrived at that age: hence in the United States the moral predominance of the young and passionate is greatest. In Ireland there are 1050 persons in every 10,000 of the population, above 50 years of age, to exercise the influence of their age and experience upon the community—an influence that will diminish with the progress of emancipation.

ordinary culinary vegetables have been known for ages, and it is remarkable that in these days, when our gardens are adorned with innumerable native plants in a cultivated state, few new grains, vegetables, or fruits have been reclaimed; the old have been produced in infinite variety, and many brought from foreign countries: yet there must exist many plants capable of cultivation, as unpromising in their wild state as the turnip or carrot.

Some families of plants are more susceptible of improvement than others, and, like man himself, can bear almost any climate. One kind of wheat grows to 62° N. latitude; rye and barley are hardier, and succeed still farther north; and few countries are absolutely without grass. The cruciform tribe abounds in useful plants: indeed that family, together with the solanum, the papilionaceous and umbelliferous tribes, furnish most of our vegetables. Many plants, like animals, are of one colour only in their wild state, and their blossoms are single. Art has introduced the variety we now see in the same species, and, by changing the anthers of the mild flowers into petals, has produced double blossoms: by art, too, many plants, natives of warm countries, have been naturalized in colder climates. Few useful plants have beautiful blossoms — but if utility were the only object, of what pleasure should we be deprived! Refinement is not wanting in the inmates of a cottage covered with roses and honeysuckle; and the little garden cultivated amidst a life of toil tells of a peaceful home.

Among the objects which tend to the improvement of our race, the flower-garden and the park adorned with native and foreign trees have no small share: they are the greatest ornaments of the British Islands; and the love of a country life, which is so strong a passion, is chiefly owing to the law of primogeniture, by which the head of a family is secured in the possession and transmission of his undivided estate, and therefore each generation takes pride and pleasure in adorning the home of its forefathers.

Animals yield more readily to man's influence than vegetables, and certain classes have greater flexibility of disposition and structure than others. Those only are capable of being perfectly reclaimed that have a natural tendency for it, without which man's endeavours would be unavailing. This predisposition is greatest in animals which are gregarious and follow a leader, as elephants, dogs, horses, and cattle do in their wild state; yet even among these some species are refractory, as the buffalo, which can only be regarded as half-reclaimed. The canine tribe, on the contrary, are capable of the greatest attachment, not the dog only, man's faithful companion, but even the wolf, and especially the hyæna, generally believed to be so ferocious. After an absence of many months, a hyæna which had been the fellow-passenger of a friend of the author's in a voyage from India, recognised his voice before he came in sight, and on seeing him showed the greatest joy, lay down like a dog and licked his hands. He had been kind to it on the voyage, and no animal forgets kindness, which is the surest way of reclaiming them. There cannot be a greater mistake than the harsh and cruel means by which dogs and horses are too commonly trained; but it is long before man learns that his power is mental, and that it is intellect alone that has given him dominion over the earth and its inhabitants, of which so many far surpass him in physical strength. The useful animals were reclaimed by the early inhabitants of Asia, and it is very remarkable, notwithstanding the enterprise and activity of the present times, that among the multitudes of animals that inhabit America, Central and Southern Africa, Australia and the Indian Archipelago, 4 only have been domesticated, yet many may be capable of becoming useful to man. Of 35 species of which we possess one or more domestic races, 31 are natives of Asia, Europe, and North Africa; these countries are far from being exhausted, and a complete hemisphere is yet unexplored. An attempt has been made to domesticate the Llama, the Dziggetai, Zebra, and some species of Indian deer, but the success is either doubtful or the attempt has not been followed up. Little has been left for modern nations but the improvement of the species, and in that they have been very successful. The variety of horses, dogs, cattle, and sheep is beyond number. The form, colour, and even the disposition, may be materially altered, and the habits engrafted are transmitted to the offspring, as instinctive properties independent of education. Domestic fowls go in flocks on their native meads when wild. There are, however, instances of solitary birds being tamed to an extraordinary degree, as the raven, one of the most sagacious.

Man's necessities and pleasure have been the cause of great changes in the animal creation, and his destructive propensity of still greater. Animals are intended for our use, and field-sports are advantageous by encouraging a daring and active spirit in young men; but the utter destruction of some races, in order to protect those destined for his pleasure, is too selfish. Animals soon acquire a dread of man, which becomes instinctive and hereditary; in newly discovered uninhabited countries, birds and beasts are so tame as to allow themselves to be caught; whales scarcely got out of the way of the ships that first navigated the Arctic Ocean, but now they universally have a dread of the common enemy: whales and seals have been extirpated in various places: sea-fowl and birds of passage are not likely to be extinguished, but many land animals and birds are vanishing before the advance of civilization. Drainage, cultivation, cutting down of forests, and even the introduction of new plants and animals, destroy some of the old, and alter the relations between those that remain. The inaccessible cliffs of the Himalaya and Andes, will afford a refuge to the eagle and condor, but the time will come when the mighty forests of the Amazon and Orinoco will disappear with the myriads of their joyous inhabitants. The lion, the tiger, and the elephant will be known only by ancient records. Man, the lord of the creation, will extirpate the noble creatures of the earth—but he himself will ever be the slave of the canker-worm and the fly. Cultivation may lessen the scourge of the insect tribe, but God's great army will ever, from time to time, appear suddenly —no one knows from whence; the grub will take possession of the ground, and the locust will come from the desert and destroy the fairest prospects of the harvest.

Though the unreclaimed portion of the animal creation is falling before the progress of improvement, yet man has been both the voluntary and the involuntary cause of the introduction of new animals and plants into countries in which they were not natives. The Spanish conquerors little thought that the descendants of the few cattle and horses they allowed to run wild, would resume the original character of their species, and roam in hundreds of thousands over the savannahs of South America. Wherever man is, civilized or savage, there also is the dog, but he too has in some places resumed his native state and habits, and hunts in packs. Domestic animals, grain, fruit, vegetables, and the weeds that grow with them, have been conveyed by colonists to all settlements. Birds and insects follow certain plants into countries in which they were never seen before. Even the inhabitants of the waters change their abode in consequence of the influence of man. Fish, natives of the rivers on the coast of the Mexican Gulf, have migrated by the canals to the heart of North America; and the mytilus polymorphus, a shell-fish brought to the London Docks in the timbers of ships from the brackish waters of the Black Sea and its tributary streams, has spread into the interior of England by the Croydon and other canals.

The influence of man on man is a power of the highest order, far surpassing that which he possesses over animate or inanimate nature. It is, however, as a collective body, and not as an individual, that he exercises this influence over his fellow-creatures. The free-will of man, nay, even his most capricious passions, neutralize each other, when large numbers of men are considered. Professor Quetelet has most ably proved, that the greater the number of individuals, the more completely does the will of each, as well as all individual peculiarities, moral or physical, disappear, and allow the series of general facts to predominate, which depend upon the causes by which society exists and is preserved. The uniformity with which the number of marriages in Belgium occurred in 20 years, places the neutralization of the free-will of the individual man beyond a doubt, and is one of many instances of the importance of average quantities in arriving at general laws.

Certainly no event in a man's life depends more upon his free will than his marriage; yet it appears from the records in Brussels, that nearly the same number of marriages take place every year, in the towns as well as in the country; and, moreover, that the same constancy prevails in each province, though the numbers of the people are so small, that accidental causes might be more likely to affect the general result than when the numbers are larger. In fact, the whole affair passes as if the inhabitants of Belgium had agreed to contract nearly the same number of marriages annually, at each stage of life. Young people may possibly be in some degree under the control of parents, but there can be no restraint on the free will of men of 30 and women of 60 years of age; yet the same number of such incongruous marriages do annually take place between men and women at those unsuitable ages — a fact which almost exceeds belief. The day fixed for a wedding is of all things most entirely dependent on the will of the parties, yet even here there is a regularity in the annual recurrence. (See Table on next page.)

With regard to crimes also, M. Quetelet observes that the same number of crimes of the same description are committed annually, with remarkable uniformity, even in the case of those crimes which would seem most likely to baffle all attempt at prediction. The same regularity occurs in the sentences passed on criminals: in France, in every hundred trials there were sixty-one convictions regularly, year after year.

Forgetfulness, as well as free-will, is under constant laws; the number of undirected letters put into the post-office in London and in Paris is very nearly the same year after year respectively — in London they amount to 2000; so that even the deviations from free-will prove the generality and the constancy of the laws that govern us.

Scientific discoveries and social combinations, which put in practice great social principles, are not without a decided influence; but these causes of action, coming from man, are placed out of the sphere of the free-will of each: so that individual impulse has less to do with the progress of mankind than is generally believed. When society has arrived at a certain point of advancement, certain discoveries will naturally be made; the general mind is directed that way, and if one individual does not hit upon the discovery, another will. Therefore in the disputes and discussions of different nations for the honour of particular inventions or discoveries, as for example the steam-engine, a narrow view of the subject is taken; they properly belong to the age in which they are made, without derogating from the merits of those benefactors of mankind who have lessened his toil or increased his comfort by the efforts of their genius. The time had come for the invention of printing, and printing was invented; and the same observation is applicable to many objects in the physical, as well as to the moral world. In the present disturbed state of society the time is come for the termina-

The following TABLE, which is one of the most curious of statistical documents, was formed by Professor QUETELET from the Register of Marriages at Brussels :—

MARRIAGES IN BELGIUM IN THE YEARS

		1841	1842	1843	1844	1845
Men of 30 years of age and under, to	Women of 30 years of age and under	12,788	12,422	12,368	13,024	13,157
	Women from 30 to 45	2,630	2,626	2,406	2,375	2,438
	Women from 45 to 60	93	121	125	129	102
	Women from 60 upwards	7	6	8	5	5
Men from 30 to 45 inclusive, to	Women of 30 and under	6,122	5,803	5,617	4,948	5,810
	Women from 30 to 45	5,531	5,396	5,100	5,205	4,981
	Women from 45 to 60	529	542	479	493	532
	Women from 60 upwards	18	12	18	21	21
Men from 55 to 60 inclusive, to	Women of 30 and under	376	346	380	355	346
	Women from 30 to 45	896	879	896	951	993
	Women from 45 to 60	461	447	433	462	460
	Women from 60 upwards	23	19	29	36	28
Men from 60 and above, to ..	Women of 30 and under	48	35	45	41	36
	Women from 30 to 45	139	147	133	119	125
	Women from 45 to 60	153	170	137	112	145
	Women from 60 upwards	62	52	48	50	31
Annual Number of Marriages		29,876	29,023	28,220	29,326	29,210

tion of the feudal system, which will be swept away by the force of public opinion, though individuality merges in these general movements.

Though each individual is accountable to God for his conduct, it is evident that the great laws which regulate mankind are altogether independent of man's will, and that liberty of action is perfectly compatible with the general design of Providence. "A more profound study of the social system will have the effect of limiting more and more the sphere in which man's free-will is exercised, for the Supreme Being could not grant him a power which tends to overthrow the laws impressed on all the parts of creation: He has traced his limits, as He has fixed those of the ocean."

Man is eminently sociable; he willingly gives up part of his free-will to become a member of a social body; and it is this portion of the individuality of each member of that body, taken in the aggregate, which becomes the directrice of the principal social movements of a nation. It may be greater or less, good or bad, but it determines the customs, wants, and the national spirit of a people; it regulates the sum of their moral statistics; and it is in that manner that the cultivation or savageness, the virtues or the vices of individuals have their influence. It is thus that private morality becomes the base of public morality.

The more man advances in civilization the greater will be his collective influence, for knowledge is power; and at no time did the mental superiority of the cultivated races produce such changes as they do at present, because they have extended their influence to the uttermost parts of the earth by emigration, colonization and commerce. In civilized society the number of people in the course of time exceeds the means of sustenance, which compels some to emigrate; others are induced by a spirit of enterprise to go to new countries, some for the love of gain, others to fly from oppression.

The discovery of the New World opened a wide field for emigration. Spain and Portugal, the first to avail themselves of it, acquired dominion over some of the finest parts of South America, which they have maintained till lately a change of times has rendered their colonies independent states. Liberal opinions have spread into the interior of that continent, in proportion to the facility of communication with the cities on the coasts, from whence European ideas are disseminated. Of this, Venezuela and Chile are instances where civilization and prosperity have advanced more rapidly than in the interior parts of South America, where the Andes are higher and the distance from the sea greater. Civilization has been impeded in many of the smaller states by war, and those broils inevitable among people unaccustomed to free institutions Brazil would have been further advanced but for slavery, that stain on the human race, which corrupts the master as much as it debases the slave.

Some of the native South American tribes have spontaneously made considerable progress in civilization in modern times; others have benefited by the Spanish and Portuguese colonists; and many have been brought into subjection by the Jesuits, who have instructed them in some of the arts of social life. But these Indians are not more religious than their neighbours, and, from the restraint to which they have been subject, have lost vigour of character without improving in intellect; so that now they are either stationary or retrograde. Extensive regions are still the abode of men in the lowest state of barbarism: some of the tribes inhabiting the silvas of the Orinoco, Amazon, and Uruguay are cannibals.

The arrival of the colonists in North America sealed the fate of the red men. The inhabitants of the Union, too late awakened to the just claims of the ancient proprietors of the land, have recently, but vainly, attempted to save the remnant.[1] The white man, like an irresistible torrent, has already reached the centre of the continent; and the native tribes now retreat towards the far west, and will continue to retreat, till the Pacific Ocean arrests them, and the animals on their hunting-grounds are exterminated. The almost universal dislike the Indian has shown for the arts of peace, has been one of the principal causes of his decline, although the Cherokee tribe, which has lately migrated to the west of the Mississippi, is a remarkable exception; the greater number of them are industrious planters or mechanics; they have a republican government, and publish a newspaper in their own language, in a character lately invented by one of that nation.

No part of the world has been the scene of greater iniquity than the West Indian islands—and that perpetrated by the most enlightened nations of Europe. The native race has long been swept away by the stranger, and a new people, cruelly torn from their homes, have been made the slaves of hard task-masters. If the odious participation in this guilt has been a stain on the British name, the abolition of slavery by the universal acclamation of the nation will ever form one of the brightest pages in their history, so full of glory; nor will it be the less so, that justice was combined with

[[1] The Republic of the United States of North America occupies the territory lying between the 25th and 49th degrees of north latitude; and the 67th and 130th degrees of west longitude. It extends from the Atlantic to the Pacific ocean, 3000 miles, and from north to south 1700 miles, embracing an area of 3,250,000 square miles. In 1850, the population was 23,257,723, of which 3,198,324 are slaves.

The total number of Indians in the United States and territories, including California and Oregon, is estimated at 388,229; and in the British North American possessions, 19,987.]

mercy, by the millions of money granted to indemnify the proprietors. It is deeply to be lamented that our brethren on the other side of the Atlantic have not followed the example of their fatherland; but in limited monarchies the voice of the people is listened to, while republican governments are more apt to become its slave. Unfortunately, the Northern States have revoked a law by which they had nobly declared every man free who set his foot on their territory; but the time will come when the Union will sacrifice interest to justice and mercy.

It seems to be the design of Providence to supplant the savage by civilized man in the continent of Australia as well as in North America, though every effort has been made to prevent the extinction of the natives. Most of the tribes in that continent are as low in the scale of mankind as the cannibal Fuegians, whom Captain FitzRoy so generously, but so ineffectually, attempted to reclaim. Some of the Australians are faithful servants for a time, but they almost always find the restraint of civilized life irksome, and return to their former habits, though truly miserable in a country where the means of existence are so scanty. Animals and birds are very scarce, and there is no fruit or vegetable for the sustenance of man.

Slavery has been a greater impediment to the improvement of Africa than even the physical disadvantages of the country—the great arid deserts and unwholesome coasts. A spontaneous civilization has arisen in various parts of Southern and Central Africa, in which there has been considerable progress in agriculture and commerce; but civilized man has been a scourge on the Atlantic coast, which has extended its baneful influence into the heart of the continent, by the encouragement it has given to warfare among the natives for the capture of slaves, and for the introduction of European vices, unredeemed by Christian virtues.

The French are zealous in improving the people in Algiers, but the constant warfare in which they have been embroiled ever since their conquest must render their success in civilizing the natives at least remote. The inhabitants of those extensive and magnificent countries in the eastern seas that have long been colonized by the Dutch have made but little progress under their rule.

The British colony at the Cape of Good Hope has had considerable influence on the neighbouring rude nations, who now begin to adopt more civilized habits. When Mr. Somerville visited Litako, the natives for the first time saw a white person and a horse, and were scantily clothed with skins. When Dr. Smith visited them 20 years afterwards, he found the chief men mounted on horseback, wearing hats made of rushes, and an attempt made to imitate European dress.

Colonization has nowhere produced such happy results as among the amiable and cultivated inhabitants of India, who are sensible of the benefits they derive from the impartial administration of just and equal laws, the foundation of schools and colleges, and the wide extension of commerce.

All the causes of emigration have operated by turns on the inhabitants of Britain, and various circumstances have concurred to make their colonies permanent. In North America, that which not many years ago was a British colony has become a great independent nation, occupying a large portion of the continent. The Australian continent and New Zealand will in after ages be peopled by a British race, and will become centres of civilization, which will spread its influence to the uttermost islands of the Pacific. These splendid islands, possessing every advantage of climate and soil, with a population in many parts far advanced in the arts of civilized life, industry, and commerce, though in others savage, will in time come in for a share of the general improvement. The success that has attended the noble efforts of Sir J. Brooke in Borneo, shows how much the influence of an active and benevolent mind can in a short time effect.

The colonies on the continent of India are already centres from which the culture of Europe is spreading over the East.

Commerce has not less influence on mankind than colonization, with which it is intimately connected; and the narrow limits of the British Islands have rendered it necessary for its inhabitants to exert their industry. The riches of our mines in coal and metals, which produce a yearly income of 24,000,000*l.* sterling, is a principal cause of our manufacturing and commercial wealth; but even with these natural advantages, more is due not only to our talents and enterprise, but to our high character for faith and honour.

Every country has its own peculiar productions, and by an unrestrained interchange of the gifts of Providence the condition of all is improved. The exclusive jealousy with which commerce has hitherto been fettered, shows the length of time that is necessary to wear out the effects of those selfish passions which separated nations when they were yet barbarous. It required a high degree of cultivation tc break down those barriers consecrated by their antiquity; and the accomplishment of this important change evinces the rate at which the present age is advancing.

A new era in the history of the world began when China was opened to European intercourse; but many years must pass before European influence can penetrate that vast empire, and eradicate those illiberal prejudices by which it has so long been governed.

Two important triumphs yet remain to be achieved over physical difficulties by the science and energy of man, namely, the junction of the Pacific and Atlantic Oceans at the American isthmus, and the union of the Red Sea with the Mediterranean at that of Suez. The first seems to be on the eve of accomplishment, and, in conjunction with the treasures with which the auriferous districts of California abound, may bring about a complete revolution in the commerce of the New World; and that country, hitherto so completely separated from the rest of the globe, and so little known, will become a new centre of civilization, whose influence will be diffused over the wide Pacific to the shores of the eastern continent; the expectation of Columbus will then be realized—of a passage to the East Indies by the Atlantic. Should the Mediterranean and Red Sea be united by a water communication, Alexandria, Venice, and the other maritime cities of southern Europe may regain, at least in part, the mercantile position which they lost by the discovery of Vasco da Gama.[1]

The advantages of colonization and commerce to the less civilized part of the world are incalculable, as well as to those at home, not only by furnishing an exchange for manufactures, important as this is, but by the immense accession of knowledge of the earth and its inhabitants, that has been thus attained.

The history of former ages exhibits nothing to be compared with the mental activity of the present. Steam, which annihilates time and space, fills mankind with schemes for advantage or defence: but however mercenary the motives for defence may be, it is instrumental in bringing nations together. The facility of communication is rapidly assimilating national character. Society in most of the capitals is formed on the same model; and as the study of modern languages is now considered a part of polite education, and every well-educated person speaks more than one modern tongue, one of the great barriers to the assimilation of character amongst nations will be removed.

Science has never been so extensively and so successfully culti-

---

[1] It is singular that the British should for years have possessed such extensive territories in Asia without having explored their mineral wealth. Perhaps the quantity of gold recently discovered in California and Africa may call the attention of the East India Company to the subject. Some of the richest mining districts are in countries where primary formation has been disturbed by igneous action; and as that is eminently the case along the eastern coast of the Bay of Bengal, from Aracan to the peninsula of Malacca, mines of the precious metals will most likely be found on that frontier, possibly in Siam and the Birman empire. The interior of the Deccan has also been greatly disturbed by ancient volcanos; and as that country is said to bear a strong analogy in structure to South Africa, it may also resemble it in the production of gold. The auriferous territory in California, which appears to be at least 400 miles long and 100 broad, is an alluvial soil, derived from the destruction of ancient sedimentary rocks of the Paleozoic period traversed by porphyries, syennites, &c.

[Great Britain has abundantly participated in the gold-production of the world, by the recent discoveries in Australia, which promise to equal, if not to surpass, even those of California.]

vated as at the present time: the collective wisdom and experience of Europe and the United States of America is now brought to bear on subjects of the highest importance in annual meetings, where the common pursuit of truth is as beneficial to the moral as to the intellectual character, and the noble objects of investigation are no longer confined to a philosophic few, but are becoming widely diffused among all ranks of society, and the most enlightened governments have given their support to measures that could not have been otherwise accomplished.[1] Simultaneous observations are made at nume-

[1] In bringing to a close a work which may in some measure be considered a kind of Résumé of Natural knowledge, it may not be either out of place here, or irrelevant to our subject, to allude more particularly to the encouragement of late years granted to scientific investigation by our own Government.

It must be confessed that Great Britain for a long time remained behind the nations of the continent in fostering scientific enterprise and research; and if England has rivalled in most branches of natural knowledge, and surpassed in some every other people, it has arisen more from individual exertion, and that spirit of association which forms so happy a characteristic of our race, and which has in our political institutions so mainly contributed to our national greatness and prosperity, than from any direct encouragement from our rulers. Whilst France and other continental nations were endowing the votaries of science, were lavishing money on scientific expeditions, and founding institutions which will hand down the names of their sovereigns to posterity as the benefactors of mankind, England had done little in the same track beyond fitting out the memorable expeditions of Cook, and subsequently those of Vancouver and Flinders, and the support granted to our great national Observatory, which, under the direction of Bradley, Maskelyne, Pond, and Airy, has attained a degree of celebrity unequalled by any Astronomical foundation in ancient or modern times.

The conclusion of a long war, in opening the scientific repositories of the continent to our countrymen, showed us how much our great institutions, with the above solitary exception, were behindhand, not only in extent and utility, but in the liberality with which they were conducted. Possessing as we did the most ample means, from our immense colonial possessions and our widely extended commerce, to add to the stock of our knowledge in natural history, our public collections were infinitely behind those of the great states of the continent, and scarcely on a par with those of the sovereigns of a second and even third rate importance. A better system was loudly called for, and a better system has been adopted. Our great national repository, the British Museum—and I here refer more particularly to its scientific and antiquarian department, for there is still much room for improvement in the literary—has in a few years, thanks to the liberality of Parliament and the exertions of its trustees and officers, become equal in every respect, and superior in many, to any similar institution on the continent. Two establishments have been created within the last dozen of years which reflect the greatest honour on the statesman, Sir F. Baring, then Chancellor of the Exchequer, and the late Earl of Besborough, as chief Commissioners of the Woods and Forests, who fostered them in their infancy, and on the talented individuals who were selected to carry out the enlightened views of the Government—the Museum of Practical Geology,

rous places in both hemispheres on electricity, magnetism, on the tides and currents of the air and the ocean, and those mysterious

a designation that conveys a very inadequate idea of the extent of its attributes or of its utility, and the Royal Botanic Gardens at Kew. To the first the public is already indebted for a geological survey and map of the empire such as never had been planned or executed in any other country —only a small instalment, however, of great services which the nation and geological science are likely to derive from the labours of Sir H. Delabeche and his collaborators. The Royal Gardens at Kew, under the direction of Sir W. J. Hooker, lose nothing when compared with the most celebrated establishments of the kind, ancient or modern; never was public money better bestowed, or in a way to convey more useful instruction and gratification to the great mass of the community. Whilst every German university had its Museum of Comparative Anatomy, when the government of revolutionary France had placed at the disposal of Cuvier ample means to lay the basis of that science of which he was considered to be the founder, an eminent surgeon, John Hunter, animated by the love of science alone, and unaided by his Government, was rendering a similar service to Great Britain in laying the foundation of that Museum which so justly bears his honoured name. Thanks to the liberality of the Government, and to the well-judged appreciation of the Royal College of Surgeons, the Hunterian Collection has become the property of the nation, and has received such additions and ameliorations as not to be behind any of those of the continent; whilst in point of arrangement, facilities granted for study, and real practical utility, it infinitely surpasses them all. To it we principally are indebted for the introduction of the study of comparative anatomy into this country, and for the possession of one of its greatest modern expositors, Professor Owen.

It may appear invidious, at a time when every department of our Government is showing itself so desirous of promoting the cause of science, to point to any in particular: still we cannot refrain from making special mention of one to which science in general, and more particularly that branch of it which forms the principal object of this work, and our best national interests owe a lasting debt of gratitude—the Hydrographic department of the Admiralty; which, under its present able chief, Sir Francis Beaufort, has attained a degree of eminence unequalled by that of any other maritime country. The Lords of the Admiralty have profited of a long period of tranquillity to extend our knowledge over almost every region of the globe, conferring thereby an immense service on geographical science, and placing in the hands of our Royal and Commercial marine a collection of charts and nautical instructions unparalleled in the history of navigation for their extent and exactitude. Another branch of inquiry, closely connected with Hydrography and Navigation, which it required the encouragement of a government to institute, the investigation of the laws of terrestrial magnetism and meteorology, has been very liberally provided for by Parliament, and most ably carried out, under the direction of Colonel Sabine, by the establishment of special observatories in our widely extended colonies, and by the publication and distribution of their results.

The several maritime expeditions undertaken since the peace in a purely scientific view reflect the highest credit on the departments of the Government with which they have originated, as they do on the eminent individuals, many of whom still live to enjoy their well-merited fame, who have carried out their country's wishes. The names of Parry, Franklin, Back, James C. Ross and Richardson will be preserved in the memory of posterity long

vicissitudes of temperature and moisture, which bless the labours of the husbandman one year, and blight them in another.

The places of the nebulæ and fixed stars, and their motions, are known with unexampled precision, and the most refined analyses embrace the most varied objects. Three new satellites and nine new planets have been discovered within four years, and one of these under circumstances the most unprecedented. In the far heavens, from disturbances in the motions of Uranus which could not be satisfactorily accounted for, an unknown and unseen body was declared to be revolving on the utmost verge of the solar system; and it was found in the very region of the heavens pointed out by mathematical analysis. On earth, though hundreds of miles apart, that invisible messenger, electricity, instantaneously conveys the thoughts of the invisible spirit of man to man — results of science sublimely transcendental.

---

after the ephemeral glory of their professional career will have been forgotten.

Although it is to the projectors of such an altered state of things, and to the statesmen who encouraged and brought it about, that our first acknowledgment is due, our thanks must be also expressed to that branch of the legislature which, holding, as it rightly does, the public purse, has so liberally come forward upon every occasion, when solicited, in granting the means to promote scientific enterprise. The votary of science therefore owes to the House of Commons the expression of his unmingled gratitude.

But, in paying that just tribute to the Ministers of the Crown and to Parliament, we must not pass over in silence the encouragement which science has in every department met with from the East India Company. Lords of an immense territory, the Court of Directors, and its servants in India, have always shown themselves ready to contribute in a most liberal spirit to the extension of our knowledge of their widely extended empire. The trigonometrical surveys of India, the establishment of observatories, the endowment of colleges and of scientific societies, the formation of collections of natural history at great expense, and which it distributes to all those who are likely to make good use of them, the publication of works on physical researches, on natural history, of astronomical observations, bestowed with so liberal a hand on men of science, the formation of such a map of its extended dominions and of charts of its coasts as would do honour to any government, must place the East India Company in the first rank of those mighty Potentates of the earth to whom science will both now and in after ages feel placed under the most lasting obligations.

Connected with our Oriental empire, it is due to some of the native sovereigns of India to state that they have not been behindhand in imitating the liberal example of their powerful protectors. Two native princes, the Rajah of Travancore and the King of Oude, have at very great expense established astronomical observatories in their territories, furnished with European instruments of the most delicate construction, and placed under the direction of European officers amply endowed and provided for. The peninsula of India at the present moment possesses four astronomical observatories little behind those of Europe as regards the means of observation: until lately there did not exist one public observatory in the whole extent of the United States of America.

Vain would be the attempt to enumerate the improvements in machinery and mechanics, the canals and railroads that have been made, the harbours that have been improved, the land that has been drained, the bridges that have been constructed; and now, although Britain is inferior to none in many things, and superior to all in some, one of our most distinguished engineers[1] declares that we are scarcely beyond the threshold of improvement. To stand still is to retrograde; human ingenuity will always keep pace with the unforeseen, the increasing wants of the age. "Who knows what may yet be in store for our use; what new discovery may again change the tide of human affairs; what hidden treasures may yet be brought to light in the air or in the ocean, of which we know so little; or what virtues there may be in the herbs of the field, and in the treasures of the earth — how far its hidden fires, or stores of ice, may yet become available: ages can never exhaust the treasures of nature or the talent of man."[2] It would be difficult to follow the rapid course of discovery through the complicated mazes of magnetism and electricity; the action of the electric current on the polarized sun-beam, one of the most beautiful of modern discoveries, leading to relations hitherto unsuspected between that power and the complex assemblage of visible and invisible influences on solar light, by one of which nature has recently been made to paint her own likeness. It is impossible to convey an idea of the rapid succession of the varied and curious results of chemistry, and its application to physiology and agriculture; moreover, distinguished works have lately been published at home and abroad on the science of mind, which has been so successfully cultivated in our own country. Geography has assumed a new character, by that unwearied search for accurate knowledge and truth that marks the present age, and physical geography is altogether a new science.

The spirit of nautical and geographical discovery, begun in the 15th century, by those illustrious navigators who had a new world to discover, is at this day as energetic as ever, though the results are less brilliant. Neither the long gloomy night of a polar winter, nor the danger of the ice and the storm, deter our gallant seamen from seeking a better acquaintance with "this ball of earth," even under its most frowning aspect; and that, for honour, which they are as eager to seek, even in the cannon's mouth. Nor have other nations of Europe and America been without their share in these bold adventures. The scorching sun and deadly swamps of the tropics as little prevent the traveller from collecting the animals and plants of the present creation, or the geologist from investigating those of ages long gone by. Man daily indicates his birthright as lord of the creation, and compels every land and sea to contribute to his knowledge.

---

[1] Sir John Rennie.

[2] Charles Babbage, Esq.

The most distinguished modern travellers, following the noble example of Baron Humboldt, the patriarch of physical geography, take a more extended view of the subject than the earth and its animal and vegetable inhabitants afford, and include in their researches the past and present condition of man, the origin, manners, and languages of existing nations, and the monuments of those that have been. Geography has had its dark ages, during which the situation of many great cities and spots of celebrity in sacred and profane history had been entirely lost sight of, which are now discovered by the learning and assiduity of the modern traveller. Of this, Italy, Egypt, the Holy Land, Asia Minor, Arabia, and the valleys of the Euphrates and Tigris, with the adjacent mountains of Persia, are remarkable instances, not to mention the vast region of the East, and the remote centres of aboriginal civilization in the New World. The interesting discoveries of Mr. Layard, who possesses every acquirement that could render a traveller competent to accomplish so arduous an undertaking, have brought to light the long-hidden treasures of the ancient Nineveh, where its own peculiar style of art had existed anterior to that of Egypt. In many parts of the world the ruins of cities of extraordinary magnitude and architecture show that there are wide regions of whose original inhabitants we know nothing. The Andes of Peru and Mexico have remains of civilized nations before the age of the Incas. Mr. Pentland has found numerous remains of Peruvian monuments in every part of the great valley of the Peru-Bolivian Andes, and many parts of the imperial capital Cusco little changed from what they were at the downfall of Atahualpa. Mr. Stephens has discovered in the woods of Central America the ruins of great cities, adorned with sculpture and pictorial writings, vestiges of a people far advanced, who had once cultivated the soil where these entangled forests now grow. Picture-writings have been discovered by Sir Robert Schomburgk on rocks in Guiana, similar to those found in the United States and in Siberia. Magnificent buildings still exist in good preservation all over eastern Asia, and many in a ruinous state belong to a period far beyond written record.

Ancient literature has furnished a subject of still more interesting research, from which it is evident that the mind of man is essentially the same under very different circumstances: every nation far advanced in civilization has had its age of poetry, the drama, romance, and philosophy, each stamped with the character of the people and times, and still more with their religious belief. Our profound Oriental scholars have made known to Europeans the refined Sanscrit literature of Hindostan, its schools of philosophy and astronomy, its dramatic writings and poetry, which are original and beautiful, and to these the learned in Greece and Italy have contributed.

The riches of Chinese literature and their valuable geography were introduced into Europe by the French Jesuits of the last century, and followed up with success by the French and English philosophers of the present: to France we also owe much of our knowledge of the poetry and letters of Persia; and from the time that Dr. Young deciphered the inscriptions on the Rosetta Stone, Egyptian hieroglyphics and picture-writing have been studied by the learned of France, England, and Italy, and we have reason to expect much new information from the more recent researches of Professor Lepsius of Berlin. The Germans, indeed, have left few subjects of ancient literature unexplored, even to the language written at Babylon and Nineveh — the most successful attempt to decipher which is due to a distinguished countryman of our own, Colonel Rawlinson.[1]

The press has overflowed with an unprecedented quantity of literature, some of standard merit, and much more that is ephemeral, suited to all ranks, on every subject, with the aim, in our own country at least, to improve the people, and to advocate the cause of morality and virtue. All this mental energy is but an effect of those laws which regulate human affairs, and include in their generality the various changes that tend to improve the condition of man.

The fine arts do not keep pace with science, though they have not been altogether left behind. Painting, like poetry, must come spontaneously, because a feeling for it depends upon innate sympathies in the human breast. Nothing external could affect us, unless there were corresponding ideas within; poetically constituted minds of the highest organization are most deeply impressed with whatever is excellent. All are not gifted with a strong perception of the beautiful, in the same way as some persons cannot see certain colours, or hear certain sounds. Those elevated sentiments which constitute genius are given to few; yet something akin, though inferior in degree, exists in most men. Consequently, though culture may not inspire genius, it cherishes and calls forth the natural perception of what is good and beautiful, and by that means improves the tone of the national mind, and forms a counterpoise to the all-absorbing useful and commercial.

Historical painting is successfully cultivated both in France and Germany. The Germans have modelled their school on the true style of the ancient masters. They have become their rivals in richness and beauty of colouring, and are not surpassed in vividness of imagination, nor in variety and sublimity of composition, which is poetry of the highest order embodied. Sculpture and architecture

[1] The most ancient forms of writing are supposed to be Himyaritic lately found in Arabia, and that of the Phœnicians, which is the origin of all the alphabets of ancient and modern Europe, and probably the form of letters in which the sacred Scriptures were written.—(Prichard.)

are also marked by that elevated and pure taste which distinguish their other works of art.[1] French artists, following in the same steps, have produced historical works of great merit. Pictures representing scenes of domestic life have been painted with much expression and beauty by our own artists; and British landscapes, like some painted by German artists, are not mere portraits of nature, but pictures of high poetical feeling, and the excellence of their composition has been acknowledged all over Europe, by the popularity of the engravings which illustrate many of our modern books. The encouragement given to this branch of art at home may be ascribed to the taste for a country life, so general in England. Water-colour painting, which is entirely of British growth, has now become a favourite style in every country, and is brought to the highest perfection in our own.

The Italians have had the merit of restoring sculpture to the pure style which it had lost, and that gifted people have produced some of the noblest specimens of modern art. The greatest genius of his time left the snows of the far North to spend his days in Rome, the head-quarters of art; and our own sculptors of eminent talents have established themselves in Rome, where they find a more congenial spirit than in their own country, in which the compositions of Flaxman were not appreciated till they had become the admiration of Europe. Munich can boast of some of the finest specimens of modern sculpture and architecture.

The Opera, one of the most refined of theatrical amusements in every capital city of Europe, displays the excellence and power of Italian melody, which has been transmitted from age to age by a succession of great composers. German music, partaking of the learned character of the nation, is rich in original harmony, which requires a cultivated taste to understand and appreciate.

Italy is the only country that has had two poetical eras of the highest order; and, great as the Latin period was, that of Dante was more original and sublime. The Germans, so eminent in every branch of literature, have also been great as poets; the power of Goethe's genius will render his poems as permanent as the language in which they are written. France is, as it long has been, the abode of the Comic Muse; and although that nation can claim great poets of a more serious cast, yet the language and the habits of the people are more suited to the gay than the grave style. Though the British

---

[1] The works of Cornelius and Kaulbach bear testimony to the justice of the observations in the text. In drawing, nothing can be more beautiful —in composition, nothing can be more varied or sublime. The 'Destruction of Jerusalem,' by Kaulbach, in which a powerful genius has combined the truth of the historian with the imagination of the poet, and executed with the hand of a master, might bear comparison even with the Italian school of colouring.

may have been inferior to other nations in some branches of the fine arts, yet poetry, immeasurably the greatest and most noble, redeems, and more than redeems us. The nation that produced the poetry of Chaucer, Spenser, Shakspeare and Milton, with all the brilliant train, down nearly to the present time, must ever hold a distinguished place, as an imaginative people. Shakspeare alone would stamp a language with immortality. The British novels stand high among works of imagination, and they have generally had the merit of advancing the cause of morality. Had French novelists attended more to this, their knowledge of the human heart and the brilliancy of their composition would have been more appreciated.

Poetry of the highest stamp has fled before the utilitarian spirit of the age; yet there is as much talent in the world, and imagination too, at the present time, as at any former period, though directed to different and more important objects, because the whole aspect of the moral world is altered. The period is come for one of those important changes in the minds of men which occur from time to time, and form great epochs in the history of the human race. The whole of civilized Europe could not have been roused to the enthusiasm which led them to embark in the Crusades by the preaching of Peter the Hermit, unless the people had been prepared for it; and men were ready for the Reformation before the impulse was given by Luther. These are the barometric storms of the human mind.

The present state of transition has been imperceptibly in progress, aided by many concurring circumstances, among which the increasing intelligence of the lower orders, and steam travelling, have been the most efficient. The latter has assisted eminently in the diffusion of knowledge, and has probably accelerated the crisis of public affairs on the continent, by giving the inhabitants of different countries opportunities of intercourse, and comparing their conditions. No invention that has been made for ages has so levelling a tendency, which accords but too well with the present disposition of the people. The spirit of emancipation, so peculiarly characteristic of this century, appears in all the relations of life, political and social. On the continent of Europe it has shaken the whole fabric of society, subverted law and order, and ruined thousands, in order to throw down the crumbling remains of the feudal system. The violence with which these changes were conducted, has naturally led to a reaction, but the present attempt to inflict upon the world political and spiritual despotism, must be ephemeral in its turn, being directly opposed to the irresistible progress of the human mind. The same emancipating spirit which has thrown young and old into a state of insubordination and rebellion abroad, has been quietly but gradually altering the relations of social and domestic life at home. Parent and child no longer stand in the same relation to one another; even at an early age boys assume the character and independence of men, which may perhaps fit them sooner for taking their share in the affairs of the world; for it must be acknowledged that, whether from early independence or some other cause, no country has produced more youthful and able statesmen than our own; but, at the same time, it places them on a less amicable and more dangerous position, by depriving them of the advice and experience of the aged, to which the same deference is no longer paid. The working man considers his interest to be at variance with that of the manufacturer, and the attachment of servants to their masters is nearly as extinct in Britain as vassalage. Ambition, to a great extent, pervades the inferior and middle grades of society, and so few are satisfied with the condition in which they were born, that the pressure upwards is enormous. The numerous instances of men rising from an inferior rank to the highest offices in the State encourage the endeavour to advance in society, which is right and natural, if pursued by legitimate means; but the levelling disposition so prevalent abroad is as pernicious as it is impracticable. So long as men are endowed with different dispositions and different talents, so long will they differ in condition and fortune, and this is as strongly marked in republics as in any other form of government; for man, with all his attempts to liberate himself from nature's ordinances, by the establishment of equal laws and civil rights, never can escape from them—inequality of condition is permanent as the human race. Hence from necessity we must fulfil the duties of the station in which we are placed, bearing in mind that, while Christianity requires the poor to endure their lot with patience, it imposes a heavy responsibility on the rich.

In Britain, respect for the labouring classes, together with active benevolence, form the counterpoise to the evil propensities of this state of transition; a benevolence which is not confined to almsgiving, but which consists in the earnest desire to contribute with energy to the sum of human happiness. In proportion as that disposition is diffused among the higher classes, and the more they can convince the lower orders that they have an ardent desire to afford them every source of happiness and comfort that is in their power, so much sooner will the transient evils pass away, and an improved state of things will commence; kindly and confident feelings will then take the place of coldness and mistrust.

The continual increase of that disinterested benevolence and liberal sentiment, which in our own country is the most hopeful and consoling feature of the age, manifests itself in the frequency with which plans for ameliorating the condition of the lower classes are brought before Parliament; in the societies formed for their relief; and in the many institutions established for their benefit and comfort.

Three of the most beneficial systems of modern times are due to the benevolence of English ladies—the improvement of prison discipline, savings-banks, and banks for lending small sums to the poor.

The success of all has exceeded every expectation, and these admirable institutions are now adopted by several foreign countries. The importance of popular and agricultural education is becoming an object of attention to the more enlightened governments; and one of the greatest improvements in education is, that teachers are now fitted for their duties, by being taught the art of teaching. The gentleness with which instruction is conveyed no longer blights the joyous days of youth, but, on the contrary, encourages self-education, which is the most efficient.

The system of infant-schools, established in many parts of Europe and throughout the United States of America, is rapidly improving the condition of the people. The instruction given in them is suited to the station of the scholars, and the moral lessons taught are often reflected back on the uneducated parents by their children. Moreover, the personal intercourse with the higher orders, and the kindness which the children receive from them, strengthen the bond of reciprocal good feeling. Since the abolition of the feudal system, the separation between the higher and the lower classes of society has been increasing; but the generous exertions of individuals, whose only object is to do good, is now beginning to correct a tendency that, unchecked, might have led to the worst consequences to all ranks. We learn from statistical reports that the pains taken by individuals and associations are not without their effect upon the character of the nation. For example, during the eleven years that preceded 1846, in which the criminal returns indicated the intellectual condition of persons accused, there were 31 counties in England and Wales, in which not one educated woman was called before a court of law, in a population of 2,617,653 females.[1]

Crime has generally decreased in proportion to the religious and moral education of the people: the improvement in the morality of the factory-children is immense since Government appointed inspectors to superintend their health and education;[2] and indeed the

[1] Twenty of these counties were in England and 11 in Wales, and so few crimes took place among educated women in the other counties during the 11 years mentioned, that the annual proportion of accusations against educated females was only 1 in 1,349,059. During the year 1846 only 48 educated persons were convicted of crimes out of the whole population of England and Wales, and none were sentenced to death. And during the years 1845 and 1846 there were 15 counties in England and 11 in Wales in which no well-educated person was convicted of any crime. The number of accusations among educated persons in Scotland is greater, because education is more general, and because the quantity of ardent spirits used in Scotland is five times greater than in England. Crime is very much below the average in the mining districts, and it is still less frequent in Wales and in the mountainous country in the North of England. The accomplishments of a *well*-educated person in these statistical records consist merely in being able to read and write fluently.—'London Statistical Journal.'

[2] Every factory-child is limited to 48 hours of labour in the week, and

improvement in the condition of the whole population appears from the bills of mortality, which unquestionably prove that the duration of human life is continually increasing throughout Great Britain.[1]

The voluntary sacrifices that were made in 1847 to relieve the necessities of a famishing nation evince the humane disposition of the age. But it is not one particular and extraordinary case, however admirable, that marks the general progress — it is not in the earthquake or the storm, but in the still small voice of consolation heard in the cabin of the wretched, that is the prominent feature of the charities of the present time, when the benevolent of all ranks seek for distress in the abodes of poverty and vice, to aid and to reform. No language can do justice to the merit of those who devote themselves to the reformation of the children who have hitherto wandered neglected in the streets of great cities; in the unpromising task they have laboured with patience, undismayed by difficulties that might have discouraged the most determined — but they have had their reward, they have succeeded.[2] The language of kindness and sympathy, never before heard by these children of crime and wretchedness, is saving multitudes from perdition. But it would require a volume to enumerate the exertions that are making for the accommodation, health, and improvement of the people, and the devotion of high and low to the introduction of new establishments and the amelioration of the old. Noble and liberal sentiments mark the proceedings of public assemblies, whether in the cause of nations or of individuals, and the severity of our penal laws is mitigated by a milder system. Happily this liberal and benevolent spirit is not confined to Britain, it is universal in the States of the American Union, and it is spreading widely through the more civilized countries of Europe.

No permanently retrograde movement can now take place in civilization; the diffusion of Christian virtues and of knowledge ensures

---

the children must by law attend school at least two hours a day for six days out of the seven, besides a Sunday school—one penny being deducted out of each shilling of wages for education. The inspectors have the power of establishing schools where wanted, and of dismissing incompetent teachers. The engagement of factory-children in Britain lasts till they are 13, in the United States it ends at 15 years of ages.—'Statistical Journal.'

[1] The average duration of the life of sovereigns is greater in modern than in ancient times, but it is still lower than any other class of mankind. The most favourable average for them is 70·05 years; for the English aristocracy it is 71·69; for the English gentry, 74·00; for the learned professions, 73·62; for English literary and scientific men it is 72·10; for the army and navy, 71·99; and for the professions of the fine arts, 71·15.—'London Statistical Journal.'

[2] There are 62 Ragged Schools in London, and Government undertakes to send annually to the colonies 150 of such of the scholars as choose to go.—'London Statistical Journal.'

the progressive advancement of man in those high moral and intellectual qualities that constitute his true dignity. But much yet remains to be done at home, especially in religious instruction and the prevention of crime; and millions of our fellow-creatures in both hemispheres are still in the lowest grade of barbarism. Ages and ages must pass away before they can be civilized; but if there be any analogy between the period of man's duration on earth and that of the frailest plant or shell-fish of the geological periods, he must still be in his infancy; and let those who doubt of his indefinite improvement compare the first revolution in France with the last, or the state of Europe in the middle ages with what it is at present. For, during the recently disturbed condition of the Continent and the mistaken means which the people employ to improve their position, crime is less frequent and less atrocious than it was in former times, and the universal indignation it now raises is a strong indication of improvement. In our own country, men who seem to have lived before their time were formerly prosecuted and punished for opinions which are now sanctioned by the legislature, and acknowledged by all. The moral disposition of the age appears in the refinement of conversation. Selfishness and evil passions may possibly ever be found in the human breast, but the progress of the race will consist in the increasing power of public opinion, the collective voice of mankind regulated by the Christian principles of morality and justice. The individuality of man modifies his opinions and belief; it is a part of that variety which is a universal law of nature; so that there will probably always be a difference of views as to religious doctrine, which, however, will become more spiritual, and freer from the taint of human infirmity; but the power of the Christian religion will appear in purer conduct, and in the more general practice of mutual forbearance, charity, and love.

APPENDIX.

TABLE OF HEIGHTS ABOVE THE SEA OF SOME REMARKABLE POINTS OF THE GLOBE.

EUROPE.

Names of Places, Mountains, &c.	Heights in English Feet.	Countries in which situated.	Authorities.
Mont Blanc	15,739	Alps, P.[1]	P. S.[2]
Monte Rosa	15,210	" L.	"
Mont Cervin	14,836	" P.	"
Finsterärhorn	14,026	" B.	Eichman
Jungfrau	13,672	" B.	"
Le Géant du M. Blanc	13,786	" P.	P. S.
Mont Combin	14,124	" P.	"
Mont Iséran	13,272	" G.	"
Monte Viso	13,599	" C.	"
Ortler Spitz	12,851	" R.	A. S.
Le Grand Rioburent	11,063	" M.	P. S.
Drey Herrn Spitz	10,122	" Car.	A. S.
Mont Terglou	9,386	" J.	"
Passes of the Alps:—			
Col du Géant	11,238[3]	" P.	Saussure
Col de St. Theodule	11,185	" P.	P. S.
Pass of Great St. Bernard	8,173	" P.	"
" La Furka	8,714	" L.	S. S.
" Mont Moro	8,937	" L.	P. S.
" Le Tavernette	9,827	" C.	"
" Mont Iséran	9,196	" G.	"
" or Col des Fenêtres	9,581	" P.	"
" of the Stelvio	9,177	" R.	A. S.

[1] The letters affixed indicate the part of the Alps to which each locality belongs—M., Maritime; C., Cottian; G., Grecian; P., Pennine; L., Lepontine; B., Bernese, or Helvetian; R., Rhetian; J., Julian; Car., Carniac.

[2] The authorities on which these heights are given are—the Piedmontese Surveys (P. S.), as published in 1845, in the work entitled 'Le Alpi che cingono l'Italia,' 1 vol. 8vo; the Austrian Survey (A. S.), as given in the splendid Maps, published by the Austrian Government, of the Regno Lombardo-Veneto, in 84 sheets; and the Swiss Trigonometrical Survey, by Eichman, 1 vol. 4to, 1846.

[3] The first eight passes are only fit for foot-passengers, and in certain seasons for mules: the remaining eleven offer carriage-roads, and are generally open at all seasons of the year, with the exception of the Stelvio.

Names of Places, Mountains, &c.	Heights in English Feet.	Countries in which situated.	Authorities.
Pass of Bernardino	7,015	Alps, R.	A. S.
" the Splugen	6,946	" R.	"
" St. Gothard	6,808	" R.	S. S.
" Mont Cenis	6,772	" G.	P. S.
" Simplon	6,578	" L.	"
" Tende	6,159	" M.	"
" Mont Genèvre	6,119	" C.	"
" Brenner	4,659	" R.	A. S.
" Pontebba	3,625	" J.	"
Malahite Peak	11,168	Pyrenees	A. B. L.[1]
Mont Perdu Peak	10,994	"	"
Maboré Cylinder of	10,899	"	"
Maladetta "	10,886	"	"
Vignemale "	10,820	"	"
Pic du Midi	9,540	"	"
Canigou	9,137	"	"
Passes of the Pyrenees:—			
Pass or Port d'Oo	9,843	"	"
" d'Estaube	8,402	"	"
" de Garvanie	7,654	"	"
" de Tourmalet	7,143	"	"
Pic de Sancy	6,188	France	"
Plomb du Cantal	6,093	"	"
Mont Mezen	5,795	"	"
Puy de Dôme	4,806	"	"
Ballon des Vosges	4,688	"	"
Mont Ventoux	6,263	"	"
Mulahaçen	11,483	Spain	"
Sierra de Gredos	10,552	"	Bory
Estrella	7,526	"	Franzini
Siete Picos	7,244	"	Bauza
Peña Laza	8,222	"	"
El Gador	6,575	"	Rojas Clemente
Monte Corno, or Gran Sasso d'Italia	9,521	Italy, Apennines	Schouw
Monte Vellino	7,851	"	De Prony
Termenillo Grande	7,212	"	Schouw
Monte Amaro di Majella	9,113	"	"
Monte Cimone	6,975	"	"
Mont Amiata	5,794	Tuscany	"
Vesuvius, Punta del Palo, Aug. 1847	3,947	Kingdom of Naples	Chev. Amanti
Monte Somma	3,869	"	"
Monte Cavi (Mons Albanus)	3,202	Campagna of Rome	Schouw
St. Oreste or Soracte	2,140	"	"

[1] Heights taken from the list published in the French 'Annuaire du Bureau des Longitudes,' converted from metres into English feet.

Names of Places, Mountains, &c.	Heights in English Feet.	Countries in which situated.	Authorities.
Passes of the Apennines:—			
Pass of Noviordi Giovi	1,550	Apennines	Schow
" La Bochetta	2,550	....	"
" Pietramala	3,294	....	"
Islands of the Mediterranean:—			
Monte Rotondo	8,767	Corsica	A. B. L.
" d'Oro	8,701	"	"
" Generargenta	6,004	Sardinia	La Marmora
Mount Etna	10,874	Sicily	W. H. Smyth
Pizzo di Cane	6,509	"	A. B. L.
Mount Eryx	3,894	"	"
Stromboli	2,687	Lipara Isles	De Borch
Greece and Morea:—			
Mount Guiona	8,538	....	Peytier[1]
Parnassus	8,068	....	"
Taygetus, Mont St. Elias	7,904	....	"
Mont Olonas	7,293	....	"
" Kelmos	7,726	....	"
" Athos	6,778	....	De Borch
" Helicon	5,738	....	Peytier
Delphi	5,725	....	"
Mont Hymettus	3,378	....	"
Central Europe:—			
Ruska Joyana	9,912	E. Carpaths.	Malte Brun
Budosch, Transylvania	9,593	"	A. B. L.
Surrul	9,593		
Mount Tatra, highest point	8,524	W. Carpaths.	Wahlenberg
" Csabi Peak	8,314	"	"
" Lomnitz Peak	8,861	"	A. B. L.
Riesenkoppe, in the Riesengeberge	5,394	Germany	Horen
Feldberg in the Schwarzwald	4,675	"	French Engineers
Belchenberg " "	4,642	"	"
Kandelberg " "	4,160	"	Bohnenberger
Schneeberg, Riesengebirge	4,784	"	
Kammkoppel "	4,265	"	Charpentier
Sonnenwerbel, in the Erzgebirge	4,124	"	
Rachelberg in the Böhmerwald	4,561	"	Sternberg
Steinberg, Moravia	3,511	"	David
Brocken, Hartz	3,658	"	Zach
Schneeberg, in the Fichtelgebirge	3,461	"	Goldfuss
Blessberg, in the Thuringerwald.	2,748	"	Zach

[1] Heights determined during the French expedition to the Morea by Captains Peytier and Boblaye, and published in the 'Connaissance des Temps' for 1839.

Names of Places, Mountains, &c.	Heights in English Feet.	Countries in which situated.	Authorities.
Glockner, in the Thuringerwald..	2,231	Germany	Zach
Gross Feldberg, in the Taunus Chain	2,775	"	Schmidt
Lowenberg, in the Siebengebirge	2,024	"	Nose
Norway and Sweden:—			
Skagtöltend............Lat 61° 24′	8,101	Scandinavian Mountains	Keilhau
Koldetind	7,224	"	"
Sognefield	7,182	"	Hagelstam
Mugnafield........... Lat. 61° 20′	7,215	"	Forsell
Schneehattan........ " 62° 20′	8,120	"	Eismark
Pighœttan .... " 62° 2′	6,788	"	Hagelstam
Sulitelma .... " 67° 5′	6,178	"	Wahlenberg
Langfield .... " 61° 53′	6,598	"	Hagelstam
Melderskin .... " 60° 0′	4,859	"	Von Buch
Lyngen Mountains.. " 69° 30′	4,300	"	"
Great Britain:—			
Ben Nevis	4,368	Scotland	Jameson
Cairntoul, Aberdeenshire	4,223	"	Playfair
Ben Avon "	3,931	"	"
Ben Lawers, Grampians	3,945	"	"
Schihallien "	3,564	"	"
Snowdon	3,557	Wales	Roy
Cader Idris	3,550	"	"
Carn Llewellyn	3,471	"	"
Cross Fell, Cumberland	3,383	England	Jameson
Helvyllen "	3,313	"	"
Skiddaw "	3,038	"	Dr. Young
Schunner Fell, Yorkshire	2,388	"	Smith
Coniston Fell, Lancashire	2,575	"	"
Cheviot Hills	2,657	"	"
Pentland Hills	1,878	Scotland	Playfair
Curran Tual, Kerry	3,412	Ireland	Nimmo
Sleib Donnard	3,146	"	"
Nephin, Mayo	2,644	"	Jameson
Mourne Mountains, Down	2,493	"	"
Ben More, Isle of Mull	3,100	Hebrides	"
Hecla, Isle of S. Uist	3,002	"	Boué
Cuchullin, Isle of Skye	2,995	"	M'Culloch
Mount Rona	3,593	Shetland	Laing
Iceland and Faroe:—			
Snœfials, Jokull	5,115	Iceland	A. B. L.
Hecla	5,210	"	"
Skalingefield, Isle Stromoe	2,172	Feroe	Stein

ASIA.

Names of Places, Mountains, &c.	Heights in English Feet.	Countries in which situated.	Authorities.
Himalaya Chain:—			
Kunchin-junga, W. part	28,178	Sikkim	Col. Waugh[1]
„ E. Peak	27,826	„	„
Dwalaghiri	28,080	Nepaul	Webb
Juwahir	25,670	Kumaöon	Herbert
Junnoo	25,312	Sikkim	Waugh
Jumnautri	25,500	Nepaul	Webb
Dhaibun	24,740	„	„
Kabroo	24,005	Sikkim	Waugh
Chumulari	23,929	Tibet	„
Powhunry, or Donkiah Lah	23,175	Sikkim	„
Kanchan-ghow	22,000	„	Dr. Hooker
Momonangli, or Gurla	23,500	Tibet	Strachey[2]
Api Peak	22,799	Nepaul	Webb
Peak No. 12	23,263	Between the Kali and E. branch of the Ganges	„
„ 13	22,313		
„ 23	22,727		
„ 25	22,277		
St. George's Peak	22,500	Between the Ganges and Sutlej	
St. Patrick's Peak	22,638		
Gungoutri Pyramid	21,219		
Jownlee Peak (highest)	21,940	Kumaöon	
Kailas Peak	21,000	Tibet	Strachey
Kohibaba	17,905	Hindoo Cush	Burnes
Peak N. of Cabul	20,232	„	„
Passes of the Himalaya:—			
Marsi Niglak Pass	19,000	Tibet	Strachey
Karokorum Pass	18,600	„	Dr. Thomson[3]
Parangla Pass	18,500	„	Cunningham
Kronbrung Pass	18,313	„	Gerard
Langpya Dhura or Doora Ghaut	17,750	„	Strachey
Lipu Lek Pass	16,884	„	Manson
Niti Ghaut Pass	16,814	„	Gerard
Paralaha Pass	16,500	„	Webb
Shatool Pass	15,500	„	„
Lachoong Pass	18,000	Sikkim	Dr. Hooker[4]
Elbrouz	18,493	Caucasus	Fuss
Kasbeck	16,530	„	A. C.[5]

[1] The heights of Sikkim Himalaya are the results of the observations of Colonel Waugh, Director of the Trigonometrical Survey of India. *See* 'Journal of As. Soc. of Bengal,' Nov. 1848.

[2] For Lieut. Strachey's observations during his very interesting Journey to the Sacred Lakes of Manasarowar, &c., *see* 'Journal of As. Soc. of Bengal,' Aug. 1848.

[3] *See* Hooker's 'Journal of Botany,' May, 1849.

[4] Journal of Geographical Society of London, Vol. xx. p. 49.

[5] The Heights followed by the letters A. C. have been taken from Humboldt's 'Asie Centrale.'

Names of Places, Mountains, &c.	Heights in English Feet.	Countries in which situated.	Authorities.
Demavend	14,695	Persia	Thomson
Ararat	17,212	"	Parrot
Argæus	13,197	Asia Minor	A. C.
Beloukha	11,062	Altaï	"
Mount Libanus	9,517	Syria	A. B. L.
Mount Horeb	8,593	"	Rüppell
" Sinai	7,498	"	"
Jebel Serbal	6,760	"	"
Kamen Peak	5,397	Ural	A. C.
Tremel Peak	5,071	"	"

AFRICA, AND ISLANDS IN THE ATLANTIC.

Names of Places, Mountains, &c.	Heights in English Feet.	Countries in which situated.	Authorities.
Mount Atlas (Miltsin)	11,400	Morocco	Washington
" Abba Jarrat... 13° 10′ N.	15,008	Abyssinia	Rüppel
" Buahat......... 13° 12′ N.	14,362	"	"
' Kilimandjaro.. 4° 0′ S. (doubtful)	20,000	" in the Mtns. of the Moon	Ans. of Phil.
Mount Woso ...... Lat. 6° 30′ N.	16,350	Southern Ethiopia	D'Abbadie[1]
" Dajan " 13° 15′ N.	15,740	Northern Ethiopia	"
" Fatra " 10° 42′ N.	14,350	Abyssinia	"
Table Mountain	3,816	Cape of Good Hope	A. B. L.
Pico Ruivo	6,056	Madeira	Vidal[2]
Peak of Teneriffe, or de Teyde...	12,172	Canaries	Von Buch
Chahorra, Teneriffe	9,885	"	"
Pico de Cruz, Palma	7,730	"	Vidal
Los Pexos, Great Canary	6,400	"	"
Alto Garaona, Gomera	4,400	"	"
San Anton, Ferro	3,907	"	"
Asses' Ears, Fuestaventura	2,770	"	"
Peak of Fogo	9,154	Cape Verde Islands	Deville
Pico, Island of San Antonio	8,815	"	Capt. King
Pico, Island of Pico	7,613	Azores	Vidal
Pico de Vara, Island of St. Michael's	3,570	"	"
Caldeira de Sta. Barbara, Terceira	3,500	"	"

[1] From a MS. list of a great number of Geographical Positions and Heights determined by M. A. T. d'Abbadie, during his travels and long residence in Abyssinia, communicated to the author through Mr. Pentland.

[2] The Heights given on Captain Vidal's authority are taken from the elaborate Surveys of Madeira, the Canaries, and Azores, executed under his direction, and published by the Admiralty.

Names of Places, Mountains, &c.	Heights in English Feet.	Countries in which situated.	Authorities.
Pico de San Jorje	3,498	Azores	Vidal
Morro Gordo, Flores	3,087	"	"
Caldeira de Corvo	2,460	"	"

AMERICA.

Names of Places, Mountains, &c.	Heights in English Feet.	Countries in which situated.	Authorities.
North America:—			
Mount St. Elias	16,775	N. America	A. B. L.
Popocatepetl	17,717	Mexico	"
Orizaba	17,374	"	Humboldt
Iztacihuatl	15,705	"	"
Nevado of Toluca	15,542	"	A. B. L.
Sierra Nevada	15,170	"	Humboldt
Perote Mountain	13,413	"	"
Fair Weather Mountain	14,925	N. America	
Jorullo	4,265	Mexico	"
Volcan de Fuego, West Peak	13,160	Guatemala	Basil Hall
" " East Peak	13,050	"	"
Irasu, or Volcano of Cartago	11,480	"	Phys. Atlas
West Indies:—			
Blue Mountains	7,277	Jamaica	
La Soufrière	5,108	Guadaloupe	
Montagne Pelée	4,432	Martinique	Monnier
Mount Garon	4,370	St. Vincent's	Chisholm
South America:—			
La Silla de Caraccas	8,600	Venezuela	Humboldt
Cerro de Duida	8,280	"	"
Roraima ....Lat. 5° 30′ N.	7,450	Guiana	Schomburgk
Plain of Bogota	8,730	"	Humboldt
Volcano of Tolima	18,020	Andes of N. Grenada	"
Volcano of Purace	17,034	"	"
" Cumbal	15,620	"	Bousingault
Cayambe	19,535	Andes of the Equator	Humboldt
Antisana	19,137		"
Cotopaxi	18,875	"	"
Pichincha	15,924	"	"
Chimborazo	21,424	"	"
Illinissa	17,380	"	Bongeur
Tunguragua	16,424	"	Humboldt
Sangaï	16,138	"	La Condamine

Names of Places, Mountains, &c.	Heights in English Feet.	Countries in which situated.	Authorities.
Vilcañota Peak	17,525	Peru	Pentland[1]
Apu-Cunuranu	17,590	"	"
Guaracoota Peak, Snow-line	16,217	"	"
Cololo.................. Lat. 14° 58′	17,930	Bolivia	"
Volcano of Arequipa	20,320	Peru	"
Queñuta.................. Lat 17° 41′	18,765	....	"
Chipicani, or Nevado of Tacora..	19,745	"	"
Pomarape	21,700	"	"
Parinacota............................	22,030	"	"
Sahama	22,350	"	"
Gualateiri............ Lat. 18° 23′	21,960	"	"
Ancohuma, S. Peak.................	21,286	Bolivian Andes	"
" N. Peak	21,043	"	"
Chachacomani, N. Peak	20,355	"	"
Angel Peak.......... Lat. 16° 10′	20,115	"	"
Supaïwasi, or Huayna Potosi	20,260	"	"
Cacaca .... Lat. 16° 25′	18,210	"	"
La Mesada, S. Peak	19,356	"	"
Illimani, S. Peak	21,140	"	"
Mount de las Litanias...............	14,500	"	"
Miriquiri Peak........................	16,100	"	"
Cerro, or Mountain of Potosi	16,152	"	"
" of Chorolque, near Tupisa.....................	16,550	"	Redhead
Aconcagua Mountain...............	23,910	Chile[2]	Beechey and Fitzroy
Peak of Doña Ana	16,070	"	Domeyko[3]
Tupungato	15,000	"	
Volcano of Antuco	8,918	"	Domeyko
Volcano of Osorno, or Llanquihue	7,550	"	Fitzroy
Yanteles	8,030	"	"
Minchinmadava Volcano	8,000	"	"
Mount Stokes	6,400	Patagonia	"

[1] The Heights given in this table on Mr. Pentland's authority have been taken from his Map of 'The Laguna of Titicaca, and of the Valleys of Yucay, Collao, and Desaguadero,' published in 1848.

[2] As stated in the text, p. 96. The height here assigned to the Peak of Anconcagua differs 700 feet from that given by Captain Fitzroy. A re-calculation, however, of his elements has led us to adopt a much greater elevation for the giant of the Chilian Andes than is given by that talented officer.

Captain Fitzroy's observations place the summit of the Peak of Aconcagua, which on his chart is incorrectly designated as a volcano, in lat. 32° 38′ 30″, long. 70° 00′ 30″ W., or 23′ 23″ N., and 100′ 45″ E. of Valparaiso, or its nearest distance about $88\frac{9}{10}$ geographical miles. From a station near Captain Fitzroy's, at Valparaiso, Captain Beechey found the angle of elevation of Aconcagua, by several very careful observations, to be 1° 55′ 45″, the distance from this station to the Peak being 88 74 geographical miles. From a discussion of all these data, the compiler of this table has deduced for the height of Aconcagua 23,910 feet above the sea.

[3] The heights given on Mr. Domeyko's authority are taken from his very interesting papers on the Geology and Mines of Chile, inserted in the Annales des Mines, 1846, '47, '48.

Names of Places, Mountains, &c.	Heights in English Feet.	Countries in which situated.	Authorities.
Mount Burney	5,800	Patagonia	Fitzroy
Mount Sarmiento	6,900	Terra del Fuego	"
Mount Darwin	6,800	"	"
Passes of the Andes:—			
Pass of Rumihuasi....................	16,160	Peru	Gaye
" Altos de Toledo	15,790	"	Pentland
" Pacuani	15,340	Bolivia	"
" Chullunquiani	15,160	"	"
" Vilcañota, or la Raya	14,520	Peru	"
" Las Gualillas.................	14,750	"	"
" Paramo d'Assüay	15,528	Equator	Humboldt
" las Guanacas.................	14,708	"	Bougeur
" Quindiu	11,502	N. Grenada	Humboldt
" el Amorsadero..............	12,850	"	"
" Come Cabello, Lat. 27° 30′ S.	14,520	Chile	Domeyko
" Dona Ana Lat 29° 36′ S.	14,829	"	"
" Portezuela de la Laguna Lat. 30° 15′ S.	15,575	"	"
" La Cumbre	12,450	"	Pentland
" las Peuquenes, E. Pass...	13,810	"	Domeyko
" el Portillo, W. "	14,370	"	"
Mountains of Brazil:—			
Itambe	5,960		Eschwege
Villarica chain, Serra da Piedade	5,830	....	"
Itacolumi	5,750	....	"
Isle of Bourbon, highest point ...	8,340	....	Phys. Atlas
Mount Ambotismene................	11,506	Madagascar	A. B. L.
Adam's Peak	6,152	Ceylon	
Mount Slamat or Tajal	11,930	Java	Junghuhn
" Sumbung	11,030	"	"
" Gounnong Pasama, or Ophir	13,840	Sumatra	Raffles
Volcano of Matua.....................	4,500	Kurile Is.	Phys. Atlas
Peak of Unimak......................	8,593	Aleutian Is.	"
Mowna Kea	13,953	Sandwich Is.	Wilkes
" Roa..............................	13,760	"	"
Tobreonou..............................	12,250	Otaheite	Phys. Atlas
Mount Wellington, or Kosciusco.	6,500	New Holland	Strelizki
" Lindsay....Lat. 28° 20′ S.	5,700	"	Mitchell
" Canobolas " 33° 25′ S.	4,551	"	"
" Edgecumbe	9,630	New Zealand	Bidwell
" Egmont	8,840	"	Dieffenbach
Tongariro Mountain	6,200	"	"
Mount Erebus	12,367	Antarctic Lands	Sir J. C. Ross
" Terror	10,884		"

LAKES AND INLAND SEAS.

Names of Places, Mountains, &c.	Heights in English Feet.	Countries in which situated.	Authorites.
Sirikol, source of the Oxus........	15,630	Pamer	Wood
Manasarowar and Raikas Thal...	15,250	Tibet	Strachey
Chumurari Lake	15,000	"	Cunningham
Titicaca	12,847	Peru-Bolivia	Pentland
Lake Ngami	2,825	S. Africa	Murray
Baikal	1,535	Asia	A. C.
Lake of Van............................	566	Turkey in Asia	"
Aral	36	Asia	"
Caspian Sea, *below* the level of the Ocean................................	82	"	Russian Survey
Dead Sea, *below* the Ocean.........	1,312	Syria	Symonds
Lake of Tiberias, *below* the Ocean	328[1]	"	"
Lake Superior.........................	596	N. America	
" of Lucerne	1,407	Switzerland	Eschman
" of Geneva	1,230	"	"

HEIGHTS OF SOME REMARKABLE INHABITED PLACES.

Rumihuasi, Post Station	15,542	Andes of Peru	Gaye
Ayavirini, Post Station	14,960	Peru	"
Pati, Post Station, Lat. 16° 5′ S.	14,400	"	Pentland
Apo " " " 16° 13′ S.	14,376	"	"
Ancochallini, farm " 17° 35′ S.	14,683	"	"
Tacora, village......" 17° 47′ S.	13,690	"	"
Calamarca............" 16° 54′ S.	13,650	Bolivia	"
Antisana, farm	13,454	Equator	Humboldt
Potosi, city............................	13,330	Bolivia	Pentland
Puno, city	12,870	Peru	"
Oruro, city	12,454	Bolivia	"
Arquaze, highest village in Ethiopia	12,235	Abyssinia	D'Abbadie
La Paz, city	12,226	"	Pentland
Miquipampa, village	11,870	Peru	Humboldt
Cusco, city	11,384	"	Pentland
Quito, capital of the Equator	9,543	....	Humboldt
Chuquisaca, capital of Bolivia...	9,343	Bolivia	Pentland

[1] The depression of the Lake of Tiberias given in the table, on the authority of Lieut. Symonds, is not to be relied upon, since it differs by some hundred feet from that determined barometrically by three different observers, de Berton, Rusegger, and M. Von Wildenbruch, the mean of which gives for its level below the Mediterranean 755 feet; the numbers given at pp. 85 and 259 will require therefore to be corrected. The mean of the barometrical measurements of the three travellers above mentioned gives for the depression of the Dead Sea 1423½ feet, or one hundred feet more than the trigonometrical measurements by Lieut. Symonds

Names of Places, Mountains, &c.	Heights in English Feet.	Countries in which situated.	Authorities.
Bogota, capital of New Grenada.	8,730	N. Grenada	Humboldt
Mexico	7,570	Mexico	"
Arequipa, city	7,852	Peru	Pentland
Highest villages on the S. side of the Himalaya	13,000	Kumäon	Strachey
Leh	11,600	Thibet	"
Niti, village	11,473	Kumäon	Webb
Darjeeling, town	7,165	Sekim Himalaya	Waugh
Cabool	6,382	Afghanistan	Burnes
Saka, capital of Enarea, 8° 11′ N.	6,050	Ethiopia	D'Abbadie
Kandahar	5,563	"	Humboldt
Teheran	4,137	Persia	A. C.
Kashmir, city	5,818	Kashmir	Hugel
Hospital of Great St. Bernard	8,110	....	A. B. L.
" of St. Gothard	6,808	Alps, P.	"
St. Veran, village	6,693	" C.	"
Breuil, village	6,584	" P.	P. S.
Barèges	4,072	Pyrenees	A. B. L.
Briançon, town	4,285	Alps, M.	"
Jerusalem	2,565	Syria	Berton and Rusegger
Madrid	1,994	Spain	A. B. L.
Santiago, capital of Chile	1,750	Chile	Pentland
Münich	1,764	Bavaria	A. B. L.
Geneva	1,450	Switzerland	"
Turin	755	Piedmont	"
Lima	520	Peru	Pentland
Vienna	436	Austria	A. B. L.
Milan	420	Lombardy	"
Paris, Observatory	213	France	"
Rome, Capitol	151	Italy	"
Brelin	131	Prussia	"

GLOSSARY.

A'BIES. Lat. A fir-tree. Specific name of a tree.

ABYSSI'NICA. Lat. Ayssinian; belonging or relating to Abyssinia.

ACA'CIA. Gr. *ake*, a point, and *akios*, not subject to worms; a thorny tree. A genus of the family Leguminósæ and order Mimósæ. About 300 species are enumerated; many of them yield gum.

ACA'CIA ARA'BICA. Arabian acacia.

ACA'CIAS. Trees belonging to the genus acacia.

A'CID. A term given by chemists to those compound bodies which unite with salifiable bases to form salts: for example, a compound of sulphur and oxygen, called sulphuric acid, unites with magnesia and forms a salt named sulphate of magnesia, or Epsom salts.

ACI'DULOUS. Sourish; possessing acid properties.

ACROCHO'RDI. Lat.; plural of acrochordus.

ACROCHO'RDUS. From the Greek *akrochordon*, a wart. A genus of non-venomous ophidians, whose bodies are entirely covered by scales resembling warts: these scales, or rather squamous tubercles, are small, numerous, rhomboidal, and surmounted by a small horn or point, more or less sharp.

ADANSO'NIA. A genus of plants named in honour of Michel Adanson, a famous French botanist, born in 1727. *Adanso'nia digita'ta.* Sour gourd, or African sour-sop. Monkeys' bread or Baobab tree of Senegal, which is considered the largest, or rather the broadest tree in the world. "Several measured by Adanson, were from sixty-eight to seventy-eight feet in circumference, but not extraordinarily high. The trunks were from twelve to fifteen feet high, before they divided into many horizontal branches, which touched the ground at their extremities; these were from forty-five to fifty-five feet long, and were so large that each branch was equal to a monstrous tree; and where the water of a neighboring river had washed away the earth, so as to leave the roots of one of these trees bare and open to the sight, they measured 110 feet long, without including those parts of the roots which remained covered. It yields a fruit which resembles a gourd, and which serves for vessels of various uses; the bark furnishes a coarse thread, which they form into ropes, and into cloth with which the natives cover their middle from the girdle to the knees; the small leaves supply them with food in times of scarcity, while the large ones are used for covering their houses, or, by burning, for the manufacture of good soap. At Sierra Leone this tree does not grow larger than an orchard apple tree." *Loudon.*

A'DIT. Lat. *adeo*, I approach. A horizontal shaft or passage in a mine, either for access, or for carrying off water.

AFRICA'NUS. Lat. African; belonging or relating to Africa.

AGAL'LOCHUM. From the Gr. *aggalomai*, to become splendid. A resinous, aromatic wood, burned by the Chinese and Japanese for the sake of its agreeable odour, from the *Excœca'ria aggal'locha.* Aloes wood.

A'GAMOUS. From the Gr. *a*, privative, and *gamos*, marriage. Having no sex.

A'GATE. A name given to all varieties of quartz which have not a vitreous aspect; are compact, semi-transparent, and whose fracture resembles that of wax. Agates are of various colours and admit of a fine polish. According to Theophrastes and Pliny, the name comes from the river Achates in Sicily, now the Drillo, on the banks of which the first agates were found.

AGLA'IA. From the Gr. *Aglaia*, beauty, elegance. A genus of plants, trees or shrubs, of which there are five or six species in the Island of Java. The *odora'ta* is one.

A'GUA. Spanish. Water.

AI'RA. From the Gr. *aira*, a tare, cockle weed. A genus of the family of Gramíneæ, or grasses, of the tribe of Avenáceæ. Hair-grass. *A. antarctica.* Antarctic hair-grass.

AIR-PLANTS. A name given to certain parasitic plants which were supposed to be nourished by the air alone, without contact with the soil. There are some species which will live many months suspended by a string in a warm apartment.

AL'BA, AL'BUS, AL'BUM, } Lat. White.

ALBI'NO. Spanish. From the Lat. *albus*, white. Applied to individuals of the human race, (and extended also to some other animals), who have white hair; the iris, pinkish or very pale; and the eyes unable to bear much light. Albinos are most frequent in the negro race; but it does not seem to be true that there are tribes of Albinos in any part of the world.

ALBU'MEN. From the Lat. *albus*, white. A chemical term, applied to an immediate organic principle, which constitutes the chief part of the white of egg. Animal and vegetable albumen are nearly the same in composition.

AL'KALINE. Having properties of an alkali.

ALCHEMI'LLAS or ALCHEMI'LLA. Arabic. A genus of plants of the family Rosáceæ. The *A. vulgaris*, common ladies' mantle.

A'LGA. Lat. Sea-weed.

ALGÆ. Plural of alga. Name of a sub-class of cryptógamous plants, which is subdivided into three families: the *Phy'ceæ*, or submerged sea-weeds; the *Lichens*, or emerged sea-weed, and the *Byssa'ceæ*, or amphibious sea-weeds. The algæ or sea-weeds are ágamous plants which live in the air, on the surface or at the bottom of fresh or salt water; they are remarkable for their cellular or filamentous structure, into which no vessels enter.

ALHA'GI. Arabic. Genus of plants of the family of Leguminósæ. The *alhagi maurorum* grows in the deserts of Egypt; a sweet, gummy substance exudes from the bark in form of small yellowish grains, which, it appears, was the manna the Hebrews ate while in the deserts of Arabia Petrea.

AL'KALI or AL'CALI. A chemical term formerly applied to potash and soda: it now embraces the oxides of potassium, sodium, lithium, barium, strontium and calcium, metals which decompose water at ordinary temperatures, and absorb, that is, combine with its oxygen, giving out heat and flame.

A'LOE. Name of a genus of plants which includes very many species. The inspissated juice of several of these species constitutes the varieties of the medicine called Soccotrine, Barbadoes aloes, &c.

ALLU'VIA. Lat. Plural of alluvium.

ALLU'VIAL. Of the nature of alluvium.

ALLU'VION, ALLU'VIUM, } From the Lat. *alluo*, I wash upon. Gravel, sand, mud, and other transported matter, washed down by rivers and floods upon lands not permanently submerged beneath water. A deposit formed of matter transported by currents of water.

ALPI'NUM, ALPI'NUS, } Lat. Alpine; belonging or relating to the Alps.

ALU'MINUM or ALUMI'NIUM. From *alu'men*, alum. The metalloid which forms the basis of alum; of alumina or pure argil.

AMARY'LLIS. From the Gr. *amarusso*, to be resplendent. A nymph in ancient mythology. Name of a genus of plants, forming the type of the family of

Amaryllídeæ, composed of about sixty species. Generally they are bulbous plants, remarkable for the size and beauty of their flowers.

AMBLYRHY'NCHUS. From the Gr. *amblus*, obtuse, and *rugchos*, snout. Name of a genus of iguanian reptiles.

A'METHYST. From the Gr. *amithustos*, not drunk. The ancients gave this name to a stone in which the wine red colour was tempered with violet. A violet variety of hyaline quartz.

AMMO'NIA. A colourless gas of a peculiar, pungent odour. It causes death when respired; and its strong alkaline reaction distinguishes it from all other elastic fluids. It is liberated from all its chemical combinations by the alkalis. Spirits of hartshorn is a solution of this gas.

AMMONI'ACAL. Of the nature of ammonia.

AM'MONITE. From the Lat. *Ammon*, a name of Jupiter. A fossil so called from a supposed resemblance to the horns engraven on the heads of Jupiter Ammon. In certain parts of England called *snake-stones*. Ammonites are fossil shells, rolled upon the same plane, consisting of a series of separate chambers, like the nautilus.

AMOR'PHOUS. From the Gr. *a*, privative, and *morphe*, form. Without definite or regular shape.

AMPE'LIDÆ. Lat. (*ampelis*), name of a family of birds in the tribe of Dentiróstres.

AMPELI'DEÆ. From the Gr. *ampelos*, a vine. Name of the family of Phaneró-gamous plants, which includes the vine.

AMPHI'BIOUS. From the Gr. *amphibios*, two-lived. Having the faculty of living in two elements.

AMPHIU'MA. From the Gr. *amphi*, both, on all sides, and *uma*, that which has been moistened. A genus of Batrachians in which lungs but no bronchiæ exist through life. *Amphiu'ma menop'oma*. A kind of Batrachian which resembles the Salamander. It is found in Louisiana.

A'MPLITUDE. In astronomy denotes the angular distance of a celestial body, at the time it rises or sets, from the east or west points of the horizon. It is sometimes used to designate the horizontal distance a projectile reaches when thrown from a gun.

AM'YRIS. From the Gr. *amuros*, not perfumed. A genus of phaneróġamous plants, which is the type of the family of Amyri'deæ, which is allied to the family of turpentines. *Am'yris gileade'nsis*. The Balm of Gilead. *Am'yris kataf*. The myrrh tree. *Am'yris opoba'lsamum*. The opobalsam, or balsam of Mecca.

ANA'NAS. Portuguese. Pine-apple. Genus of the family Bromeliáceæ, and type of the tribe Ananáceæ.

ANDRO'MEDA. Mythological name of a constellation. Genus of the family Ericáceæ, and type of the tribe Andromédeæ or Andromedas.

ANGE'LICA ARCHENGE'LICA. Garden Angelica. Roots and seeds used in medicine as an aromatic stimulant.

ANGUI'NUS. Lat. Of the nature of a snake; belonging or relating to a snake.

A'NEROID. From the Gr. *a* or *an*, privative, without, and *reō*, to flow. A name given to a kind of barometer which is constructed without a liquid to counterpoise the air. The Aneroid barometer consists of a cylinder of copper with a very thin and corrugated end, partially exhausted of air, and hermetically sealed. The effect of the varying pressure of the atmosphere on the thin end is magnified by a system of levers, so as to affect the index of a dial like that of a watch or clock. This is a French invention, but was patented in England, in the year 1844. See Barómeter.

ANIMA'LCULA. Lat. Plural of Animálculum.

ANIMA'LCULE. A diminutive animal. A term used to designate animals so small that they cannot be seen by the unassisted eye.

ANIMA'LCULUM. Lat. Animalcule.

ANISA'TUM. Lat. Belonging or relating to aniseed. Specific name of the tree which produces star-aniseed.

ANO'LIS. A kind of Saurian, called *anoli* in the Antilles. Also called, *long-toed lizard*, or *dactyloa*.

AN'NUAL. From the Lat. *annus*, a year. Yearly. A plant which rises from the seed, reaches perfection, and perishes within a year, is termed an annual.
ANTA'RCTICA. Lat. Antarctic.
ANTILO'PUS MONTA'NA. Mountain Antelope. *A. rupicapra.* Chamois. *A. cervicapra.* Common Antelope. *A. dorcas.* Gazelle. *A. gazella.* Algazel. *A. mhorr.* Mhorr.
A'NTHER. From the Gr. *anthera*, a flowery herb. In botany: the essential part of the stamen. The small yellowish body, compared to a diminutive leaf folded on itself, which crowns the stamen, and in which the pollen is formed.
ANTIQUO'RUM. Lat. Of the ancients.
APHE'LION. From the Greek *apo*, from, and *élios*, the sun. That point of a planet's orbit most distant from the sun; opposed to *perihelion.*
A'PHIDES. Plural of aphis.
A'PHIS. Gr. A plant-louse; a vine-fretter.
APOCY'NEÆ. From the Gr. *apo*, far from, and *kuon*, dog. Having the virtue of driving away dogs; the plant which kills dogs. Botanical name of a family of which the genus *apo'cynum* is the type.
A'PTENODY'TES. From the Gr. *apten*, without wings, and *dutes*, diver. A genus of birds. *A'ptenody'tes patagonica.* A species of penguin.
A'PTERYX. From the Gr. *apteros*, without wings. Name of a genus of birds.
A'QUEOUS ROCKS. Are those formed by deposits from water.
A'QUILA. Lat. An eagle. *Aquila albicilla.* The fishing eagle.
ARAUCA'RIA. From *Arauco.* Name of a department or district of Chile where the first species was seen. Name of a genus of the family of Conifers. *Arauca'ria excelsa.* The Norfolk Island pine.
AR'BUTUS. Lat. A shrub. A genus of plants.
ARCTOCE'PHALUS. From the Gr. *arktos*, a bear, *kephale*, head. Name of a genus of mammals.
AR'DEA. Lat. A Heron. Name of a genus of birds. *Ar'dea helias.* The Sun Bird.
A'REA OF SUBSIDENCE. A geological expression used to designate a space which has settled.
ARE'CA. Cabbage-tree. A genus of plants of the family of Palmæ. *Are'ca catechu.* The medicinal or betel-nut palm.
ARENA'CEOUS. From the Lat. *are'na*, sand. Sandy; of the nature of sand.
ARGEN'TEUM. Lat. Silvery; relating to silver.
ARGENTI'FEROUS. From the Lat. *argentum*, silver, and *fero*, I bear. Containing silver.
ARGILLA'CEOUS. From the Lat. *argilla*, clay or argil. Of the nature of clay.
ARGONAU'TA. Lat. From the Gr. *argo*, name of a vessel, and *nautes*, a navigator. Name of a genus of cephalo'podous mollusks.
ARMADI'LLO. Spanish. Diminutive of *armado*, armed. Name of a mammal of the family of edentáta or edentates.
AROMA'TICUS. Lat. Aromatic; spicy.
AR'SENIC. A metal of a shining, steel gray colour. Heated in contact with atmospheric air, it rapidly absorbs oxygen, and forms *arsenious acid*, which is the poison commonly called arsenic, or *rat's bane.* Arsenic is found in its metallic state, in the form oxide or arsenious acid, or white arsenic; and combined with sulphur, forming orpiment, and realgar.
ARTEMI'SIÆ. } A tribe of plants, of which the genus Artemísia is the type.
ARTEMISIAS. } Many of them are used in medicine.
ARTE'SIAN. From *Artois*, name of a province of France where especial attention has been given to a means of obtaining water, which consists in boring vertical perforations of small diameter in the exterior crust of the earth, frequently of great depth. These are termed Artesian wells.
ARTICULA'TA. Lat. From *articulus*, a joint or articulation. Articulated; having joints or articulations.
ASCLE'PIAS. A name of Escalapius. A genus of phanerógamous plants. *Ascle'pias gigante'a.* Mudar of the Hindoos. The milky juice is very caustic; the bark of the root as well as the juice are used in medicine by the Asiatics

A'SPHALT.
ASPHA'LTUM. } From the Gr. *a*, privative, and *sphalto*, I slip, or *asphaltos*, bitumen. Used anciently as a cement. A black, brittle bitumen, found on the surface and banks of the Dead Sea, hence called the Asphaltic lake.

ASPHODE'LEÆ. Name of a family of phanerógamous plants.

ASPHYX'IA. From the Gr. *a*, privative, and *sphuxis*, pulse. Without pulse. Seeming death from suspended respiration, from any cause, such as drowning, strangulation, or suffocation.

ASSI'MILATE. From the Lat. *ad*, and *similare*, to render similar. Assimilation is the act by which living bodies appropriate and transform into their own substance, matters with which they may be placed in contact. In man, assimilation is a function of nutrition.

ASPLE'NIFO'LIA. Compound of *asplenium*, a genus of ferns, and *folia*, leaves. Having leaves resembling those of the asplénium.

A'STER. From the Gr. *aster*, a star. A name given to the plant by the Greeks in allusion to the radiate form of the flowers. Name of a genus of plants which forms the type of the *asteroides* or asters—literally, *star-flowers*.

ASTRA'GALI. Lat. Plural of Astragalus.

ASTRA'GALUS. Lat. Name of a genus of phanerógamous plants of the family of leguminósæ.

A'TOLL. A chaplet or ring of coral, enclosing a lagoon or portion of the ocean in its centre.

AUCU'BA.
AUKU'BA. } A genus of plants of the family of Rhamnoides. There is but one species, which grows in Japan. *Aucuba Japonica*.

AUCHE'NIA. From the Gr. *auchenios*, belonging to the head or neck. Lat. Name of a genus of mammals, the Llama. Also, a genus of celeópterous insects.

AURI'CULA. Lat. Little ear. A genus of phanerógamous plants of the family of Primuláceæ.

AURI'FEROUS. From the Lat. *aurum*, gold, and *fero*, I bear. Gold-bearing, containing gold.

AU'ROCHS. An alteration of the German *Auerochs*, wild-bull. Their race is now almost extinct; a few individuals are found in the forests of Lithuania, &c.

AUSTRA'LE.
AUSTRA'LIS. } Lat. Belonging or relating to the south.

AZA'LEA. From the Gr. *azalea*, burned. A genus of phanerógamous plants of the family of Ericáceæ.

AZE'DARACH. From the Arab. *Azadaracht*, a name given by Avicenna to a plant.

A'ZOTE.
AZO'TIC GAS. } From the Gr. *a*, privative, and *zo'on*, life. The name given by chemists to a gas, now also called nitrogen, which will support neither respiration nor combustion. It constitutes seventy-nine per cent. of the atmosphere, and enters into the composition of all animal matter, except fatty substances, and into a certain number of proximate vegetable principles.

BACCI'FERUM. Lat. Compound of *bacca*, a berry, and *fero*, I bear. Berry-bearing. Specific name of a plant.

BALANCE OF TORSION, or TORSION BALANCE. A machine invented by Coulomb for measuring the intensities of electric or magnetic forces, by establishing an equilibrium between them and the force of torsion.

BALÆ'NA. Lat. A whale. Name of a genus of mammals, belonging to the order Cetácea. *Balæ'na mystece'tus*. The common whale. *Balæ'na gibbo'sa*. A kind of whale which has five or six protuberances on its back.

BAN'KSIA. A genus of phanerógamous plants of the family of Proteáceæ.

BA'OBAB. See Adansonia.

BA'RIUM. From the Gr. *barus*, heavy. A metal obtained from bary'tes by Sir H. Davy.

BARO'METER. From the Gr. *baros*, weight, and *metron*, a measure. An instrument for measuring the weight of atmospheric air.

BAROME'TRIC.
BAROME'TRICAL. } Belonging or relating to the barometer.

Barringto'nia. A genus of phanerógamous plants of the family of Myrtáceæ, and the type of the tribe of Barringtóniæ.

Basa'lt. An Ethiopian word. A black or bluish gray rock, harder than glass, very tenacious, and consequently difficult to break; it is homogeneous in appearance, although essentially composed of pyroxene and feldspar, with a large proportion of oxide of iron or titanium. Basalt is considered by all geologists to be a product of igneous formation.

Basa'ltic. Belonging or relating to basalt.

Batrac'hian. From the Gr. *ba'trachos*, a frog. The name given by naturalists to those reptiles which resemble frogs in their organization. Batrachians form the fourth order in the class of Reptiles.

Beaufo'rtia. Name of a genus of the family of Myrtáceæ, named in honour of Mary, the Duchess of Beaufort, who encouraged the study of Botany.

Be'lemnites. From the Gr. *be'lemnon*, a dart. A genus of dibranchiate cephalopods, the shells of which are chambered and perforated by a siphon, but internal. They are long, straight, and conical; and commonly called "thunder stones."

Benjami'na. Lat. Benjamin. A genus of plants; also the specific name of a plant.

Berni'cla. Generic name of a kind of goose, having a short beak. *Berni'cla cyana'ptera.* The goose of Shoa.

Ber'yl. A mineral allied to the emerald. It is transparent, of a pale green colour, and in Brazil it is sometimes sold under the name of emerald.

Be'tel. The leaf of the betel or Siriboa pepper.

Be'tula. Lat. Birch. Name of a genus of plants. *Be'tula nana.* Dwarf birch.

Betulöi'des. From *betula*, a birch-tree, and Gr. *eidos*, resemblance. Specific name of a plant.

Bigno'nia. A genus of plants named in honour of the Abbé Bignon, the Librarian of Louis XIV.

Bis'muth. From the Germ. *Wismuth.* A brittle, yellowish white metal.

Bitu'men. A combustible mineral, composed of carbon, hydrogen, and oxygen.

Bi'xa Orlea'na. A plant which produces a colouring matter, called *annotto.*

Bo'a. Name of a genus of non-venomous reptiles.

Bohe'a. Specific name of a tea-plant.

Bom'bax. From *bombux*, one of the Greek names of cotton. A genus of plants of the family Malváceæ. *Bombax heptaphyllum.* A kind of cotton-tree. *Bombax ceiba.* The cotton-wood tree, much valued for making canoes.

Bon'duc. A synonym of the *Guilandi'na.* Specific name of a plant.

Bora'cic acid. An acid obtained from borax, consisting of boron and oxygen.

Bora'ssus. From the Gr. *borassos*, a date. A genus of the family of Palms. *Borassus flabellifórmis.* The fan-leaved palm.

Bo'rate. The salt resulting from a combination of boracic acid and a salifiable base, as the borate of soda.

Bo'rax. Tinkal. A natural compound of soda and boracic acid.

Bore. A high-crested wave where the water is shallow, as on a sand-bar.

Bo'ron. A simple or undecomposable substance, the basis of boracic acid and borax.

Bori'chthys. From the Fr. *borgne*, one-eyed or blind, and the Gr. *ichthus*, a fish.

Borragi'neæ. Name given by Jussieu to a group of plants.

Borrer'ia. From Borrera, name of a man. A genus of phanerógamous plants of the family of Rubiáceæ.

Boswe'llia. A genus named in honour of Dr. John Boswell. *Boswellia serrata.* The olibanum tree.

Bos. Lat. An ox. A genus of ruminating mammals, embracing several species. *Bos urus.* The Urus. *Bos caffer.* Cape buffalo. *Bos buba'los.* Common buffalo. *Bos America'nus.* The Bison. *Bos moscha'tus.* The Musk Ox. *Bos gru'niens.* The Yak.

Bo'tany. From the Gr. *botane*, plant. The branch of natural history which embraces the knowledge and study of plants.

Bota'nic. Belonging or relating to botany.

Boul'ders, or Bowl'ders. Rounded masses of stone lying upon the surface, or loosely imbedded in the soil.

Boulder formation, or Erratic block formation. A geological term applied to a part of the diluvial drift. See Ruschenberger's Natural History.

Brac'teæ. Lat. Bracts. Floral leaves, different in colour from other leaves.

Bra'nichia. Lat. A gill.

Bra'nchiæ. Lat. From the Gr. *bragchos*, the throat. The gills of fishes. They are the breathing organs of fishes; they differ from lungs both in their form and structure.

Bras'sica. Lat. Cabbage.

Brec'cia. Italian. A rock composed of an agglutination of angular fragments. When the fragments are rolled pebbles, it constitutes a conglomerate rock, called *pudding-stone*.

Brevise'tum. Lat. *Brevis*, short, and *setum*, a bristle. A specific name.

Brex'ia. From the Gr. *brexis*, rain; in allusion to the protection from rain afforded by its ample foliage. A genus of plants of the family of Brexiáceæ.

Bu'bo. Lat. An owl. A specific as well as generic name. *Bubo maximus*. A kind of owl.

Bu'fo. Lat. A toad. *Bufo Agua*. A Brazilian toad.

Bu'phaga. Lat. From the Gr. *bos*, an ox, and *phago*, I eat. A genus of birds, which includes the African beef-eater.

Bur'sa. Lat. A sack, a purse, or pouch.

Bu'tea. A genus of the family of Papilionáceæ, named in honour of John, Count of Bute, a cultivator of botanic science. *Butea fro'ndosa* yields a gum (*butea*) which has been confounded with Kino.

Cac'ti. Lat. Plural of cactus.

Cac'tus. From the Gr. *kaktos*, spiny plant. Name of a genus of the family of Cactáceæ. *Cactus coccine'llifer*. The cochineal cactus. *Cactus opu'ntia*. Indian fig.

Caca'lia. Name of a genus of phanerógamous plants of the family of Compósitæ. Several species are useful as condiments.

Cacha'lot, or Cache'lot. Fr. Name of the spermaceti whale. Used to designate a variety of the order of Cetáceans, which has teeth in both jaws.

Cad'mium. A white metal, much like tin. Its ores are associated with those of zinc. Discovered in 1818.

Cadu'cous. From the Lat. *cado*, I fall. In Botany, when a part is temporary, and soon disappears or falls off, it is said to be caducous.

Cæci'liæ. From the Lat. *cæcus*, blind. A tribe of Batrachians.

Cæspito'sa. Lat. From *cæspes*, turf or sod. Belonging or relating to turf.

Ca'feine. Fr. In chemistry the name of the proximate principle of coffee.

Ca'japute, Cajapu'ta, } A Malay name for a greenish, volatile oil used as a remedy in rheumatism, &c.

Ca'lamus. A genus of phanerógamous plants of the family of Palms. *Ca'lamus draco*. An East Indian plant which yields an astringent substance called Dragon's blood. *Ca'lamus rotan*. The rattan plant.

Calca'reous. From the Lat. *calx*, *calcis*, lime. Belonging to or relating to lime. Calcareous rocks are those of which lime forms a principal part.

Calceola'ria. From the Lat. *calce'olus*, a little shoe. A remarkable genus of phanerógamous plants of the family of Scrophulariáceæ.

Cal'cium. From the Lat. *calx*, *calcis*, lime. A metal discovered by Sir H. Davy, in 1807, which, united with oxygen, forms oxide of calcium or lime.

Callit'riche. From the Gr. *kallithrix*, having luxuriant hair. A genus of aquatic plants. Also the name of a genus of American monkeys.

Calo'ric. From the Lat. *caleo*, I am warm. The term used by chemists to designate *the matter of heat*.

CALORI'FIC. Belonging or relating to caloric.
CALYCA'NTHUS. From the Gr. *kalux*, a calyx, and *anthos*, flower. A genus of the family of Calycanthácea.
CALYPTO'MENE. From the Gr. *kaluptos*, concealed, and *meno*, I remain. Name of a genus of birds.
CAM'BRIAN SYSTEM. From Cambria in Wales. A name given by geologists to the lowest sedimentary rocks, characterized by fossil remains of animals lowest in the scale of organization, such as corallines, &c. It is also called the Schistose system, on account of its slaty nature.
CAMEL'LIA. A genus of the family of Aurantiácea, named in honour of Kamel, a botanist. It contains the tea plants. *Camel'lia sasanqua.* Lady Banks' Camellia. *Camel'lia odorifera.* Sweet-smelling Camellia.
CAMPA'NULA. From the Lat. *campana*, a bell, from the shape of its corolla. A genus of phanerógamous plants of the family of Campanulácea, of which it is the type. 182 species are described.
CAMPHORO'SMA. From the Lat. *campho'ra*, camphor, and the Gr. *osme*, odour. A genus of plants of the family of Chenopodácea.
CAM'PHORA. Lat. Camphor. Belonging or relating to camphor.
CANARIE'NSIS. Lat. Belonging or relating to the Canary Islands.
CANDELA'BRUM. Lat. A candlestick.
CA'NINE. From the Lat. *canis*, a dog. Teeth which resemble those of a dog are so called; the canine teeth of the upper jaw in man are commonly called the eye-teeth.
CAOU'TCHOUC. Gum-elastic; India-rubber, a substance obtained from the *Jatro'pha ela'stica*, the *Ficus indica* and the *Urce'ola ela'stica*.
CAPE'NSIS. Lat. Belonging or relating to the Cape of Good Hope.
CARAGA'NA. A genus of plants of the family of Papilionácea.
CAR'BON. From the Lat. *carbo*, charcoal. A chemical element or undecomposed body. The diamond is pure carbon. It is the basis of anthracite, and of all the varieties of mineral coal, and is one of the principal constituents of all organic bodies.
CAR'BONATE. Any compound of carbonic acid and a salifiable base, as *carbonate of lime, carbonate of soda.*
CARBO'NIC ACID. A compound of carbon and oxygen.
CARBONI'FEROUS. From the Lat. *carbo*, coal, and *fero*, I bear, coal-bearing; containing carbon. In geology the term is applied to those strata which contain coal, and to the period when the coal-measures were formed.
CARDAMI'NE. Gr. Name of a plant. A genus of the family of Crucíferæ. Lady's smock. *Cardami'ne hirs'uta.* Hairy Cardamine.
CAR'DUI. Lat. Genitive case of *carduus*, a thistle. Specific name of a butterfly.
CARNI'VORA. From the Lat. *caro, carnis*, flesh, and *voro*, I eat. Name of a family of Mammals.
CARTILA'GINOUS FISHES. A term used to designate that division of the class of fishes which includes only those having cartilaginous instead of bony skeletons.
CARYOPHY'LLUS. Lat. A garden pink. A genus of plants of the family of Caryophy'lleæ. *Caryophy'llus aroma'ticus.* The clove-tree.
CARYO'TA. A genus of Palms of equatorial Asia. The *caryota urens* derives its specific name from a burning sensation its fruit imparts when eaten.
CA'SPIA. Lat. Belonging or relating to the Caspian Sea.
CAS'SIA. From the Gr. *kassia*, cinnamon. A genus of plants of the family of Papilionácea. The genus contains more than 300 species.
CA'STANOSPE'RMUM. From the Gr. *kastanon*, chestnut, and *sperma*, fruit. A genus of the family of Papilionácea.
CASUARI'NÆ. A family of plants separated from that of the Cónifers. The *casuari'næ* are found in New Holland, and in India, and are remarkable for the absence of leaves.
CATA'LPA. A genus of plants of the family of Bignoniácea.
CAT'ECHU. An astringent extract, used in medicine.

CAT'S EYE. A beautiful silicious mineral, penetrated by fibres of asbestos, which, when polished, reflects an effulgent, pearly light, much resembling the mutable reflections from the eye of a cat.

CAULE'RPA. From the Gr. *kaulos*, a stem, and *erpo*, I creep. A genus of algæ of the family of Zoosper'meæ. There are about 35 species of caulérpa, which inhabit equatorial seas. The *caule'rpa proli'fera* belongs to the Mediterranean.

CA'VIA. Genus of mammals of the family of rodents, including the guinea-pig.

CE'BUS. Lat. Name of a genus of monkeys; the marmoset.

CECRO'PIS. A genus of birds.

CEDRE'LA. Genus of plants of the family of Cedreláceæ.

CEI'BA. Synonym of *Bombax*, cotton. Specific name of a kind of cotton.

CENTAURE'A. A genus of plants of the family of Synanthéreæ-Cyanáreæ, and type of the tribe of Centaurieæ.

CERATI'TES. From the Gr. *keratites*, horned. A generic name of certain insects.

CERATO'DES. From the Gr. *keratodes*, formed of horns. A genus of mollusks.

CER'EAL. From the Lat. *ceres*, corn. Applied to grasses which produce the bread corns; as wheat, rye, barley, oats, rice, &c.

CEREA'LIA. Lat. Name of a tribe of grasses.

CEREO'PSIS. From the Gr. *keros*, wax, and *opsis*, aspect. A genus of birds of the order of Palmípedes and family of Lamelliróstres. It is marked by a wax-like membrane on the beak. *Cereopsis striata.* A kind of goose.

CE'RIUM. Named after the planet Ceres. A white brittle metal discovered in 1803, by Hisinger and Berzelius.

CER'THIA. Lat. Name of a genus of passerine birds, commonly called creepers.

CER'VUS. Lat. A stag. A genus of mammals.

CETA'CEA. From the Gr. *ketos*, a whale. A genus of pisciform mammals that have fins in place of feet, and inhabit the sea. Name of an order of aquatic mammals.

CHALK. Earthy carbonate of lime.

CHAMBERED SHELLS. A term used to designate those shells of mollusks which are divided internally into cells or chambers by partitions.

CHAMÆ'ROPS. From the Gr. *chamai*, on the ground, and *rops*, a brush. Name of a genus of palms. *Chamæ'rops hu'milis.* The dwarf fan palm.

CHEIRO'PTERA. From the Gr. *cheir*, hand, and *pteron*, a wing; signifying the hand has become a wing. Name of a family of mammals, including the bats.

CHEIROSTE'MON. From the Gr. *cheir*, hand, and *stemon*, filament. A genus of plants of the family of Sterculiáceæ, and tribe of Bombáceæ.

CHELO'NIAN. From the Gr. *chelone*, a tortoise. Applied to reptiles resembling tortoises.

CHEL'YDÆ. From the Gr. *chelus*, a tortoise. A tribe of reptiles of the family Emy'des.

CHLAM'YPHORE. From the Gr. *chlamus*, a cloak, and *phero*, I bear. A genus of mammals of the tribe of armadillos.

CHLENA'CEÆ. From the Gr. *chlaina*, a cloak. A tribe of plants, native in Madagascar.

CHLOA'NTHES. From the Gr. *chloros*, greenish yellow, and *anthos*, flower. A genus of plants of the family of Chloantháceæ.

CHLORI'TIC. From the Gr. *chloros*, green. Belonging or relating to chlorite, an earthy mineral found in the cavities of slate rocks.

CHROME, CHRO'MIUM. From the Gr. *chroma*, colour. A whitish brittle metal, discovered by Vauquelin in 1797. In union with oxygen it forms chromic acid.

CICHORA'CEÆ. From the Gr. *kichora*, chichory. A tribe of plants of the family of Compósitæ.

CI'RRI. Plural of *cirrus*.

CI'RRO-CU'MULUS. A sondercloud; a kind of cloud. The cirro-cumulus is intermediate between the cirrus and cumulus, and is composed of small well defined masses, closely arranged.

CI'RRO-STRA'TUS. A wanecloud. The cirro-stratus, intermediate between the cirrus and stratus, consists of horizontal masses separated into groups, with which the sky is sometimes so mottled as to suggest the idea of resemblance to the back of a mackerel.

CI'RRUS. Lat. A tendril. A kind of cloud. Applied to certain appendages of animals; as the beard from the end and sides of the mouth of certain fishes. The cirrus cloud consists of fibres or curling streaks which diverge in all directions. It occupies the highest region, and is frequently the first cloud which is seen after a continuance of clear weather.

CI'STUS. A genus of plants of the family of Cistáceæ.

CLA'RKIA. Proper name. A genus of plants of the tribe of Epilóbiæ.

CLAY-SLATE. A rock which resembles clay or shale, but is generally distinguished by its structure; the particles having been re-arranged, and exhibiting what is called slaty cleavage. It is one of the metamorphic rocks.

CLAYTO'NIA. A genus of plants of the family of Portuláceæ-calandríneæ.

CLEAVAGE. The mechanical division of the laminæ of rocks and minerals, to show the constant direction in which they may be separated.

CLERODE'NDRON. From the Gr. *kleros*, accident, and *dendron*, tree. In allusion to its accidental effects in medicine. A genus of plants of the family of Verbenáceæ-Lantáneæ.

COAL MEASURES. The geological formation in which coal is found.

CO'BALT. From the Germ. *kobold*, a devil. A brittle metal of a reddish gray colour. Its ores are always associated with arsenic.

CO'BRA CAPEL'LO. Portu. *cobra*, snake, and *capello*, a cawl or hood. Hood snake, a venomous serpent.

CO'CA. Quechua, an aboriginal Peruvian word. Specific name of the genus Erythróxylum.

COCCINE'LLA. From the Gr. *kokkinos*, scarlet. A genus of celeopt'erous insects; commonly called Lady birds.

COCCINE'LLIFER. From *coccinella* (the diminutive of the Lat. *coccinus*, crimson), a genus of celeopterous insects, and *fero*, I bear. A specific name.

COC'CUS. From the Gr. *kokkos*, a seed which dyes scarlet. A genus of insects of the order Hermip'tera. *Coccus la'cca.* A species of cochineal insect. *Coccus i'licus.* Green oak cochineal.

CO'COS. Gr. A genus of palms; the cocoanut. *Cocos olera'cia.* The oil cocoanut.

CO'DIUM. From the Gr. *kodion*, a fleece. A genus of plants of the tribe Siphóneæ. *Codium bu'rsa* and *Codium flabellifo'rme* are species.

COLO'BUS. From the Gr. *kolobos*, mutilated. A genus of monkeys which belong to the old world. *Colobus como'sus.* A hairy monkey.

COLU'BRIFORM. From the Lat. *co'luber*, a serpent, an adder, and *forma*, shape. Adder-shape.

COLUM'BA. Lat. A pigeon. A genus of birds. *Columba migrato'ria.* Wild pigeon.

COLUM'BIUM. A metal discovered in a mineral found in Massachusetts by Mr Hachett, in 1801.

COLU'MNAR. In the form of columns.

COMBU'STION. The combination of two bodies accompanied by the extrication of heat and light. When a body rapidly combines with oxygen, for example, with a disengagement of heat and light, it is said to undergo combustion.

COMPARATIVE ANATOMY. The comparative study of the various parts of the bodies of different animals.

COMPO'SITÆ. A family of Monopetalous plants.

CONDUC'TOR. Those substances which possess the property of transferring caloric or heat, and electricity, are termed conductors of heat or caloric, and conductors of electricity.

CONFE'RVÆ. Tribe of plants of the family of Zoospérmeæ. It includes many sea-weeds.

CON'GENER. From the Lat. *con*, with, and *genus*, race. Species belonging to the same genus, are termed congeners, or congeneric.

CONGLO'MERATE. From the Lat. *conglomero*, I heap together. Any rock composed of pebbles cemented together by another mineral substance, either calcareous, silicious, or argillaceous.
CO'NIFER. From the Lat. *conus*, a cone, and *fero*, I bear. A tree or plant which bears cones, such as pines, fir-trees, &c.
CONI'FERÆ. A family of plants which includes the conifers.
CO'RAL. From the Gr. *koreo*, I ornament, and *als*, the sea. The hard calcareous support formed by certain polypi.
CO'RALLINE. Belonging or relating to coral.
CORALLI'NEÆ. The corallines, a tribe of calciferous polypi.
COREO'PSIS. From the Gr. *koris*, a bug, and *opsis*, aspect. A genus of plants.
COR'DIA. A genus of plants of the family of Cordiáceæ. It contains about 150 species.
CORIA'CEOUS. From the Lat. *corium*, the hide of a beast. Leathery.
CORO'NA. Lat. A crown. A genus of plants.
CORO'NÆ. Plural of corona.
CORU'NDUM. A crystallized or massive mineral of extreme hardness, almost opaque, and of a reddish colour. It is allied to the sapphire, and is composed of nearly pure alúmina.
COT'TUS. A genus of fishes.
COTYLE'DON. From the Gr. *kotule'don*, a seed-lobe.
COTYLE'DONOUS. Belonging or relating to a cotyle'don or seed-lobe.
CRA'TER. Lat. A great cup or bowl. The mouth of a volcano.
CRA'TERIFORM. In form of a crater.
CRATERI'FEROUS. Containing craters.
CRETA'CEOUS. From the Lat. *creta*, chalk. Of the nature of chalk, relating to chalk.
CRINOI'DEÆ. From the Gr. *krinon*, a lily, and *eidos*, resemblance. A family of radiate animals.
CROP OUT. When a rock, in place, emerges on the surface of the earth, it is said to crop out.
CRO'TON. A genus of plants of the family of Euphorbiáceæ.
CRUCI'FERÆ. From the Lat. *crux*, *crucis*, a cross, and *fero*, I bear. A family of plants which have flowers in form of a Maltese cross.
CRU'CIFORM. In shape of a cross.
CRUSTA'CEA. From the Lat. *crusta*, a crust. A class of articulated animals.
CRUSTA'CEAN. An animal of the class of crustacea; a crab.
CRYPTOGA'MIA. From the Gr, *kruptos*, concealed, and *gamos*, marriage. A class of plants, which are propagated without apparent seeds.
CRYPTO'GAMOUS. Belonging or relating to the cryptogámia.
CRYP'TONYX. From the Gr. *kruptos*, concealed, and *onux*, a nail. A genus of birds; also, a genus of insects.
CRYST'AL. From the Gr. *krustallos*, ice. This term was originally applied to those beautiful transparent varieties of silica or quartz known under the name of *rock-crystal*. When substances pass from the fluid to the solid state, they frequently assume those regular forms which are generally termed crystals. A crystal is any inorganic solid of homogeneous structure, bounded by natural planes and right lines, symmetrically arranged.
CRYS'TALLINE. Relating to, or resembling crystals.
CRYSTALLIZA'TION. The process by which crystals are formed.
CUCI'FERA THEBA'ICA. A palm of Egypt which grows to the height of 20 feet. Also known as the genus *Hyphæne*, from the Gr. *Huphaino*, I entwine. A fan-leaf palm of the tribe of Borassíneæ.
CU'CULUS. Lat. A cuckoo. A genus of passerine birds.
CU'LEX. Lat. A gnat. A genus of insects of the family of Dip'tera, and type of the tribe of Culícides: *culex pi'piens*, the common gnat.
CU'MULI. Plural of cumulus.
CU'MULO-STRA'TUS. Twain cloud: it partakes of the appearance of the cumulus and stratus.
CU'MULUS. A form of cloud. A convex aggregate of watery particles, increasing upwards from a horizontal base, and assuming more or less of a conical figure

CUR'VIDENS. Lat. *Curvus*, bent, and *dens*, tooth. Having a bent tooth.
CUSPA'RIA. A genus of plants, named after the tree which yields the Angustura bark.
CYANAP'TERA. From the Gr. *kuanos*, blue, and *pteron*, wing. A specific name.
CYANEROI'DES. From the Gr. *kuanos*, blue, and *eidos*, resemblance. A family of medusæ.
CY'CAS. A genus of plants, the type of the family cycádeæ. *Cy'cas revolu'ta.* Narrow-leaved cycas.
CYCA'DÆ. A family of plants allied to the cónifers.
CYCADA'CEOUS. Belonging or relating to the cycádeæ.
CY'CLAS. From the Gr. *kuklos*, a circle. A genus of gasteropods.
CYG'NUS. Lat. A swan. A genus of birds. *Cyg'nus mu'sicus.* The whistling swan.
CYNOCE'PHALUS. From the Gr. *kuon*, a dog, and *kephale*, head. A genus of mammals. Dog-headed monkey or baboon.
CYPERA'CEÆ. Name of a family of herbaceous plants.
CYPRÆ'A. From the Gr. *kupris*, Venus. A cowry. A genus of mollusks. *Cypræ'a mone'ta.* The money cowry.
CYPRI'NIDÆ. From the Gr. *kuprinos*, a carp. Name of a family of fishes.
CYSTOSEI'RIÆ. From the Gr. *kustis*, a vesicle, and *seira*, a chain. A tribe of sea-weeds.

DAC'TYLIS. From the Gr. *daktulos*, a finger. A genus of the family of Gramíneæ. *Dactylis cæspitosa.* Tussock grass.
DAH'LIA. After Dahl, a Swedish botanist. Genus of plants of the family of Compósitæ.
DALBE'RGIA. After Dalberg, a Swedish botanist. A genus of plants of the family of Papilionáceæ, and of the tribe of Dalbergiæ.
DA'MAN. Alteration of the Arabic word *Ghannem*, the name of an animal. Specific name of a mammal.
DANA'IS. Genus of plants of the family of Rubiáceæ.
DAPH'NE. A genus of plants of the family Daphnáceæ.
DARWI'NII. The name of Darwin latinized. Belonging or relating to Darwin.
DASYU'RIDÆ. From the Gr. *dasus*, thick, hairy, and *oura*, tail. A family of mammals.
DE'BRIS. Fr. Wreck, ruins, remains. In geology the term is applied to large fragments, to distinguish them from *detritus*, or those which are pulverized.
DECI'DUOUS. From the Lat. *decido*, I fall off. Applied to plants whose leaves fall off in autumn, to distinguish them from evergreens.
DECLINA'TION of any celestial body, is the angular distance of the body, north or south, from the equator.
DEINOTHE'RIUM. From the Gr. *deinos*, terrible, and *ther*, wild beast. A genus of fossil pachyderms.
DELESSE'RIÆ. Proper name. Tribe of plants of the family of Flori'deæ.
DELPHI'NUS. Lat. Dolphin. A genus of aquatic mammals.
DEL'TA. The Gr. letter Δ. The triangular deposits, shoals or islands, at the mouths of rivers are called deltas.

DEL'TOID. From the Gr. letter Δ and *eidos*, resemblance. Resembling the letter delta.

DENUDA'TION. From the Lat. *denudo*, a strip. A removal of a part of the land, so as to lay bare the inferior strata.

DEODA'R. A kind of pine tree.

DEPOSI'TION. From the Lat. *depono*, I let fall. In geology the falling to the bottom of matters suspended or dissolved in water.

DEVO'NIAN SYSTEM. So called because it is largely developed in Devonshire, England. It is synonymous with the old red sand-stone formation. It is composed at first of pudding-stone, and then passes into sandstone, with which it alternates at different places.

DE'TINENS. Lat. Detaining; that which has the power to detain.

DETRI'TUS. A geological term applied to deposits composed of various substances which have been comminuted by attrition. The larger fragments are usually termed *débris;* those which are pulverized, as it were, constitute *detritus.* Sand is the detritus of silicious rocks.

DIAMAGNE'TIC. If a bar of iron be suspended between the poles of an electro-magnet, it will be attracted by both poles on the line of force. But if a bar of bismuth be suspended in the same manner, it will be repelled by both poles, and rest at right angles to the line of force. Substances which are attracted by both poles of an electro-magnet are said to be *magnetic,* and those which are repelled by both poles are termed *diamagnetic.*

DICHO'TOMA, DICHO'TOMUM, DICHO'TOMUS, } From the Gr. *dichotomos,* equally divided. In zoology this term is applied to a species of the genus Iris, the body of which is bifurcate. In botany it is applied to the stem, branches, peduncles, leaves, hairs, styles, &c., when they are bifurcated in form.

DICOTYLE'DON. From the Gr. *dis,* two, and *kotuledon,* seed-lobe. A double seed-lobe.

DICOTYLE'DONOUS. Relating to dicotyle'don; having a double seed-lobe.

DIDEL'PHOUS. From the Gr. *dis,* double, and *delphus,* womb. Applied to opossums and other marsupial mammals.

DIDEL'PHIS. A genus of marsupial mammals.

DIDEL'PHIDÆ. A tribe of marsupial mammals.

DIDY'MIUM. A metal discovered recently by Mosander.

DIGITA'TA. Lat. Di'gitate; spread out like the fingers.

DINO'RNIS. From the Gr. *deinos,* great, terrible, and *ornis,* a bird. A genus of fossil, or extinct birds.

DIO'TIS. From the Gr. *diótos,* having two ears: referring to the flower. A genus of plants of the family of helianthácecæ.

DISLOCA'TION. Displacement. In geology where strata or veins have been displaced from the position where first deposited or formed, they are said to be dislocated.

DI'SA. A genus of plants of the family of Orchi'deæ. *Di'sa grandiflo'ra.* Large-flowered Disa.

DIO'SMA. From the Gr. *dios,* divine, and *osme,* smell. A genus of plants of the family of Dios'meæ.

DILLENIA'CEÆ. Proper name. A family of plants.

DIONÆ'A. One of the names of Venus. A genus of plants of the family of Droserácea. *Dionæ'a musci'pula.* Venus' Fly-trap.

DIP'TERYX. From the Gr. *dis,* double, and *pterux,* a wing, in allusion to the two appendages of the calyx. Tonquin Bean. A genus of plants of the family of Leguminósæ. *Dip'teryx odora'ta.* Sweet-scented Tonquin Bean.

DIC'TYOTA. From the Gr. *dictuon,* a net. A genus of plants of the family of Phy'ceæ, and tribe of dictyóteæ.

DICTYONE'MA. From the Gr. *dictuon,* a net, and *nema,* a filament. A genus of plants of the family of Phy'ceæ.

DIP. In geology, direction of the inclination of strata. "To take a dip," is to measure the degree that a stratum inclines or dips from a horizontal line.

DISIN'TEGRATE. From the Lat. *de,* privative, *integer,* a whole. To separate or break up an aggregate into parts.

DO'LOMITE. Magnesian marble, or granular magnesian carbonate of lime. Named after Dolomieu.

DOMBE'YA. In honour of Joseph Dombey. A genus of plants of the family of Byttneriácea: it is found in Madagascar and the Isle of Bourbon.

DORSI'GERA. Lat. From *dorsum,* the back, a ridge, and *gero,* I carry or wear. A specific name.

DORYA'NTHES. From the Gr. *doru, doratos,* a lance, and *anthesis,* a flowering. A genus of plants of the family of Amaryllidáceæ.

DRA'BA. A genus of plants of the family of Cruci'feræ.

DRACÆ'NA. Lat. A genus of Saurians.

DRACÆ'NÆ. Plural of Dracæ'na.

DRYOBA'LANOPS. From the Gr. *drus, os*, an oak, *balanos*, an acorn, and *ops*, aspect. A genus of plants of the family of Dip'terocárpeæ. *Dryoba'lanops ca'mphora.* The camphor tree of Sumatra.

DYNA'MIC. From the Gr. *dunamis*, power, force. Belonging or relating to dynamics.

DYNA'MICS. The doctrine of forces as exhibited in moving bodies which are at liberty to obey the impulses communicated to them. The motions of celestial bodies in their orbits, or of a stone falling freely through the air, are embraced in the study of dynamics.

EARTHS. Formerly chemists, believing them to be simple bodies, included the following substances under the name of earths: Baryta, Strontia, Lime, Magnesia, Alumina or clay, Silica, Glucina, Zirconia, and Yttria. Research has shown that all have metallic or metalloid bases.

ECHID'NA. Greek name of a monster, supposed to have the body of a beautiful woman, and the tail of a serpent. A genus of mammals of the family of Monotre'mata.

E'CHIMYS. From the Gr. *echinos*, spiny, and *mus*, a rat. A genus of mammals; a sort of rat found in South America.

ECLIP'TIC. In Astronomy the great circle of the heavens which the sun appears to describe in his annual revolution.

EDENTA'TA. From the Lat. *e*, without, and *dens*, tooth: without teeth. An order of mammals which are destitute of teeth.

EDU'LIS. Lat. Eatable; that which may be eaten.

EFFLORE'SCENCE. The pulverulent covering formed on the surface of saline substances from which the atmosphere has removed the water of crystallization. When saline substances give up their water of crystallization to the air, they are said to effloresce.

ELA'IS. }
ELÆ'IS. } From the Gr. *elaia*, the olive. A genus of plants of the family of Palms. The *Elais Guinea'nensis* yields the Palm oil.

E'LAPS. Gr. Name of a serpent. A genus of ophidians.

ELECTRI'CITY. From the Gr. *elektron*, amber, the substance in which it was first observed. The property acquired by glass and resin from friction to attract light substances. Electricity exists in all bodies, and becomes manifest, at least partially, whenever the natural state of equilibrium of their molecules is disturbed by any cause.

ELE'CTRO-MA'GNETISM. The phenomena produced when a current of electricity is traversing any substance, or when electricity is in motion, magnetism is at the same time developed.

ELEC'TRO-MAG'NET. An apparatus for exhibiting the phenomena of electro-magnetism.

ELEC'TRICUS. Lat. Electric. Belonging to, or relating to electricity.

ELLIP'TICA. Lat. Elliptic.

E'LEPHAS. Lat. Gr. name of the elephant. A genus of mammals of the order of pachydermata.

ELEPHAN'TINA. Lat. Belonging or relating to an elephant; elephantine.

EM'BRYO. From the Gr. *embruon*, from *bruō*, I bud forth. A germ at the early stages of development.

E'MERALD. A mineral of a beautiful green colour, much valued for ornamental jewelry. It consists of silica, alumina, glucina, oxide of chromium, which is the colouring matter, and a trace of lime.

E'MYS. Lat. From the Gr. *emus*, a water tortoise. A genus of reptiles of the family of emy'dians.

EMY'DIANS. A family of reptiles of the order of Chelónia.

ENCRINI'TES. From the Gr. *krinon*, a lily. A genus of fossil *Echi'noderms.* The skeleton of this animal is said to consist of not less than 26,000 separate pieces.

E'OCENE. From the Gr. *eōs*, dawn, and *kainos*, recent. In geology a name for the older tertiary formation, in which the first dawn, as it were, of existing species, appears.

EPACRI'DEÆ. From the Gr. *epi*, upon, and *akros*, an elevated place, a hill. A family of plants.
EP'IPHYTE. From the Gr. *epi*, upon, and *phutos*, a plant. Applied to plants which grow upon other plants.
EQUINOCTIA'LIS. Lat. Equinoctial.
EQUISE'TUM. From the Lat. *equus*, a horse, and *seta*, hair. A genus of plants of the family of Equisitáceæ.
E'QUUS. Lat. A horse. A genus of mammals.
ER'BIUM. A metal, recently discovered.
ERI'CA. A genus of plants of which there are 429 species.
ERIOCAU'LON. From the Gr. *erion*, wool, and *kaulon*, stem or stalk. A genus of plants of the family of Eriocaulóneæ.
ERYTHRI'NA. From the Gr. *eruthros*, red. A genus of plants of the family of Papilionáceæ.
ERYTHROX'YLON. From the Gr. *eruthros*, red, and *xulon*, wood. A genus of plants.
ESCARP'MENT. From the Ital. *scarpa*, sharp, formed from the Lat. *carpere*, to cut. The steep face often presented by the abrupt termination of strata where subjacent beds crop out from beneath them.
ESCULEN'TA. Lat. Esculent.
ETHNO'GRAPHER. From the Gr. *ethnos*, a nation, and *graphô*, I write. One who cultivates ethnography: an ethno'logist.
ETHNO'GRAPHY. A department of knowledge which treats of the different natural races and families of men. A treatise on the subject.
EUCALY'PTI. Lat. Plural of eucalyptus.
EUCALY'PTUS. From the Gr. *eu*, well, and *kaluptos*, covered. A genus of plants of the family of Myrtáceæ.
EUPHO'RBIA. Gr. Name of a plant. A genus of plants of which there are 300 species.
EXCE'LSA. Lat. Noble, tall, stately.
EXCO'RTICA. Lat. Without bark.
EXO'GENOUS. From the Gr. *ex*, from, and *geinomai*, I grow. Applied to plants which grow by successive external additions to their wood.
EXTEN'SILE. Having the power to extend itself.
EXU'VIÆ. Lat. The sloughs or cast skins, or cast shells of animals.

FA'GUS. Lat. Beech. A genus of plants of the family of Amentáceæ.
FA'LCO. Lat. Falcon. A genus of birds. *Falco isla'ndicus.* The Gerfalcon.
FA'MILY. In natural history the term is applied to an assemblage of several genera which resemble each other in many respects.
FARI'NA. Lat. Meal.
FAR'INHA. Portu. Meal, flour.
FARINO'SA. Lat. Mealy; belonging or relating to meal.
FAU'NA. All animals of all kinds peculiar to a country constitute the *fauna* of that country.
FE'LIS. Lat. A cat. A genus of mammals of the family of carni'vora. *Felis irbis.* The panther.
FENESTRA'LIS. Lat. Belonging or relating to a window or opening.
FER'BIUM. A recently discovered metal.
FERNS. The filices; an order of cryptogámic plants.
FI'CUS. Lat. A fig. A genus of plants of the family of Moræ'ceæ.
FICOI'DE. A genus of plants of the family of *Mesembryanthe'meæ*, of which there are about 200 species.
FICOI'DES, FICOI'DEÆ. The family of Mesembryanthémeæ. *Ficoides* is applied also as a specific name.
FIORD. A frith, firth, or furth; a rocky chasm penetrated by the sea; a rock-bound strait.
FLACOUR'TIA. Proper name. A genus of plants of the family of Flacourtiáceæ.
FLABELLIFO'RME. From the Lat. *flabellum*, a fan, and *forma*, form. Fan-shaped.

FLO'RA. Lat. Name of the Goddess of Flowers. All the plants of all kinds belonging to a country constitute the *flora* of that country.
FLO'RIDA. Belonging or relating to flowers; or relating to the State of Florida.
FO'CI. Lat. Plural of focus.
FO'CUS. Lat. A hearth. In optics the term describes the point or space where the rays of light are concentrated by a lens. The apex of a cone of rays of light, or of heat, formed by a lens, or concave mirror.
FOLIA'CEOUS. From the Lat. *folium*, a leaf. Leafy. Having the form of leaves.
FOO'TSTALKS. In botany the stalks of flowers, or of leaves.
FO'SSIL. From the Lat. *fodio*, I dig. Any organic body, or the traces of any organic body, whether animal or vegetable, which has been buried in the earth by natural causes.
FOSSILI'FEROUS. Containing fossils: fossil-bearing.
FORMI'CIDÆ. From the Lat. *fo'rmica*, an ant, and the Gr. *eidos*, resemblance. A family of insects of the family of Hymenóptera.
FROND. Also, *frons*. A name applied to the leaves of palms, and of cryptóga-mous plants.
FRONDO'SA. Lat. Full of green leaves.
FRA'GRANS. Lat. Fragrant; odorous.
FRA'GILIS. Lat. Fragile; easily broken.
FRINGI'LLÆ. Lat. *fringilla*, a chaffinch. A family of birds, the most numerous of the group of conirostres, or thick-billed birds.
FU'CCA. Name of a genus of aquatic plants.
FU'CI. Lat. Plural of fucus.
FU'CUS. Lat. Sea-weed. A genus of aquatic plants.
FUCH'SIA. After Leonard Fuchs, a physician of the 16th century. A genus of plants.
FUNC'TION. From the Lat. *fungor*, I act. The action of an organ, or system of organs.
FUN'GI. Lat. Plural of fungus.
FUN'GUS. Lat. A mushroom.
FUNE'REUS. Lat. Funereal: belonging to a dead body.

GA'DUS. Lat. A codfish.
GALLINA'CEOUS. From the Lat. *galli'na*, a hen. Relating to birds of the order of Gallináceæ.
GALE'NA. From the Gr. *galene*, lead ore. A mineral composed of sulphur and lead: a natural sulphuret of lead.
GAL'VANISM. From *Galvani*, a distinguished Italian philosopher. That branch of electrical science in which electricity is made manifest by the mediate contact of different metals. Also, the phenomena exhibited by living animal matter when placed between the poles or extremities of an apparatus for showing electricity by the mediate contact of different metals.
GALVA'NIC. Belonging or relating to galvanism.
GANGEA'TICUS. Lat. Gangeatic; belonging or relating to the river Ganges.
GARDE'NIA. After a proper name. A genus of plants of the family of Rubiá-ceæ; it contains some forty species. The *Garde'nia grandiflo'ra* is the Cape Jasmin.
GAR'NET. A mineral consisting of silicates of alumina, lime, iron, and manganese. It occurs imbedded in mica-slate, granite, and gneiss, and occasionally in limestone, chlorite-slate, serpentine, and lava. There are several varieties of garnet.
GAS. From the Germ. *geist*, spirit. The name given to all permanently elastic fluids, or airs, different from the atmospheric air.
GAS'EOUS. Of the nature of gas.
GENRE. Fr. Genus, kind, manner, style. In painting it is applied to signify the representation of certain kinds of objects, as landscapes, views, animals, plants, flowers, scenes in common life. Pictures of *genre*, then, are pictures of a genus or kind as to subject; as landscapes, marine views, flower pieces, still-life, &c.

GE'NERA. Lat. Plural of genus.
GE'NUS. Lat. A kindred, breed, race or family.
GEO'LOGY. From the Gr. *ge*, the earth, and *logos*, discourse. That branch of natural history which treats of the structure of the terrestrial globe. It is divided into *descriptive* geology; *dynamic* geology, which treats of the forces by which the surface of the earth has been modified; *practical* and *economic* geology, embracing the application of geological science to mining, road-making, architecture, and agriculture.
GEOTHER'MAL. From the Gr. *ge*, the earth, and *thermos*, heat, temperature. Relating to temperature of the earth.
GERA'RDIA. Proper name. A genus of plants of the family of Scrophulariáceæ.
GERMINA'TION. The process of the development of the seed, and the embryo which it contains.
GEY'SERS. From an Icelandic word, signifying raging or roaring. Celebrated spouting fountains of boiling water in Iceland.
GIBBO'SA. Lat. Gibbous; having protuberances or bunches.
GIBRALTA'RICA. Lat. Belonging or relating to Gibraltar.
GIGANTE'A. } GIGANTE'US. } Lat. Gigantic, huge.
GILEADE'NSIS. Lat. Belonging or relating to Gilead.
GLA'CIAL. Belonging or relating to ice.
GLA'CIERS. Fr. Masses or beds of ice formed in high mountains, derived from the snows or lakes frozen by the continued cold of those regions.
GLADIO'LUS. A genus of plants of the family of Iri'deæ.
GLAND. An organ formed for the purpose of secreting a peculiar fluid.
GLAU'COUS. From the Gr. *glaukos*, blue. Applied to the bluish and pulverulent aspect which certain plants present, such as the leaves of cabbages, &c. Also used to signify the bloom of the color of cabbage leaves, sometimes observed on polished bodies.
GLEDI'TSCHIA. A genus of plants of the family of Leguminósæ, named in honor of J. G. Gleditsch, a German botanist. It includes the Honey and Swamp locust trees among its species.
GLOBA'RIA. From the Lat. *globum*, a ball. A genus of insects: also a specific name.
GLUCI'NUM. A metal discovered in glucina in 1798 by Vauquelin.
GLU'TEN. Lat. The viscid elastic substance which remains when wheat flour is wrapped in a coarse cloth, and washed under a stream of water, so as to carry off the starch and soluble matters. It exists in many plants and in animals. It is the basis of glue.
GLYCE'RIA. A genus of plants of the family of grami'neæ, and the tribe Festucáceæ.
GLY'CINE. From the Gr. *glukus*, sweet. A genus of plants of the family of Papilionáceæ.
GNAPHA'LIUM. From the Gr. *gnaphalion*, the cotton tree. A genus of plants of the family of Compósitæ.
GNEISS. Germ. A rock resembling granite. It is composed chiefly of feldspar and mica, and is more or less slaty in its structure. Gneiss is used for building and flagging.
GOLD. The most valuable and longest known of the metals.
GOODE'NIA. Proper name. A genus of plants of the family of Goodeniáceæ.
GORDO'NIA. Proper name. A genus of plants of the family of Gordoniáceæ.
GRANDIFLO'RA. Lat. Large-flowered.
GRAMI'NEÆ. Lat. Grasses. A family of monocotylédonous plants, containing about 3000 species.
GRANI'VOROUS. } GRANI'VORA. } Applied to animals which feed upon grains, especially to passerine birds.
GRÆ'CA. Lat. Greek.
GRA'NULAR. Composed of grains.
GRA'NITE. A rock which is a crystalline aggregate of quartz, feldspar, and mica.
GRANI'TIC. Of the nature of granite.
GREEN'STONE. A rough variety of trap-rock, consisting chiefly of hornblende.

GRIT. A coarse-grained sandstone.
GUILANDI'NA. A proper name. A genus of plants of the family of Leguminósæ. *Guilandina Bonduc*, the oval-leaved Nicker-tree.
GUINEANEN'SIS. Lat. Belonging or relating to Guinea.
GUM. A vegetable product, which is tasteless and inodorous, and is distinguished by being soluble in water, and insoluble in alcohol: gum-arabic, for example.
GYMNO'TUS. From the Gr. *gumnos*, naked, and *nôtos*, back. A genus of fishes.
GYP'SUM. Native sulphate of lime. It is converted into plaster of Paris by heat.
GYRO'PHORA. From the Gr. *guros*, a circle, and *pherô*, I bear. A genus of cryptógamous plants.

HA'BITAT. Lat. He inhabits. Used to designate the place in which animals and plants are naturally found.
HALIO'TIS. From the Gr. *als*, the sea, and *ous*, the ear. A genus of mollusks.
HA'LCYON. From the Gr. *alkuo'n*, a king-fisher. A genus of birds.
HELIÁ'NTHUS. From the Gr. *elios*, the sun, and *anthos*, flower; sunflower.
HELI'ACAL. From the Gr. *elios*, the sun. Relating to the sun. When a star rises so as to be visible in morning twilight before the appearance of the sun, it is said to rise *heliacally*.
HEPTAPHY'LLUM. From the Gr. *epta*, seven, and *phulion*, a leaf. Seven-leaved. A specific name.
HERBA'CEOUS. In botany, Herb-like; that perishes every year. An annual stem. Not woody.
HERBI'VORA. Lat. Herbivorous.
HERBIV'OROUS. From the Lat. *herba*, a plant, and *vorare*, to eat. Plant eating. Applied to animals which feed chiefly or exclusively on plants or herbs.
HERITIE'RA. Proper name. A genus of plants of the family of Sterculiáceæ.
HI'BERNATE. From the Lat. *hibernare*, to winter. Animals which retire and sleep throughout the winter, are said to hibernate.
HIBI'SCUS. A genus of plants of the family of Malváceæ.
HIERO'CHLOA. From the Gr. *ieros*, sacred, and *chloa*, herb. A genus of plants of the family of Grami'neæ.
HIPPOPO'TAMUS. From the Gr. *ippos*, a horse, and *potamos*, river. River Horse. A genus of mammals.
HIRSU'TA. Lat. Hirsute; covered with soft hairs.
HOL'CUS. A genus of plants of the family of Grami'neæ.
HO'PEA, or HO'PPEA. Proper name. A genus of plants.
HO'RRIDA. Lat. Horrid; spiny.
HO'RARY. From the Lat. *hora*, an hour. The motion of a celestial body, or the space it moves through in an hour, is termed its *horary motion*.
HORSE'SHOE MAG'NET. A magnet in form of a horse-shoe.
HUMI'RIA. A genus of plants of the family of Humoriáceæ. They inhabit tropical America.
HYDRAN'GEA. From the Gr. *udôr*, water, and *aggos*, a vessel. A genus of plants of the family of Saxifragáceæ, and tribe of Hydrangéæ.
HYDRAU'LIC. From the Gr. *udôr*, water, and *aulos*, a pipe. Relating to liquids in motion. Hydraulics is that branch of natural philosophy or physics which treats of the force of water and other liquids in motion.
HYDROSTA'TIC. From the Gr. *udôr*, water, and *staô*, I stand. Relating to water in a state of rest. Hydrostatics is the science which treats of the equilibrium and pressure of water and other liquids.
HY'DROGEN. From the Gr. *udôr*, water, and *gennaein*, to generate. A colorless, tasteless, inodorous gas, one part of which, by weight, combined with eight parts of oxygen forms water; — combined with sulphur it constitutes *sulphuretted* Hydrogen; — and with carbon, carburetted Hydrogen, the gas used for illumination.
HYDROGE'TON. A synonym of *Ouvirau'dra*. A genus of aquatic plants.
HY'LA. From the Gr. *ule*, wood, a tree. A tree-frog.

HYMENÆ'A. A genus of plants of the family of Papilionáceæ. A resinous tree of tropical America.
HYMENO'PTERA. From the Gr. *umen*, a membrane, and *pteron*, wing. Systematic name of a class of insects, characterized by membranous wings.
HY'RAX. From the Gr. *urax*, a shrew mouse. A genus of mammals.

IA'NTHINA. See Janthina.
I'BEX. Lat. A wild goat. A genus of mammals.
I'BIS. A genus of birds.
IG'NEOUS ROCKS. Are those rocks whose structure is attributable to the influence of heat, such as granite and basalt. They are distinct from stratified rocks, or those formed by deposits from water.
IGUA'NA. A reptile of the lizard tribe.
IGUA'NIAN. Applied to Saurians which resemble the iguana.
IGUA'NODON. From *iguana*, and the Gr. *odous*, tooth. A genus of extinct or fossil reptiles of gigantic size discovered in the south of England.
I'LEX. Lat. The Holly.
IL'ICIS. Lat. Of the Holly; belonging or relating to the holly.
ILLI'CIUM. From *illicio*, to attract; from its agreeable perfume. The anniseed tree. A genus of plants of the family of Magnoliáceæ.
IM'BRICATE. Laid one over another like tiles.
INCONSPIC'UOUS. Lat. Not conspicuous or remarkable.
INCI'SOR. From the Lat. *incido*, I cut. Applied to those teeth which occupy the anterior or centre of the upper and lower jaws, because they are used for cutting the food.
IN'CA. Designation of the aboriginal Peruvian princes; used as a specific name. Also, a genus of insects.
IN'DICA—IN'DICUS. Lat. Indian: Belonging or relating to India.
INDICA'TOR. Lat. Indicator; one who points out. A genus of birds.
INFUSO'RIA. Animals of infusions; microscopic animalcules.
INFUSO'RIAL. Belonging or relating to the Infusoria.
INORGA'NIC. Without organs or organization.
IN'SECT. From the Lat. *in*, into, *seco*, I cut. Applied to animals whose bodies are cut, as it were, into three parts—head, thorax, and abdomen.
IRI'DEÆ. A family of monocotylédonous plants.
IRI'DIUM. From the Lat. *iris*, the rainbow. A grey, brittle, very infusible metal, which is found associated with the ores of platinum.
ISA'TIS. From the Gr. *isazô*, I render equal. Woad. A genus of plants of the family of Cruci'feræ. Also the name of a species of dog.
ISLA'NDICUS. Lat. Belonging or relating to Iceland.
ISOCHI'MENAL. From the Gr. *isos*, equal, and *cheima*, winter. Isochimenal lines pass through all places where the mean winter temperature is the same.
IS'OGEOTHE'RMAL. From the Gr. *isos*, equal, *ge*, the earth, and *thermos*, heat. Applied to lines which are supposed to pass through all parts of the earth's structure on the surface where the mean heat is the same.
ISOTHE'RMAL. From the Gr. *isos*, equal, and *thermos*, heat. Isothermal lines are supposed to pass through all places where the mean temperature of the air is the same.
ISOTHE'RIAL. From the Gr. *isos*, equal, and *thereios*, having the heat of summer. Isotherial lines are supposed to be drawn through all places having the same mean summer temperature.
I'XIA. A genus of plants of the family of Iri'deæ.

JA'NTHINA. From the Gr. *ianthinos*, violet. A genus of mollusks.
JAPO'NICA—JAPO'NICUS. Belonging or relating to Japan.
JAS'PER. A silicious mineral of various colours; sometimes spotted, banded or variegated. It takes a fine polish.
JER'BOA. A genus of mammals of the family of Rodents, or gnawers. The jumping mouse.
JURA'SSIC. Belonging or relating to the Jura mountains. Applied to a system of rocks of the middle secondary geological period. Also termed oolitic.

KA'LMIA. A genus of plants of the family of Ericáceæ.
KER'RIA. Proper name. A genus of plants of the family of Rosáceæ.
KEUR'VA. Synonym of Pandanus.
KING'IA. Proper name. A genus of plants of the family of Joncáceæ, found in New Holland. *Kin'gia austra'lis;* the grass tree.

LABIA'TÆ. From the Lat. *labium,* lip; in allusion to the form of the corolla. A family of dicotylédonous plants.
LAGO'PUS. From the Gr. *lagôs,* a hare, and *pous,* foot: hare-footed. A genus of birds of the order of Gallináceæ.
LAM'ANTIN. The manatus. A genus of mammals of the order of Cetácea.
LAMINA'RIA. A genus of aquatic plants of the family of Phy'ceæ.
LAMPRATO'RNIS. A genus of birds. *Lamprato'rnis super'ba.* A kind of raven.
LANA'TA. Lat. Woolly.
LANCEOLA'TUS. Lat. Lanceolate; lance-shaped.
LAND'SLIP, or LAND'SLIDE. In geology, the removal of a portion of land down an inclined surface, from its attachment being loosened by the action of water beneath, or by an earthquake.
LANTA'NIUM. A metal discovered in 1840 by Mosander.
LA'PIS LA'ZULI. A mineral belonging to the aluminous silicates, of an azure blue colour.
LAURE'OLA. Specific name of a plant.
LAURI'NEÆ. } LAUREA'CEÆ. } From *laurus,* laurel, one of the genera. A family of plants.
LATENT HEAT. Heat not indicated by the thermometer; that heat upon which the liquid and aëriform conditions of bodies depend, and which becomes *sensible* duringthe conversion of vapour into liquids, and of liquids into solids.
LA'RVA. Lat. A mask. The first state of an insect after leaving the egg.
LA'RVÆ. Lat. Plural of larva.
LA'VA. In geology, substances which flow in a melted state from a volcano. Lavas vary in consistence and texture.
LEGUMINO'SÆ. From the Lat. *legu'men,* a bean. A family of plants.
LEGU'MINOUS. Belonging or relating to the Leguminosæ.
LEONI'NA. Lat. Belonging or relating to a lion.
LEPORI'NA. Lat. Belonging or relating to a hare.
LEPIDO'PTERA. From the Gr. *lepis,* a scale, and *pteron,* a wing, scaly wings. An order of insects characterized by scaly wings.
LESSO'NIA. Proper name. A genus of plants; also a genus of birds.
LEUCADE'NDRON. From the Gr. *leukos,* white, and *dendron,* tree. A genus of plants of the family of Proteáceæ.
LI'AS. Provincial corruption of the word *layers.* In geology, a division of the secondary formation. It is also called the Liassic, Jurassic, and Oolitic system of rocks.
LI'CHENS. An order of cryptógamous plants. They include various mosses.
LILIA'CEOUS. Belonging or relating to the lily.
LILIA'CEÆ. A family of plants.
LIMO'NIA. A genus of plants of the family of Aurantiáceæ.
LIMB. In botany, the spreading part or border of a leaf or petal. In astronomy, the outermost edge of the sun or moon.
LI'RIODE'NDRON. From the Gr. *leirion,* a lily, and *dendron,* a tree. The tulip tree. A genus of plants of the family of Magnoliáceæ.
LI'THIUM. A metal.
LLA'NOS. Span. Planes.
LO'ASA. A genus of plants of the family of Loasáceæ.
LOBE. A term applied in botany to the more or less profound divisions of a leaf, corolla, or other part of a plant.
LOBELIA'CEÆ. In honour of Lobel, a botanist. A family of dicotylédonous plants.
LON'GIFRONS. Lat. Having a long front or forehead.
LOPHOBRA'NCHES. From the Gr *lophos,* a tuft, or crest, and *branchia,* gills. An order of fishes.

Lopho'phorus. From the Gr. *lophos*, a tuft, and *phoros*, bearer. A genus of birds of the order of Gallináceæ.

Lora'nthus. From the Gr. *lôron*, a leather strap, and *anthos*, flower. Loranth. A genus of plants of the family of Loranthaceæ.

Lo'tus. A genus of plants of the family of Leguminósæ.

Lo'xia. A genus of birds.

Lu'teum. Lat. Yellow; dirty; made of clay. A specific name.

Macroce'phalus. From the Gr. *makros*, large, and *kephale*, head. A genus of insects. The specific name of a mammal.

Macrocy'stis. From the Gr. *makros*, large, and *kustis*, bladder. A genus of aquatic plants of the family of Phy'ceæ. Gigantic sea-weeds found in the southern hemisphere.

Macrou'rus. From the Gr. *makros*, great, and *oura*, tail. Having a long or large tail.

Mag'net. Loadstone is the natural magnet, which has the property of attracting iron. Artificial magnets are prepared so as to possess the peculiar attractive properties of the loadstone.

Mag'netism. The science which investigates the phenomena presented by natural and artificial magnets, and the laws by which they are connected.

Magne'sium. A silvery white metal obtained from magnesia.

Magne'sian. Containing magnesia.

Magno'lia. Name of Magnol, a French botanist. A genus of plants of the family of Magnoliáceæ.

Ma'lachite. A mineral; native green carbonate of copper.

Mal'lotus. A genus of fishes of the family of Salmones. A genus of plants of the family of Euphorbiáceæ. A synonym of the genus *Rottlera*.

Malu'rus. A genus of passerine birds.

Mam'mal. Any animal that suckles its young.

Mamma'lia. From the Lat. *mamma*, a breast. The name of the class of mammals or animals which suckle their young.

Mammi'feræ. Same as mammalia.

Mana'ti. Lat. Plural of manatus.

Mana'tus. A genus of mammals. The Lamantin.

Mangane'se. A metal.

Mari'tima. Lat. Maritime; relating to the sea.

Marl. Argillaceous carbonate of lime. There are several varieties of marl.

Marsu'pial. From the Lat. *marsupium*, a pouch. Any animal having a peculiar pouch in front or on the abdomen.

Mas'todon. From the Gr. *mastos*, a nipple, and *odous*, a tooth. A genus of extinct mammals allied to the elephant.

Ma'trix. In geology, the stony substance or bed in which metallic ores and crystalline minerals are embedded. The *gangue*.

Mauri'tia. Belonging to the island of Mauritius.

Mauro'rum. Lat. Of the Moors.

Ma'xima.
Ma'ximum. } Lat. The greatest.
Ma'ximus.

Medu'sa. A genus of marine animals of the class Aca'lepha.

Megathe'rium. From the Gr. *megas*, great, and *therion*, beast. Name of a fossil quadruped.

Melaleu'ca. From the Gr. *melas*, black, and *leukos*, white. A genus of plants of the family of Myrtáceæ.

Mela'stoma. From the Gr. *melas*, black, and *stoma*, opening. A genus of plants of the family of Melastomáceæ.

Mel'ia. A genus of plants of the family of Meliáceæ.

Melofo'rmis. From the Lat. *melo*, a melon, and *forma*, shape. Melon-shaped.

Menopo'ma. From the Gr. *menos*, strong, and *poma*, cover. A genus of reptiles of the family of Salamanders. Specific name of a batrachian.

Menu'ra. A genus of passerine birds. The *Menu'ra supe'rba*, the lyre-bird.

Mercury. Quicksilver. A metal which is liquid at ordinary temperatures.

MESEMBRYAN'THEMUM. From the Gr. *mesembria*, the mid-day, and *anthemum*, flowering; so called because the flowers usually expand at mid-day. The fig marygold. A genus of plants of the family of Fico'ides.
ME'SA. Span. A table.
MES'PILUS. From the Gr. *mesos*, half, and *pilos*, bullet, the fruit resembling a half ball. The medlar. A genus of plants of the family of Rosáceæ.
MET'ALLOID. Literally, resembling metal. The metals obtained from the alkalis and earths are called metalloids.
METALLI'FEROUS. Containing metal, or metals.
METAMOR'PHIC. From the Gr. *meta*, indicating change, and *morphe*, form. Metamorphic rocks are those which are evidently of mechanical origin, but owing to the presumed action of heat, have undergone change. Altered rocks.
METROSI'DEROS. From the Gr. *metra*, heart of a tree, and *sideron*, iron; in allusion to the hardness of its wood. A genus of plants of the family of Myrtáceæ.
ME'TUR. A species of wild corn which grows in Iceland.
MIA'SMA. } From the Gr. *miainô*, I contaminate. Applied to any emana-
MIA'SMATA. } tion from animal or vegetable substances, or from the earth, which may prejudicially influence the health of those persons who may be exposed to it.
MI'CA. From the Lat. *mico*, I shine. A mineral, generally found in thin elastic laminæ, soft, smooth, and of various colors and degrees of transparency. It is one of the constituents of granite.
MI'CA-SCHIST. Germ. (Gr. *schistos*, slaty, easily split.) Mica-slate. A lamellar rock composed of quartz, ordinarily grayish, and a great quantity of brilliant lamellæ of mica arranged in scales, or extended leaves.
MI'DAS. Name of a genus of monkeys; also, of a genus of reptiles.
MIGRATO'RIA. Lat. Migrating.
MILLINGTO'NIA. Proper name. A genus of plants of the family of Bignoniáceæ.
MILLE'PORA. From the Lat. *mille*, a thousand, and *pori*, holes. A genus of stony polyps, or corallines
MIMO'SA. From the Lat. *mimus*, a comedian; in allusion to its numerous varieties. A genus, and a tribe of plants.
MI'OCENE. From the Gr. *meiôn*, less, and *kainos*, recent. In geology, a name of a group of rocks of the tertiary period.
MI'NIMUM. Lat. The least.
MIRA'GE. Fr. A kind of natural optical illusion, arising from the unequal and irregular refraction of light by the lower strata of the atmosphere. The illusive appearance of water in deserts is explained in this manner.
MISODEN'DRON. A genus of plants of the family of Loranthâceæ.
MITE'LLA. A genus of plants of the family of Saxafragáceæ.
MOL'LUSK. From the Lat. *mollis*, soft. Applied to certain soft animals which inhabit shells, as oysters.
MOLLU'SCA. Lat. Mollusks. A branch of the animal kingdom.
MOLLU'SCOUS. Belonging or relating to mollusks.
MOLYBDE'NUM. A white, brittle metal.
MONOCOTYLE'DON. From the Gr. *monos*, single, and *kotuledon*, seed-lobe. A single seed-lobe.
MONOCOTYLE'DONOUS. Relating to monocotylédons.
MONO'CEROS. From the Gr. *monos*, single, and *keras*, horn. Having one horn.
MO'NODON. From the Gr. *monos*, single, and *odous*, tooth. Name of a genus of aquatic mammals. The Narwhal.
MONE'TA. Lat. Belonging or relating to money.
MONI'LIFORM. From the Lat. *monile*, a necklace. In form of a string of beads; necklace-like.
MO'NITOR. A genus of Saurian reptiles.
MONOPH'YSITE. From the Gr. *monos*, only, and *phusis*, nature. One of the sect of the ancient church who maintained that the human and divine natures in Jesus Christ became so blended and confounded as to constitute but one nature.

MONOSPE'RMA. From the Gr. *monos*, single, and *sperma*, seed. One-seeded. A specific name.

MON'TIA. A genus of plants of the family of Portuláceæ.

MORI'NDA. Indian Mulberry. A genus of plants of the family of Rubiáceæ.

MORAI'NES. Fr. The name given by geologists to longitudinal deposits of stony detritus found at the bases, and along the edges of all the great glaciers.

MO'RUS. Mulberry. A genus of plants of the family of Urti'ceæ.

MOSASAU'RUS. From *Meuse*, name of a river, and the Gr. *sauros*, a lizard. A genus of fossil reptiles.

MOS'CHUS. Lat. from the Gr. *moschos*, musk. Name of a genus of mammals.

MOSCHI'FERUS. Lat. Musk-bearing; containing musk.

MO'SSES. Cryptógamous parasites of the family of Lycopodeáceæ.

MU'CILAGE. A mixture of gum and water.

MU'RAL. Belonging or relating to a wall.

MU'SA. The banana. A genus of plants of the family of Musáceæ.

MUS'CHELKALK. German. Shell limestone.

MUSCI'PULA. Lat. A fly-trap or mouse-trap. A name of a plant.

MU'SICUS. Lat. Relating to music; musical.

MYCE'TUS. Name of a genus of monkeys.

MYOPO'TAMUS. From the Gr. *mus*, a rat, and *potamos*, a river. Name of a genus of gnawing mammals.

MYRI'STICA. A genus of plants of the family of Myrista'ceæ; *Myris'tica moscha'ta*, the nutmeg tree.

MYR'TUS. Myrtle. A genus of plants of the family of Myrta'ceæ.

NA'NA. From the Lat. *nanus*, a dwarf. A specific name.

NA'PHTHA. A limpid bitumen.

NARCI'SSUS. Name of a genus of plants of the family of Amarylli'deæ.

NA'TRIUM. A metal.

NA'TRON. A subcarbonate of soda.

NEC'TARY. That part of a flower which secretes *nectar* or honey.

NELUM'BIUM. A genus of plants of the family of Nymphæa'ceæ. Sacred Bean.

NES'TOR. An extinct bird.

NEURO'PTERA. From the Gr. *neuron*, a nerve, and *pteron*, wing. An order of insects.

NEW RED SAND'STONE. In geology, a system of rocks of the secondary formation.

NICK'EL. A white metal. It is the basis of "German Silver."

NI'GRA. Lat. Black.

NIM'BUS. A rain cloud.

NI'TIDA. Lat. Neat, clean, bright.

NI'TROGEN. A simple, permanently elastic fluid or gas, also called azote, which constitutes four-fifths of the atmosphere, and is the basis of nitric acid.

NIVA'LIS. Lat. Snowy.

NON-CONDUC'TOR. Applied to substances which do not possess the property of transmitting electricity, or heat.

NOTACAN'THUS. From the Gr. *nôtos*, back, and *akantha*, a spine. Name of a genus of fishes.

NOTOR'NIS. Name of an extinct bird.

NOTOTHE'RIUM. A fossil genus of marsupial mammals.

NUMMULA'RIA. From the Lat. *nummus*, a coin. A family of Mollusks. Nummuli'tes.

NYMPHÆ'A. Name of a genus of plants of the family of Nymphæa'

NYS'SA. From the Gr. *nussô*, I prick. Name of a genus of plan'

OBSI'DIAN. A glassy lava. Volcanic glass.

OCEAN'ICA. Lat. Relating to the ocean.

ODORA'TA. Lat. Odorous.

ODORATIS'SIMA. Lat. Very, or most odorous.

ODORI'FERA. Lat. Odoriferous.

OENOTHE'RA. From the Gr. *oinos*, wine, and *therô*, I hunt. A genus of plants of the family of Oenothera'ceæ. Synonym of onagrariæ.

O'LEA. Lat. Olive. A genus of plants of the family of Olea'ceæ.

OLD RED SAND'STONE. A system of rocks of the secondary formation.

O'OLITE. From the Gr. *ôon*, an egg, and *lithos*, stone. A granular variety of carbonate of lime, frequently called *roe-stone*.

O'PAL. A brittle mineral, characterized by its iridescent reflection of light. It consists of *silica* with about ten per cent. of water.

OPALES'CENT. Resembling opal.

OPHI'DIAN. From the Gr. *ophis*, a serpent; applied to reptiles of the order of Ophidia.

OPHICE'PHALUS. From the Gr. *ophis*, serpent, and *kephale*, head. Serpent-head. A genus of acánthoptery'gian, or bony-finned fishes.

OR'CHIS. A genus of plants of the family of Orchide'æ, named from most of the species being marked by two tubercles.

ORCHID'EOUS. Relating to the genus orchis.

OR'GAN. From the Gr. *organon*, an instrument. Part of an organized being, destined to exercise some particular function; for example, the ears are the organs of hearing, the muscles are the organs of motion.

ORGA'NIC. Relating to an organ. *Organic remains*, are the fossil remains of organized beings.

ORGANIZA'TION. The mode or manner of structure of an organized being.

OR'GANIZED. Composed of organs; having a mode of structure.

ORIENTA'LE. }
ORIENTA'LIS. } Lat. Eastern. Belonging to the East.

ORNITHORYN'CHUS. From the Gr. *ornis*, *ornithos*, a bird, and *rugchos*, a beak. A genus of mammals, having the beak of a duck.

OR'TYGIS. From the Gr. *ortux*, a quail. A genus of birds.

OS'MIUM. From the Gr. *osme*, odour. A metal discovered in 1803, by Tennant.

OSCILLA'TION. The act of moving backwards and forwards like a pendulum.

OTA'RIA. From the Gr. *ôtarion*, a small ear. A genus of amphibious mammals, of the tribe of seals.

OUT'CROP. In geology, the emergence of a rock in place, at the surface.

O'VARY. In botany, that part of a flower in which the young seeds are contained.

OX'ALIS. A genus of plants of the family of Oxali'deæ.

OXLE'YA. A genus of plants of the family of Cedrela'ceæ.

OX'YGEN. The vivifying gas which constitutes about one-fifth of the atmosphere, the presence of which is essential to life.

O'ZONE. From the Gr. *ozô*, I smell of something. The odorous matter perceived when electricity passes from pointed bodies into the air.

PACHYDER'MATA. Lat. from the Gr. *pachus*, thick, and *derma*, skin. An order of mammals—Pachyderms.

PADI'NA. Same as Zonária, a beautiful marine plant. *Padi'na pavo'nia*, or *Zona'ria pavo'nia*. Turkey feather.

PALAP'TERYX. From the Gr. *palaios*, ancient, and *apteryx*, (formed from the Gr. *a*, privative, and *pteron*, wing, wingless. Name of a genus of fossil birds, discovered recently in New Zealand.

PALÆOTHE'RIUM. From the Gr. *palaios*, ancient, and *therion*, beast. A fossil genus of pachyder'matous mammals.

PALÆONTO'LOGY. From the Gr. *palaios*, ancient, and *on*, a being or creature, and *logos*, discourse. That branch of zoological science, which treats of fossil organic remains.

PALÆOZO'IC. From the Gr. *palaios*, ancient, and *zoe*, life. Relating to ancient life; belonging or relating to fossils.

PALMA REA'L. Spanish. Royal Palm.

PALLA'DIUM. A white, hard, very malleable and ductile metal, which is susceptible of a fine polish. It is more difficult to melt than gold.

PALMEL'LA. A genus of plants of the family of Confervácеæ. *Palme'lla niva'lis*, a plant of the snowy regions, which gives color to the snow amidst which it grows. *Protoco'ccus* is the red snow plant.

PANDA'NA. Relating to, or resembling the Screw-pines.
PANDA'NUS. From the Malay name of the tree, *pandang*. Screw-pine. A genus of plants, of the family of Pandaneæ. *Pandanus candelabrum*. Candlestick screw-pine.
PA'NICUM. Panic-grass. A genus of plants of the family of Grami'neæ. *Pa'nicum milia'ceum*, millet, a grain used for feeding poultry in England.
PA'PA. Spanish. Pope, Specific name of a vulture.
PAPY'RUS. A genus of plants of the family of Cypera'ceæ. The *Papy'rus antiquo'rum* yields the substance used as paper by the ancient Ægyptians.
PAPYRI'FERA. From *papyrus*, a sort of paper, and *fero*, I bear. Paper-bearing.
PARHE'LIA. Plural of parhelion.
PARHE'LION. From the Gr. *para*, alongside of, and *elios*, the sun. A mock sun. A meteor which consists in the simultaneous appearance of several suns, "fantastic images of the true one."
PARADISA'ICA. Lat. Belonging or relating to Paradise. A specific name.
PARNA'SSUS. A genus of lepidopterous insects of the tribe of Parna'ssidæ.
PASS'ERINES. / PASS'ERINE BIRDS. / PASS'ERES. } From the Lat. *Passer*, a sparrow, name of a varied and extensive order of birds, not easily characterized.
PASSIFLO'RA. Abbreviation of *flos*, flower, and *passionis*, of the passion. Passion-flower, so called from a supposed resemblance between its floral organs, and the instruments of the Passion of our Saviour. An extensive and beautiful genus of plants.
PATAGO'NICA. Lat. Relating to Patagonia. Specific name of a penguin.
PAVO'NIA. Formed from the Lat. *pavo*, a peacock. A specific name.
PEAT. The natural accumulation of vegetable matter on the surface of lands not in a state of cultivation; always moist to a greater or less degree, varying, according to the kind of plants to the decay of which the formation of peat is due.
PELARGO'NIUM. From the Gr. *pelargos*, a stork. Stork's bill. A genus of plants of the family of Geraniáceæ.
PELA'GIC. From the Lat. *pelagus*, the sea. Relating to the sea.
PELO'PIUM. A metal discovered by Prof. H. Rose.
PEN'DULUM. From the Lat. *pendo*, I hang. A weight suspended at the end of a rod, so that it may vibrate from side to side in a plane, is called a pendulum.
PENNISE'TUM. From the Lat. *penna*, a feather or pen, and *seta*, a bristle. A genus of plants of the family of Grami'neæ.
PE'PLIS. Gr. Water-purslane. A genus of plants of the family of Salicáriæ.
PER'MIAN. After the ancient kingdom of Permia. A name applied by Mr. Murchison to a system of rocks, consisting of an extensive group of fossiliferous strata, intermediate, in their geological position, between the Carboniferous and Triassic systems, the latter being the upper portion of the New Red Sandstone formation.
PERTURBA'TION. In astronomy, the deviation of a celestial body from the elliptic orbit which it would describe, if acted upon by no other attractive force than that of the sun, or central body about which it revolves.
PE'TAL. From the Gr. *petalon*, a leaf. A part of the corolla of a flower analogous to a leaf.
PETRO'LEUM. From the Gr. *petros*, a rock, and the Lat. *oleum*, oil. Rock-oil, often called *Barbadoes tar*. A brown, liquid bitumen, found in the West Indies, Europe, &c.
PHACOCHÆ'RE. Fr. / PHACOCHÆ'RUS. Lat. } From the Gr. *phake*, a wart, and *choiros*, a hog. A genus of mammals of the order of pachydermata; allied to the hogs.
PHALAN'GER. From the Gr. *phalagx*, a phalanx. A genus of marsupial or pouch-bearing mammals.
PHANEROGA'MIA. From the Gr. *phaneros*, evident, and *gamos*, marriage. Phanerógamous plants. Applied to plants having distinct flowers.
PHI'LEDON. / PHI'LEMON. } Name of a genus of birds.
PHLE'UM. Cat's-tail grass. A genus of plants of the family of Grami'neæ.

PHLOX. Gr. Flame. A genus of beautiful plants of the family of Polemoniáceæ.

PHO'CA. Lat. A seal. A genus of aquatic mammals, embracing the common seal or *Pho'ca vituli'na;* the Harp seal or *P. ocea'nica;* the Hare-tailed seal or *P. lagura;* the sea-lion; sea-wolf; sea-elephant; sea-cow; &c., &c.

PHO'CÆ. Lat. Plural of phoca.

PHO'NOLITE. From the Gr. *phoneô*, I resound, and *lithos*, a stone. Clink-stone. A kind of compact basalt which is sonorous when struck.

PHOR'MIUM. From the Gr. *phormos*, a basket. Flax-lily. A genus of plants of the family of Asphodéleæ. *Pho'rmium te'nax*, Iris-leaved flax-lily of New Zealand.

PHOS'PHORUS. From the Gr. *phos*, light, and *pherô*, I bear. A simple substance which is highly inflammable.

PHOSPHO'RIC A'CID. A compound of phosphorus and oxygen, having the properties of acids.

PHOS'PHATES. Compounds of phosphoric acid with salifiable bases, as soda, are termed phosphates; Phosphate of soda, for example.

PHOSPHORE'SCENCE. Emission of light from substances at common temperatures, or below a red heat.

PHOSPHORE'SCENT. Having the property of emitting light without sensible heat.

PHOTO'METER. From the Gr. *phôs*, light, and *metron*, a measure. An instrument for measuring the intensity of light.

PHYSA'LIA. } PHY'SALIS. } From the Gr. *phuse*, a vesicle. A genus of animals of the family of Acalepha. The Portuguese man-of-war belongs to this genus.

PHY'SALIS. A genus of plants of the family of Solanáceæ. *Physa'lis e'dulis*, the Cape gooseberry.

PHYSE'TER. A blower. Name of a genus of mammals of the family of Ceta'cea.

PIME'NTO. Allspice; Jamaica pepper.

PIN'NATE. From the Lat. *pinnatus*, feathered. Having leaflets arranged along each side of a common petiole, like the feather of a quill.

PINNATI'FIDA. Lat. Pinna'tifid. A leaf is so called when it is divided into lobes from the margin nearly to the midriff.

PI'NUS. Lat. A pine-tree. A genus of plants of the family of Coniferæ. *Pi'nus a'bies.* The Norway Spruce. *Pi'nus canarie'nsis.* The Canary pine. *Pi'nus ce'mbra.* The Riga balsam tree; the Cembran or Siberian pine. *Pi'nus exce'lsa.* The lofty or Nepal pine. *Pi'nus marit'ima.* The maritime pine. *Pi'nus pi'nea.* The Stone pine.

PI'PIENS. Lat. Peeping like a chicken.

PI'PA. A genus of batrachian reptiles. A kind of toad.

PLA'TINA. } PLA'TINUM. } The diminutive of the Spanish *plata*, silver. A metal of a steel gray colour, approaching to the white colour of silver, to which resemblance it owes its name. It was found in Choco, one of the provinces of Colombia, and brought to Europe in 1741, by Don Antonio de Ulloa.

PLEI'OCENE. } PLI'OCENE. } From the Gr. *pleion*, more, and *kainos*, recent. A term applied by geologists to the newer tertiary formation, because there is found fossilized in it a greater number of existing than of extinct species.

PLUMB-LINE. } PLU'MMET. } From the Lat. *plumbum*, lead. An instrument, consisting of a string with a weight, usually of lead, attached to a straight staff, for the purpose of ascertaining the direction of gravitation, or the perpendicular to the horizon.

PLUTO'NIC ROCKS. Unstratified crystalline rocks, probably formed at great depths beneath the surface by igneous fusion. *Volcanic rocks* are formed near the surface.

PODOCA'RPUS. From the Gr. *pous, podos*, the foot, and *karpos*, fruit. A genus of plants of the family of Coniferæ.

Po'larized light. Light so modified as to possess poles, or sides, having opposite properties. Light by reflection or refraction, when passed through crystals possessing the power of double refraction, becomes modified, so that it does not present the same phenomena of transmission and reflection, as light which had not been polarized.

Polariza'tion. The process by which light is polarized.

Polyg'onum. From the Gr. *polus*, many, and *gonu*, a knee or joint. A genus of plants of the family of Polygonáceæ. *Polygonum viriparum*, Alpine Bistort.

Polymo'rpha. Lat. From the Gr. *polus*, many, and *morphe*, form. Many-shaped. A specific name.

Po'lypi. Lat. Plural of polypus.

Polyple'ctron. Name of a genus of birds.

Pol'ypus. From the Gr. *polus*, many, and *pous*, foot. A genus of radiate animals.

Pon'tica. Lat. From *pontus*, the sea. Belonging or relating to the sea.

Pontoppida'na. Synonym of Couroupita. A genus of plants of the family of Myrtáceæ, Lecythi'deæ. A large tree of Guiana.

Ponto'phidan. From the Lat. *pontus*, the sea, and the Gr. *ophis*, a serpent. The sea-serpent.

Portulaca'ria. A genus of plants of the family of Portula'ceæ. The Purslane-tree. *Portulaca'ria a'fra*. The African purslane-tree.

Porphyrit'ic. Of the nature of porphyry.

Por'phyry. From the Gr. *porphura*, purple. Originally applied to a *red rock* found in Egypt. A compact feldspathic rock containing disseminated crystals of feldspar, the latter when polished forming small angular spots, of a light color, thickly sprinkled over the surface. The rock is of various colors, dark green, red, blue, black, &c.

Port'land bed. A name given by geologists to the superior division of the upper óolite or lias system. The "Portland stone" is a kind of lime-stone found in the south of England, and more particularly in the Isle of Portland. In this series of strata is a silicious sand known as the "Portland Sand."

Potas'sium. A metal discovered in potash by Sir H. Davy, in 1807.

Potenti'lla. A genus of plants of the family of Rosáceæ. Cinquefoil. *Potenti'lla tridenta'ta*. Trifid-leaved cinquefoil.

Prehen'sile. From the Lat. *prehendere*, to lay hold of. Having the faculty to lay hold of. Applied to the tails of those monkeys, for example, which have the power to suspend themselves by the tail.

Preda'ceous. Living on prey.

Pri'mary forma'tion. A term applied by geologists to designate the different rocks which were formed prior to the creation of plants and animals.

Primige'nius. Lat. Original; first of its kind.

Pri'mum mo'bile. That which first imparts motion.

Prim'ula. Lat. A primrose. A genus of plants of the family of Primuláceæ. *Prim'ula farino'sa*, the Bird's-eye primrose.

Prism. A solid bounded by three planes, two of which are equal.

Prisma'tic. Belonging or relating to a prism.

Probosci'dian. From the Gr. *proboskis*, a proboscis or trunk. Applied to mammals of the family which includes the elephant.

Procella'ria. From the Lat. *procella*, a tempest at sea. A genus of birds of the family of Palmi'pedes.

Proli'fera. Lat. Formed from *proles*, a race or stock, and *fero*, I bear. Prolific.

Pro'tea. A genus of plants of the family of Proteáceæ. *Pro'tea cyanero'ides*, Artichoke-flowered protea.

Pro'teus. Lat. A genus of reptiles.

Pteroca'rpus. From the Gr. *pteron*, a wing, and *karpos*, fruit. So called because the pods have membranous wings. A genus of plants of the family of Leguminóseæ. *Pteroca'rpus santali'nus*. The red saunders tree.

Pte'ris. Gr. Name of Fern. A genus of cryptógamous plants. Brake. *Pte'ris escule'nta*. Edible fern.

PTER'OPUS. From the Gr. *pteron*, wing, and *pous*, foot. A genus of mammals of the tribe of bats, termed Roussettes.
PU'MA. A name of the couguar or American Lion.
PYROG'ENOUS. From the Gr. *pur*, fire, and *geinomai*, I beget. Applied to rocks which owe their origin to the action of fire, as granite.
PY'RUS. Lat. A pear-tree. A genus of plants of the family of Rosáceæ.
PYRI'FERA. Lat. From *pyrus*, a pear, and *fero*, I bear. Pear-bearing.
PY'THON. A genus of reptiles.

QUADRU'MANA. Formed from the Lat. *quatuor*, four, and *manus*, hand. An order of mammals characterized by having four hands.
QUARTZ. Germ. Rock crystal.
QUA'RTZITE. A mineral resembling quartz. Granular quartz.
QUARTZ'OSE. Of the nature of quartz.
QUI'CKSILVER. Mercury. A metal which is fluid at ordinary temperatures.

RADIA'TA. Lat. Radiates; the name of a class of zóophytes.
RA'DIATE. From the Lat. *radius*, a ray. Furnished with rays; having rays.
RADIA'TION. The emission of the rays of light, or of heat, from a luminous or a heated body.
RAFFLE'SIA. After Sir T. Raffles. A genus of plants of the family of Rafflesiáceæ, which are parasites, growing on the roots of dicotyle'donous plants. The flowers of some of them are enormously large; those of the Rafflesia Arnoldi are said to be three feet in diameter.
RAINE'TTE. Fr. A tree-frog.
RA'NA. Lat. A frog. A genus of reptiles.
RANUN'CULI. Lat. Plural of ranunculus.
RANUN'CULUS. From the Lat. *rana*, a frog, because the species inhabit humid places. Crow-foot. A genus of plants of the family of Ranunculáceæ.
REFRA'CTION. From the Lat. *refractus*, broken. The deviation of a ray of light from its rectilinear course, caused by passing through a transparent substance. The degree of refraction depends upon the density of the medium through which the ray of light passes.
REFU'LGENS. Lat. Shining brightly; refulgent.
RE'PTILE. From the Lat. *repere*, to crawl. A term applied to any animal that moves naturally upon its belly, or on very short legs, as serpents, &c.
REPTI'LIA. Lat. The class of reptiles: it comprises those vertebrate animals which have cold blood, an aërial respiration, and an incomplete circulation.
RESINI'FERA. Lat. Containing resin.
REVOLU'TA. Lat. Turned back; tumbled.
RHE'A. Synonym of *Struthio*, an ostrich.
RHO'DIUM. From the Gr. *rodon*, a rose, on account of the rose-red color of some of its salts. A metal discovered in the year 1803, by Wollaston.
RHODODE'NDRON. From the Gr. *rodon*, a rose, and *dendron*, a tree. A genus of plants of the family of Ericáceæ.
RHODOME'LIA. From the Gr. *rodon*, a rose, and *melas*, black. A genus of plants of the family of Phy'ceæ.
RHUS. A genus of plants of the family of Terebintháceæ. *Rhus vernix.* The varnish Sumach.
RHY'NCHOPS. From the Gr. *rugchops*, a beak. A genus of birds: the skimmers or scissor-bills.
ROCK-SALT. Common salt found in masses or beds in the new red sandstone.
RODEN'TIA. From the Lat. *rodere*, to gnaw. An order of mammals.
RO'DENTS. Animals of the order of Rodentia.
ROR'QUAL. A kind of whalebone whale.
RO'SA. Lat. Rose. A genus of plants of the family of Rosáceæ. *Rosa sinensis.* The Chinese rose.
RU'BER. Lat. Red.
RUBIA'CEÆ. A family of plants.
RU'BY. A crystallized gem of various shades of red.
RUFIM'ANUS. Lat. Red-handed.

RU'MINANT. An animal that chews the cud.
RUMINA'NTIA. An order of mammals which are characterized by chewing the cud.
RU'MINATE. To chew the cud.

SA'LINES. Natural deposits of salt; salt springs.
SA'LIX. Lat. Willow. A genus of plants of the family of Salici'neæ. *Sa'lix lana'ta.* Woolly willow.
SALT. A combination of an acid with one or more bases.
SA'MOLUS. From the Celtic, *san*, salutary, and *mos*, pig. Salutary to pigs. Brook-weed. A genus of plants of the family of Primulắceæ. *Sa'molus vale-ra'ndi.* Common brook-weed.
SAN'DARACH. A name given by the Arabs to an odorous resin.
SANDALI'NUS. Lat. Sandal-like.
SA'NDSTONE. Any rock consisting of aggregated grains of sand.
SAPA'JOU. Fr. A genus of monkeys.
SAPI'NDUS. Abbreviation of *sapo*, soap, and *indicus*, Indian soap. Soap-berry. A genus of plants of the family of Sapindắceæ. *Sapi'ndus sapona'ria.* Common soap-berry.
SAPONA'RIA. Lat. Soapy.
SAPOTA'CEÆ. A family of plants.
SA'PPHIRE. A very hard gem consisting essentially of crystallized alumina. It is of various colors; the *blue* variety being usually called sapphire; the *red*, the oriental ruby; the *yellow*, the oriental topaz.
SA'KIS. } SA'QUIS. } A genus of monkeys.
SARGA'SUM. From the Span. *sarga'zo*, sea-lentils. A genus of plants of the family of Phy'ceæ.
SARRACE'NIA. After Dr. Sarrazin. The side-saddle flower, or pitcher plant. A genus of plants of the family of Sarracéneæ.
SAU'RIAN. From the Gr. *sauros*, a lizard. Applied to animals of the lizard tribe.
SAU'ROID. From the Gr. *sauros*, a lizard, and *eidos*, resemblance. Resembling a lizard.
SCA'NDENS. Lat. Climbing.
SCHIST. From the Gr. *schistos*, split. Slate.
SCHISTO'SE. Slaty.
SCHOT'IA. After Schott, a Dutch gardener. A genus of plants of the family of Leguminósæ. *Scho'tia specio'sa.* Small-leaved Schotia.
SCITAME'NEÆ. A family of plants.
SCLE'RIA. From the Gr. *skleros*, hard. A genus of plants of the family of Cyperắceæ.
SCO'LOPAX. Lat. A genus of birds: a heron.
SCOPA'RIA. From *scopa*, a broom. A genus of plants of the family of Scrophulari'neæ.
SCO'RIÆ. Volcanic cinders. Cinders and slags of basaltic lavas of a reddish brown and black color.
SCORIA'CEOUS. Of the nature of scoriæ.
SCO'RIFORM. In form of scoriæ.
SCY'THROPS. From the Gr. *skuthrops*, sad. A genus of birds of the order of climbers. A cuckoo.
SEAMS. In geology, thin layers of strata interposed between others.
SEBI'FERA. Lat. Containing tallow.
SE'CONDARY FORMA'TION. In geology, the formation which is next in order to the transition formation.
SE'CULAR. From the Lat. *seculum*, a century. *Secular elevations* are those which take place gradually and imperceptibly, through a long period of time. *Secular tides* are those which are dependent upon the secular variation of the moon's mean distance from the earth.

SE'DIMENT. From the Lat. *sedeo,* I sit. That which subsides or settles to the bottom of any liquid.
SEDIME'NTARY. Belonging or relating to sediment.
SEE'D-LOBE. The envelope in which the seed of a plant is formed.
SE'LENITE. A variety of gypsum, or sulphate of lime.
SEMI'TIC. Applied to the languages of the descendants of *Shem,* or the Orientals.
SE'PAL. That part of the calyx of a flower which resembles a leaf.
SE'PIA. A kind of paint prepared from the cuttle-fish. A genus of mollusks.
SER'PENTINE. A magnesian rock of various colors, and often speckled like a serpent's back. It is generally dark green.
SEPTENTRIONA'LIS. Lat. Northern.
SER'RATE. From the Lat. *serra,* a saw. Toothed like a saw.
SERRA'TA. Lat. Serrate.
SHAFT. A cylindrical hollow space, or pit, in mines, made for the purpose of extracting ores, &c.
SHALE. An indurated slaty clay, or clay-slate.
SHI'NGLE. Loose, water-worn gravel and pebbles.
SHORE'A. Synonym of *Vatica.* A genus of plants of the family of Dip'tero-cárpeæ.
SIER'RA. Span. A mountain chain.
SI'LEX. From the Gr. *chalix,* a pebble. The principal constituent of quartz, rock-crystal, and other *sili'cious* minerals.
SI'LICA. Silicious earth: the oxide of *silicon* (the elementary basis of Silica,) constituting almost the whole of *silex* or flint. It combines with many of the metallic oxides, and is for this reason sometimes called *sili'cic* acid.
SI'LICATE. A compound of silicic acid and a base. *Plate-glass* and *window-glass* are silicates of soda and potassa; and *flint-glass* is a similar compound with a large addition of silicate of lead.
SILI'CIOUS. Containing silicia.
SILI'CIFIED. Petrified or mineralized by silicious earth.
SILT. The name given to the sand, clay, and earth, which accumulate in running waters.
SILI'CIUM. The metalloid which forms the basis of silica.
SILU'RIAN SYS'TEM. A series of rocks formerly known as the *greywacke series.* So called after the *Silures* or *Siluri,* the ancient Britons who inhabited the region where these strata are most distinctly developed. They are entirely of marine origin.
SILU'RUS. Lat. A genus of fishes of the family of Siluridæ.
SIL'VA. A forest, or woods.
SIMU'LIUM. From the Lat. *simulo,* I feign. A genus of insects of the order of Diptera.
SINE'NSIS. Lat. Chinese; belonging or relating to China.
SIN'TER. Germ. A scale. *Calcareous sinter* is a variety of carbonate of lime composed of successive concentric layers. *Silicious sinter* is a variety of common opal.
SIPHO'NIA. A genus of plants of the family of Euphorbiáceæ.
SLATE. A well known rock which is divisible into thin plates or layers.
SMYN'THUS, or SMIN'THUS. From the Gr. *sminthos,* rat. A genus of rodent mammals.
SMYRNE'NSIS. Lat. Belonging, or relating to Smyrna.
SO'LAR SPE'CTRUM. Lat. *Spectrum,* an image. In optics, the name given to an elongated image of the sun formed on a wall or screen by a beam of undecomposed light, received through a small hole, and refracted by a prism.
SOLFATA'RA. Italian. A volcanic vent emitting sulphur and sulphurous compounds.
SOLIDA'GO. Golden-rod. A genus of plants of the family of Compósitæ.
SOMATE'RIA. Synonym with *platypus.* A genus of birds.
SOPHO'RA. A genus of plants of the family of Leguminósæ.
SPAR. (Germ. *Spath.*) Applied to certain crystallized mineral substances, which easily break into cubic, prismatic, or other forms.
SPAR'RY. Of the nature of spar.

SPE'CIES. A kind; a subdivision of genus. According to Dr. Morton, a primordial type. "An animal," says Mr. John Cassin, "which constantly perpetuates its kind, or in other words, produces itself either exactly or within a demonstrable range of variation, is a species." Extinct species is a term applied to those kinds of organized beings, whether plants or animals, which are not found living upon the face of the earth.

SPECI'FIC. Relating to species.

SPECI'FIC WEIGHT, or SPECI'FIC GRA'VITY. The relative weight of one body with that of another of equal volume.

SPECIO'SA. / SPECIO'SUM. / SPECIO'SUS. } Lat. Handsome. A word used as a specific name.

SPE'CULAR IRON. A kind of iron ore of granular structure, and metallic lustre, sometimes shining.

SPECTA'BILIS. Lat. Visible, remarkable, notable.

SPICA'TA. Lat. Having spikes; eared like corn.

SPINE'LLE, or SPINE'L. Fr. A sub-species of ruby.

SPIRÆ'A. A genus of plants of the family of Rosáceæ.

SPORES. The seeds of lichens, and cryptógamous plants.

SPOR'ULES. The diminutive of spores.

SPUMA'CEOUS. From the Lat. *spuma*, foam. Foamy.

STAGMA'RIA. From the Gr. *stagma*, a drop. A genus of plants of the family of Anacardiáceæ. *Stagma'ria vernici'flua*, a tree of Sumatra, from the bark of which exudes an extremely acrid juice. This juice quickly dries in the air, becomes black, and is sold at a high price; it is employed in the preparation of a varnish. The Sumatrans consider it dangerous to sit or sleep in the shade of this tree.

STA'MEN. Lat. The male apparatus of a flower.

STAPE'LIA. Proper name. A genus of plants of the family of Asclepiádeæ.

STARCH. A vegetable substance which exists in many tuberous roots, the stalks of palms, and in the seeds of the cereal grasses.

STEPPE. Fr., from the Lat. *stipes*, a landmark. A term applied to the savannas of Tartary, of the Crimea, &c., and salt deserts of Northern Asia. A level waste, destitute of trees: a prairie.

STI'GMA. The superior, terminating part of the pistil of a flower.

STILLIN'GIA After Dr. Stillingfleet. A genus of plants of the family of Euphorbiáceæ. *Stilli'ngia sebi'fera*. The tallow tree of China.

STRA'TA. Lat. Plural of *stratum*, a layer, a bed.

STRATIFICA'TION. An arrangement in beds or layers.

STRA'TIFIED. Arranged in strata.

STRA'TUM. Lat. In Geology, a bed of sedimentary rock.

STRA'TUS. A kind of cloud: it consists of horizontal layers, and includes fogs and mists; its under surface usually rests upon the land or sea, and it is therefore the lowest of the clouds.

STRELIT'ZIA. After Queen Charlotte, of the family of Mecklenburgh Strelitz. A genus of plants of the family of Musáceæ.

STRI'Æ. Lat. Diminutive channels or creases.

STRIA'TA. Lat. Striated; marked with striæ.

STRON'TIUM. A metalloid found in the earth called strontia.

STRU'THIO. Lat. An ostrich. A genus of birds.

STRU'THIOUS. Of the nature of an ostrich.

STRYCH'NOS. A genus of plants of the family of Apocy'neæ. *Strych'nos toxica'ria*. The poison strychnos. The Nux Vomica is the seed of a plant of this genus.

STYLE'DIUM. From the Gr. *stulos*, a column. A genus of plants of the family of Stylideæ, found in New Holland.

SUBLIMA'TION. The process by which volatile substances are raised by heat, and again condensed into the solid form. The substances so obtained are called *su'blimates*.

SUB'SOIL. An under soil.

SUBSTRA'TA. Lat. Plural of *substratum*, an under layer or bed.

SUL'PHURET. A compound of sulphur with another solid, as with iron, forming *sulphuret of iron.*
SUL'PHURETTED. Containing sulphur; as, hydrogen containing sulphur, is called sulphuretted hydrogen.
SUPE'RBA. Lat. Superb, elegant.
SURIA'NA. A genus of plants of the family of Suriáneæ; it was formerly of the Rosáceæ.
SYCOMO'RUS. Lat. The Sycamore; applied also as a specific name.
SY'ENITE and SI'ENITE. A granite rock from *Syene* or *Siena* in Egypt. It consists of quartz, feldspar, and hornblende. It is tougher than granite.
SYL'VIA. Name of a genus of birds.
SYNGENE'SIA. From the Gr. *sun*, together, and *geinomai*, to grow. Linnean name of a class of plants.

TAC'CA. Malay. A genus of plants of the family of Aroïdeæ. *Tac'ca pinnati'fida.* The Salep tree.
TANG'HINIA. From the Madagascar name, *Tanghing.* A genus of plants of the family of Apocy'neæ. *Tang'hinia veneni'flua* yields an active poison which is used to cause death, under judicial sentence, by the natives of Madagascar.
TANTA'LIUM. A metal, remarkable for its insolubility in acids.
TA'NTALUS. A genus of birds of the family of Herons.
TARTA'RICA. Lat. Belonging or relating to Tartary.
TELESÇO'PIC. Relating to the telescope; telescopic objects are those which may be seen by the aid of a telescope.
TELLU'RIUM. A rare metal, found in the gold mines of Transylvania.
TEM'PERATURE. A definite degree of sensible heat.
TEMPORA'RIA. Lat. Temporary; relating to time.
TE'NAX. Lat. Tenacious.
TER'MES. A genus of insects of the order of Neuroptera, and family of Termitidæ. White ants.
TER'RA JAPO'NICA. An astringent medicinal gum, obtained from the Acácia ca'techu.
TER'TIARY FORMA'TION. A series of sedimentary rocks which are superior to the primary and secondary, and distinguished by the fossil remains found in them.
TESTA'CEÆ. From *testa*, a shell. Testáceans; animals provided with an external shelly cover, composed chiefly of carbonate of lime.
TESTA'CEOUS. Consisting of carbonate of lime and animal matter.
TESTU'DO. Lat. Tortoise. A genus of reptiles of the order of Chelo'nians.
TETRA'CERA. From the Gr. *tettares*, four, and *keras*, a horn. A genus of plants of the family of Dillenáceæ.
TETRA'GONA. From the Gr. *tettares*, four, and *gonu*, a knee. Having four angles; applied as a specific name. A genus of plants of the family of Portuláceæ.
THE'A. A genus of plants of the tribe of Came'lleæ. *The'a bo'hea*, Bohea tea; *The'a vi'ridis*, Green tea.
THE'INE. The proximate principle of tea.
THER'MAL. From the Gr. *thermos*, heat. Warm; belonging or relating to heat.
THO'RIUM. A metal obtained from Thori'na, an earthy substance.
THU'IA, also THU'JA. A genus of plants of the family of Coni'feræ. *Thu'ia articula'ta.* Jointed arbor vitæ. *T. orienta'lis;* Chinese arbor vitæ. *T. sa'ndarach*, Shittim wood.
TI'DAL. Relating to tides. *Tidal wave* is the elevation of the water of the ocean, produced by the attraction of the moon.
TILLAND'SIA. A genus of plants of the family of Bromeliáceæ.
TITA'NIUM. A metal discovered in 1781, by W. Gregor, in a ferruginous sand.
TO'DUS. Lat. A tôdy. Name of a genus of birds of the order of Passeri'næ.
TO'PAZ. A crystallized pellucid mineral, harder than quartz; commonly of a yellow wine color, but it also occurs white, blue, and brown.
TO'RSION BA'LANCE. See BALANCE.
TOU'RMALINE. A mineral substance consisting of a Bo'ro-si'licate of a'lumine, harder than quartz, but not as hard as topaz.

TOURNFO'RTIA. After Tournefort. A genus of plants of the family of Borragi'neæ.

TRA'CHYTE. From the Gr. *trachus*, rough. A variety of lava. A feldspathic rock, which often contains glassy feldspar and hornblende. When the feldspar crystals are thickly and uniformly disseminated, it is called *trachy'tic por'phyry*.

TRAP. From the Swedish *trappa*, a flight of stairs, because *trap rocks* frequently occur in large tabular masses rising one above another like the successive steps of a stair-case. Applied to certain igneous rocks composed of feldspar, augite, and hornblende.

TRA'PPEAN. Belonging to trap rocks.

TREMA'NDRA. A genus of plants of the family of Tremándrea.

TRI'AS. From the Lat. *tres*, three. Synonym of the triássic system of rocks, consisting of the *Bunter sandstein*, the *Muschelkalk* and *Keuper*, a group of sandy marls of variegated colors.

TRICHO'MANES. From the Gr. *trichos*, hair, and *mania*, madness, excess. A genus of plants of the class of Cryptoga'mia. *Tricho'manes brevise'tum.* Short-styled trichómanes.

TRICY'RTIS. From the Gr. *treis*, three, and *kurtis*, a sack or pouch. A genus of plants of the family of Melantháceæ.

TRIDENTA'TA. Lat. Three-toothed; having three teeth.

TRIGONOCE'PHALUS. From the Gr. *treis*, three, *gonos*, an angle, and *kephale*, head. A genus of very venomous serpents. *Trigonoce'phalus lanceola'tus.* Lance-head viper.

TRI'LOBITE. From the Lat. *tres*, three, and *lobus*, lobe. A genus of fossil crustáceans.

TRI'ONYX. From the Gr. *treis*, three, and *onux*, a nail. A genus of Chelonians.

TRIO'STEUM. From the Gr. *treis*, three, and *osteon*, a bone, a nut. A genus of plants of the family of Caprifoliáceæ.

TU'FA. Italian. A volcanic rock, composed of an agglutination of fragmented scoriæ.

TUNG'STEN. Swedish. *Heavy stone.* A metal which is hard, white, brittle, and difficult to fuse.

TU'RDUS. Lat. A thrush. Name of a genus of birds.

TU'RQUOISE. A blue mineral found in Persia; its color depends on the presence of oxide of copper.

UM'BEL. A form of inflorescence, in which several peduncles expand so as to produce a flower somewhat resembling a parasol when open.

UMBELLI'FERÆ. From *umbélla*, a sun-shade, and *fero*, I bear. Name of a family of plants.

UMBELLI'FEROUS. Bearing umbels. Belonging or relating to the Umbelli'feræ.

UNCINA'TA. Lat. From *uncus*, a hook. Hooked; having hooks.

UPHEAV'AL. The elevation of land by earthquakes.

URANIUM. A metal discovered by Klaproth, in 1789.

URSI'NUS. Lat. Belonging or relating to bears.

USNEOI'DES. From *u'sne*, a kind of lichen, and the Gr. *eidos*, resemblance. Resembling the *u'sne*.

U'RENS. Lat. Burning.

VA'CUUM. From the Lat. *vacuus*, empty. A portion of space void of matter.

VAGINA'LIS. Lat. From *vagina*, a sheath. A genus of birds.

VANA'DIUM. A silvery white metal, discovered originally by Del Rio, in 1801, but not admitted until 1830.

VANE'SSA. A genus of butterflies. *Vane'ssa ca'rdui*, the painted lady butterfly.

VERONI'CA. A genus of plants of the family of Scrophularínæ.

VE'RTEBRA. From the Lat. *vertere*, to turn. A joint or bone of the spine. *Ve'rtebral co'lumn*, is the spine or back-bone.

VER'TEBRATE. Having vertebræ, or a spine.

VER'TICOSE. Whorl-like.

VENENI'FLUA. Lat. Flowing with poison.

VERNICI'FLUA. Lat. Flowing with varnish.
VER'NIX. Lat. Varnish.
VILLO'SUS. Lat. Velvety.
VI'RIDIS. Lat. Green.
VI'TEX. Chaste-tree. A genus of plants of the family of Verbenáceæ.
VIT'RIFIED. From the Lat. *vitrum*, glass. Converted into glass.
VITULI'NA From the Lat. *vitulus*, a sea-calf. Belonging or relating to seals.
VIV'ERRINE. From the Lat. *viverra*, a ferret or civet. Belonging or relating to a civet.
VIVI'PARUM. Lat. Vivi'parous.
VOLCA'NIC. Belonging or relating to volcanoes.
VOLTA'IC. Applied to electricity produced after the manner of Volta, an Italian philosopher.
VUL'TUR PA'PA. The king of vultures.

WA'TERSHED. The general declivity of the face of a country which determines the direction of the flowing of water.
WEALD. Name of a part of Kent and Surrey in England. The *Wealden clay* and *Wealden deposit* are found in this part of England.

XANTHOX'YLUM. From the Gr. *xanthos*, yellow, *xulon*, wood. Tooth-ache tree. A genus of plants of the family of Rutáceæ.
XERA'NTHEMUM. From the Gr. *xeros*, dry, and *anthos*, flower. A genus of plants of the family of Compo'sitæ.

YER'BA MATÈ. Spanish name of the Ilex Paraguensis.
YTT'RIUM. A metal discovered by Wöhler, in 1828; it is of a dark gray color and brittle.
YUC'CA. Adam's needle. A genus of plants of the family of Liliáceæ. It yields an esculent root.

ZA'MIA. A genus of plants of the family of Cycádeæ.
ZANNICHE'LLIA. After Zannichella, a Venetian apothecary. Pond weed. A genus of plants of the family of Nai'ades.
ZIRCO'NIUM. A metal found in *zirconia*, an earth, discovered by Klaproth in 1789.
ZI'ZYPHUS. A genus of plants of the family of Rhamni. *Z. ju'juba*, yields the jujube fruit.
ZOSTE'RA. From the Gr. *zoster*, a riband. Sea-wrack-grass. A genus of plants of the family of Fluviales.
ZOOL'OGY. From the Gr. *zo'on*, an animal, and *logos*, a discourse. That branch of Natural History which treats of animals.
ZO'OPHYTE. From the Gr. *zo'on*, an animal, and *phuton*, a plant. An animal without vertebræ, or extremities, that attaches itself to solid bodies, and seems to live and vegetate like a plant.

INDEX.

THE END.

CATALOGUE
OF
BLANCHARD & LEA'S PUBLICATIONS.

CAMPBELL'S LORD-CHANCELLORS. New Edition. (Just Issued.)

LIVES OF THE LORD CHANCELLORS
AND
KEEPERS OF THE GREAT SEAL OF ENGLAND

FROM THE EARLIEST TIMES TO THE REIGN OF KING GEORGE IV.

BY LORD CHIEF-JUSTICE CAMPBELL, A.M., F.R.S.E.

Second American, from the Third London Edition.

Complete in seven handsome crown 8vo. volumes, extra cloth, or half morocco.

This has been reprinted from the author's most recent edition, and embraces his extensive modifications and additions. It will therefore be found eminently worthy a continuance of the great favor with which it has hitherto been received.

Of the solid merit of the work our judgment may be gathered from what has already been said. We will add that, from its infinite fund of anecdote, and happy variety of style, the book addresses itself with equal claims to the mere general reader, as to the legal or historical inquirer; and while we avoid the stereotyped commonplace of affirming that no library can be complete without it, we feel constrained to afford it a higher tribute by pronouncing it entitled to a distinguished place on the shelves of every scholar who is fortunate enough to possess it.—*Frazer's Magazine.*

A work which will take its place in our libraries as one of the most brilliant and valuable contributions to the literature of the present day.—*Athenæum.*

BY THE SAME AUTHOR—TO MATCH—(Now Ready).

LIVES OF THE CHIEF-JUSTICES OF ENGLAND,

From the Norman Conquest to the Death of Lord Mansfield.

SECOND EDITION.

In two very neat vols., crown 8vo., extra cloth, or half morocco.
To match the "Lives of the Chancellors" of the same author.

MEMOIRS OF THE LIFE OF WILLIAM WIRT.—By JOHN P. KENNEDY, Esq. In two handsome royal 12mo. volumes, extra cloth, with a Portrait. Also, a handsome Library Edition, in two octavo volumes.

GRAHAM'S HISTORY OF THE UNITED STATES; FROM THE FOUNDING OF THE BRITISH COLONIES TILL THEIR ASSUMPTION OF INDEPENDENCE.—Revised Edition, from the Author's MSS. With a Portrait, and a Memoir by PRESIDENT QUINCY. In two large and handsome octavo volumes, extra cloth.

WILLIAM PENN, AN HISTORICAL BIOGRAPHY. With an Extra Chapter on the Macauley Charges. By W. HEPWORTH DIXON.—In one neat royal 12mo. volume, extra cloth.

GUIZOT'S OLIVER CROMWELL. (Now ready.)

HISTORY OF OLIVER CROMWELL AND THE ENGLISH COMMONWEALTH, from the Execution of Charles I. to the Death of Cromwell. In two large and handsome royal 12mo. volumes, extra cloth.

RUSSELL'S LIFE OF FOX. (Just issued).

MEMORIALS AND CORRESPONDENCE OF CHARLES JAMES FOX. Edited by LORD JOHN RUSSELL. In two handsome royal 12mo. volumes, cloth.

THE ENCYCLOPÆDIA AMERICANA;

A POPULAR DICTIONARY OF ARTS, SCIENCES, LITERATURE, HISTORY, POLITICS, AND BIOGRAPHY. In fourteen large octavo volumes of over 600 double-columned pages each. For sale very low, in various styles of binding.

MURRAY'S ENCYCLOPÆDIA OF GEOGRAPHY.

THE ENCYCLOPÆDIA OF GEOGRAPHY, comprising a Complete Description of the Earth, Physical, Statistical, Civil, and Political; exhibiting its Relation to the Heavenly Bodies, its Physical Structure, the Natural History of each Country, and the Industry, Commerce, Political Institutions, and Civil and Social State of all Nations. Revised, with Additions, by THOMAS G. BRADFORD. In three large octavo volumes, various styles of binding.

YOUATT AND SKINNER ON THE HORSE.

THE HORSE.—BY WILLIAM YOUATT. A New Edition, with numerous Illustrations. Together with a General History of the Horse; a Dissertation on the American Trotting Horse; how Trained and Jockeyed; an Account of his Remarkable Performances; and an Essay on the Ass and the Mule. By J. S. SKINNER, Assistant Postmaster-General, and Editor of the Turf Register. In one handsome octavo volume, extra cloth.

This edition of Youatt's well-known and standard work on the Management, Diseases, and Treatment of the Horse, embodying the valuable additions of Mr. Skinner, has already obtained such a wide circulation throughout the country, that the Publishers need say nothing to attract to it the attention and confidence of all who keep Horses, or are interested in their improvement.

YOUATT AND LEWIS ON THE DOG.

THE DOG. By WILLIAM YOUATT. Edited by E. J. LEWIS, M.D. With numerous and beautiful Illustrations. In one very handsome volume, crown 8vo., crimson cloth, gilt.

ADVICE TO YOUNG SPORTSMEN IN ALL THAT RELATES TO GUNS AND SHOOTING. By LIEUT.-COL. P. HAWKER. Second American, from the Ninth London Edition. Edited, with numerous Additions and Illustrations, by W. T. PORTER, Editor of the New York Spirit of the Times. In one very handsome octavo volume, extra crimson cloth, with plates.

A DICTIONARY OF MODERN GARDENING. By G. W. JOHNSON, Esq. With numerous Additions, by David Landreth. With one hundred and eighty wood-cuts. In one very large royal 12mo. volume, of about 650 double-columned pages. This work is now offered at a very low price.

THE YOUNG MILLWRIGHT AND MILLER'S GUIDE. By OLIVER EVANS. With Additions and Corrections by THOMAS P. JONES; and a Description of an Improved Merchant Flour-Mill, by C. and O. EVANS. With twenty-eight Plates. Fourteenth Edition. In one neat Octavo volume, leather.

ACTON'S MODERN COOKERY, IN ALL ITS BRANCHES, REDUDED TO A SYSTEM OF EASY PRACTICE. Edited by MRS. S. J. HALE. In one neat royal 12mo. volume, extra cloth, with numerous Illustrations.

THOMSON'S DOMESTIC MANAGEMENT OF THE SICK-ROOM. With additions by R. E. GRIFFITH, M.D. In one royal 12mo. volume, extra cloth.

WILSON ON HEALTHY SKIN. (Now Ready.)

HEALTHY SKIN; A POPULAR TREATISE ON THE MANAGEMENT OF THE SKIN AND HAIR IN RELATION TO HEALTH. Second American, from the Fourth and Revised London Edition. By ERASMUS WILSON. In one handsome royal 12mo. volume, with numerous cuts, extra cloth; also, paper covers, price 75 cts.

NEW AND IMPROVED EDITION.

LIVES OF THE QUEENS OF ENGLAND,

FROM THE NORMAN CONQUEST. WITH ANECDOTES OF THEIR COURTS. Now first published from Official Records, and other Authentic Documents, Private as well as Public. New Edition, with Additions and Corrections. By AGNES STRICKLAND. In six volumes, crown octavo, extra crimson cloth, or half morocco, printed on fine paper and large type. Copies of the Duodecimo Edition, in twelve volumes, may still be had.

A valuable contribution to historical knowledge, to young persons especially. It contains a mass of every kind of historical matter of interest, which industry and resource could collect. We have derived much entertainment and instruction from the work.—*Athenæum*.

The execution of this work is equal to the conception. Great pains have been taken to make it both interesting and valuable.—*Literary Gazette*.

A charming work—full of interest, at once serious and pleasing.—*Monsieur Guizot*.

TO BE HAD SEPARATE.

LIVES OF THE QUEENS OF HENRY VIII., and of his mother, Elizabeth of York. By MISS STRICKLAND. Complete in one handsome crown octavo volume, extra cloth. (Just Issued.)

MEMOIRS OF ELIZABETH, Second Queen Regnant of England and Ireland. By MISS STRICKLAND. Complete in one handsome crown octavo volume, extra cloth. (Just Issued.)

INGERSOLL'S HISTORY OF THE LATE WAR. In two octavo volumes—sold separate. Volume I. containing 1812–13; and volume II., 1814.

MRS. MARSH'S ROMANTIC HISTORY OF THE HUGONOTS. In two handsome royal 12mo. volumes, extra cloth.

NIEBUHR'S ANCIENT HISTORY.

LECTURES ON ANCIENT HISTORY; FROM THE EARLIEST TIMES TO THE TAKING OF ALEXANDRIA BY OCTAVIANUS. Comprising the History of the Asiatic Nations, the Egyptians, Greeks, Macedonians, and Carthagenians. By B. G. NIEBUHR. Translated from the German Edition of DR. MARCUS NIEBUHR, by DR. LEONHARD SCHMITZ; with Additions and Corrections from his own MS. Notes. In three handsome crown octavo volumes, extra cloth.

PRINCIPLES OF THE MECHANICS

OF MACHINERY AND ENGINEERING. By PROFESSOR JULIUS WEISBACH. Translated and edited by PROFESSOR GORDON. Edited, with American additions, by PROFESSOR WALTER R. JOHNSON. In two very handsome octavo volumes, of nearly 900 pages, with about 900 superb illustrations.

CHEMICAL TECHNOLOGY;

OR, CHEMISTRY APPLIED TO THE ARTS AND TO MANUFACTURES. By DR. F. KNAPP. Edited, with numerous notes and additions, by DR. EDMUND RONALDS and DR. THOMAS RICHARDSON. With American additions, by PROF. WALTER R. JOHNSON. In two handsome octavo volumes, of about 1000 pages, with nearly five hundred splendid illustrations.

GRAHAM'S ELEMENTS OF CHEMISTRY;

INCLUDING THE APPLICATIONS OF THE SCIENCE IN THE ARTS. Edited, with Notes, by ROBERT BRIDGES, M.D. Part I., handsome 8vo., of 430 pages, with 185 illustrations. Part II. preparing.

THE LAWS OF HEALTH IN RELATION TO MIND AND BODY. A Series of Letters from an Old Practitioner to a Patient. By LIONEL S. BEAL, M.D. In one royal 12mo. volume.

HUMBOLDT'S ASPECTS OF NATURE, IN DIFFERENT LANDS AND DIFFERENT CLIMATES. Translated by Mrs. Sabine. In one large royal 12mo. volume, extra cloth.

NARRATIVE OF THE UNITED STATES' EXPEDITION TO THE DEAD SEA AND RIVER JORDAN.

BY W. F. LYNCH, U. S. N. (COMMANDER OF THE EXPEDITION).

In one very large and handsome octavo volume, with twenty-eight beautiful Plates, and two Maps.

This book, so long and anxiously expected, fully sustains the hopes of the most sanguine and fastidious. It is truly a magnificent work. The type, paper, binding, style, and execution, are all of the best and highest character, as are also the maps and engravings. It will do more to elevate the character of our national literature than any work that has appeared for years. The intrinsic interest of the subject will give it popularity and immortality at once. It must be read to be appreciated: and it will be read extensively, and valued, both in this and other countries.—*Lady's Book.*

Also, to be had—

CONDENSED EDITION, one neat royal 12mo. volume, extra cloth, with a Map.

DON QUIXOTE DE LA MANCHA. Translated from the Spanish of MIGUEL DE CERVANTES SAAVEDRA, by CHARLES JARVIS, Esq. Carefully revised and corrected, with a memoir of the Author and a notice of his works. With numerous Illustrations, by Tony Johannot. In two beautifully-printed volumes, crown octavo, various styles of binding.

The handsome execution of this work, the numerous spirited illustrations with which it abounds, and the very low price at which it is offered, render it a most desirable library edition for all admirers of the immortal Cervantes.

PICCIOLA, THE PRISONER OF FENESTRELLA; or, CAPTIVITY CAPTIVE. By X. B. SAINTINE. New Edition, with Illustrations. In one very neat royal 12mo. volume, paper covers, price 50 cents, or extra cloth.

THE LANGUAGE OF FLOWERS, with Illustrative Poetry. To which are now added the CALENDAR OF FLOWERS, and the DIAL OF FLOWERS. Ninth American, from the Tenth London Edition. Revised by the editor of "Forget-me-Not." In one elegant royal 18mo. volume, extra crimson cloth, gilt, with beautiful colored Plates.

HALE'S ETHNOGRAPHY AND PHILOLOGY OF THE U. S. EXPLORING EXPEDITION. In one large royal quarto volume, extra cloth.

DANA ON ZOOPHYTES. Being part of the publications of the UNITED STATES EXPLORING EXPEDITION. One large royal quarto volume of letter-press, and an Atlas in imperial folio, of 60 plates, containing many hundred figures, exquisitely engraved on steel, and coloured after nature.

JOHNSTON'S PHYSICAL ATLAS.

THE PHYSICAL ATLAS OF NATURAL PHENOMENA, FOR THE USE OF COLLEGES, ACADEMIES, AND FAMILIES. By ALEXANDER KEITH JOHNSTON, F.R.G.S., &c. In one large imperial quarto volume, strongly and handsomely bound in half morocco. With twenty-six Plates, engraved and colored in the best style, together with over one hundred pages of descriptive letter-press, and a very copious Index.

A work which should be in every family and every school-room, for consultation and reference. By the ingenious arrangement adopted by the author, it makes clear to the eye every fact and observation relative to the present condition of the earth, arranged under the departments of Geology, Hydrography, Meteorology, and Natural History. The letter-press illustrates this with a body of information, nowhere else to be found condensed into the same space; while a very full Index renders the whole easy of reference.

Now Ready,

PRINCIPLES OF COMPARATIVE PHYSIOLOGY.

BY W. B. CARPENTER, M.D., F.R.S, &c.

A new American, from the Fourth and revised London Edition. In one large and handsome octavo volume, of 750 pages, with 309 beautiful illustrations.

The present edition of this work will be found in every way worthy of its high reputation as the standard text-book on this subject. Thoroughly revised and brought up by the author to the latest date of scientific investigation, and illustrated with a profusion of new and beautiful engravings, it has been printed in the most careful manner, and forms a volume which should be in the possession of every student of natural history.

By the same author (Just Issued),

ON THE USE AND ABUSE OF ALCOHOLIC LIQUORS IN HEALTH AND DISEASE. In one neat royal 12mo. volume, extra cloth. Also, in [illegible]ble cloth, for mailing, price 50 cents, free by post.

BUSHNAN'S POPULAR PHYSIOLOGY. (Now Ready.)

THE PRINCIPLES OF ANIMAL AND VEGETABLE PHYSIOLOGY, A POPULAR TREATISE ON THE FUNCTIONS AND PHENOMENA OF ORGANIC LIFE. To which is prefixed an Essay on the Great Departments of Human Knowledge. By J. STEVENSON BUSHNAN, M.D. In one handsome royal 12mo. volume, with over one hundred illustrations.

Though cast in a simple and unpretending form, this little work is based upon the latest scientific developments. It is therefore admirably adapted for the private reader, or for a text-book in those academies and seminaries which include this subject in their course of studies.

OWEN ON THE SKELETON AND TEETH. (Now Ready.)

THE PRINCIPAL FORMS OF THE SKELETON AND OF THE TEETH. By PROFESSOR R. OWEN, author of "Comparative Anatomy," &c. In one handsome royal 12mo. volume, extra cloth, with numerous illustrations.

Written as a popular introduction to his favourite science, by the most distinguished osteologist of the age, this work cannot fail to find favour with all students of Geology, Zoology, and Comparative Anatomy, of which its subject may be considered as the foundation.

DE LA BECHE'S GEOLOGY.

THE GEOLOGICAL OBSERVER. By SIR HENRY T. DE LA BECHE, F.R.S., &c. In one large and handsome octavo volume, with over 300 illustrations.

ABEL AND BLOXAM'S CHEMISTRY. (Now Ready.)

A HAND-BOOK OF CHEMISTRY, THEORETICAL, PRACTICAL, AND TECHNICAL. By F. A. ABEL and C. L. BLOXAM. In one large and handsome octavo volume, of over 650 pages, with numerous illustrations.

The department of theoretical chemistry has been amply elucidated in many late publications, but a want has been felt of a work which should afford a guide to the practical student in the numerous and complicated processes required in the laboratory, while the operative chemist has had no recent manual detailing the new and valuable improvements which are daily being made. It has been the aim of the authors to supply this vacancy, and the applause which their labours have received from competent judges is a sufficient evidence of their success.

PRINCIPLES OF PHYSICS AND METEOROLOGY.

By PROFESSOR J. MULLER. Revised, and illustrated with over five hundred engravings on wood, and two handsome coloured plates. In one large and beautiful octavo volume of nearly 650 pages.

Now Complete.

HANDBOOKS OF NATURAL PHILOSOPHY AND ASTRONOMY.

BY DIONYSIUS LARDNER, D.C.L.,

Formerly Professor of Natural Philosophy and Astronomy in University College, London.

This valuable Series is now complete, consisting of three Courses, as follows:—

FIRST COURSE,

MECHANICS, HYDROSTATICS, HYDRAULICS, PNEUMATICS, SOUND, & OPTICS,

In one large royal 12mo. volume, of 750 pages, with 424 Illustrations.

SECOND COURSE,

HEAT, MAGNETISM, COMMON ELECTRICITY, AND VOLTAIC ELECTRICITY,

In one royal 12mo. volume, of 450 pages, with 244 Illustrations.

THIRD COURSE,

ASTRONOMY AND METEOROLOGY.

In one very large royal 12mo. volume, of nearly 800 pages, with 37 Plates, and over 200 Illustrations.

These volumes can be had either separately or in uniform sets, containing about 2000 pages, and nearly 1000 Illustrations on steel and wood.

To accommodate those who desire separate treatises on the leading departments of Natural Philosophy, the First Course may also be had, divided in three portions, viz:

Part I. Mechanics.—Part II. Hydrostatics, Hydraulics, Pneumatics, and Sound.—Part III. Optics.

It will thus be seen that this work furnishes either a complete course of instruction on these subjects, or separate treatises on all the different branches of Physical Science. The object of the author has been to prepare a work suited equally for the collegiate, academical, and private student, who may desire to acquaint himself with the present state of science, in its most advanced condition, without pursuing it through its mathematical consequences and details. Great industry has been manifested throughout the work to elucidate the principles advanced by their practical applications to the wants and purposes of civilized life, a task to which Dr. Lardner's immense and varied knowledge, and his singular felicity and clearness of illustration render him admirably fitted. This peculiarity of the work recommends it especially as the text-book for a practical age and country such as ours, as it interests the student's mind, by showing him the utility of his studies, while it directs his attention to the further extension of that utility by the fulness of its examples. Its extensive adoption in many of our most distinguished colleges and seminaries is sufficient proof of the skill with which the author's intentions have been carried out.

BIRD'S NATURAL PHILOSOPHY.

ELEMENTS OF NATURAL PHILOSOPHY; being an Experimental Introduction to the Physical Sciences. Illustrated with over 300 wood-cuts. By Golding Bird, M.D., Assistant Physician to Guy's Hospital. From the Third London edition. In one neat volume, royal 12mo.

We are astonished to find that there is room in so small a book for even the bare recital of so many subjects. Where everything is treated succinctly, great judgment and much time are needed in making a selection and winnowing the wheat from the chaff. Dr. Bird has no need to plead the peculiarity of his position as a shield against criticism, so long as his book continues to be the best epitome in the English language of this wide range of physical subjects.—*North American Review.*

For those desiring as extensive a work, I think it decidedly superior to anything of the kind with which I am acquainted.—*Prof. John Johnston, Wesleyan Univ., Middletown, Ct.*

ARNOT'S ELEMENTS OF PHYSICS.

ELEMENTS OF PHYSICS; or, Natural Philosophy, General and Medical, written for Universal Use in Plain or Non-technical Language. By Neil Arnot, M.D. In one octavo volume, with about two hundred Illustrations.

A COMPLETE COURSE OF NATURAL SCIENCE. (Just Issued.)

THE BOOK OF NATURE;

An Elementary Introduction to the Sciences of Physics, Astronomy, Chemistry, Mineralogy, Geology, Botany, Zoology, and Physiology. By FREDERICK SCHOEDLER, PH. D., Professor of the Natural Sciences at Worms. First American Edition, with a Glossary, and other Additions and Improvements; from the Second English Edition, translated from the Sixth German Edition, by HENRY MEDLOCK, F.C.S., &c. Illustrated by 679 engravings on wood. In one handsome volume, crown octavo, of about 700 large pages, extra cloth.

To accommodate those who desire to use the separate portions of this work, the publishers have prepared an edition in parts, as follows, which may be had singly, by mail or otherwise, neatly done up in flexible cloth.

NATURAL PHILOSOPHY	114 pages,	with 149	Illustrations.
ASTRONOMY	64	" 51	"
CHEMISTRY	110	" 48	"
MINERALOGY AND GEOLOGY	104	" 167	"
BOTANY	98	" 176	"
ZOOLOGY AND PHYSIOLOGY	106	" 84	"
INTRODUCTION, GLOSSARY, INDEX, &c.	96		

As a work for popular instruction in the Natural and Physical Sciences, it certainly is unrivalled, so far as my knowledge extends. It admirably combines perspicuity with brevity; while an excellent judgment and a rare discrimination are manifest in the selection and arrangement of topics, as well as in the description of objects, the illustration of phenomena, and the statement of principles. A more careful perusal of those departments of the work to which my studies have been particularly directed has been abundantly sufficient to satisfy me of its entire reliableness — that the object of the author was not so much to *amuse* as really to *instruct*.—*Prof. Allen, Oberlin Institute, Ohio.*

I do not know of another book in which so much that is important on these subjects can be found in the same space.—*Prof. Johnston, Wesleyan University, Conn.*

Though a very comprehensive book, it contains about as much of the details of natural science as general students in this country have time to study in a regular academical course; and I am so well pleased with it that I shall recommend its use as a text-book in this institution.—*W. H. Allen, President of Girard College, Philadelphia.*

I am delighted with Dr. Schœdler's "Book of Nature;" its tone of healthful piety and reverence for God's word add a charm to the learning and deep research which the volume everywhere manifests.—*Prof. J. A. Spencer, N. Y.*

BROWNE'S CLASSICAL LITERATURE. (Now Complete.)

A HISTORY OF GREEK CLASSICAL LITERATURE.

BY THE REV. R. W. BROWNE, M.A.,

Professor of Classical Literature in King's College, London.

In one very handsome crown octavo volume.

By the same Author, to match. (Now ready.)

A HISTORY OF ROMAN CLASSICAL LITERATURE.

In one very handsome crown octavo volume.

These two volumes form a complete course of Classical Literature, designed either for private reading or for collegiate text-books. Presenting, in a moderate compass and agreeable style, the results of the most recent investigations of English and continental scholars, it gives, in a succession of literary biographies and criticisms, a body of information necessary to all educated persons, and which cannot elsewhere be found in so condensed and attractive a shape.

New and much improved Edition.—(Lately Issued.)

PHYSICAL GEOGRAPHY.

BY MARY SOMERVILLE.

A new American, from the third and revised London edition.

WITH NOTES AND A GLOSSARY,

BY W. S. W. RUSCHENBERGER, M.D., U. S. NAVY.

In one large royal 12mo. volume, of nearly six hundred pages.

The subject of Physical Geography is one of which the acknowledged importance is rapidly forcing its introduction into all systems of education which pretend to keep themselves on a level with the improvements and requirements of the age. It is no longer considered sufficient to drill the scholar into a mechanical knowledge of the names of rivers and mountains, and the territorial divisions of the earth's surface. A want is now felt of an acquaintance with the structure of the globe, externally and internally, and of the causes and effects of the variations of land and water, forest and desert, heat and cold, tides, currents, rain, wind, and all the other physical phenomena occurring around us, which have so direct and immense an influence upon the human race. This is all summed up in "Physical Geography," which may be regarded as the résumé of all that is known on the natural history and present state of the earth and its inhabitants — the practical application of the principles which are elucidated by the minute investigations of the scientific observer. This vast and interesting subject has been successfully grappled by Mrs. Somerville, who in the present volume has set forth, in a picturesque and vivid style, a popular yet condensed account of the globe, in its relations with the Solar System; its geological forces; its configuration and divisions into land and water, mountain, plain, river, and lake; its meteorology, mineral productions, vegetation, and animal life; estimating and analyzing the causes at work, and their influence on plants, animals, and mankind. A study such as this, taken in conjunction with ordinary political geography, lends to the latter an interest foreign to the mere catalogue of names and boundaries, and, in addition to the vast amount of important information imparted, tends to impress the whole more strongly on the mind of the student.

Eulogy is unnecessary with regard to a work like the present, which has passed through three editions, on each side of the Atlantic, within the space of a few years. The publishers therefore only consider it necessary to state that the last London edition received a thorough revision at the hands of the author, who introduced whatever improvements and corrections the advance of science rendered desirable; and that the present issue, in addition to this, has had a careful examination on the part of the editor, to adapt it more especially to this country. Great care has been exercised in both the text and the glossary to obtain the accuracy so essential to a work of this nature; and in its present improved and enlarged state, with no corresponding increase of price, it is confidently presented as in every way worthy of a continuation of the striking favor with which it has been everywhere received.

BUTLER'S ANCIENT ATLAS.

AN ATLAS OF ANCIENT GEOGRAPHY. By SAMUEL BUTLER, D.D., late Lord Bishop of Litchfield. In one handsome octavo volume, containing twenty-one coloured quarto Maps, and an accentuated Index.

The very low price at which this work is now offered, and the authoritative position which it has so long maintained, render it a very desirable reference book for all institutions where this branch of study is pursued.

BUTLER'S ANCIENT GEOGRAPHY.

GEOGRAPHIA CLASSICA; or, THE APPLICATION OF ANCIENT GEOGRAPHY TO THE CLASSICS. By SAMUEL BUTLER, D.D., late Lord Bishop of Litchfield. Sixth American, from the last and revised London edition. In one neat royal 12mo. volume, half bound.

NOW COMPLETE.

SCHMITZ AND ZUMPT'S CLASSICAL SERIES.

By the completion of this series, the classical student is now in possession of a thorough and uniform course of Latin instruction, on a definite system. Besides the advantages which these works possess in their typographical accuracy and careful adaptation to educational purposes, the exceedingly low price at which they are offered is a powerful argument in favor of their general introduction, as removing a barrier to the general diffusion of classical education in the size and costliness of the text-books heretofore in use.

The series consists of the following volumes, clearly and handsomely printed, on good paper, in a uniform large 18mo. size, strongly and neatly bound, and accompanied with notes, historical and critical introductions, maps, and other illustrations.

SCHMITZ'S ELEMENTARY LATIN GRAMMAR AND EXERCISES, extra cloth, price [illegible]
KALTSCHMIDT'S SCHOOL LATIN DICTIONARY, in two Parts, Latin-English, and English-Latin, nearly 900 pages, strongly bound in leather .... $[illegible]
Part I., Latin-English, about 500 pages, " " " .... .90
Part II., English-Latin, nearly 400 pages, " " " .... .75
SCHMITZ'S ADVANCED LATIN GRAMMAR, 318 pages, ... half bound, .60
ADVANCED LATIN EXERCISES, WITH SELECTIONS FOR READING, extra cloth, .50
CORNELII NEPOTIS LIBER DE EXCELLENTIBUS DUCIBUS, &c., extra cloth, .50
CÆSARIS DE BELLO GALLICO, LIBRI IV., 232 pages, ... extra cloth, .50
C. C. SALLUSTII CATILINA ET JUGURTHA, 168 pages, ... extra cloth, .50
EXCERPTA EX P. OVIDII NASONIS CARMINIBUS, 246 pages, ... extra cloth, .60
Q. CURTII RUFI DE ALEXANDRI MAGNI QUÆ SUPERSUNT, 326 pp., ex. cloth, .70
P. VIRGILII MARONIS CARMINA, 438 pages, ... extra cloth, .75
ECLOGÆ EX Q. HORATII FLACCI POEMATIBUS, 312 pages, ... extra cloth, .60
T. LIVII PATAVINI HISTORIARUM LIBRI I. II. XXI. XXII., 350 pp., ex. cloth, .70
M. T. CICERONIS ORATIONES SELECTÆ XII., 300 pages, ... extra cloth, .60

Also, uniform with the Series,

BAIRD'S CLASSICAL MANUAL OF ANCIENT GEOGRAPHY, ANTIQUITIES, CHRONOLOGY, &c., ... extra cloth, .50

The volumes in cloth can also be had, strongly half-bound in leather, with cloth sides, at an extra charge of five cents per volume.

The very numerous recommendations of this series from classical teachers of the highest standing, and their adoption in many of our best academies and colleges, sufficiently manifest that the efforts of the editors and publishers have not been unsuccessful in supplying a course of classical study suited to the wants of the age, and adapted to the improved modern systems of education.

With your Classical Series I am well acquainted, and have no hesitancy in recommending them to all my friends. In addition to your Virgil, which we use, we shall probably adopt other books of the series as we may have occasion to introduce them.—*Prof. J. J. Owen, N. Y. Free Academy.*

I regard this series of Latin text-books as decidedly superior to any others with which I am acquainted. The Livy and Horace I shall immediately introduce for the use of the college classes.—*Prof. A. Rollins, Delaware College.*

Having examined several of them with some degree of care, we have no hesitation in pronouncing them among the very best extant.—*Prof. A. C. Knox, Hanover College, Indiana.*

I can give you no better proof of the value which I set on them than by making use of them in my own classes, and recommending their use in the preparatory department of our institution. I have read them through carefully, that I might not speak of them without due examination; and I flatter myself that my opinion is fully borne out by fact, when I pronounce them to be the most useful and the most correct, as well as the cheapest editions of Latin Classics ever introduced in this country. The Latin and English Dictionary contains as much as the student can want in the earlier years of his course; it contains more than I have ever seen compressed into a book of this kind. It ought to be the student's constant companion in his recitations. It has the extraordinary recommendation of being at once portable and comprehensive.—*Prof. R. N. Newell, Masonic College, Tenn.*

That invaluable little work, the Classical Manual, has been used by me for some time. I would not, on any account, be without it. You have not perhaps been informed that it has recently been introduced in the High School of this place. Its typographical accuracy is remarkable.—*Reginald H. Chase, Harvard University.*

Shaw's English Literature—Lately Published.

OUTLINES OF ENGLISH LITERATURE.

By Thomas B. Shaw, Professor of English Literature in the Imperial Alexander Lyceum, St. Petersburg, Second American Edition. With a Sketch of American Literature, by Henry T. Tuckerman, Esq. In one large and handsome volume, royal 12mo., of about five hundred pages.

The object of this work is to present to the student, within a moderate compass, a clear and connected view of the history and productions of English Literature. To accomplish this, the author has followed its course from the earliest times to the present age, seizing upon the more prominent "Schools of Writing," tracing their causes and effects, and selecting the more celebrated authors as subjects for brief biographical and critical sketches, analyzing their best works, and thus presenting to the student a definite view of the development of the language and literature, with succinct descriptions of those books and men of which no educated person should be ignorant. He has thus not only supplied the acknowledged want of a manual on this subject, but by the liveliness and power of his style, the thorough knowledge he displays of his topic, and the variety of his subjects, he has succeeded in producing a most agreeable reading-book, which will captivate the mind of the scholar, and relieve the monotony of drier studies.

Its merits I had not now for the first time to learn. I have used it for two years as a text-book, with the greatest satisfaction. It was a happy conception, admirably executed. It is all that a text-book on such a subject can or need be, comprising a judicious selection of materials, easily yet effectively wrought. The author attempts just as much as he ought to, and does well all that he attempts; and the best of the book is the *genial spirit*, the genuine love of genius and its works which thoroughly pervades it, and makes it just what you want to put into a pupil's hands.—*Professor J. V. Raymond, University of Rochester.*

Of "Shaw's English Literature" I can hardly say too much in praise. I hope its adoption and use as a text-book will correspond to its great merits.—*Prof. J. C. Pickard, Ill. College.*

BOLMAR'S COMPLETE FRENCH SERIES.

Blanchard and Lea now publish the whole of Bolmar's Educational Works, forming a complete series for the acquisition of the French language, as follows:

BOLMAR'S EDITION OF LEVIZAC'S THEORETICAL AND PRACTICAL GRAMMAR OF THE FRENCH LANGUAGE. With numerous Corrections and Improvements, and the addition of a complete Treatise on the Genders of French Nouns and the Conjugation of the French Verbs, Regular and Irregular. Thirty-fifth edition. In one 12mo. volume, leather.

BOLMAR'S COLLECTION OF COLLOQUIAL PHRASES, on every topic necessary to maintain conversation; arranged under different heads; with numerous remarks on the peculiar pronunciation and use of various words. The whole so disposed as considerably to facilitate the acquisition of a correct pronunciation of the French. In one 18mo. volume, half bound.

BOLMAR'S EDITION OF FENELON'S AVENTURES DE TELEMAQUE. In one 12mo. volume, half bound.

BOLMAR'S KEY TO THE FIRST EIGHT BOOKS OF TELEMAQUE, for the literal and free translation of French into English. In one 12mo. volume, half bound.

BOLMAR'S SELECTION OF ONE HUNDRED OF PERRIN'S FABLES, accompanied with a Key, containing the text and a literal and a free translation, arranged in such a manner as to point out the difference between the French and the English idiom; also, a figured pronunciation of the French. The whole preceded by a short treatise on the Sounds of the French language as compared with those of English. In one 12mo. volume, half bound.

BOLMAR'S BOOK OF FRENCH VERBS, wherein the Model Verbs, and several of the most difficult, are conjugated Affirmatively, Negatively, Interrogatively, and Negatively and Interrogatively, containing also numerous Notes and Directions on the Different Conjugations, not to be found in any other book published for the use of English scholars; to which is added a complete list of all the Irregular verbs. In one 12mo. volume, half bound.

The long and extended sale with which these works have been favoured, and the constantly increasing demand which exists for them, renders unnecessary any explanation or recommendation of their merits.

HERSCHELL'S ASTRONOMY.

OUTLINES OF ASTRONOMY.

BY SIR JOHN F. W. HERSCHEL, BART., F.R.S., &c.

A New American, from the Fourth and Revised London Edition.

In one handsome crown octavo volume, with numerous plates and wood-cuts.

The present work is reprinted from the last London Edition, which was carefully revised by the author, and in which he embodies the latest investigations and discoveries. It may therefore be regarded as fully on a level with the most advanced state of the science, and even better adapted than its predecessors as a full and reliable text-book for advanced classes.

A few commendatory notices are subjoined, from among a large number with which the publishers have been favored.

A rich mine of all that is most valuable in modern Astronomy.—*Professor D. Olmsted, Yale College.*

As a work of reference and study for the more advanced pupils, who yet are not prepared to avail themselves of the higher mathematics, I know of no work to be compared with it.—*Prof. A. Caswell, Brown University, R. I.*

This treatise is too well known, and too highly appreciated in the scientific world, to need new praise. A distinguishing merit in this, as in the other productions of the author, is, that the language in which the profound reasonings of science are conveyed is so perspicuous that the writer's meaning can never be misunderstood.—*Prof. Samuel Jones, Jefferson College, Pa.*

I know no treatise on Astronomy comparable to "Herschel's Outlines." It is admirably adapted to the necessities of the student. We have adopted it as a text-book in our College.—*Prof. J. F. Crocker, Madison College, Pa.*

As far as I am able to judge, it is the best work of its class in any language.—*Prof. James Curley, Georgetown College.*

It would not become me to speak of the scientific merits of such a work by such an author; but I may be allowed to say, that I most earnestly wish that it might supersede every book used as a text-book on Astronomy in all our institutions, except perhaps those where it is studied mathematically.—*Prof. N. Tillinghast, Bridgewater, Mass.*

CHEMICAL TEXT-BOOK FOR STUDENTS. (Just Issued.)

ELEMENTARY CHEMISTRY,

THEORETICAL AND PRACTICAL.

BY GEORGE FOWNES, PH. D., &c.

With Numerous Illustrations.

A NEW AMERICAN, FROM THE LAST AND REVISED LONDON EDITION. EDITED, WITH ADDITIONS,

BY ROBERT BRIDGES, M.D.

In one large royal 12mo. volume, containing over 550 pages, clearly printed on small type, with 181 Illustrations on Wood.

We know of no better text-book, especially in the difficult department of Organic Chemistry, upon which it is particularly full and satisfactory. We would recommend it to preceptors as a capital "office-book" for their students who are beginners in Chemistry. It is copiously illustrated with excellent wood-cuts, and altogether admirably "got up."—*N. J. Medical Reporter.*

A standard manual, which has long enjoyed the reputation of embodying much knowledge in a small space. The author has achieved the difficult task of condensation with masterly tact. His book is concise without being dry, and brief without being too dogmatical or general.—*Virginia Medical and Surgical Journal.*

The work of Dr. Fownes has long been before the public, and its merits have been fully appreciated as the best text-book on Chemistry now in existence. We do not, of course, place it in a rank superior to the works of Brande, Graham, Turner, Gregory, or Gmelin, but we say that, as a work for students, it is preferable to any of them.—*London Journal of Medicine.*

A TEXT-BOOK OF SCRIPTURE GEOGRAPHY AND HISTORY. (Just Issued.)

OUTLINES OF SCRIPTURE GEOGRAPHY AND HISTORY;

Illustrating the Historical Portions of the Old and New Testaments.

DESIGNED FOR THE USE OF SCHOOLS AND PRIVATE READING.

BY EDWARD HUGHES, F.R.A.S., F.G.S.,

Head Master of the Royal Naval Lower School, Greenwich, &c.

BASED UPON COLEMAN'S HISTORICAL GEOGRAPHY OF THE BIBLE.

With twelve handsome Colored Maps.

In one very neat royal 12mo. volume, extra cloth.

The intimate connection of Sacred History with the geography and physical features of the various lands occupied by the Israelites, renders a work like the present an almost necessary companion to all who desire to read the Scriptures understandingly. To the young, especially, a clear and connected narrative of the events recorded in the Bible, is exceedingly desirable, particularly when illustrated, as in the present volume, with succinct but copious accounts of the neighboring nations, and of the topography and political divisions of the countries mentioned, coupled with the results of the latest investigations, by which Messrs. Layard, Lynch, Olin, Durbin, Wilson, Stephens, and others, have succeeded in throwing light on so many obscure portions of the Scriptures, verifying its accuracy in minute particulars. Few more interesting class-books could therefore be found for schools where the Bible forms a part of education, and none, perhaps, more likely to prove of permanent benefit to the scholar. The influence which the physical geography, climate, and productions of Palestine had upon the Jewish people will be found fully set forth, while the numerous maps present the various regions connected with the subject at their most prominent periods.

We have given it considerable examination, and have been very favorably impressed with it as a work of rare excellence, and as well calculated to answer a demand, which, so far as our knowledge extends, has never yet been fully accomplished.—*Evangelical Repository.*

We have read it with care, and can recommend it with confidence. Indeed, we do not know of a more convenient and reliable handbook for a pastor, Sunday-school teacher, or a general student to refer to for information in regard to Palestine, whether as to its physical features or its geography, its climate or its productions, its past history or its present condition.—*Southern Presbyterian.*

It appears to contain, in a compressed form, a vast deal of important and accurate geographical and historical information. I hope the book will have the wide circulation which its merits entitle it to. I shall not fail to recommend it so far as opportunity offers.—*Prof. Samuel H. Turner, N. Y. Theological Seminary.*

We have long needed just such a book, and as soon as possible we shall make it one of the text-books of our college. It should be a text-book in all our theological institutions.—*Rev. Samuel Findley, President of Antrim College, Ohio.*

Few more interesting class-books, where the Bible is used in schools, can be found than the "Outlines of Scripture Geography and History," and it will prove, in families where the Bible is read, a valuable auxiliary to the understanding of that blessed volume. It is therefore to be hoped that it will receive that patronage which it so richly deserves.—*Rev. Eliphalet Nott, President of Union College, N. Y.*

I have studied the greater portion of it with care, and find it so useful as a book of reference, that I have placed it on the table with my Bible, as an aid to my daily Scripture readings. It is a book which ought to be in the hands of every biblical student, and I cannot but hope that it will have a wide circulation. To such as desire to borrow, I answer, "I cannot loan it, for I am obliged to refer to it daily!"—*Prof. E. Everett, New Orleans.*

It comprises the fullest and most instructive, as well as the most attractive course of lessons on its particular subjects that has hitherto been offered in the compass of a single volume.—*William Russell, N. E. Normal Institute, Mass.*

Its thoroughness and comprehensiveness, combined with conciseness and portable size, and especially its neat and beautiful maps, render it peculiarly adapted to Bible-classes and Sabbath-Schools, and, indeed, to every religious family and every reader of the Bible. It is also very valuable to the student of Ancient History, whether sacred or profane. I have seen no work of the kind which pleases me so well.—*Prof. Sturtevant, Illinois College.*

www.ingramcontent.com/pod-product-compliance
Lightning Source LLC
LaVergne TN
LVHW021226110826
845150LV00002B/257